JEAN LERAY '99 CONFERENCE PROCEEDINGS

MATHEMATICAL PHYSICS STUDIES

Jean Leray
'99 Conference Proceedings

The Karlskrona Conference
in Honor of Jean Leray

Edited by

Maurice de Gosson
Department of Health, Science and Mathematics,
Blekinge Institute of Technology,
Karlskrona, Sweden

KLUWER ACADEMIC PUBLISHERS
DORDRECHT / BOSTON / LONDON

A C.I.P. Catalogue record for this book is available from the Library of Congress.

ISBN   978-90-481-6316-8

Published by Kluwer Academic Publishers,
P.O. Box 17, 3300 AA Dordrecht, The Netherlands.

Sold and distributed in North, Central and South America
by Kluwer Academic Publishers,
101 Philip Drive, Norwell, MA 02061, U.S.A.

In all other countries, sold and distributed
by Kluwer Academic Publishers,
P.O. Box 322, 3300 AH Dordrecht, The Netherlands.

Printed on acid-free paper

To Jean Leray (1906–1998)

TABLE OF CONTENTS

Preface xi
Introduction 1
Acknowledgements..3

PART I: HYPERBOLIC SYSTEMS AND EQUATIONS 5

Caractère holonôme d'une solution élémentaire
L. Boutet de Monvel 7

Necessary conditions for hyperbolicity of first order systems
A. Bove and T. Nishitani 17

On the Cauchy problem for hyperbolic operators with non-regular
coefficients
F. Colombini, D. Del Santo, and T. Kinoshita 37

Multiple points of the characteristic manifold of a diagonalizable operator
H. Delquié and J. Vaillant 53

Large temps de vie des solutions d'un problème de Cauchy non linéaire
D. Gourdin and M. Mechab 65

Une remarque sur un prolongement analytique de la solution du problème
de Cauchy
Y. Hamada 75

Conormality and lagrangian properties along diffractive rays
P. Laubin 83

Caractérisation des opérateurs différentiels hyperboliques
Y. Ohya 97

A dependence domain for a class of micro-differential equations with
involutive double characteristics
Y. Okada and N. Tose 109

Ramification non abélienne
C. Wagschal 115

PART II: SYMPLECTIC MECHANICS AND GEOMETRY 123

Extension du calcul différentiel et application à la théorie des groupes de
Lie en dimension infinie
P. Dazord 125

viii

The cohomological meaning of Maslov's lagrangian path intersection index
M. de Gosson and S. de Gosson 143

A Kähler structure on the punctured cotangent bundle of the Cayley projective plane
K. Furutani 163

On mechanical systems with a Lie group as configuration space
C.-M. Marle 183

Dirac fields on asymptotically simple space–times
J.–P. Nicolas 205

An embedding result for some general symbol classes in the Weyl calculus
J. Toft 219

The lagrangian in symplectic mechanics
G. M. Tuynman 235

Geometry of solution spaces of spaces of Yang–Mills equations
J. Śniatycki 249

PART III: SHEAVES AND SPECTRAL SEQUENCES 267

La theórie des résidus sur un espace analytique complexe
V. Ancona and B. Gaveau 269

Derivation of exact triples and Leray–Koszul spectral sequences
B. Bendiffalah 281

PART IV: ELLIPTIC OPERATORS; INDEX THEORY 303

Le noyau de la chaleur des opérateurs sous-elliptiques des groupes d'Heisenberg
R. Beals, B. Gaveau, and P. Greiner 305

The geometry of Cauchy data spaces
B. Booss–Bavnbek, K. Furutani, and K.-P. Wojciechowski 321

On the Cauchy problem for Kirchhoff equations of p-laplacian type
K. Kajitani 355

A remark on surgery in index theory of elliptic operators
V.-E. Nazaikinskii and B.-Y. Sternin 363

The η invariant and elliptic operators in subspaces
A. Savin, B.-W. Schulze, and B. Sternin 373

Regularisation of mixed boundary problems
B.-W. Schulze, A. Shlapunov, and N. Tarkhanov 389

PART V: MATHEMATICAL PHYSICS 411

Covariant method for solution of Cauchy's problem based on Lie group
analysis and Leray's form
N. Ibragimov 413

Liouville forms, parallelisms and Cartan connections
P. Libermann 423

A two-dimensional non-linear shell model of Koiter's type
P.-G. Ciarlet 437

A model of the process of thinking based on the dynamics of bundles of
branches and sets of bundles of p-adic trees
A. Khrennikov 451

Global wave maps on black holes
Y. Choquet–Bruhat 469

Entanglement, parataxy, and cosmology
E. Binz and W. Schempp 483

Sur le contrôle des équations de Navier–Stokes
J.-L. Lions 543

Addresses 559

Preface

Jean Vaillant

L'oeuvre de Jean Leray est originale et profonde; ses théorèmes et ses théories sont au coeur des recherches mathématiques actuelles: la beauté de chacun de ses travaux ne se divise pas. Son cours de Princeton, sous forme de notes en anglais (et d'une traduction en russe) en est une belle illustration: ce cours présente les équations aux dérivées partielles à partir de la transformation de Laplace et du théorème de Cauchy–Kowaleska et contient l'essentiel de nombreuses recherches modernes. Leray avait pour but de résoudre un problème, souvent d'origine mécanique ou physique — qui se pose, et non qu'on se pose —, de démontrer un théorème; il construit alors son oeuvre de façon complète et essentiellement intrinsèque. En fait, Leray construit une théorie dont l'extension tient à son origine naturelle, l'acuité, la perfection, la profondeur d'esprit de son auteur; en même temps il domine les calculs, qu'il mène avec plaisir et élégance: «Il n'y a pas de mathématiques sans calculs» disait-il.

La science était au centre de la vie de Jean Leray. Il s'inquiétait de sa sauvegarde. Rappelons quelques phrases de ses textes de 1974:

> «D'ailleurs la science ne s'apprend pas: elle se comprend. Elle n'est pas lettre morte et les livres n'assurent pas sa pérennité; elle est une pensée vivante. Pour la maîtriser notre esprit doit, habilement guidé, la redécouvrir de même que notre corps à dû revivre dans le sein maternel, toute l'évolution qui créa notre espèce. Aussi n'y a-t-il qu'une façon efficace d'enseigner les sciences et les techniques: transmettre l'esprit de recherche.»

> «La vie scientifique est concentrée dans quelques institutions de recherche et quelques Universités. Elle est alimentée par quelques fonds privés et une très modeste partie des fonds publics. Ses besoins sont donc appréciés par des profanes. Or elle ne peu survivre qu'en se diversifiant et en suivant ses inspirations sans contrainte aucune: faire une découverte réellement originale, c'est faire ce que nul n'a encore fait, ce dont nul ne peut prévoir l'utilité; on ignore nécessairement l'intérêt qu'acquerra ce qu'on trouve.»

> «Il est vain d'effectuer des réformes formelles, ignorant le détail des difficultés, les compétences authentiques, les traditions fécondes. Il est dangereux d'appliquer à la vie intellectuelle les critères valables dans la vie économique: en simplifiant et uniformisant les structures, on détruit la féconde émulation des institutions; en planifiant, en ignorant l'importance, pour la recherche, des fortes personnalités, en jugeant les chercheurs par équipes, on traite l'activité intellectuelle comme doit l'être l'activité industrielle: en prétendant jauger le rendement d'un chercheur de réputation

*mondiale par la statistique des doctorats qu'il décerne et celle des cita-
tions qui sont faites de ses travaux on se targue de juger mécaniquement
l'activité intellectuelle.»*

*«La primauté de l'enseignement mathématique sur les autres enseigne-
ments scientifiques ne peut résulter que de ses qualités interdisciplinaires.»*

Jean Leray manifestait une grande curiosité et une vraie bienveillance pour
les recherches nouvelles et originales. Les textes rassemblés ici tentent de
respecter cette ouverture d'esprit, ils concernent les équations aux dérivées partielles de tous types, les problèmes holomorphes, la géométrie et la mécanique,
la topologie et l'analyse, la physique mathématique.

Cet hommage réalisé à Karlskrona est un témoignage international de
l'immense renom de Jean Leray, pour nous, le plus grand mathématicien de
ce siècle.

Introduction

The Conference in the Honor of Jean Leray was held at the University of Karlskrona, in the south east of Sweden during the last week of August 1999. There were around fifty participants from all around the world: from France, of course, but also from Belgium, Germany, Italy, Poland, Russia, Scandinavia, India, Japan, and North America. All these participants had come to honor the memory of one of the twentieth century's greatest mathematicians, probably even the greatest. An overwhelming majority of the talks were held in Leray's language, French. This, and that there were so many participants from Paris, led some people to say, jokingly, that this was Sweden's first *conférence parisienne* ...

Jean Vaillant, who knew Leray for much longer than I, tells us in the preface to these Proceedings about the decisive influence Leray had on contemporary mathematics. I want to say in this Introduction a few words about the *person* Jean Leray.

My first direct contact with Leray's work was in 1984, when I discovered his treatise *Lagrangian Analysis, and Quantum Mechanics*. I was at that time still working on the theory of pseudo-differential operators, and I immediately was captivated by Leray's style, both rigorous and concise, but at the same time of an incredible clarity. This was very different from what I had been accustomed to; in the eighties it was still trendy in France to pretend that there was no life for partial differential equation outside the rather new calculus of pseudo-differential operators developed in the early 70's (and by which Leray was only moderately impressed ...). What struck me most was the geometric set-up of Lagrangian analysis, which I perceived as both deep, elegant, and efficient. In particular, I was captivated by Leray's construction of the Maslov index, and I started trying to develop his idea. However, it required very hard work to understand the underlying machinery and I soon realized that I would be rather quickly stuck without help. I decided to ask the Master himself for some advice and I wrote him a polite letter, apologizing for coming with such stupid and trivial questions. Just a few days later came an answer. Not only did the great man assure me that my questions were not stupid at all, but he encouraged me to pursue my efforts: «*Votre jeune enthousiasme me plaît*»[1] he said, adding «*Mon Analyse Lagrangienne vous intéresse? Elle m'a passionnée!*»[2] My 'youth' was actually very relative; I was almost thirty five, but this was typical of Leray: once he called to tell me that he would defer for some days the proof of theorem he wanted to send me because he was receiving a young Japanese mathematician. This young mathematician, Professor Hamada (who has contributed to this volume) was over sixty!

---

[1] «I like your youthful enthusiasm»

[2] «You like my Lagrangian Analysis? It is my passion!»

There have been many articles listing, commenting, and explaining the impact of his mathematics. But no one, to the best of my knowledge, has ever mentioned his *human qualities*: kindness, warmth, and carefulness. Here is one example. We were working in 1990 on some extensions of the Maslov index. There were some delicate points that needed to be carefully analyzed and discussed. I was in Nice at that moment, Leray was in Paris, so we exchanged letters and phone calls. As soon as he had received one of my letters, he would give me a phone call, always late in the afternoon (some times even *very late!*) and give me some valuable advice on this or that refinement of a particular point. Our conversations often lasted more than one hour. When Leray had decided that he had enough, he abruptly hung up, just saying «*Bonsoir, et mes hommages à Madame de Gosson . . .*»: he was also a *gentleman*, definitely a rare quality today!

My wife and I used to visit the Leray's each summer at the lovely summer resort La Baule, in Bretagne, where he had retired (he was living in a flat, just a few miles away from his family home in La Turballe). We met Leray there for the last time in July 1997. He looked tired, but was still working on some questions of Lagrangian Analysis (this was heroic because he did not have access to his books anymore: he had donated his private library to the University of Paris VI). During our conversation, I asked him which, of all theories he had created or significantly contributed to, he preferred. He answered, with a little smile:

«*Monsieur de Gosson, j'aime autant tous mes enfants!*»[3]

This is the Leray I knew, and will always remember.

Because Jean Leray contributed to so many different areas of pure and applied mathematics, I found it not only fair, but also necessary, to divide this Volume into five parts, each devoted to one aspects of Jean Leray's work. This division is for, some texts, somewhat artificial because some areas of course overlap, especially since mathematics is becoming more and more interdisciplinary. I have tried to respect as much as possible the wishes of the authors concerning the category under which they appear, but I am of course, in the end, the only one responsible for any errors or misinterpretations.

Maurice de Gosson

---

[3] 'Monsieur de Gosson, I love all my children equally'

Acknowledgments

The immense success of the Conference in Honor of Jean Leray is owed to the enthusiasm and generosity, both in time, efforts and money, of many great people and institutions. It was indeed a challenge to organize an international conference in the memory of a French mathematician (even of the calibre of Jean Leray!) in a small local Swedish University. These difficulties were overcome thanks to the encouragements and the logistical help of the local academic authorities, especially those of Dr. Per Ericsson, President of the University, and Professor Claes Jogréus, Dean of Science.

But I wish above all to express my warmest thanks to Professor Jean Vaillant, University of Paris VI, who was co-organizer of the event. Without his invaluable help and efforts, this conference would not have reached this international dimension. My warmest thanks also, to the Members of the Scientific Committee: Professor Y. Choquet–Bruhat, B. Booss–Bavnbek, K. Furutani, B.-W. Schulze, ... I want to express my very special gratitude to a key person, my beloved wife Charlyne. Thanks to her efficiency in organizing and solving all the practical details, and her kindness, she succeeded in making this Conference also a successful social event!

The major part of the financing from the Swedish side came from the University of Karlskrona–Ronneby and from Wenner-Grén Foundation (Stockholm), but the City of Karlskrona (Telecom City) and Regional Council (Länsstyrelsen) also contributed significantly. Travel grants were given by the University of Paris VI, and also by the French Embassy in Stockholm, whose Scientific Attaché, Monsieur B. Gustavino, honored us with his presence at the Conference. Air France helped us significantly by arranging a group fare from Paris to Copenhagen for about thirty participants, who were then taken by chartered bus to Karlskrona.

Finally, I wish to express my gratitude to Prof. A. Boutet de Monvel for her encouragement, kindness and support for the publication of these Proceedings by Kluwer Academic Publishers. Last, but not least, thanks are due to Serge de Gosson, who turned a fluctuating compilation of author files into this book. Those who have produced a Proceedings volume before know how much hard work this requires.

Maurice de Gosson

PART I

HYPERBOLIC SYSTEMS AND EQUATIONS

CARACTÈRE HOLONÔME D'UNE SOLUTION ÉLÉMENTAIRE

L. BOUTET DE MONVEL

Contents

1   Introduction   9
2   Opérateurs à coefficients constants   10
3   Opérateurs à coefficients variables   12
4   Conclusions. $\mathcal{E}$-modules et $\mathcal{D}$-modules   14

M. de Gosson (ed.), Jean Leray '99 Conference Proceedings, 7-16.
© *2003 Kluwer Academic Publishers.*

Résumé. *Nous décrivons un système holonôme régulier d'équations aux dérivées partielles analytiques simple et naturel que devrait idéalement satisfaire la solution élémentaire d'un opérateur différentiel $P(x, d)$ à coefficients analytiques et à caractéristiques simples. Ce système est effectivement vérifié par la solution élémentaire avancée de P lorsque P est hyperbolique, ou mod les fonctions analytiques lorsque P est elliptique. Ceci permet de retrouver de façon un peu plus «algébrique» les théorèmes de Leray qui expriment que la solution élémentaire a un prolongement holomorphe ramifié, ainsi que le lieu de ramification et la monodromie. Ce travail a été développé par D. Meyer dans sa thèse [27] (Paris 6, déc 1998).*

Abstract. *Holonomic character of an elementary solution. We describe a regular holonomic system of partial differential equations which should ideally be satisfied by an elementary solution of a differential operator $P(x, d)$ with simple characteristics and analytic coefficients. In many concrete cases this is exact, and gives the correct ramification locus and monodromy as described in the theory of J. Leray. This work was developed by D. Meyer in his thesis [27] (Paris 6, Dec. 1998).*

1. Introduction

Pour décrire notre sujet nous aurons besoin de rappeler deux notions. La première est celle des fonctions de classe de Nilsson, qui généralisent les fonctions à singularité du type de Fuchs (cf. [28], [23]). Une fonction de classe de Nilsson sur un ouvert $X \subset \mathbf{C}^n$ (ou plus généralement U variété complexe) est une fonction holomorphe f sur X ramifiée le long d'une hypersurface Γ de U, i.e., une fonction holomorphe sur le revêtement universel de $U - \Gamma$ telle que:

(i)   les différentes branches γf, $\gamma \in \Pi_1(U - \Gamma)$ engendrent un espace vectoriel de dimension finie (sur $\mathbf{C}$);

(ii) f est à croissance modérée le long de Γ;

exemple type: Γ est définie par une équation $u = 0$ et $f = u^s \operatorname{Log}^k u$: $s \in \mathbf{C}$, k entier ≥ 0 (on montre que f est de classe de Nilsson, ramifiée le long de Γ, si et seulement si au voisinage de chaque point lisse de Γ elle est combinaison linéaire finie, à coefficients holomorphes, de fonctions ramifiées élémentaires du type ci-dessus).

La deuxième est la notion de $\mathscr{D}$-module holonôme, ou système holonôme d'équations différentielles, resp. holonôme régulier, introduite par M. Sato, T. Kawai et M. Kashiwara dans leurs théorie des $\mathscr{D}$-modules et des systèmes d'équations différentielles (cf. [11], [15], [16]; $\mathscr{D}$ désigne le faisceau des opérateurs différentiels à coefficients analytiques). Un système (fini) d'équations différentielles à coefficients analytiques (M) sur U:

$$P_j(x, d)f = 0$$

est holonôme si sa variété caractéristique $car\ M \subset T^*U$ est de dimension minimum n. Rappelons que $car\ M$ est sous-variété analytique du fibré cotangent définie par les équations $\sigma_{P_j}(x, \xi) = 0$ (plus exactement $car\ M$ est la variété des zéros communs des σ_P lorsque P parcourt l'idéal engendré par les P_j; σ_P désigne le symbole de $P(x, d)$ i.e., la somme des termes de plus haut degré du polynôme $P(x, \xi)$). Dans le cas d'un système M à coefficients analytiques $car\ M$ est involutive (cf. [16], [26]), i.e., la forme symplectique canonique $\sum d\xi_j\, dx_i$ (ou la forme de Liouville $\sum \xi_j\, dx_j$) s'y annule, de sorte que sa dimension est $\geq n$ (ceci est faux pour les systèmes à coefficients C^∞, dont il ne sera jamais question ici). Parmi les systèmes holonômes on distingue les systèmes «holonômes réguliers» (cf. [15]), dont les solutions sont de classe de Nilsson (à croissance modérée), et qui sont la géneralisation naturelle à plusieurs variables des équations du type de Fuchs.

Il est notoire et remarquable que la solution élémentaire des opérateurs différentiels les plus usuels et les plus utiles (équations des ondes et de Maxwell, équations de Laplace et bien d'autres) sont «holonômes régulières», i.e., s'obtiennent en raccordant par morceaux de fonctions analytiques de classe de Nilsson. La solution élémentaire de l'équation de la chaleur $(4\pi t)^{-n/2}\exp{-x^2/4t}$ est aussi holonôme, irrégulière (on a perdu le caractère «à croissance modérée»). De nombreux auteurs (Gårding [6], [7], Hadamard [8] et Petrowskii [29], [30]) ont contribué à mettre cela en évidence, et J. Leray a expliqué cela de façon générale pour les opérateurs hyperboliques à coefficients analytiques dans ses articles sur le problème de Cauchy (cf. [18], [20], [21], [22]).

Dans leurs théorie des $\mathcal{D}$-modules et des systèmes d'équations différentielles (cf. [11], [15], [16]), M. Sato, T. Kawai et M. Kashiwara ont introduit la notion de $\mathcal{D}$-module holonôme resp. holonôme régulier, qui correspondent (avec des précisions supplémentaires) aux systèmes d'équations dont les solutions sont les fonctions de classes de Nilsson: les solutions élémentaires usuelles des équations classiques vérifient un système holonôme d'équations différentielles. Il est alors naturel, pour une équation donnée, d'essayer de construire un système holonôme que doit satisfaire la solution élémentaire. L'existence d'un tel système redonnerait de façon plus algébrique et plus simple le théorème de J. Leray.

2. Opérateurs à coefficients constants

Dans [1] I.M. Bernstein a montré que si P est un opérateur différentiel à coefficients constants sur $\mathbb{R}^n$ il admet toujours une solution élémentaire E holonôme (peut-être pas à croissance modérée—i.e., «régulière»); E a pour transformée de Fourier la distribution $P(\xi)^{-1}$, construite en utilisant le prolongement méromorphe de l'application $s \rightarrow f^\lambda$, si f est un polynôme positif. Mais sa démonstration, bien que remarquablement simple, est trop implicite pour donner commodément des renseignements sur le lieu de ramification, le mode de croissance le long de celui-ci, ou la monodromie (entre autres parce que le lien entre les propriétés géométriques de E et de sa transformée de Fourier reste compliqué); en outre cette démonstration ne se généralise pas aux opérateurs à coefficients variables, pour lesquels le résultat est en général faux.

Au contraire il est facile de former un système d'équations différentielles qui exprime idéalement qu'une distribution $E(x, y)$ est le noyau de «l'inverse» de P: E est un noyau de convolution, de la forme $T(x - y)$ où T est une distribution sur $\mathbb{R}^n$; autrement dit on a

$$\left(\frac{\partial}{\partial x_i} - \frac{\partial}{\partial y_i} \right) E = 0 \quad \text{pour } i = 1, \ldots, n. \tag{2.1}$$

En outre l'inverse de P est une fonction de P, donc la transformée de Fourier $\widehat{T}$ devrait vérifier les équations

$$P(\xi)\widehat{T} = 1 \tag{2.2}$$

$$\partial \widehat{T} \wedge \partial P = 0, \quad \text{i.e.,} \quad P_i(\xi) : \frac{\partial \widehat{T}}{\partial \xi_j} - P_j(\xi) \frac{\partial \widehat{T}}{\partial \xi_j} \quad \text{pour } i, j = 1, \ldots, n \tag{2.3}$$

où ∂ désigne la différentielle extérieure par rapport à ξ, et on a posé

$$P_i = \frac{\partial P(\xi)}{\partial \xi_i}$$

ce qui équivaut, en plus des équations (2.1) à

$$P(d)T = \delta \tag{2.4}$$

$$\left(x_i P_j(d) - x_j P_i(d) \right) T = 0 \quad \text{pour } i, j = 1, \ldots, n. \tag{2.5}$$

Ce système d'équations résulte aussi du système suivant (qui lui est équivalent si P est à caractéristiques simples):

$$\text{on a} \quad B \circ E = E \circ A \tag{2.6}$$

pour tout couple d'opérateurs différentiels $(A(x, d), B(x, d))$ tel que

$$A \circ P = P \circ B$$

que devrait évidemment satisfaire l'inverse de P si tout se passait dans une algèbre associative (on a utilisé la notation des opérateurs et noté par la même lettre E et l'opérateur de noyau E).

En général le système (2.6) est insuffisant pour décrire une solution élémentaire; mais si P est à caractéristiques simples il est immédiat de vérifier que c'est un bon système d'équations différentielles:

Theorem 2.1. *Si P est à caractéristiques simples, le système (2.6) est équivalent à (2.1), (2.4), (2.5); il est holonôme régulier, bien défini dans le domaine complexe (sur $\mathbf{C}^n \times \mathbf{C}^n$); sa variété caractéristique est la réunion de la diagonale et du flot bicaractéristique de P issu de la diagonale.*

Lorsque P est hyperbolique, sa solution élémentaire avancée satisfait (2.6) (tout a un sens dans l'algèbre des opérateurs à noyau dans le cône d'ondes d'avenir). De même si P est elliptique la solution élémentaire satisfait (2.6) mod une fonction analytique comme on voit en se plaçant dans l'algèbre des opérateurs pseudo-différentiels analytiques.

En général l'argument d'algèbre associative invoqué plus haut ne marche pas et il n'y a plus vraiment de bonne relation entre le système (2.6) et la solution élémentaire. Par exemple si P est à partie principale complexe, il a une solution élémentaire holonôme mais qui en général ne vérifie pas (2.6) (qui est un système trop exigeant). Si P est à caractéristiques multiples, le système (2.6) n'est en général pas holonôme (il n'y a pas assez d'équations); dans le cas de l'opérateur de la chaleur

$$P = \frac{\partial}{\partial t} - \Delta$$

le système (2.6) est bien holonôme (pas régulier) mais cela ne se voit pas sur les symboles (parties dominantes) des opérateurs.

3. Opérateurs à coefficients variables

Soit P un opérateur différentiel à coefficents analytiques. P est d'abord défini sur une variété analytique réelle Y et se prolonge analytiquement à un voisinage complexe X de Y. La solution élémentaire (si elle existe) a pour noyau une distribution sur la variété réelle $Y \times Y$ (en fait nous nous limiterons à un petit voisinage de la diagonale). Mais cela a un sens de dire qu'elle satisfait à un système d'équations différentielles (ou pseudo-différentielles) analytiques, et un tel système, comme ceux que nous aurons à considérer ici, se prolongent à un petit voisinage complexe de la diagonale de Y dans $X \times X$.

Dans le cas des opérateurs à coefficients variables, la situation pour la solution élémentaire est encore pire que dans le cas des opérateurs à coefficients constants. Tout d'abord s'il ne s'agit pas d'un opérateur à caractéristiques simples, la géométrie des bicaractéristiques peut être trop compliquée et trop mauvaise pour que le flot bicaractéristique définisse une variété, ce qui ne laisse peu d'espoir qu'il existe une solution élémentaire holonôme. Pour ces raisons nous nous limitons exclusivement aux opérateurs à caractéristiques simples (il s'agit de toutes les caractéristiques, réelles ou complexes; nous nous limitons aussi exclusivement aux opérateurs à coefficients analytiques pour pouvoir utiliser la théorie des $\mathcal{D}$-modules cohérents).

Une conséquence géométrique immédiate de l'hypothèse «à caractéristiques simples» est la suivante:

Proposition. On suppose P à caractéristiques simples sur X complexe. Alors il existe un voisinage de la diagonale au-dessus duquel le flot bicaractéristique de P issu de la diagolale est une variété Λ.

Rappelons que les bicaractéristiques de P sont les courbes intégrales du champ hamiltonien

$$\sum \frac{\partial p}{\partial \xi_i} \frac{\partial}{\partial x_i} - \frac{\partial p}{\partial x_i} \frac{\partial}{\partial \xi_i}$$

où p désigne le symbole de P. $\Lambda \subset T^*X \times T^*X$ est l'ensemble des couples $((x, \xi), (y, \eta))$, extrémités d'un arc même bicaractéristique au dessus d'un voisinage convenable assez petit de la diagonale ($\xi, \eta \neq 0$). La proposition résulte du fait que si P est à caractéristiques simples, le champ hamiltonien est homogène et sans point fixe (pour $\xi \neq 0$).

Le système (2.6) en général ne contient pas assez d'équations (il n'y a pas assez d'opérateurs qui commutent à P), mais il est naturel de regarder ce qu'il devient en microlocalisant. Notons que l'équation $AP = PB$ suppose A et B définis au même point, donc le système (2.6) définit au mieux un germe de faisceau de $\mathcal{E}$-modules au voisinage de la diagonale (ou, puisqu'il s'agira de modules cohérents, un $\mathcal{E}$-module défini dans un petit voisinage de la diagonale). $\mathcal{E}$ désigne le faisceau des opérateurs microdifférentiel, qui est un faisceau sur T^*X ou $T^*(X \times X)$ privé de sa section nulle (X complexe).

Ceci dit, comme P est à caractéristiques simples, il est microlocalement équivalent au voisinage de chaque point à $(\partial/\partial x_1) \times$ elliptique. Comme le théorème est vrai pour l'opérateur $\partial/\partial x_1$, il est aussi vrai, microlocalement, pour P. Autrement dit.

Proposition. Si P est à caractéristiques simples, le système (2.6) définit un germe de $\mathcal{E}$-module holonôme $\mathcal{M}$ sur la diagonale de $T^*X \times T^*X$ (qui se prolonge à un petit voisinage complexe de celle-ci).

Le $\mathcal{E}$-module $\mathcal{M}$ n'est pas simple, en fait il y a une suite exacte $0 \to \mathcal{N} \to \mathcal{M} \to \delta \to 0$ où les deux extrémités sont de multiplicité 1; δ est le $\mathcal{E}$-module correspondant à l'opérateur identité; c'est un quotient de $\mathcal{M}$ parce que le système (2.6) contient les équations $(x_i - y_i)P(x, d_x)E = 0$ essentiellement équivalentes à $P(x, d_x)E = \delta(x - y)$. $\mathcal{N}$ est de multiplicité 1, et sa variété caractéristique est exactement le flot bicaractéristique de P issu de la diagonale. (Rappelons que ces objets sont définis seulement au dessus d'un voisinage de la diagonale de $X \times X$; le fait qu'il se prolonge au dessus de $X \times X$ tout entier (i.e., que $\mathcal{N}$ se prolonge) dépend alors essentiellement uniquement de la géométrie globale des bicaractéristiques de P).

4. Conclusions. $\mathcal{E}$-modules et $\mathcal{D}$-modules

Pour pouvoir tirer de cette construction des conclusion pour la solution élémentaire, il faut encore montrer que notre $\mathcal{E}$-module $\mathcal{M}$ correspond en fait à un $\mathcal{D}$-module (ce sont les $\mathcal{D}$ modules holonômes réguliers qui sont bien reliés aux fonctions de classes de Nilsson, pas les $\mathcal{E}$-modules). Dans sa thèse, en utilisant le théorème de cohérence des images directes non caractéristiques (non propres) de Houzel et Schapira [9], D. Meyer montre en effet que le germe $p_* \mathcal{M}_x$ de l'image de $\mathcal{M}$ par la projection $p : T^* X \times X \to X \times X$ en point (x, x) de la diagonale est un $\mathcal{D}$-module holonôme, dont les sections engendrent $\mathcal{M}$.

Ceci pose plus généralement la question du lien entre $\mathcal{D}$-modules et $\mathcal{E}$-modules sur une variété complexe, en particulier dans le cas holonôme (régulier ou non). M. Kashiwara et T. Kawai [15] ont démontré qu'il y a une bonne relation en un point «générique» $x \in X$, i.e., tel que la fibre Λ_ξ du support (Lagrangien) Λ soit réduite à un seul rayon (ou à un nombre fini de rayons); les thèses de D. Meyer [27] et M. Carette [5] donnent d'autres exemples utiles. On peut espérer qu'il y a toujours une bonne relation dans le cas holonôme, mais ce n'est certainement pas vrai en général, et on peut essayer de comprendre les obstructions (la première obstruction vient des fibres de dimension 2 de la variété caractéristique).

Ceci étant, sachant que le module microlocal $\mathcal{M}$ correspond à un vrai système d'équations différentielles, reste à savoir si l'opérateur P possède bien une solution élémentaire satisfaisant à ces équations. C'est encore moins probable que dans le cas des opérateurs à coefficients constants. Par exemple l'opérateur de H. Lewy, qui est à caractéristiques simples, n'a pas de solution élémentaire du tout. Mais ici encore le système microlocal (2.6) est bien satisfait par la solution élémentaire avancée de P si P est hyperbolique, et on retrouve ainsi les théorèmes de J. Leray. Il l'est aussi (mod une fonction analytique) si P est elliptique, ce qui explique le caractère holonôme de la solution élémentaire dans ce cas.

Références

[1]   Bernstein, I.N., Modules over rings of differential operators. An investigation of the fundamental solution of equations with constant coefficients, *Funkc. Anal i Prilozen* **5**(2) (1971), 1–16 & *Funct. Anal. appl.* **5** (1971), 89–101.

[2]   _____, Modules over a ring of differential operators, *Funct. An. and Appl.* **6** (1972), 273–285.

[3]   _____, The analytic continuation of generalized functions with respect to a parameter, *Funct. Anal. and Appl.* **6** (1972), 273–285.

[4]   Boutet de Monvel, L., On the holonomic character of the elementary solution of a partial differential operator, in Bony, J.M. and Morimoto, M. (eds), *New trends in Microlocal Analysis*, Springer–Verlag Tokyo (1997), 171–178.

[5]   Carette, M., *E-modules microlocalement libres et connexions non-intégrables en dimension 2*, thèse Paris **6**, nov. 1999.

[6]   Gårding, L., Linear hyperbolic partial differential equations with constant coefficients, *Acta Math.* **85** (1951), 1–62.

[7]   _____, Solution directe du problème de Cauchy pour les équations hyperboliques, *Coll. Int. CNRS, Nancy* (1956), 71–90.

[8]   Hadamard, J., *Oeuvres*, 4 vol., C.N.R.S., Paris, 1968.

[9]   Houzel, Ch. and Schapira, P., Images directes de modules différentiels, *C. R. Acad. Sc.* **298** (1984), 461–464.

[10]   John, F., *Plane waves and spherical means applied to partial differential equations*, Interscience, New York 1955.

[11]   Kashiwara, M., On the maximally overdetermined systems of linear differential equations I, *Publ. RIMS, Kyoto University* **10** (1975), 563–5.

[12]   _____, *Faisceaux constructibles et systèmes holonômes d'équations aux dérivées partielles linéaires à points singuliers réguliers*. Sém. Goulaouic–Schwartz 1979–80, exposé no. 19.

[13]   _____, *Systems of microdifferential equations*, Cours à l'Université Paris Nord, Progress in Math. **34**, Birkhäuser 1983.

[14]   Kashiwara, M., Kawai, T., and Kiwura, T., *Foundations of algebraic analysis*, Princeton Math. Series 37, Princeton University Press, Princeton N.J. 1986.

[15]   Kashiwara, M. and Kawai, T., *On holonomic systems of micro-differential equations III—systems with regular singularities*, publ. RIMS, Kyoto University **17** (1981), 813–979.

[16]   Kashiwara, M., Kawai, T., and Sato, M., *Microfunctions and pseudo-differential equations*, Lecture Notes **287** (1973), 265–524, Springer–Verlag.

[17]   Leray, J., Le problème de Cauchy pour une équation linéaire à coefficients polynomiaux, *C. R. Acad. Sc. Paris* **242** (1956), 1483–1488.

[18]   _____, Uniformisation de la solution du problème linéaire analytique de Cauchy près de la variété qui porte les données de Cauchy (problème de Cauchy I), *Bull. Soc. Math. France* **85** (1957), 389–429.

[19]   _____, La solution élémentaire d'un opérateur différentiel linéaire, *Bull. Soc. Math. France* **86** (1958), 389–429.

[20]   _____, Le calcul différentiel et intégral sur une variété complexe (problème de Cauchy III), *Bull. Soc. Math. France* **87** (1959), 81–180.

[21]   _____, Un prolongement de la transformée de Laplace qui transforme la solution unitaire d'un opérateur hyperbolique en sa solution élémentaire (problème de Cauchy IV), *Bull. Soc. Math. France* **90** (1962), 39–156.

[22]   Garding, L., Kotake, T., and Leray, J., Uniformisation et développement asymptotique de la solution du problème de Cauchy linéaire, à données holomorphes; analogie avec la théorie des ondes asymptotiques et approchées (problème de Cauchy Ibis et VI), *Bull. Soc. Math. France* **92** (1964), 263–361.

[23] Leray, J., Un complément au théorème de N. Nilsson sur les intégrales de formes différentielles à support singulier algébrique, *Bull. Soc. Math. France* **95** (1967), 313–374.

[24] ———, *Solutions asymptotiques et groupe métaplectique*, Séminaire sur les équations aux dérivées partielles 1973–74, Collège de France, Paris.

[25] ———, *Analyse Lagrangienne et mécanique quantique*, Séminaire sur les équations aux dérivées partielles 1976–77, Collège de France, Paris.

[26] Malgrange, B., *L'involutivité des caractéristiques des systèmes différentiels et micro-différentiels*, Séminaire Bourbaki 1977–78, no. 552.

[27] Meyer, D., *D*-modules et *E*-modules associés à un opérateur différentiel à caractéristiques simples, *C. R. Acad. Sci. Paris* **328**, Série I, (1999), 489–494; thèse, Paris 6, déc. 1998 et article en préparation.

[28] Nilsson, N., Some growth and ramification properties of certain integrals on algebraic manifolds, *Ark. Mat.* **5** (1965), 463–476.

[29] Petrowsky, I.G., Über das Cauchysche Problem für ein System linearer partieller Differentialgleichungen, *Mat. Sb.* **2**(44) (1937), 815–870.

[30] ———, Über das Cauchysche Problem für Systeme von partiellen Differentialgleichungen im Gebiete der nichtanalytischen Funktionen, *Bull. Univ. Moscow Sér. Int.* **1**, no. 7 (1938), 1–74.

[31] Schapira, P. and Leray, J., *Microdifferential systems in the complex domain*, Grundlehren der mathematischen Wissenschaften **269**, Springer 1985.

NECESSARY CONDITIONS FOR HYPERBOLICITY OF FIRST ORDER SYSTEMS

ANTONIO BOVE AND TATSUO NISHITANI

Contents

1   Introduction... 19
2   Inductive lemmas 20
3   Proof of Theorem 1.1 31

17

M. de Gosson (ed.), Jean Leray '99 Conference Proceedings, 17-35.

1. Introduction

In this note some necessary conditions for the well posedness of the Cauchy problem for hyperbolic systems of arbitrary order will be studied. In the scalar case Ivrii and Petkov [7] has shown that the well posedness of the Cauchy problem implies that, near a multiple characteristic point, a set of vanishing conditions on the homogeneous parts of the lower order terms must be satisfied. Evidently in the case of hyperbolic differential systems the situation is much more complex and the vector structure playing a relevant role. As a consequence the above mentioned result is not true any more.

Our purpose here is to study a truly vectorial case and to find necessary conditions for the well posedness which correspond to Ivrii–Petkov conditions mentioned above. Let m, N be positive integers and Ω an open subset of $\mathbb{R}^{n+1}$ containing the origin. We denote by $P(x, D)$ a differential operator of order m with coefficients in $C^\infty(\Omega, M_N(\mathbb{C}))$, the space of all $N \times N$ matrices depending smoothly on $x \in \Omega$. We shall also write

$$P(x, D) = P_m(x, D) + P_{m-1}(x, D) + \cdots + P_0(x), \qquad (1.1)$$

where $P_{m-j}(x, D)$ is the homogeneous part of order $m - j$. We shall always assume that the polynomial $h(x, \xi_0, \xi') = \det P_m(x, \xi_0, \xi')$ has only real roots with respect to ξ_0. We also assume that the hyperplane $\{x_0 = 0\}$ is non-characteristic for P, i.e., it is non-characteristic for h. Let $\rho \in T^*\Omega \setminus 0$ be a characteristic point of P_m of multiplicity r and define $P_{m-j,\rho}$ by

$$P_{m-j}\left(\rho + \lambda^{-1}\delta z\right) = \lambda^{-s_j}\left(P_{m-j,\rho}(\delta z) + O\left(\lambda^{-1}\right)\right), \quad \lambda \longrightarrow \infty,$$

where $P_{m-j,\rho}(\delta z)$ does not vanish identically. Then we have

Theorem 1.1. *Let P be as in (1.1) and ρ be a characteristic point of P_m such that $d^j P_m(\rho) = 0$ for $j = 0, 1, \ldots, r - 1$. Denote by s_j the degree of the homogeneous polynomial $P_{m-j,\rho}$, $j = 1, \ldots, [r/2]$. Assume that:*

    i) There exists at least one $j \in \{1, \ldots, [r/2]\}$ for which $s_j < r - 2j$. Define then

$$\theta_0 = \min_{\substack{j \in \{1,\ldots,[r/2]\} \\ s_j < r-2j}} \frac{j}{r - s_j}, \qquad \widehat{P}(x, \xi) = \sum_{\substack{m-r\theta_0=m-j-s_j\theta_0 \\ s_j < r \text{ when } j > 0}} P_{m-j,\rho}(x, \xi). \quad (1.2)$$

    ii) The polynomial $\xi_0 \mapsto \det \widehat{P}(x, \xi_0, \xi')$ has non-real roots at least one of which has multiplicity at most 3.

Then the Cauchy problem for P is not well posed.

In the subsequent sections we give a sketch of the proof of Theorem 1.1 when $\det \widehat{P} = 0$ has a non-real double root. The detailed proof, including the case of non-real triple root, will be given in [3].

2. Inductive lemmas

Let now θ_0 be defined by (1.2). Due to assumption i) in Theorem 1.1, we obtain that

$$\theta_0 < \frac{1}{2}.$$

Define $\sigma_0 = 1 - 2\theta_0$. For sake of simplicity we shall assume that $\rho = (0, e_n)$. Let us set

$$P_\lambda(x, D) = \sum_{j=0}^{m} P_{m-j} \left(\lambda^{-\theta_0} x, \lambda e_n + \lambda^{\theta_0} D \right)$$

$$= \sum_{j=0}^{m} \lambda^{m-j} P_{m-j} \left(\lambda^{-\theta_0} x, e_n + \lambda^{-\theta_0-\sigma_0} D \right)$$

$$= \sum_{j=0}^{m} \sum_{\substack{k \geq 0 \\ m-j-s_j\theta_0-k\theta_0 > -M}} \lambda^{m-j-s_j\theta_0-k\theta_0}$$

$$\times \sum_{|\alpha+\beta|=s_j+k} \frac{1}{\alpha!\beta!} P_{m-j(\beta)}^{(\alpha)} (0, e_n) x^\beta \left(\lambda^{-\sigma_0} D \right)^\alpha + O\left(\lambda^{-M} \right),$$

where M is an arbitrarily large positive integer and by $O(\lambda^{-M})$ we denote a differential operator whose coefficients are bounded by λ^{-M} on any bounded preassigned open set U in $\mathbb{R}^{n+1}$. Let us denote $U^\pm = \{x \in U \mid \pm(x_0 - t) < 0\}$ for an open set U in $\mathbb{R}^{n+1}$. Here we recall (see Proposition 2.2 in [8]).

Proposition 2.1. Assume that $0 \in \Omega$ and the Cauchy problem for $P(x, D)$ is well posed in both Ω^t and Ω_t for every small t. Then for every compact set $W \subset \mathbb{R}^{n+1}$ and for every positive $T > 0$ we can find $C > 0$, $\bar\lambda > 0$ and $p \in \mathbb{N}$ such that

$$\|u\|_{C^0(W^t)} \leq C\lambda^{(\theta_0+1)p} \|P_\lambda u\|_{C^p(W^t)},$$

$$\|u\|_{C^0(W_t)} \leq C\lambda^{(\theta_0+1)p} \|P_\lambda u\|_{C^p(W_t)}$$

for every $u \in C_0^\infty(W)$, $\lambda \geq \bar\lambda$, $|t| < T$.

The method of proof is just to construct an asymptotic solution u_λ to the equation $P_\lambda u_\lambda \equiv 0$ depending on a large parameter λ and violating the a priori estimate in Proposition 2.1. Let

$$\widetilde{G}^{(0)}(x, \xi; \lambda) = \sum_{j=0}^{m} \sum_{\substack{k \geq 0 \\ m-j-s_j\theta_0-k\theta_0 > -M}} \lambda^{-j+(r-s_j)\theta_0-k\theta_0}$$

$$\times \sum_{|\alpha+\beta|=s_j+k} \frac{1}{\alpha!\beta!} P_{m-j(\beta)}^{(\alpha)}(0, e_n) x^\beta \xi^\alpha \tag{2.1}$$

then it is clear that

$$P_\lambda(x, D) = \lambda^{m-r\theta_0} \widetilde{G}^{(0)}\left(x, \lambda^{-\sigma_0} D; \lambda\right) + O\left(\lambda^{-M}\right).$$

It is useful to rewrite $\widetilde{G}^{(0)}(x, \xi; \lambda)$ in the following way

$$\widetilde{G}^{(0)}(x, \xi; \lambda) = \sum_{j\geq 0} \lambda^{-\delta_j(\widetilde{G}^{(0)})} \widetilde{G}_j^{(0)}(x, \xi), \tag{2.2}$$

where

$$0 = \delta_0\left(\widetilde{G}^{(0)}\right) < \delta_1\left(\widetilde{G}^{(0)}\right) < \cdots < \delta_j\left(\widetilde{G}^{(0)}\right) < \cdots$$

and it is understood that the sum in (2.2) is finite. Furthermore from (2.1) we obtain

$$\widetilde{G}_0^{(0)}(x, \xi) = \sum_{\substack{j\geq 0, \, s_j < r \\ j-(r-s_j)\theta_0=0}} P_{m-j,\rho}(x, \xi) = \widehat{P}(x, \xi).$$

Note that $\delta_1(\widetilde{G}^{(0)}) \geq \theta_0$. We say that a differential operator $P(x, D; \lambda)$, depending on a large positive parameter λ is in the class $\mathfrak{R}$ if there exists a positive rational number κ and differential operators $P_j(x, D)$, $j = 0, 1, \ldots, L$ for some $L \in \mathbb{N}$ whose coefficients are C^∞ in some open set such that

$$P(x, D; \lambda) = \sum_{j=0}^{L} \lambda^{-\kappa j} P_j(x, D).$$

Our next step is to prove the following lemma.

Lemma 2.2. *Let $G(x, D)$ be a differential operator with smooth coefficients defined in U and let σ, θ be rational numbers with $\sigma \geq \theta > 0$. Denote by $\varphi(x)$ a function in $C^\infty(U)$. Then:*

i)

$$e^{-i\lambda^\theta \varphi(x)} G\left(x, \lambda^{-\sigma} D\right) e^{i\lambda^\theta \varphi(x)}$$
$$= G\left(x, \lambda^{-(\sigma-\theta)}\left(\varphi_x(x) + \lambda^{-\theta} D\right)\right) + \lambda^{-\theta} r\left(x, \lambda^{-\theta} D; \lambda\right), \quad (2.3)$$

where $r \in \mathcal{R}$.

ii) If $G(x, \xi) = O(|\xi|^q)$ *as* $|\xi| \to 0$, *uniformly with respect to* x, *for some positive integer* q, *then*

$$e^{-i\lambda^\theta \varphi(x)} G\left(x, \lambda^{-\sigma} D\right) e^{i\lambda^\theta \varphi(x)} = G\left(x, \lambda^{-(\sigma-\theta)}\left(\varphi_x(x) + \lambda^{-\theta} D\right)\right)$$
$$+ \lambda^{-(\sigma-\theta)q-\theta} r\left(x, \lambda^{-\theta} D; \lambda\right), \quad (2.4)$$

where again $r \in \mathcal{R}$.

Remark 2.3. It is important to remark that in the notation above the quantity $G(x, \lambda^{-(\sigma-\theta)}(\varphi_x(x) + \lambda^{-\theta} D))$ does not contain the terms in which the derivatives land on $\varphi_x(x)$, as will be clear from the proof; those terms are pushed into the 'error' term r and thus $G(x, \lambda^{-(\sigma-\theta)}(\varphi_x(x) + \lambda^{-\theta} D))$ is to be thought of as a commutative expression.

In order to prove Theorem 1.1 we prove first a more general inductive lemma in the following. From the assumption we may start off assuming that

$$\xi_0 \longmapsto \det \widetilde{G}_0^{(0)}\left(x, \xi_0, \xi'\right) \text{ has a non-real root of multiplicity } q_0, \text{ i.e.,}$$

$$\det \widetilde{G}_0^{(0)}(x, \xi) = \left(\xi_0 - \tau_0\left(x, \xi'\right)\right)^{q_0} \Delta_0(x, \xi), \quad \Delta_0\left(x, \tau_0\left(x, \xi'\right), \xi'\right) \neq 0 \quad (2.5)$$

in some open set $U \times V$ in $\mathbb{R}_x^{n+1} \times \mathbb{R}_{\xi'}^n$. In the sequel U and V stands for an open set in $\mathbb{R}^{n+1}$ and in $\mathbb{R}^n$, respectively, which may differ from line to line but the subsequent one will be contained in the preceding one.

Denote by $\varphi^{(0)}(x)$ a complex-valued smooth (i.e., real analytic) function in U such that

$$\partial_{x_0} \varphi^{(0)}(x) = \tau_0\left(x, \partial_{x'} \varphi^{(0)}(x)\right). \quad (2.6)$$

Then

$$e^{-i\lambda^{\sigma_0} \varphi^{(0)}(x)} \widetilde{G}^{(0)}\left(x, \lambda^{-\sigma_0} D; \lambda\right) e^{i\lambda^{\sigma_0} \varphi^{(0)}(x)}$$
$$= \widetilde{G}^{(0)}\left(x, \varphi_x^{(0)}(x) + \lambda^{-\sigma_0} D; \lambda\right) + \lambda^{-\sigma_0} R^{(0)}\left(x, \lambda^{-\sigma_0} D; \lambda\right),$$

where $R^{(0)}$ is a symbol in the class.

In order to construct an asymptotic solution for the $N \times N$ matrix-valued operator $\widetilde{G}^{(0)}(x, \lambda^{-\sigma_0} D; \lambda)$ we first prove a general inductive lemma enabling us to construct the phase functions required.

Lemma 2.4 (First inductive step). *Let σ_p, $\theta_p \in \mathbb{Q}^+$, $r_p \leq N$, $\varphi^{(p)}(x)$ be a smooth function defined in U and consider the $r_p \times r_p$ matrix-valued differential operator*

$$\widetilde{G}^{(p)}\left(x, \varphi_x^{(p)}(x) + \lambda^{-\sigma_p} D; \lambda\right) + \lambda^{-\sigma_p} R^{(p)}\left(x, \lambda^{-\sigma_p} D; \lambda\right), \qquad (2.7)$$

where:

$i)_p$

$$\widetilde{G}^{(p)}(x, \xi; \lambda) = \sum_{j \geq 0} \lambda_j^{-\delta_j(\widetilde{G}^{(p)})} \widetilde{G}^{(p)}(x, \xi),$$

$$0 = \delta_0\left(\widetilde{G}^{(p)}\right) < \delta_1\left(\widetilde{G}^{(p)}\right) < \cdots$$

the sum being finite and $\widetilde{G}_j^{(p)}$ denoting differential operators with real analytic coefficients;

$ii)_p$

$$\det \widetilde{G}_0^{(p)}(x, \xi) = \left(\xi_0 - \tau_p\left(x, \xi'\right)\right)^{q_p} \Delta_p(x, \xi), \quad \Delta_p\left(x, \tau_p\left(x, \xi'\right), \xi'\right) \neq 0$$

in some open set $U \times V$ in $\mathbb{R}_x^{n+1} \times \mathbb{R}_{\xi'}^n$.

$iii)_p$ *The function $\varphi^{(p)}(x)$, defined in U, satisfies*

$$\partial_{x_0} \varphi^{(p)}(x) = \tau_p\left(x, \partial_{x'} \varphi^{(p)}(x)\right).$$

$iv)_p$

$$\text{rank } \widetilde{G}_0^{(p)}\left(x, \varphi_x^{(p)}(x)\right) = r_p - r_{p+1} \qquad (2.8)$$

in U for a suitable positive integer $r_{p+1} \leq r_p$.

Then we can find a $r_{p+1} \times r_{p+1}$ matrix-valued differential operator

$$F^{(p)}\left(x, \lambda^{-\sigma_p} D; \lambda\right) \qquad (2.9)$$

such that

$$F^{(p)}(x, \xi; \lambda) = \sum_{j \geq 0} \lambda^{-\delta_j(F^{(p)})} F_j^{(p)}(x, \xi),$$

$$0 = \delta_0\left(F^{(p)}\right) < \delta_1\left(F^{(p)}\right) < \cdots$$

the sum being finite and

$$\sigma_p, \theta_p, \delta_j\left(F^{(p)}\right), \quad j \geq 1, \; \text{belong to } \frac{\mathbb{N}}{k'(p)}$$

for a suitable positive integer $k'(p)$.

$$F_0^{(p)}(x, 0) = 0. \tag{2.10}$$

$$\det F_0^{(p)}\left(x, \lambda^{-\sigma_p}\xi\right) = \frac{1}{E\left(x, \lambda^{-\sigma_p}\xi\right)} \det \widetilde{G}_0^{(p)}\left(x, \varphi_x^{(p)}(x) + \lambda^{-\sigma_p}\xi\right)$$

$$+ O\left(\lambda^{-(M+1)\delta_p^*}\right), \tag{2.11}$$

where M is an arbitrarily large positive integer and $\delta_p^* > 0$, suitable and $E(x, \xi)$ is an elliptic symbol. Moreover the construction of an asymptotic solution for the operator (2.7) is reduced to the construction of an asymptotic solution for the operator (2.9).

Set

$$G^{(p)}(x, \xi; \lambda) = \widetilde{G}^{(p)}\left(x, \varphi_x^{(p)}(x) + \xi; \lambda\right) + \lambda^{-\sigma_p} R^{(p)}(x, \xi; \lambda).$$

Then $G^{(p)}$ can be written as a finite sum of differential operators:

$$G^{(p)}(x, \xi; \lambda) = \sum_{j \geq 0} \lambda^{-\delta_j(G^{(p)})} G_j^{(p)}(x, \xi); \tag{2.12}$$

we remark that in the above (finite) sum

$$\delta_0\left(G^{(p)}\right) = \delta_0\left(\widetilde{G}^{(p)}\right) = 0, \qquad G_0^{(p)}(x, \xi) = \widetilde{G}_0^{(p)}\left(x, \varphi_x^{(p)} + \xi\right).$$

By (2.8) we have

$$\text{rank } G_0^{(p)}(x, 0) = r_p - r_{p+1}. \tag{2.13}$$

Thus we can find two non-singular smooth matrices, $M_p(x)$, $N_p(x)$, defined in U, such that

$$M_p(x) G_0^{(p)}(x, 0) N_p(x) = \left[\begin{array}{c|c} I_{r_p - r_{p+1}} & 0 \\ \hline 0 & 0 \end{array}\right],$$

where $I_{r_p - r_{p+1}}$ denotes the $(r_p - r_{p+1}) \times (r_p - r_{p+1})$ identity matrix. Then we have

$$\widehat{G}^{(p)}\left(x, \lambda^{-\sigma_p} D; \lambda\right) = M_p(x) G^{(p)}\left(x, \lambda^{-\sigma_p} D; \lambda\right) N_p(x)$$
$$= \sum_{j \geq 0} \lambda^{-\delta_j(\widehat{G}^{(p)})} \widehat{G}_j^{(p)}\left(x, \lambda^{-\sigma_p} D\right)$$

whose proof follows from the next straightforward lemma:

Lemma 2.5. *Let*

$$A\left(x, \lambda^{-\sigma} D; \lambda\right) = \sum_{j \geq 0} \lambda^{-\delta_j(A)} A_j\left(x, \lambda^{-\sigma} D\right),$$

$$B\left(x, \lambda^{-\sigma} D; \lambda\right) = \sum_{\ell \geq 0} \lambda^{-\delta_\ell(B)} B_\ell\left(x, \lambda^{-\sigma} D\right)$$

be operators of the form (2.12). Then

$$(A \circ B)\left(x, \lambda^{-\sigma} D; \lambda\right) = \sum_{j \geq 0} \lambda^{-\delta_j(C)} C_j\left(x, \lambda^{-\sigma} D\right),$$

where $\delta_j(C) = \delta_i(A) + \delta_\ell(B) + k\sigma$ for suitable i, ℓ and k.

Since M_p and N_p are non-singular matrices then the construction of an asymptotic solution for $G^{(p)}(x, \lambda^{-\sigma_p} D; \lambda)$ is equivalent to the construction of an asymptotic solution for $\widehat{G}^{(p)}(x, \lambda^{-\sigma_p} D; \lambda)$ and moreover we have that

$$\widehat{G}_0^{(p)}(x, \xi) = M_p(x) G_0^{(p)}(x, \xi) N_p(x)$$

so that

$$\widehat{G}_0^{(p)}(x, 0) = \left[\begin{array}{c|c} I_{r_p - r_{p+1}} & 0 \\ \hline 0 & 0 \end{array} \right]. \tag{2.14}$$

Here $\widehat{G}^{(p)}(x, \xi; \lambda)$ has the same properties as $G^{(p)}(x, \xi; \lambda)$ where (2.13) has to be replaced by (2.14).

From now on we switch back to the $G^{(p)}$ notation, dropping the caps. Let us write $G^{(p)}$ in block form:

$$G^{(p)}\left(x, \lambda^{-\sigma_p} D; \lambda\right) = \left[\begin{array}{cc} G_{11}^{(p)}\left(x, \lambda^{-\sigma_p} D; \lambda\right) & G_{12}^{(p)}\left(x, \lambda^{-\sigma_p} D; \lambda\right) \\ G_{21}^{(p)}\left(x, \lambda^{-\sigma_p} D; \lambda\right) & G_{22}^{(p)}\left(x, \lambda^{-\sigma_p} D; \lambda\right) \end{array} \right],$$

where the blocking corresponds to that of (2.14). We have

$$G_{11}^{(p)}(x, \xi; \lambda) = \sum_{j \geq 0} \lambda^{-\delta_j(G^{(p)})} G_{11,j}^{(p)}(x, \xi)$$

$$= I - \sum_{j \geq 0} \lambda^{-\delta_j(G^{(p)})} B_j^{(p)}(x, \xi) \qquad (2.15)$$

$$= I - B^{(p)}(x, \xi; \lambda),$$

where $B_j^{(p)}(x, \xi)$ are $(r_p - r_{p+1}) \times (r_p - r_{p+1})$ matrices and

$$B_0^{(p)}(x, 0) = 0. \qquad (2.16)$$

Define

$$\mathcal{R}^{(p)}\left(x, \lambda^{-\sigma_p} D; \lambda\right) = \sum_{k=0}^{M} \left(B^{(p)}\left(x, \lambda^{-\sigma_p} D; \lambda\right)\right)^k, \qquad (2.17)$$

where M is a positive integer that, in the sequel, will be chosen suitably large. We may then write

$$\mathcal{R}^{(p)}\left(x, \lambda^{-\sigma_p} D; \lambda\right) = \sum_{j \geq 0} \lambda^{-\delta_j(\mathcal{R}^{(p)})} \mathcal{R}_j^{(p)}\left(x, \lambda^{-\sigma_p} D\right),$$

where the sum is a finite sum whose number of terms depends on M and, using again Lemma 2.5, the $\delta_j(\mathcal{R}^{(p)})$, $j \geq 1$ are an increasing sequence of rational numbers. From (2.17) we obtain that

$$G_{11}^{(p)}\left(x, \lambda^{-\sigma_p} D; \lambda\right) \mathcal{R}^{(p)}\left(x, \lambda^{-\sigma_p} D; \lambda\right) = I - \left(B^{(p)}\left(x, \lambda^{-\sigma_p} D; \lambda\right)\right)^{M+1}.$$

We want to show that $(B^{(p)}(x, \lambda^{-\sigma_p} D; \lambda))^{M+1}$ becomes negligible provided M is chosen large enough.

Lemma 2.6. *Let $B^{(p)}$ be defined as in (2.15). Then*

$$\partial_\xi^\gamma \sigma\left(\left[B^{(p)}\left(x, \lambda^{-\sigma_p} D; \lambda\right)\right]^{M+1}\right) = O\left(\lambda^{-(M+1)\delta_p^*}\right),$$

where $\delta_p^* = \min\{\sigma_p, \delta_1(G^{(p)})\} > 0$ and $\sigma(\cdot)$ denotes the symbol of a given differential operator.

Remark 2.7. From the proof we can also deduce that

$$\partial_\xi^\gamma \sigma\left(\left(B^{(p)}\left(x, \lambda^{-\theta} \xi; \lambda\right)\right)^{M+1}\right) = O\left(\lambda^{-(M+1)\theta}\right)$$

for any θ with $0 < \theta \leq \min\{\sigma_p, \delta_1(G^{(p)})\}$.

Now define

$$\Lambda^{(p)}\left(x,\lambda^{-\sigma_p}D;\lambda\right) = \begin{bmatrix} I & -\mathcal{R}^{(p)}\left(x,\lambda^{-\sigma_p}D;\lambda\right)G_{12}^{(p)}\left(x,\lambda^{-\sigma_p}D;\lambda\right) \\ 0 & I \end{bmatrix}$$

in block form notation, the blocks correspond to those in (2.14). We have

$$G^{(p)}\left(x,\lambda^{-\sigma_p}D;\lambda\right)\Lambda^{(p)}\left(x,\lambda^{-\sigma_p}D;\lambda\right)$$

$$= \begin{bmatrix} G_{11}^{(p)}\left(x,\lambda^{-\sigma_p}D;\lambda\right) & \left(I-G_{11}^{(p)}\mathcal{R}^{(p)}\right)\left(x,\lambda^{-\sigma_p}D;\lambda\right)G_{12}^{(p)}\left(x,\lambda^{-\sigma_p}D;\lambda\right) \\ G_{21}^{(p)}\left(x,\lambda^{-\sigma_p}D;\lambda\right) & \left(G_{22}^{(p)}-G_{21}^{(p)}\mathcal{R}^{(p)}G_{12}^{(p)}\right)\left(x,\lambda^{-\sigma_p}D;\lambda\right) \end{bmatrix}$$

$$(2.18)$$

and define

$$F^{(p)}\left(x,\lambda^{-\sigma_p}D;\lambda\right) = \left(G_{22}^{(p)}-G_{21}^{(p)}\mathcal{R}^{(p)}G_{12}^{(p)}\right)\left(x,\lambda^{-\sigma_p}D;\lambda\right). \quad (2.19)$$

Furthermore we have that

$$\left(I-G_{11}^{(p)}\mathcal{R}^{(p)}\right)\left(x,\lambda^{-\sigma_p}D;\lambda\right)G_{12}^{(p)}\left(x,\lambda^{-\sigma_p}D;\lambda\right)$$

$$= \left(B^{(p)}\left(x,\lambda^{-\sigma_p}D;\lambda\right)\right)^{M+1}G_{12}^{(p)}\left(x,\lambda^{-\sigma_p}D;\lambda\right).$$

Here $S^{(p)}T^{(p)}(x,\lambda^{-\sigma_p}D;\lambda)$ stands for $S^{(p)}(x,\lambda^{-\sigma_p}D;\lambda)T^{(p)}(x,\lambda^{-\sigma_p}D;\lambda)$. The proof of the following proposition is then a straightforward consequence of Lemma 2.6.

Proposition 2.8. *The $(1,2)$-block of the matrix in (2.18), as a differential operator, is $O(\lambda^{-(M+1)\delta_p^*})$. In particular if M is chosen suitably large, since $\Lambda^{(p)}$ is non singular, the construction of an asymptotic solution for the $r_p \times r_p$ matrix of differential operators $G^{(p)}(x,\lambda^{-\sigma_p}D;\lambda)$ reduces to the construction of an asymptotic solution for the $r_{p+1} \times r_{p+1}$ matrix of differential operators $F^{(p)}(x,\lambda^{-\sigma_p}D;\lambda)$, defined in (2.19).*

Using the proof of Lemma 2.6 we can write

$$F^{(p)}\left(x,\lambda^{-\sigma_p}D;\lambda\right) = \sum_{j\geq 0}\lambda^{-\delta_j(F^{(p)})}F_j^{(p)}\left(x,\lambda^{-\sigma_p}D\right), \quad (2.20)$$

where the exponents $\delta_j(F^{(p)})$ are obtained summing a number of $\delta_j(G^{(p)})$ to integer multiples of σ_p. First remark that (2.10) is immediate because of the construction of $F^{(p)}$. Due to Lemma 2.6 we can see that the (operator-valued) matrix in (2.18) is lower triangular modulo terms that are $O(\lambda^{-(M+1)\delta_p^*})$ and therefore

$$F_0^{(p)}(x,\xi) = G_{22,0}^{(p)}(x,\xi) - G_{21,0}^{(p)}(x,\xi)\mathcal{R}_0^{(p)}(x,\xi)G_{12,0}^{(p)}(x,\xi)$$

$$\det F_0^{(p)}\left(x,\lambda^{-\sigma_p}\xi\right) = \frac{1}{\det G_{11,0}^{(p)}\left(x,\lambda^{-\sigma_p}\xi\right)} \det G_0^{(p)}\left(x,\lambda^{-\sigma_p}\xi\right) \qquad (2.21)$$

$$+ O\left(\lambda^{-(M+1)\delta_p^*}\right).$$

This proves (2.11), since $\det G_{11,0}^{(p)}(x,0) \neq 0$.

Let us consider the operator $F^{(p)}$ as given by (2.20), defined in U, and denote by $s_j^{(p)}$ the vanishing order with respect to the variable ξ as $|\xi| \to 0$ of $F_j^{(p)}, j = 0, 1, \ldots$:

$$F_j^{(p)}\left(x,\lambda^{-\theta}\xi\right) = \lambda^{-\theta s_j^{(p)}}\left[\widehat{F}_j^{(p)}(x,\xi) + O\left(\lambda^{-\theta}\right)\right],$$

where θ is any positive real number. Here $\widehat{F}_j^{(p)}$ is an $r_{p+1} \times r_{p+1}$ matrix valued homogeneous polynomial in the variable ξ of degree $s_j^{(p)}$. Define

$$\theta_{p+1} = \min_{\substack{j \geq 1 \\ s_j^{(p)} < s_0^{(p)}}} \left\{ \frac{\delta_j\left(F^{(p)}\right)}{s_0^{(p)} - s_j^{(p)}}, \theta_p \right\} \qquad (2.22)$$

so that, in particular, $\theta_{p+1} \leq \theta_p$ and let

$$\sigma_{p+1} = \sigma_p - \theta_{p+1}.$$

For our present purpose we shall assume that $\sigma_{p+1} > 0$. If $\sigma_{p+1} \leq 0$ we make a different argument in the following.

Let now $\varphi^{(p+1)}(x)$ be a real analytic function defined in some open set U; in the following we shall precise this function. Applying Lemma 2.2 we compute

$$e^{-i\lambda^{\sigma_{p+1}}\varphi^{(p+1)}(x)} F^{(p)}\left(x,\lambda^{-\sigma_p}D;\lambda\right) e^{i\lambda^{\sigma_{p+1}}\varphi^{(p+1)}(x)}$$

$$= \sum_{j \geq 0} \lambda^{-\delta_j(F^{(p)})} F_j^{(p)}\left(x,\lambda^{-\theta_{p+1}}\left(\varphi_x^{(p+1)}(x) + \lambda^{-\sigma_{p+1}}D\right)\right)$$

$$+ \sum_{j \geq 0} \lambda^{-\delta_j(F^{(p)})-\theta_{p+1}s_j^{(p)}-\sigma_{p+1}} \widetilde{R}_j^{(p+1)}\left(x,\lambda^{-\sigma_{p+1}}D;\lambda\right),$$

where $\widetilde{R}_j^{(p+1)} \in \mathcal{R}$. Defining $\widetilde{G}^{(p+1)}(x, \xi; \lambda)$ and $R^{(p+1)}(x, \xi; \lambda)$ by

$$F^{(p)}\left(x, \lambda^{-\theta_{p+1}}\xi; \lambda\right) = \lambda^{-\theta_{p+1}s_0^{(p)}}\widetilde{G}^{(p+1)}(x, \xi; \lambda)$$

$$= \lambda^{-\theta_{p+1}s_0^{(p)}}\sum_{j \geq 0}\lambda^{-\delta_j(\widetilde{G}^{(p+1)})}\widetilde{G}_j^{(p+1)}(x, \xi), \quad (2.23)$$

$$\sum_{j \geq 0}\lambda^{-\delta_j(F^{(p)})-\theta_{p+1}s_j^{(p)}}\widetilde{R}_j^{(p+1)}(x, \xi; \lambda) = \lambda^{-\theta_{p+1}s_0^{(p)}}R^{(p+1)}(x, \xi; \lambda),$$

the right hand side of the above equality can be written as

$$\lambda^{-\theta_{p+1}s_0^{(p)}}\left[\widetilde{G}^{(p+1)}\left(x, \varphi_x^{(p+1)}(x)+\lambda^{-\sigma_{p+1}}D\right)+\lambda^{-\sigma_{p+1}}R^{(p+1)}\left(x, \lambda^{-\sigma_{p+1}}D; \lambda\right)\right]$$

$$= \lambda^{-\theta_{p+1}s_0^{(p)}}\left[\sum_{j \geq 0}\lambda^{-\delta_j(\widetilde{G}^{(p+1)})}\widetilde{G}_j^{(p+1)}\left(x, \varphi_x^{(p+1)}(x)+\lambda^{-\sigma_{p+1}}D\right)\right.$$

$$\left.+\lambda^{-\sigma_{p+1}}R^{(p+1)}\left(x, \lambda^{-\sigma_{p+1}}D; \lambda\right)\right].$$

We want to show that, provided an additional assumption is made, the operator $\widetilde{G}^{(p+1)}$ satisfies the conditions i)$_{p+1}$–iv)$_{p+1}$ of Lemma 2.2, thus enabling us to start over again the induction process.

i)$_{p+1}$ is obvious. From (2.23) we obtain

$$\widetilde{G}_0^{(p+1)}(x, \xi) = \sum_{\substack{\theta_{p+1}s_0^{(p)}=\theta_{p+1}s_j^{(p)}+\delta_j(F^{(p)}) \\ s_j^{(p)}<s_0^{(p)} \text{ if } j>0}} \widehat{F}_j^{(p)}(x, \xi), \quad (2.24)$$

where each $\widehat{F}_j^{(p)}$ being homogeneous of degree $s_j^{(p)}$ with respect to the variable ξ.

We make the following assumption:

$$q_p = r_{p+1}s_0^{(p)}. \quad (2.25)$$

From (2.21) and Remark 2.7 we have

$$\det F_0^{(p)}\left(x, \lambda^{-\theta}\xi\right) = \frac{1}{\det G_{11,0}^{(p)}\left(x, \lambda^{-\theta}\xi\right)}\det G_0^{(p)}\left(x, \lambda^{-\theta}\xi\right)$$

$$+O\left(\lambda^{-(M+1)\theta}\right)$$

for a suitable $\theta > 0$. But

$$F_0^{(p)}\left(x, \lambda^{-\theta}\xi\right) = \lambda^{-\theta s_0^{(p)}}\left[\widehat{F}_0^{(p)}(x, \xi) + O\left(\lambda^{-\theta}\right)\right]$$

so that

$$\begin{aligned}
\det G_0^{(p)}\left(x, \lambda^{-\theta}\xi\right) &= \det G_{11,0}^{(p)}\left(x, \lambda^{-\theta}\xi\right)\lambda^{-\theta q_p}\\
&\quad\times\left[\det \widehat{F}_0^{(p)}(x, \xi) + O\left(\lambda^{-\theta}\right)\right] + O\left(\lambda^{-(M+1)\theta}\right)\\
&= \lambda^{-\theta q_p}\left[E_p(x) + O\left(\lambda^{-\theta}\right)\right]\left[\det \widehat{F}_0^{(p)}(x, \xi) + O\left(\lambda^{-\theta}\right)\right]\\
&\quad + O\left(\lambda^{-(M+1)\theta}\right).
\end{aligned}$$

(2.26)

On the other hand

$$\begin{aligned}
\det G_0^{(p)}\left(x, \lambda^{-\theta}\xi\right) &= c_p(x)\det \widetilde{G}_0^{(p)}\left(x, \varphi_x^{(p)}(x) + \lambda^{-\theta}\xi\right)\\
&= \lambda^{-q_p\theta}c_p(x)\sum_{|\alpha|=q_p}\frac{1}{\alpha!}\left(\det \widetilde{G}_0^{(p)}\right)^{(\alpha)}\\
&\quad\times\left(x, \varphi_x^{(p)}(x)\right)\xi^\alpha + o\left(\lambda^{-q_p\theta}\right),
\end{aligned}$$

(2.27)

where c_p denotes a non-zero smooth function defined in U. Thus from (2.25) and (2.28) we obtain

$$\det \widehat{F}_0^{(p)}(x, \xi) = c_p(x)\sum_{|\alpha|=q_p}\frac{1}{\alpha!}\left(\det \widetilde{G}_0^{(p)}\right)^{(\alpha)}\left(x, \varphi_x^{(p)}(x)\right)\xi^\alpha. \quad (2.28)$$

The above relation allows us to conclude that:

a) The right hand side of (2.28) is a non-characteristic polynomial with respect to the variable ξ with real analytic coefficients. Then one can find a real analytic $\tau_{p+1}(x, \xi')$ defined in $U \times V$ such that

$$\begin{aligned}
\det \widetilde{G}_0^{(p+1)}(x, \xi) &= \left(\xi_0 - \tau_{p+1}\left(x, \xi'\right)\right)^{q_{p+1}}\Delta_{p+1}(x, \xi),\\
\Delta_{p+1}\left(x, \tau_{p+1}\left(x, \xi'\right), \xi'\right) &\neq 0,
\end{aligned}$$

(2.29)

where $q_{p+1} \leq q_p$.

b) We can find a real analytic function $\varphi^{(p+1)}(x)$, defined in U such that

$$\partial_{x_0}\varphi^{(p+1)}(x) = \tau_{p+1}\left(x, \partial_{x'}\varphi^{(p+1)}(x)\right).$$

c) There exists a non-negative integer r_{p+2} such that

$$\text{rank } \widetilde{G}_0^{(p+1)}\left(x, \varphi_x^{(p+1)}(x)\right) = r_{p+1} - r_{p+2}$$

in U.

These statements easily imply conditions ii)$_{p+1}$–iv)$_{p+1}$. Thus far we proved the following

Lemma 2.9 (Second half of the induction step). *Let $F^{(p)}(x, \xi; \lambda)$ be as in Lemma 2.4 and assume that hypothesis (2.25) holds. Then we can find a positive rational number $\theta_{p+1} \leq \theta_p$ with $\sigma_{p+1} = \sigma_p - \theta_{p+1} > 0$, a real analytic function $\varphi^{(p+1)}(x)$ defined in U and $r_{p+1} \times r_{p+1}$ matrix valued differential operators $\widetilde{G}^{(p+1)}(x, D; \lambda)$, $R^{(p+1)}(x, D; \lambda)$ with $R^{(p+1)} \in \mathcal{R}$ such that the construction of an asymptotic solution for the operator (2.9) is reduced to the construction of an asymptotic solution for*

$$\widetilde{G}^{(p+1)}\left(x, \varphi_x^{(p+1)}(x) + \lambda^{-\sigma_{p+1}} D; \lambda\right) + \lambda^{-\sigma_{p+1}} R^{(p+1)}\left(x, \lambda^{-\sigma_{p+1}} D; \lambda\right). \quad (2.30)$$

Furthermore conditions i)$_{p+1}$–iv)$_{p+1}$ hold.

As a consequence of Lemmas 2.4 and 2.9 we have the following

Lemma 2.10. *Assume (2.25) holds. Then the construction of an asymptotic solution for (2.7) satisfying i)$_p$–iv)$_p$ is reduced to the construction of an asymptotic solution for (2.30) satisfying i)$_{p+1}$–iv)$_{p+1}$.*

3. Proof of Theorem 1.1

This section is devoted to the proof of Theorem 1.1. We limit ourselves to the case $q_0 \leq 2$. By (2.5), (2.6) and (2.22) we may start our induction process with $\widetilde{G}^{(0)}$, σ_0, θ_0 and apply Lemma 2.10. We point out explicitly here that (2.25) is not assumed to hold. However by (2.29) for each p either $q_p = 2$ or $q_p = 1$ and $q_p \geq r_{p+1} s_0^{(p)}$ in general. This implies that if $q_p = 1$ necessarily $r_{p+1} = s_0^{(p)} = 1$, i.e., we are in a scalar case and (2.25) is verified.

If the former case holds, i.e. , if $q_p = 2$, then either $r_{p+1} = 2$ and (2.25) holds or $r_{p+1} = 1$. But in this case we are again in a scalar case and it cannot occur then that $q_p > r_{p+1} s_0^{(p)}$. Summing up if $q_p \leq 2$ then necessarily (2.25) holds true. Next we show that the induction process for computing the phase function ends after finitely many steps.

Proposition 3.1. *Under the assumptions of Theorem 1.1 the iteration procedure of Lemma 2.10 occurs only a finite number of times before reaching a point where*

$$\sigma_{\bar{p}+1} = \sigma_0 - \sum_{i=1}^{\bar{p}+1} \theta_i \le 0$$

for a suitable integer $\bar{p}$.

In order to prove Proposition 3.1 we need a preliminary lemma:

Lemma 3.2. *Assume that there exists $\bar{p} \in \mathbb{N}$ such that*

$$q_{\bar{p}} = q_{\bar{p}+1} = \cdots = q, \qquad r_{\bar{p}} = r_{\bar{p}+1} = \cdots = r.$$

Under the hypotheses of Theorem 1.1 there exists a $k = k(\bar{p})$ such that

$$\sigma_p, \theta_p, \delta_j\left(\widetilde{G}^{(p)}\right), \quad j \ge 1, \text{ belong to } \frac{\mathbb{N}}{k}$$

for every $p \ge \bar{p}$.

Proof. By (2.29) we have $q \le 2$. Let us start considering the case $q = 2$. If $q = 1$ the argument is of the same kind and easier.

By what has been said above if $q = 2$ then either $s_0^{(p)} = 1$ or $r = 1$.

i) $s_0^{(p)} = 1$. The property that $q_{p+1} = q_p$ implies that there are no roots of the equation $\det \widetilde{G}_0^{(p+1)}(x, \xi) = 0$ with respect to ξ_0 with uniform multiplicity less than q_p. $\widetilde{G}_0^{(p+1)}(x, \xi)$ is given by (2.24). Two cases may occur: either the sum in (2.24) has $\widehat{F}_0^{(p)}$ as the only summand or there are also other summands. In the former case we have that $\theta_{p+1}s_0^{(p)} < \theta_{p+1}s_j^{(p)} + \delta_j(F^{(p)})$, for every $j \ge 1$, which implies, if $s_j^{(p)} < s_0^{(p)}$, that

$$\theta_{p+1} < \frac{\delta_j\left(F^{(p)}\right)}{s_0^{(p)} - s_j^{(p)}}$$

or, because of (2.22),

$$\theta_{p+1} = \theta_p. \tag{3.1}$$

Assume now that there are terms other than $\widehat{F}_0^{(p)}$, corresponding to $j > 0$, in the sum in (2.24). Since $s_0^{(p)} = 1$ the condition defining the sum implies that

$$\delta_1\left(F^{(p)}\right) = \theta_{p+1}. \tag{3.2}$$

ii) $r = 1$. We are then in a scalar case. Again considering the sum in (2.24) we conclude (3.1) if $\widehat{F}_0^{(p)}$ is the only summand. Let us assume that there are also other summands different from $\widehat{F}_0^{(p)}$.

Now $s_0^{(p)} = q$ and the assumption of the lemma implies that there is a $j \geq 1$ such that

$$\delta_j \left(F^{(p)} \right) = \theta_{p+1} \tag{3.3}$$

because of the following lemma.

Lemma 3.3. Let $p(\tau) = \sum_{j=0}^s a_j \tau^{q_j}$ where $0 = q_0 < q_1 < \cdots < q_s$ and $a_s \neq 0$, $a_0 \neq 0$. Then the roots of $p(\tau) = 0$ have multiplicity at most s.

We skip the proof of Lemma 3.3 and go back to the proof of Lemma 3.2. Summing up in both cases we conclude that either (3.1) or (3.2) hold. In particular this implies that $k(p + 1) = k(p)$, since the $\delta_j(\widetilde{G}^{(p+1)})$ are obtained summing and multiplying rational numbers whose denominator is $k(p)$. This ends the proof of Lemma 3.2. $\square$

Proof of Proposition 3.1. By contradiction. If one could go through infinitely many iteration steps then necessarily the assumption of Lemma 3.2 must hold and hence θ_p, $p \geq \overline{p}$ has the same denominator. Thus after a finite number of iteration steps we get a negative σ_t for a suitable positive integer t. $\square$

In order to complete the proof of Theorem 1.1, by Proposition 3.1, we may assume that for a certain positive integer t, $\sigma_t > 0$ but

$$\sigma_{t+1} = \sigma_t - \theta_{t+1} \leq 0.$$

Therefore

$$\theta_{t+1} = \min_{\substack{j > 0 \\ s_j^{(t)} < s_0^{(t)}}} \left\{ \frac{\delta_j \left(F^{(t)} \right)}{s_0^{(t)} - s_j^{(t)}}, \theta_t \right\} \geq \sigma_t.$$

Our purpose is to construct an asymptotic null solution for the operator

$$\widetilde{G}^{(t)} \left(x, \varphi_x^{(t)}(x) + \lambda^{-\sigma_t} D; \lambda \right) + \lambda^{-\sigma_t} R^{(t)} \left(x, \lambda^{-\sigma_t} D; \lambda \right), \tag{3.4}$$

where $R^{(t)} \in \mathfrak{R}$, in a neighborhood of the origin.

At this stage of the construction we can still apply Lemma 2.4 in order to possibly reduce the rank of the matrix in (3.5). Hence we wind up with the construction of an asymptotic solution for the operator

$$F^{(t)} \left(x, \lambda^{-\sigma_t} D; \lambda \right) = \sum_{j \geq 0} \lambda^{-\delta_j(F^{(t)})} F_j^{(t)} \left(x, \lambda^{-\sigma_t} D \right) \tag{3.5}$$

of size $r_{t+1} \times r_{t+1}$. If

$$F_j^{(t)}\left(x, \lambda^{-1}\xi\right) = \lambda^{-s_j^{(t)}}\left[\widehat{F}_j^{(t)}(x, \xi) + O\left(\lambda^{-1}\right)\right]$$

when $j \geq 0$ and $\lambda \to +\infty$, the operator in (3.5) can be written as

$$F^{(t)}\left(x, \lambda^{-\sigma_t}D; \lambda\right) = \sum_{j \geq 0} \lambda^{-\delta_j(F^{(t)})-s_j^{(t)}\sigma_t}\left[\widehat{F}_j^{(t)}(x, D) + O\left(\lambda^{-\sigma_t}\right)\right],$$

where $O(\lambda^{-\sigma_t})$ stands for a (matrix-valued) differential operator of order $\leq s_j^{(t)}$ whose coefficients are $O(\lambda^{-\sigma_t})$ uniformly in U. The fact that $\theta_{t+1} \geq \sigma_t$ implies that for every $j \geq 1$

$$\delta_j\left(F^{(t)}\right) + \sigma_t s_j^{(t)} \geq \sigma_t s_0^{(t)}$$

so that

$$F^{(t)}\left(x, \lambda^{-\sigma_t}D; \lambda\right)$$

$$= \lambda^{-\sigma_t s_0^{(t)}}\left[\sum_{\sigma_t s_0^{(t)}=\sigma_t s_j^{(t)}+\delta_j(F^{(t)})} \widehat{F}_j^{(t)}(x, D) + \lambda^{-\varepsilon_t}\sum_{j \geq 0}\lambda^{-\tilde{\delta}_j(F^{(t)})}\widetilde{F}_j^{(t)}(x, D)\right],$$

where ε_t is a positive rational number whose denominator can be chosen to be the same as the denominator of σ_t, $\delta_j(F^{(t)})$, $j \geq 1$. Moreover $0 = \tilde{\delta}_0(F^{(t)}) < \tilde{\delta}_1(F^{(t)}) < \cdots$ and the terms with $j > 0$ in the first sum have order $s_j^{(t)}$, $j > 0$, with $s_j^{(t)} < s_0^{(t)}$.

Arguing as in the proof of Lemma 2.9 we can show that the principal part of the differential operator in the first sum, $\widehat{F}_0^{(t)}$, is non-characteristic.

Disposing of the power of λ in front of the operator in square brackets we are left, in the end, with the task of constructing an asymptotic solution for an operator of the form

$$P_{s_0^{(t)}}(x, D) + \widetilde{P}(x, D) + \lambda^{-\ell/k}\sum_{j \geq 0}\lambda^{-j/k}P_j(x, D), \tag{3.6}$$

where $k \in \mathbb{N}$, $\ell \in \mathbb{N}$, ord $\widetilde{P} < s_0^{(t)}$ and $P_{s_0^{(t)}}$ is a non-characteristic homogeneous differential operator of order $s_0^{(t)}$. One can then seek an asymptotic solution for (3.6) in the form

$$\sum_{j \geq 0}\lambda^{-j/k}u_j(x)$$

and this is a well known procedure. This ends the proof of Theorem 1.1.

References

[1]   Benvenuti, S. and Bove, A., On a class of hyperbolic systems with multiple characteristics, *Osaka J. Math.* **35** (1998), 313–356.

[2]   Benvenuti, S., Bernardi, E., and Bove, A., The Cauchy problem for hyperbolic systems with multiple characteristics, *Bull. Sci. Math.* **122** (1998), 603–634.

[3]   Bove, A. and Nishitani, T., Necessary conditions for the well posedness of the Cauchy problem for hyperbolic systems, *Osaka J. Math.* **39** (2002), 149–189.

[4]   Hörmander, L., The Cauchy problem for differential equations with double characteristics, *J. Analyse Math.* **32** (1977), 118–196.

[5]   _____, *The analysis of linear partial differential operators, voll. I–IV*, Springer–Verlag, Berlin-Heidelberg 1985.

[6]   Ivrii, Ya.V., *Linear hyperbolic equations*, Encyclopædia of Mathematical Sciences, vol. 33, Yu.V. Egorov and M.A. Shubin (eds), Springer–Verlag, Berlin 1993, 149–235.

[7]   Ivrii, V. and Petkov, M., Necessary conditions for the correctness of the Cauchy problem for non-strictly hyperbolic equations, *Uspehi Mat. Nauk* **29** (1974), 3–70.

[8]   Nishitani, T., Necessary conditions for strong hyperbolicity of first order systems, *Journal d'Analyse Math.* **61** (1993), 181–229.

[9]   _____, *Strongly hyperbolic systems of maximal rank*, vol. 33, Publ. RIMS Kyoto Univ. 1997, 765–773.

ON THE CAUCHY PROBLEM FOR HYPERBOLIC OPERATORS WITH NON-REGULAR COEFFICIENTS

FERRUCCIO COLOMBINI, DANIELE DEL SANTO,
AND TAMOTU KINOSHITA

To the memory of Professor J. Leray

Contents

1   Introduction and results   39
2   Proofs of Theorems 1.1, 1.2, 1.5   43
3   Construction of the counter examples   48

M. de Gosson (ed.), Jean Leray '99 Conference Proceedings, 37-52.
© *2003 Kluwer Academic Publishers.*

1. Introduction and results

In this work we collect some new results on the well posedness of the Cauchy problem for a class of strictly hyperbolic operators. Let $T > 0$. We are concerned with the equation

$$u_{tt} - \sum_{i,j=1}^{n} a_{ij}(t)u_{x_i x_j} + \sum_{i=1}^{n} b_i(t)u_{x_i} + c(t)u = 0 \quad \text{in } [0, T] \times \mathbb{R}^n, \quad (1.1)$$

with initial data

$$u(0, x) = u_0(x), \qquad u_t(0, x) = u_1(x) \quad \text{in } \mathbb{R}^n, \quad (1.2)$$

where (a_{ij}) is a real symmetric matrix such that

$$a(t, \xi) = \sum_{i,j=1}^{n} a_{ij}(t)\xi_i \xi_j / |\xi|^2 \geq \lambda_0 > 0, \quad (1.3)$$

for all t and for all $\xi \neq 0$, and the coefficients b_i and c are measurable and bounded.

Denoting by $\mathcal{B}$ a class of infinitely differentiable functions on $\mathbb{R}^n$, we will say that the Cauchy problem (1.1), (1.2) is $\mathcal{B}$ well posed if for all pair of functions $u_0, u_1 \in \mathcal{B}$ there exists a unique solution u to (1.1), (1.2) in $\mathscr{C}^1([0, T]; \mathcal{B})$.

It is well known that if the coefficients a_{ij} are Lipschitz continuous then the Cauchy problem (1.1), (1.2) is $\mathscr{C}^\infty$ well posed. Precisely, under the hypothesis of Lipschitz continuity of the coefficients of the principal part of (1.1), one can prove that for all $u_0 \in \mathscr{H}^s(\mathbb{R}^n)$, $u_1 \in \mathscr{H}^{s-1}(\mathbb{R}^n)$ there is a unique solution in $\mathscr{C}([0, T]; \mathscr{H}^s(\mathbb{R}^n)) \cap \mathscr{C}^1([0, T]; \mathscr{H}^{s-1}(\mathbb{R}^n))$ (see e.g. [6, Ch. 9]).

We are interested in the following question: *what well posedness results hold when the mentioned Lipschitz continuity assumption for the a_{ij} is replaced by a weaker hypothesis?*

Let us recall some know results of this type. Supposing that the coefficients of the principal part of (1.1) are log Lipschitz continuous (a function f is log Lipschitz continuous if

$$\sup_{0 < |t-s| < 1/2} \frac{|f(t) - f(s)|}{|t - s||\log|t - s||} < +\infty)$$

one can see that the Cauchy problem (1.1), (1.2) is still $\mathscr{C}^\infty$ well posed, even if in this case, considering the well posedness in Sobolev spaces, the following fact holds: there exists $\delta > 0$ (depending on the log Lipschitz norm of the coefficients a_{ij}) such that for all $u_0 \in \mathscr{H}^s(\mathbb{R}^n)$, $u_1 \in \mathscr{H}^{s-1}(\mathbb{R}^n)$ there is a unique solution in $\mathscr{C}([0, T]; \mathscr{H}^{s-\delta}(\mathbb{R}^n)) \cap \mathscr{C}^1([0, T]; \mathscr{H}^{s-1-\delta}(\mathbb{R}^n))$ (this behavior goes under the name of loss of derivatives). Moreover it turns out that the $\mathscr{C}^\infty$ well posedness in the case of log Lipschitz continuous coefficients is an optimal result in the following sense: for every function μ such that $\lim_{\sigma \to 0^+} \mu(\sigma) = +\infty$ it is possible to find a positive function a with

$$\sup_{0 < |t-s| < 1/2} \frac{|a(t) - a(s)|}{|t - s|| \log |t - s|| \mu (|t - s|)} < +\infty$$

and two $\mathscr{C}^\infty$ functions u_0, u_1 such that the Cauchy problem

$$\begin{cases} u_{tt} - a(t)u_{xx} = 0 \\ u(0, x) = u_0(x), \quad u_t(0, x) = u_1(x) \end{cases} \tag{1.4}$$

does not have any solution in $\mathscr{C}^1([0, T]; \mathscr{C}^\infty(\mathbb{R}))$ (actually there will be no solution not even if one looks for a solution which is only a distribution in the x variable) (see [1], [3]).

Similarly if one consider the case of the coefficients a_{ij} which are Hölder-continuous of exponent $\alpha < 1$, it is possible to prove that the Cauchy problem is well posed in the Gevrey space $\gamma^{(s)}(\mathbb{R}^n)$ for all $s < 1/(1-\alpha)$ (we recall that, given an open set $\Omega \subset \mathbb{R}^n$, a $\mathscr{C}^\infty$ function f is in $\gamma^{(s)}(\Omega)$ for some $s \geq 1$, if for all compact sets $K \subset \Omega$ there exist $C, M > 0$ such that

$$|\partial_x^\beta f(x)| \leq C M^{|\beta|} (\beta!)^s,$$

for all $x \in K$ and $\beta \in \mathbb{N}^n$). Also this result is optimal since, fixing $s_0 = 1/(1 - \alpha_0)$, there exists a positive function $a \in \mathscr{C}^{0,\alpha_0}(\mathbb{R})$ and there exist u_0, $u_1 \in \gamma^{(s)}(\mathbb{R})$ for all $s > s_0$ such that the Cauchy problem (1.4) does not have any solution in $\mathscr{C}^1([0, T]; \mathscr{D}'^{(s)}(\mathbb{R}))$ for all $s > s_0$, where $\mathscr{D}'^{(s)}(\mathbb{R})$ denotes the space of the Gevrey-ultradistributions of order s on $\mathbb{R}$ (see [1], [5] and [7] for related results).

In the present paper we want to show some results obtained in weakening the Lipschitz continuity assumption on the coefficients of the principal part of the operator in a different way: we suppose that there exist $\bar{t} \in [0, T]$ and there exist $q, C > 0$ such that $a_{ij} \in \mathscr{C}^1([0, T] \setminus \{\bar{t}\})$ for all $i, j = 1, \ldots, n$, and

$$|a'(t, \xi)| \leq C |t - \bar{t}|^{-q}, \tag{1.5}$$

for all $(t, \xi) \in ([0, T] \setminus \{\bar{t}\}) \times (\mathbb{R}^n \setminus \{0\})$ (here and in the following $'$ will denote the derivative with respect to the variable t).

We want to point out that if $q \geq 1$ then the condition (1.5) does not have any relation with the regularity of the a_{ij}'s on the whole $[0, T]$, as we can see from the following examples. Consider on the interval $[0, 1/2]$ the function

$$a_1(t) = \begin{cases} 1 + \dfrac{\sin(\log t)}{\log t} & \text{for } t \neq 0 \\ 1 & \text{for } t = 0. \end{cases}$$

We have that $\sup_{t\in]0,1/2]} t|a_1'(t)| < +\infty$ but it is easy to see that for all $0 < \alpha < 1$, $a_1 \notin \mathscr{C}^{0,\alpha}$. On the other hand defining on the same interval the function

$$a_2(t) = \begin{cases} 1 + \dfrac{t^2 \sin(\exp(\exp(1/t)))}{\exp(\exp(1/t))} & \text{for } t \neq 0 \\ 1 & \text{for } t = 0, \end{cases}$$

one can prove that a_2 is a log Lipschitz continuous function and, for all $q \geq 1$, $\sup_{t\in]0,1/2]} t^q |a_2'(t)| = +\infty$.

We can now state the first result of the paper.

Theorem 1.1. *Consider the equation (1.1) and suppose that the condition (1.5) holds with $q = 1$.*

Then the Cauchy problem (1.1), (1.2) is $\mathscr{C}^\infty$ well posed.

As it will be clear from the proof, considering in the case of Theorem 1.1 the well posedness in Sobolev spaces, there will be a loss of derivatives.

Our second result concerns the well posedness in Gevrey spaces.

Theorem 1.2. *Let $q > 1$ and $0 < p < 1$, with $p \leq q-1$. Consider the equation (1.1) and suppose that the condition (1.5) holds. Suppose, moreover, that there exists $C' > 0$ such that*

$$|a(t, \xi)| \leq C' |t - \bar{t}|^{-p}, \tag{1.6}$$

for all $(t, \xi) \in ([0, T] \setminus \{\bar{t}\}) \times (\mathbb{R}^n \setminus \{0\})$.

Then (1.1), (1.2) is $\gamma^{(s)}$ well posed for all $s < (q - p)/(q - 1)$.

Theorem 1.2 has two easy corollaries. If $1 < q < 2$ then the condition (1.6) with $p = q - 1$ is implied by (1.5). So that we can state the following.

Corollary 1.3. *Let $1 < q < 2$. Suppose that the condition (1.5) holds.*

Then (1.1), (1.2) is $\gamma^{(s)}$ well posed for all $s < 1/(q - 1)$.

On the other hand if the coefficients a_{ij} are bounded then the condition (1.6) is verified with $p = 0$. Consequently we have the following:

Corollary 1.4. *Consider the equation (1.1) and suppose that the condition (1.5) holds. Suppose moreover that there exists $C' > 0$ such that*

$$|a(t, \xi)| \leq C',$$

for all $(t, \xi) \in ([0, T] \setminus \{\bar{t}\}) \times (\mathbb{R}^n \setminus \{0\})$.

Then (1.1), (1.2) is $\gamma^{(s)}$ well posed for all $s < q/(q - 1)$.

Our third theorem gives a link between the recalled result for operators having Hölder-continuous coefficients in the principal part and the Corollary 1.4.

Theorem 1.5. *Let $q > 1$ and $0 < \alpha < 1$. Consider the equation (1.1) and suppose that $a_{ij} \in \mathscr{C}^{0,\alpha}([0, T])$ for all $i, j = 1, \ldots, n$. Suppose, moreover, that the condition (1.5) holds.*

Then (1.1), (1.2) is $\gamma^{(s)}$ well posed for all $s < (q/(q-1))(1/(1-\alpha))$.

One may ask what happens when s reaches the limit value, i.e., if $s = (q - p)/(q-1)$, under the hypotheses of Theorem 1.2 or $s = (q/(q-1))(1/(1-\alpha))$ under the hypotheses of Theorem 1.5. In these situations it is possible to prove that for all pair of functions $u_0, u_1 \in \gamma^{(s)}(\mathbb{R}^n)$ there exists $T' \in]0, T]$ such that the Cauchy problem

$$\begin{cases} u_{tt} - \displaystyle\sum_{i,j=1}^{n} a_{ij}(t)u_{x_i x_j} + \sum_{i=1}^{n} b_i(t)u_{x_i} + c(t)u = 0 & \text{in } [0, T'] \times \mathbb{R}^n, \\ u(0, x) = u_0(x), \ u_t(0, x) = u_1(x) & \text{in } \mathbb{R}^n, \end{cases}$$

has a unique solution in $\mathscr{C}^1([0, T']; \gamma^{(s)}(\mathbb{R}^n))$.

Similarly to the cited works on log Lipschitz continuous and Hölder-continuous coefficients, the sharpness of the results of the Theorems 1.1, 1.2 and 1.5 is assured by some counter examples.

Theorem 1.6. *There exists a function $a(t)$ satisfying*

$$a : [0,1] \longrightarrow \left[\frac{1}{2}, \frac{3}{2}\right] \quad \text{and} \quad a \in \mathscr{C}^\infty([0,1[) \cap \mathscr{C}^{0,\alpha}([0,1]) \text{ for all } \alpha < 1, \quad (1.7)$$

$$\sup_{t \in [0,1[} (1 - t)^q |a'(t)| < +\infty \quad \text{for all } q > 1, \quad (1.8)$$

and there exist $u_0, u_1 \in \mathscr{C}^\infty(\mathbb{R})$ such that the Cauchy problem (1.4) has no solution in $\mathscr{C}^1([0, 1]; \mathscr{D}'(\mathbb{R}))$.

Theorem 1.7. *Let $q_0 > 1$ and $0 \le \alpha_0 < 1$. There exists a function $a(t)$ satisfying*

$$a : [0, 1] \longrightarrow \left[\frac{1}{2}, \frac{3}{2}\right] \quad \text{and} \quad a \in \mathscr{C}^\infty([0, 1[) \cap \mathscr{C}^{0,\alpha_0}([0, 1]) \quad (1.9)$$

$$\sup_{t \in [0,1[} (1 - t)^{q_0} |a'(t)| < +\infty, \quad (1.10)$$

and there exist $u_0, u_1 \in \gamma^{(s)}(\mathbb{R})$ for all $s > s_0 = (q_0/(q_0 - 1))(1/(1 - \alpha_0))$ such that the Cauchy problem (1.4) has no solution in $\mathscr{C}^1([0, 1]; \mathscr{D}'^{(s)}(\mathbb{R}))$, for all $s > s_0$.

We point out that whilst Theorem 1.6 and Theorem 1.7 in the case of $\alpha_0 > 0$ are the converse of Theorem 1.1 and 1.5, respectively, Theorem 1.7 for $\alpha_0 = 0$

(in this case $\mathscr{C}^{0,0}$ means simply $\mathscr{C}$) gives only a counter example to the Corollary 1.4.

Finally, it is worth recording that a slightly more general non-well posedness results related to Theorem 1.6 and 1.7 can be found in [2]; in the same work the well posedness is obtained for operators depending on one space variable and in this case the coefficient c may depend also on x.

2. Proofs of Theorems 1.1, 1.2, 1.5

In this section we outline the main points of the proofs of the Theorems 1.1, 1.2 and 1.5. A precise and slightly different demonstration of the same results can be found in [2].

Let us begin with the proof of the Theorem 1.1 in the case of $\bar{t} = T$, i.e., we suppose that $a_{ij} \in \mathscr{C}^1([0, T[)$ for all $i, j = 1, \ldots, n$ and there exists $C > 0$ such that

$$|a'(t, \xi)| \le C(T - t)^{-1}, \tag{2.1}$$

for all $(t, \xi) \in [0, T[\times \mathbb{R}^n \setminus \{0\}$. Remarking that

$$|a(t, \xi) - a(0, \xi)| \le \int_0^t |a'(\sigma, \xi)| \, d\sigma \le C \log\left(\frac{T}{T - t}\right), \tag{2.2}$$

we deduce that the function $t \mapsto a(t, \xi)$ is in $\mathscr{L}^1([0, T])$ for all $\xi \in \mathbb{R}^n \setminus \{0\}$ and its $\mathscr{L}^1$-norm is uniformly bounded with respect to ξ. From this, by using the symmetry of the matrix (a_{ij}), we have that all the coefficients a_{ij} are in $\mathscr{L}^1([0, T])$. An important consequence of this fact is the so called finite speed of propagation (see [1, Th. 6]), hence instead of the $\mathscr{C}^\infty$ well posedness it will be sufficient to prove the well posedness in $\mathscr{C}_0^\infty$. We know (see [1, Th. 3]) that if $u_0, u_1 \in \mathscr{C}_0^\infty$ then the Cauchy problem (1.1), (1.2) has a unique solution $u \in \mathscr{C}^1([0, T]; \mathscr{A}'(\mathbb{R}))$, where $\mathscr{A}'(\mathbb{R})$ is the space of the analytic functionals on $\mathbb{R}$.

Our goal is to show that under the hypothesis (2.1) u actually belongs to $\mathscr{C}^1([0, T]; \mathscr{C}_0^\infty(\mathbb{R}))$; in doing this the main tools will be the estimate of an approximate energy of the solution and the Paley–Wiener theorem. We denote by v the Fourier transform of u with respect to x and from (1.1) we deduce

$$v''(t, \xi) = -a(t, \xi)|\xi|^2 v(t, \xi) - i \sum_{j=1}^n b_j(t)\xi_j v(t, \xi) - c(t)v(t, \xi). \tag{2.3}$$

The Paley–Wiener theorem ensures that for all $p > 0$ there exists $C_p > 0$ such that

$$|v(0, \xi)|^2 + |v'(0, \xi)|^2 \le C_p |\xi|^{-p},$$

for all $\xi \in \mathbb{R}^n$, $|\xi| \geq 1$. On the other hand, again by the Paley–Wiener theorem, it will be sufficient to show that for all $q > 0$ there exists $C_q > 0$ such that

$$|v(t, \xi)|^2 + |v'(t, \xi)|^2 \leq C_q |\xi|^{-q},$$

for all $t \in [0, T]$ and for all $\xi \in \mathbb{R}^n$, $|\xi| \geq 1$, to have $u \in \mathscr{C}^1([0, T]; \mathscr{C}_0^\infty(\mathbb{R}))$.
We define

$$a_\varepsilon(t, \xi) = \begin{cases} a(t, \xi) & \text{for } t \in [0, T - \varepsilon] \\ a(T - \varepsilon, \xi) & \text{for } t \in [T - \varepsilon, T], \end{cases} \tag{2.4}$$

and we set

$$E_\varepsilon(t, \xi) = a_\varepsilon(t, \xi)|\xi|^2 |v(t, \xi)|^2 + |v'(t, \xi)|^2, \tag{2.5}$$

the approximate energy of the solution u. Utilizing (2.3) in the differentiation of E_ε we obtain

$$E'_\varepsilon \leq |a'_\varepsilon| |\xi|^2 |v|^2 + 2\mathfrak{Re}\left(v'\bar{v}\right)\left(|a_\varepsilon - a| |\xi|^2 + \sum_{j=1}^n |b_j| |\xi_j| + |c|\right)$$

$$\leq \frac{1}{\lambda_0}\left(|a'_\varepsilon| + |a_\varepsilon - a| |\xi| + K\right) E_\varepsilon,$$

where $K = \sup_{t \in [0,T]}\{\sum_j |b_j(t)| + |c(t)|\}$. By Gronwall's lemma we deduce

$$E_\varepsilon(t, \xi) \leq E_\varepsilon(0, \xi)$$

$$\times \exp\left(\frac{1}{\lambda_0}\left(\int_0^T \left(|a'_\varepsilon(t, \xi)| + |\xi| |a_\varepsilon(t, \xi) - a(t, \xi)|\right) dt + KT\right)\right),$$

for all $t \in [0, T]$. From (2.1) and (2.4) we have that there exist $C', C'' > 0$ such that

$$\int_0^T |a'_\varepsilon(t, \xi)| \, dt \leq \int_0^{T-\varepsilon} C(T - t)^{-1} dt \leq C'\left(1 + |\log \varepsilon|\right),$$

and

$$\int_0^T |a_\varepsilon(t, \xi) - a(t, \xi)| \, dt \leq \int_{T-\varepsilon}^T \left(|a(T - \varepsilon, \xi)| + |a(t, \xi)|\right) dt$$

$$\leq C''\varepsilon\left(1 + |\log \varepsilon|\right).$$

Hence there exist $D, \delta > 0$ such that

$$E_\varepsilon(t, \xi) \leq D E_\varepsilon(0, \xi) \exp\left(\delta |\log \varepsilon| \left(1 + \varepsilon|\xi|\right)\right), \tag{2.6}$$

for all $t \in [0, T]$ and for all $\xi \in \mathbb{R}^n$. We remark now that

$$E_\varepsilon(0, \xi) = a(0, \xi)|\xi|^2 |v(0, \xi)|^2 + |v'(0, \xi)|^2, \qquad (2.7)$$

and

$$E_\varepsilon(t, \xi) \geq \lambda_0|\xi|^2 |v(t, \xi)|^2 + |v'(t, \xi)|^2, \qquad (2.8)$$

for all $\varepsilon \leq 1$, for all $t \in [0, T]$ and for all $\xi \in \mathbb{R}^n$, $|\xi| \geq 1$. Taking $\varepsilon = |\xi|^{-1}$ in (2.6) we obtain

$$\begin{aligned} \lambda_0|\xi|^2 |v(t, \xi)|^2 &+ |v'(t, \xi)|^2 \\ &\leq D\left(a(0, \xi)|\xi|^2 |v(0, \xi)|^2 + |v'(0, \xi)|^2\right) |\xi|^{2\delta}, \end{aligned} \qquad (2.9)$$

for all $t \in [0, T]$ and for all $\xi \in \mathbb{R}^n$, $|\xi| \geq 1$. In view of the recalled Paley–Wiener theorem the conclusion follows. Let us remark that (2.9) gives also the well posedness in Sobolev spaces with a loss of δ derivatives.

If $\bar{t} = 0$ we define

$$a_\varepsilon(t, \xi) = \begin{cases} a(\varepsilon, \xi) & \text{for } t \in [0, \varepsilon], \\ a(t, \xi) & \text{for } t \in [\varepsilon, T]. \end{cases}$$

With an argument similar to the previous one, using also the fact that in this case (1.5) implies

$$|a(\varepsilon, \xi)| \leq |a(T, \xi)| + C|\log \varepsilon| + \log T,$$

and consequently

$$E_\varepsilon(0, \xi) \leq C'(1 + |\log \varepsilon|) |\xi|^2 |v(0, \xi)|^2 + |v'(0, \xi)|^2,$$

for some C' independent from ε and ξ. Taking $\varepsilon = |\xi|^{-1}$ we deduce that there exists $D, \delta > 0$ such that

$$\begin{aligned} \lambda_0|\xi|^2 |v(t, \xi)|^2 &+ |v'(t, \xi)|^2 \\ &\leq C\left((1 + \log|\xi|) |\xi|^2 |v(0, \xi)|^2 + |v'(0, \xi)|^2\right) |\xi|^{2\delta}, \end{aligned}$$

for all $t \in [0, T]$ and for all $\xi \in \mathbb{R}^n$, $|\xi| \geq 1$, and the $\mathscr{C}_0^\infty$ well posedness follows.

The conclusion of the proof is easily reached remarking that if $\bar{t} \in {]0, T[}$ it will be sufficient to solve first the Cauchy problem in $[0, \bar{t}]$, then to solve the same problem in $[\bar{t}, T]$ (with initial data $u(\bar{t}, \cdot)$, $u_t(\bar{t}, \cdot)$ obtained from the previous problem) and finally to glue together the two solutions.

The proof of the Theorem 1.2 is similar to the previous one. Let us describe only the case of $\bar{t} = T$. We set

$$a_\varepsilon(t, \xi) = \begin{cases} a(t, \xi) & \text{for } t \in [0, T - \varepsilon^\beta] \\ a\left(T - \varepsilon^\beta, \xi\right) & \text{for } t \in [T - \varepsilon^\beta, T], \end{cases}$$

where $\beta = 1/(q - p)$. Defining E_ε as in (2.5), Gronwall's lemma gives

$$E_\varepsilon(t, \xi) \leq E_\varepsilon(0, \xi)$$

$$\times \exp\left(\frac{1}{\lambda_0}\left(\int_0^T \left(|a'_\varepsilon(t, \xi)| + |\xi|\,|a_\varepsilon(t, \xi) - a(t, \xi)|\right) dt + KT\right)\right),$$

for all $t \in [0, T]$. By (1.5) we have

$$\int_0^T |a'_\varepsilon(t, \xi)|\, dt \leq C \int_0^{T-\varepsilon^\beta} (T - t)^{-q}\, dt \leq \frac{C}{q-1} \varepsilon^{-(q-1)/(q-p)},$$

and from (1.6) we deduce

$$\int_0^T |a_\varepsilon(t, \xi) - a(t, \xi)|\, dt \leq \int_{T-\varepsilon^\beta}^T \left(|a_\varepsilon(t, \xi)| + |a(t, \xi)|\right) dt$$

$$\leq \frac{(2-p)}{1-p} C' \varepsilon^{(1-p)/(q-p)}.$$

Consequently we have that there exist $D, \delta > 0$ such that

$$E_\varepsilon(t, \xi) \leq D E_\varepsilon(0, \xi) \exp\left(\delta\left(\varepsilon^{-(q-1)/(q-p)} + |\xi|\varepsilon^{(1-p)/(q-p)}\right)\right), \quad (2.10)$$

for all $t \in [0, T]$ and for all $\xi \in \mathbb{R} \setminus \{0\}$. By (2.7) and (2.8), taking $\varepsilon = |\xi|^{-1}$ in (2.10) we deduce that there exists $D' > 0$ such that

$$\lambda_0|\xi|^2\,|v(t, \xi)|^2 + |v'(t, \xi)|^2$$

$$\leq D'\left(a(0, \xi)|\xi|^2|v(0, \xi)|^2 + |v'(0, \xi)|^2\right) \exp\left(2\delta|\xi|^{(q-1)/(q-p)}\right),$$

for all $t \in [0, T]$ and for all $\xi \in \mathbb{R}^n$, $|\xi| \geq 1$. The well posedness in $\gamma^{(s)}$, for $s < (q - p)/(q - 1)$ follows by applying the Paley–Wiener theorem.

Let us come briefly to the proof of the Theorem 1.5, in the case of $\bar{t} = T$. We consider a real non-negative $\mathscr{C}^\infty$ function ρ defined on $\mathbb{R}$ such that $\text{supp}(\rho) \subset [-1, 1]$ and $\int \rho(x)\, dx = 1$. We extend the domain of a to $\mathbb{R} \times \mathbb{R}^n \setminus \{0\}$, setting, for $t \leq 0$, $a(t, \xi) = a(0, \xi)$ and, for $t \geq T$, $a(t, \xi) = a(T, \xi)$. We define

$$a_\varepsilon(t, \xi) = \int a(t - \varepsilon\tau, \xi)\rho(\tau)\, d\tau.$$

As before we find

$$E_\varepsilon(t, \xi) \le E_\varepsilon(0, \xi)$$

$$\times \exp\left(\frac{1}{\lambda_0}\left(\int_0^T (|a_\varepsilon'(t, \xi)| + |\xi| \, |a_\varepsilon(t, \xi) - a(t, \xi)|) \, dt + KT\right)\right),$$

for all $t \in [0, T]$. We have

$$\int_0^T |a_\varepsilon'(t, \xi)| \, dt = \int_0^{T-\varepsilon^\beta} |a_\varepsilon'(t, \xi)| \, dt + \int_{T-\varepsilon^\beta}^T |a_\varepsilon'(t, \xi)| \, dt,$$

where $\beta = (1 - \alpha)/q$. From (1.5) we obtain that

$$\int_0^{T-\varepsilon^\beta} |a_\varepsilon'(t, \xi)| \, dt \le \int_0^{T-\varepsilon^\beta} \frac{1}{\varepsilon^2} \int |a(t - \tau, \xi) - a(t, \xi)| \left|\rho'\left(\frac{\tau}{\varepsilon}\right)\right| d\tau \, dt$$

$$\le \int_0^{T-\varepsilon^\beta} \frac{1}{\varepsilon} \int C \, (T - (t + \varepsilon))^{-q} \left|\rho'\left(\frac{\tau}{\varepsilon}\right)\right| d\tau dt$$

$$\le C \int_0^{T-\varepsilon^\beta} (T - (t + \varepsilon))^{-q} \, dt \int |\rho'(\sigma)| \, d\sigma$$

$$\le \frac{C}{q - 1} \varepsilon^{(1-\alpha)((1-q)/q)}.$$

On the other hand, from the fact that $a \in \mathscr{C}^{0,\alpha}$ we deduce that

$$\int_{T-\varepsilon^\beta}^T |a_\varepsilon'(t, \xi)| \, dt \le \int_{T-\varepsilon^\beta}^T \frac{1}{\varepsilon^2} \int |a(t - \tau, \xi) - a(t, \xi)| \left|\rho'\left(\frac{\tau}{\varepsilon}\right)\right| d\tau dt$$

$$\le \int_{T-\varepsilon^\beta}^T \frac{1}{\varepsilon^2} \int \|a\|_{\mathscr{C}^{0,\alpha}} |\tau|^\alpha \left|\rho'\left(\frac{\tau}{\varepsilon}\right)\right| d\tau dt$$

$$\le \|a\|_{\mathscr{C}^{0,\alpha}} \int_{T-\varepsilon^\beta}^T \varepsilon^{\alpha-1} dt \int |\sigma|^\alpha \, |\rho'(\sigma)| \, d\sigma$$

$$\le \|a\|_{\mathscr{C}^{0,\alpha}} \varepsilon^{(1-\alpha)((1-q)/q)} \int |\sigma|^\alpha \, |\rho'(\sigma)| \, d\sigma.$$

Hence there exists $C' > 0$ such that

$$\int_0^T |a_\varepsilon'(t, \xi)| \, dt \le C' \varepsilon^{(1-\alpha)((1-q)/q)}.$$

Similarly it is possible to prove that there exists $C'' > 0$ such that

$$\int_0^T |a_\varepsilon(t, \xi) - a(t, \xi)| \, dt \le C'' \varepsilon^{\alpha+((1-\alpha)/q)}.$$

Putting together these inequalities we obtain that

$$E_\varepsilon(t,\xi) \le DE_\varepsilon(0,\xi) \exp\left(\delta\left(\varepsilon^{-(1-\alpha)((q-1)/q)} + |\xi|\varepsilon^{1-(1-\alpha)((q-1)/q)}\right)\right), \quad (2.11)$$

for some $D, \delta > 0$ not depending on t or ξ. We take $\varepsilon = |\xi|^{-1}$ and the well posedness in $\gamma^{(s)}$, for $s < (q/(q-1))(1/(1-\alpha))$ follows via the Paley–Wiener theorem since by (2.7), (2.8) and (2.11) we deduce that there exists $D' > 0$ such that

$$\lambda_0|\xi|^2 |v(t,\xi)|^2 + |v'(t,\xi)|^2 \le D'\left(a(0,\xi)|\xi|^2 |v(0,\xi)|^2 + |v'(0,\xi)|^2\right)$$
$$\times \exp\left(2\delta|\xi|^{((q-1)/q)(1-\alpha)}\right),$$

for all $t \in [0,T]$ and for all $\xi \in \mathbb{R}^n$, $|\xi| \ge 1$. This concludes the sketch of the proofs.

3. Construction of the counter examples

The proofs of the Theorems 1.6 and 1.7 are based on the technique of construction of counter examples to the existence of solutions in the Cauchy problem for hyperbolic operators developed in [1], [4] and [3]. We start fixing a $\mathscr{C}^\infty$ real non-negative 2π-periodic function v such that $v(\tau) = 0$ for all τ in a neighborhood of 0 and

$$\int_0^{2\pi} v(\tau)\cos^2\tau\,d\tau = \pi.$$

We define, for all $\tau \in \mathbb{R}$ and for all $\varepsilon \in {]}0,\bar\varepsilon]$,

$$\alpha_\varepsilon(\tau) = 1 + 4\varepsilon v(\tau)\sin 2\tau - 2\varepsilon v'(\tau)\cos^2\tau - 4\varepsilon^2 v^2(\tau)\cos^4\tau,$$
$$\widetilde{w}_\varepsilon(\tau) = \cos\tau \exp\left(-\varepsilon\tau + 2\varepsilon\int_0^\tau v(s)\cos^2 s\,ds\right),$$
$$w_\varepsilon(\tau) = e^{\varepsilon\tau}\widetilde{w}_\varepsilon(\tau).$$

The functions α_ε and $\widetilde{w}_\varepsilon$ are 2π-periodic; moreover there exists $M > 0$, such that for all $\tau \in \mathbb{R}$ and for all $\varepsilon \in {]}0,\bar\varepsilon]$,

$$|\alpha_\varepsilon(\tau) - 1| \le M\varepsilon, \qquad |\alpha'_\varepsilon(\tau)| \le M\varepsilon, \qquad (3.1)$$

and finally the function w_ε is the solution of the following Cauchy problem

$$\begin{cases} w''_\varepsilon + \alpha_\varepsilon(\tau)w_\varepsilon = 0 & \text{in } \mathbb{R}, \\ w_\varepsilon(0) = 1, & w'_\varepsilon(0) = 0. \end{cases}$$

We consider three monotone sequences $\{h_k\}$, $\{\rho_k\}$, $\{\varepsilon_k\}$ of positive real numbers such that

$$h_k \longrightarrow +\infty, \qquad \varepsilon_k \longrightarrow 0, \qquad \rho_k \longrightarrow 0; \tag{3.2}$$

$$\varepsilon_k \leq \min\left\{\bar{\varepsilon}, (2M)^{-1}\right\}, \quad \text{for all } k \in \mathbb{N}; \tag{3.3}$$

$$h_k, h_k \rho_k (4\pi)^{-1} \in \mathbb{N}, \quad \text{for all } k \in \mathbb{N}; \tag{3.4}$$

$$\sum_{k=0}^{+\infty} \rho_k \leq 1. \tag{3.5}$$

We define, for all $k \in \mathbb{N}$,

$$t_k = 1 - \left(\frac{\rho_k}{2} + \sum_{j=k+1}^{+\infty} \rho_j\right),$$

and

$$I_k = \left[t_k - \frac{\rho_k}{2},\ t_k + \frac{\rho_k}{2}\right].$$

We set

$$a(t) = \begin{cases} \alpha_{\varepsilon_k}\left(h_k(t - t_k)\right) & \text{for } t \in I_k, \\[2mm] 1 & \text{for } t \in [0, 1] \setminus \displaystyle\bigcup_{k=0}^{+\infty} I_k. \end{cases}$$

The function a is identically equal to 1 in a neighborhood of the boundary of each interval I_k, consequently $a \in C^\infty([0, 1[)$. Easily from (3.1) and (3.3) we deduce that

$$|a(t) - 1| < \frac{1}{2},$$

for all $t \in [0, 1]$. Supposing that

$$\sup_{k \in \mathbb{N}} \varepsilon_k h_k^\alpha < +\infty \quad \text{for all } \alpha < 1, \tag{3.6}$$

or

$$\sup_{k \in \mathbb{N}} \varepsilon_k h_k^{\alpha_0} < +\infty, \tag{3.7}$$

we have that (1.7) or (1.9) are satisfied, respectively, (let us remark that if $\alpha_0 = 0$ then (3.7) is implied by (3.2)). Finally if

$$\sup_{k \in \mathbb{N}} \left(\sum_{j=k}^{+\infty} \rho_j\right)^q \varepsilon_k h_k < +\infty, \tag{3.8}$$

for all $q > 1$, or

$$\sup_{k \in \mathbb{N}} \left(\sum_{j=k}^{+\infty} \rho_j \right)^{q_0} \varepsilon_k h_k < +\infty, \tag{3.9}$$

then (1.8) or (1.10) holds, respectively. We come to the construction of the functions u_0, u_1 and u. We denote by φ_k the solution of the following ordinary Cauchy problem

$$\begin{cases} \varphi_k'' + h_k^2 a(t) \varphi_k = 0, \\ \varphi_k(t_k) = \eta_k, \quad \varphi_k'(t_k) = 0. \end{cases}$$

The functions φ_k are in $\mathscr{C}^\infty([0, 1[) \cap \mathscr{C}^2([0, 1])$; moreover,

$$\varphi_k(t) = w_{\varepsilon_k}(h_k(t - t_k)) = e^{\varepsilon_k h_k(t - t_k)} \widetilde{w}_{\varepsilon_k}(h_k(t - t_k))$$

for all $t \in I_k$. Since $\widetilde{w}_\varepsilon$ is 2π-periodic, we deduce from (3.4) that

$$\varphi_k\left(t_k - \frac{\rho_k}{2}\right) = \exp\left(-\frac{1}{2}\varepsilon_k h_k \rho_k\right), \qquad \varphi_k'\left(t_k - \frac{\rho_k}{2}\right) = 0, \tag{3.10}$$

$$\varphi_k\left(t_k + \frac{\rho_k}{2}\right) = \exp\left(\frac{1}{2}\varepsilon_k h_k \rho_k\right), \qquad \varphi_k'\left(t_k + \frac{\rho_k}{2}\right) = 0. \tag{3.11}$$

We introduce the energy

$$E_{\varphi_k}(t) = h_k^2 a(t) |\varphi_k(t)|^2 + |\varphi_k'(t)|^2, \tag{3.12}$$

and applying Gronwall's lemma, from (3.10) we obtain that

$$E_{\varphi_k}(t) \le E_{\varphi_k}\left(t_k - \frac{\rho_k}{2}\right) \exp\left(\int_t^{t_k - \rho_k/2} \frac{|a'(t)|}{a(t)} dt\right)$$

$$\le h_k^2 \exp\left(-\varepsilon_k h_k \rho_k + 2M \sum_{j=0}^{k-1} \varepsilon_j h_j \rho_j\right), \tag{3.13}$$

for all $t \in [0, t_k - \rho_k/2[$. Suppose now

$$4M \sum_{j=0}^{k-1} \varepsilon_j h_j \rho_j \le \varepsilon_k h_k \rho_k, \tag{3.14}$$

for all $k \in \mathbb{N}$; we deduce from (3.13) and (3.14) that

$$E_{\varphi_k}(t) \le h_k^2 \exp\left(-\frac{1}{2}\varepsilon_k h_k \rho_k\right) \tag{3.15}$$

for all $k \in \mathbb{N}$ and for all $t \in [0, t_k - \rho_k/2[$. We define

$$u_0(x) = \sum_{k=0}^{+\infty} \varphi_k(0) e^{i h_k x}, \qquad u_1(x) = \sum_{k=0}^{+\infty} \varphi_k'(0) e^{i h_k x},$$

and

$$u(t, x) = \sum_{k=0}^{+\infty} \varphi_k(t) e^{i h_k x}.$$

Finally, we suppose that

$$\lim_{k \to +\infty} p \log h_k - \varepsilon_k h_k \rho_k = -\infty \tag{3.16}$$

for all $p > 0$. From (3.12), (3.15) and (3.16) we deduce that for all $p > 0$ there exists $C_p > 0$, such that

$$|\varphi_k(t)| + |\varphi_k'(t)| \leq C_p h_k^{-p},$$

for all $k \in \mathbb{N}$ and for all $t \in [0, t_k - \rho_k/2[$. In view of the Paley–Wiener theorem for Fourier series this last inequality implies that $u_0, u_1 \in \mathscr{C}^\infty(\mathbb{R})$ and $u \in \mathscr{C}^1([0, 1 - \delta]; \mathscr{C}^\infty(\mathbb{R}))$ for all $\delta \in]0, 1[$; moreover u is the solution to

$$\begin{cases} u_{tt} - a(t) u_{xx} = 0 \\ u(0, x) = u_0(x), \quad u_t(0, x) = u_1(x). \end{cases} \tag{3.17}$$

On the other hand (3.11) and (3.16) imply that $\varphi_k(t_k + \rho_k/2)$ increases faster than any power of h_k and consequently $u \notin \mathscr{C}([0, 1]; \mathscr{D}'(\mathbb{R}))$ (remark that in view of [1, Th. 4 and Th. 6] u is the unique solution to (3.17) in $\mathscr{C}^1([0, 1]; \mathscr{D}'^{(s)}(\mathbb{R}))$ for all $s > 1$).

Similarly, in the case of Theorem 1.7 it will be enough to suppose that for all $s > s_0 = (q_0/(q_0 - 1))(1/(1 - \alpha_0))$,

$$\lim_{k \to +\infty} h_k^{1/s} - \varepsilon_k h_k \rho_k = -\infty. \tag{3.18}$$

Then for all $s > s_0$ there exists $C_s > 0$ such that

$$|\varphi_k(t)| + |\varphi_k'(t)| \leq C_s \exp\left(-h_k^{1/s}\right),$$

for all $k \in \mathbb{N}$ and for all $t \in [0, t_k - \rho_k/2]$. Consequently $u_0, u_1 \in \gamma^{(s)}(\mathbb{R})$ and $u \in \mathscr{C}^1([0, 1 - \delta]; \gamma^{(s)}(\mathbb{R}))$ for all $\delta \in]0, 1[$; u is the unique solution to (3.17) and, as a consequence of (3.11), (3.18), $u \notin \mathscr{C}([0, 1]; \mathscr{D}'^{(s)}(\mathbb{R}))$ for all $s > s_0$.

To conclude the construction of the counter examples we choose the sequences $\{h_k\}$, $\{\rho_k\}$ and $\{\varepsilon_k\}$ in the following way: in the case of Theorem 1.6,

$$h_k = N^{2k+1} \left(\left[\log(k+3)\right]\right)^k,$$

$$\rho_k = \frac{4\pi}{N} \left(\left[\log(k+3)\,(\log(\log(k+3)))^{-1}\right]\right)^k \left(\left[\log(k+3)\right]\right)^{-k},$$

$$\varepsilon_k = \frac{N^k L}{h_k \rho_k} = \frac{L}{4\pi N^k} \left(\left[\log(k+3)\,(\log(\log(k+3)))^{-1}\right]\right)^{-k},$$

where $[x]$ denotes the maximum integer $\leq x$ and N, $1/L$ are integers sufficiently large. We let to the reader to verify that (3.2), (3.3), (3.4), (3.5), (3.6), (3.8), (3.14), (3.16) are satisfied. In the case of Theorem 1.7, if $\alpha_0 = 0$,

$$h_k = N^{k+1}, \qquad \rho_k = \frac{4\pi}{N^{k+1}} \left[N^{(1-1/q_0)k}\right], \qquad \varepsilon_k = \frac{L}{k^{q_0}},$$

and if $\alpha_0 > 0$,

$$h_k = N^{k+1}, \qquad \rho_k = \frac{4\pi}{N^{k+1}} \left[N^{(1-(1-\alpha_0)/q_0)k}\right], \qquad \varepsilon_k = \frac{L}{N^{\alpha_0 k}},$$

where N and $1/L$ are integers sufficiently large. Again it will be sufficient to verify, with these choices, the conditions (3.2), (3.3), (3.4), (3.5), (3.7), (3.9), (3.14), (3.18).

References

[1] Colombini, F., De Giorgi, E., and Spagnolo, S., Sur les équations hyperboliques avec des coefficients qui ne dépendent que du temps, *Ann. Sc. Norm. Sup. Pisa* **6** (1979), 511–559.

[2] Colombini, F., Del Santo, D. and Kinoshita, T., *Well posedness of the Cauchy problem for a hyperbolic equation with non-Lipschitz coefficients*, to appear in Ann. Sc. Norm. Sup. Pisa.

[3] Colombini, F. and Lerner, N., Hyperbolic operators with non-Lipschitz coefficients, *Duke Math. J.* **77** (1995), 657–698.

[4] Colombini, F. and Spagnolo, S., Some examples of hyperbolic equations without local solvability, *Ann. scient. Éc. Norm. Sup.* 4e série, **22** (1989), 109–125.

[5] Jannelli, E., Regularly hyperbolic systems and Gevrey classes, *Ann. Mat. Pura. Appl.* **140** (1985), 133–145.

[6] Hörmander, L., *Linear partial differential operators*, Springer–Verlag, Berlin 1963.

[7] Nishitani, T., Sur les équations hyperboliques à coefficients höldériens en t et de classe de Gevrey en x, *Bull. Sci. Math.* **107** (1983), 113–138.

MULTIPLE POINTS OF THE CHARACTERISTIC MANIFOLD OF A DIAGONALIZABLE OPERATOR

HÉLÈNE DELQUIÉ AND JEAN VAILLANT

Contents

1   Notations. Statement of theorem   55
2   Proof of the theorem   56

M. de Gosson (ed.), Jean Leray '99 Conference Proceedings, 53-64.
© *2003 Kluwer Academic Publishers.*

The authors have stated in a recent paper [1] that if a 4×4 matrix-valued linear differential operator $I D_0 + a(D') + a_0(x)$ with real coefficients is such that its principal part $a(\xi)$ is diagonalizable with respect to $(1, 0, \ldots, 0)$ and is of reduced dimension superior or equal to $4(4+1)/2 - 2 = 8$, its characteristic manifold admits at least a triple point; more exactly there exists a $\xi' \neq 0$ such that the largest eigenvalue of $a(\xi')$ is triple. They propose to generalize this result to the case of a $m \times m$ operator and to a point of order $m - p$, $1 \leq p$. In this statement they obtain the result in the case $m \leq 5$; the case $m > 5$ will be studied in another publication. The existence of a point of order $m - 1$ is particularly useful in order to prove that the system is presymmetric, i.e., symmetrizable by a matrix with constant coefficients [6], [8], [9] and [10]. It is also useful for the study of systems with variable coefficients where the existence of a regular symmetrizer is also obtained [5].

The corresponding result in the case of the symmetric matrices has been obtained in [2]. We follow the main lines of their work, but the assumption of diagonalizability implies more complicated calculations than that of symmetry.

1. Notations. Statement of theorem

Let E be a $(n + 1)$-dimensional vector space and F an m-dimensional vector space. We denote ξ an element of E. Let a be a linear mapping of E into the vector space $\mathscr{L}(F, F)$ of linear mappings of F in F: $a(\xi) \in \mathscr{L}(F, F)$, $a \in \mathscr{L}[E, \mathscr{L}(F, F)]$.

Definition. *The reduced dimension d of a is the rank of a.*

i)   Choose a basis of F, then d is also the dimension of the vector subspace of the space of $m \times m$ matrices formed by the matrices $a_j^i(\xi)$, $\xi \in E$.

ii) d is the dimension of the vector subspace of the space of linear forms of E spanned by the forms:

$$\xi \longmapsto a_j^i(\xi), \quad 1 \leq i, j \leq m.$$

Definition. *N is a non-characteristic vector in E if $\det a(N) \neq 0$.*

We assume there exists a non-characteristic vector $N \in E$, then we replace $a(\xi)$ by $a^{-1}(N)a(\xi)$, i.e., we can assume that there exists N such that:

$$a(N) = I.$$

Let N be the first vector of a basis in E; we denote by ξ' an element with components $(0, \xi_1, \xi_2, \ldots, \xi_n)$. An element ξ of E is also written:

$$\xi = \xi_0 N + \xi'$$

and:

$$a(\xi) = \xi_0 I + a\left(\xi'\right) = \xi_0 I + \sum_{1 \leq k \leq n} a_k \xi_k, \quad \left(\xi_0, \xi'\right) \in \mathbb{R}^{n+1}.$$

Definition. *a is diagonalizable with respect to N if and only if:*

i) *the zeros in τ of $\det a(\tau N + \xi) = \det(\tau I + a(\xi))$ are all real, for all ξ.*

ii) *if μ is the multiplicity of a zero, then the dimension of the corresponding proper subspace is μ.*

If we choose N as the first vector of a basis in E, the previous definition is equivalent to:
i) the zeros in τ of $\det(\tau I + a(\xi'))$ are all real, i.e., the eigenvalues of $a(\xi')$ are real.
ii) the dimension of the proper subspace corresponding to an eigenvalue is equal to its multiplicity.

Let now N be the first vector in a basis in E.

Proposition 1.1. [7] If a is diagonalizable with respect to N and if:

$$a_j^i(\xi) = \xi_0 I_j^i + a_j^i\left(\xi'\right)$$

then for $i < j$, a_j^i belongs to the $(d-1)$-dimensional vector subspace spanned by the forms a_j^i, $i \geq j$.

Theorem 1.2. *If $a(\xi)$ is diagonalizable with respect to N, if:*

$$d[a] \geq m(m+1)/2 - (p+1)(p+2)/2 + 1, \quad 0 \leq p \leq m-1, \text{ if } m \leq 5, \quad (*)$$

then there exists a point of multiplicity $m - p$ different from zero in the characteristic manifold: $\det a(\xi) = 0$; more exactly, there exists $\xi' \neq 0$ such that the largest eigenvalue of $a(\xi')$ has multiplicity $m - p$. The condition $(*)$ can not be improved.

Remark. If $m > 5$, we obtain the result by imposing the additional condition:

$$m(m+1)/2 - (p+1)(p+2)/2 - 1 > m(m-p-2).$$

2. Proof of the theorem

If $p = m - 1$ the result is obvious. If $p = 0$, it follows from Proposition 1.1 that the elements $a(\xi')_j^i$, $i < j$, depend linearly on the elements $a(\xi')_j^i$, $i \geq j$; then by cancelling them we obtain a zero ξ_0 of multiplicity m.

We will first prove a weaker result than that in the theorem.

Lemma 2.1. *If $d \geq m(m+1)/2 - (p+1)(p+2)/2 + 2$, $0 \leq p \leq m-1$, then it exists $\xi' \neq 0$ such that the largest eigenvalue of $a(\xi')$ is at least of multiplicity $m - p$ (we can choose it equal to 1).*

Proof. We prove it by induction on $m - p$, m is fixed.   □

The result is obviously true for $m - 1$. Assume it true at the rank $m - p - 1$ and prove it at the rank $m - p$. Then we have:

$$d \geq m(m+1)/2 - (p+1)(p+2)/2 + 2;$$

hence:

$$d \geq m(m+1)/2 - (p+2)(p+3)/2 + 2.$$

From the induction assumption we infer that there exists $\eta' \neq 0$ such that the largest eigenvalue of $a(\eta')$ is at least of multiplicity $m - p - 1$ and equal to 1, and therefore such that:

$$\lambda_1 \left[a \left(\eta' \right) \right] = \cdots = \lambda_{m-p-1} \left[a \left(\eta' \right) \right] = 1 \geq \lambda_{m-p} \left[a \left(\eta' \right) \right].$$

If this eigenvalue is of multiplicity $m - p$, the result is obtained. We suppose then that:

$$\lambda = \lambda_1 \left[a \left(\eta' \right) \right] > \lambda_{m-p} \left[a \left(\eta' \right) \right].$$

Then we can choose a basis of F such that in this basis:

$$a \left(\eta' \right) = \mathrm{diag} \left(\lambda_1 \left[a \left(\eta' \right) \right], \lambda_2 \left[a \left(\eta' \right) \right], \ldots, \lambda_m \left[a \left(\eta' \right) \right] \right).$$

we pose:

$$a \left(\eta' \right) = b.$$

Let (X_i), $1 \leq i \leq m - p - 1$, the normalized proper vectors corresponding to the eigenvalue 1:

$$X_i = \text{columns } (0, \ldots, 1, 0, \ldots, 0), \quad 1 \text{ at the place } i;$$

we have:

$$b X_i = X_i.$$

We consider now the system in η':

$$a \left(\eta' \right) X_i = 0, \quad 1 \leq i \leq m - p - 1, \tag{2.1}$$

and we propose to show it admits a non-trivial solution.
We define the following sets:

$$\mathcal{T} = \left\{ \Phi^i_j / 1 \leq i < j \leq m - p - 1 \right\},$$

$$\mathcal{I}_1 = \left\{ \Phi^i_j / 1 \leq j \leq i \leq m \text{ and } j \leq m - p - 1 \right\},$$

$$\mathcal{I}_2 = \left\{ \Phi^i_j / m - p \leq j \leq i \leq m \right\}.$$

We know that $\{\Phi^i_j / i \leq j\}$ spans the space of linear forms of the matrix Φ by Proposition 1.1; thus there exists a basis $\mathcal{B}$ with elements of $\mathcal{I}_1$ and $\mathcal{I}_2$.

The total number of elements of $\mathcal{I}_1$ is:

$$m(m+1)/2 - (p+1)(p+2)/2$$

and that of $\mathcal{I}_2$ is:

$$(p+1)(p+2)/2.$$

There exists an integer $k \geq 0$ such that the number of elements in $\mathcal{I}_1$ belonging to the basis $\mathcal{B}$ is:

$$m(m+1)/2 - (p+1)(p+2)/2 - k, \quad k \geq 0;$$

the number of elements in $\mathcal{I}_2$ belonging to the basis $\mathcal{B}$ is then larger than or equal to $k+1$, because the reduced dimension of Φ is larger or equal to:

$$m(m+1)/2 - (p+1)(p+2)/2 + 1.$$

Assume that $\Phi^i_j(\eta') = 0$ for $\Phi^i_j \in \mathcal{I}_1$; $\xi_0 = 0$ is an eigenvalue of multiplicity $m - p - 1$ of the obtained matrix; its minors of order $p + 2$ are equal to zero. If the minor constituted from $p + 1$ lines and columns vanished, then $\xi_0 = 0$ would be an eigenvalue of order $m - p$ of $a(\eta')$, which is excluded; we deduce from this that elements of triangular $\mathcal{T}$ all vanish, this implies then that elements of $\mathcal{T}$ depend linearly on elements of $\mathcal{I}_1$.

The system (2.1) has then rank at most:

$$m(m+1)/2 - (p+1)(p+2)/2.$$

The number of unknowns is larger than or equal to:

$$m(m+1)/2 - (p+1)(p+2)/2 + 1.$$

Then there exists a non-trivial solution χ' of the system (2.1) such that $a(\chi') \neq 0$; assume that $a(\chi') = c$; we have:

$$cX_i = 0, \quad 1 \leq i \leq m - 1;$$

we can choose c such that:

$$\lambda(c) > 0.$$

Then, as in [2], we consider the matrix:

$$c(\alpha) = b + \alpha c;$$

we have:

$$c(\alpha)X_i = X_i, \quad 1 \leq i \leq m - p - 1.$$

For α small, eigenvalues of $c(\alpha)$ are near those of b. For α large:

$$\lambda_1[c(\alpha)] = \alpha\lambda_1(c) + 0(1).$$

We deduce from this that exists $\alpha^* > 0$ such that:

$$\lambda_{m-p}\left[c\left(\alpha^*\right)\right] = 1,$$

consequently we have a contradiction, and the lemma is proved.

Lemma 2.2. *Supposing that*

$$d \geq m(m+1)/2 - (p+1)(p+2)/2 + 1, \quad 0 \leq p \leq m - 2,$$

and, moreover, that for all $\xi' \neq 0$

$$\lambda_1\left[a(\xi')\right] > \lambda_{m-p}\left[a(\xi')\right], \tag{2.2}$$

let $Y_1, Y_2, \ldots, Y_{m-p-1}$ be a system of $m - p - 1$ linearly independent vectors. We consider the system in (λ, ξ'):

$$a\left(\xi'\right)Y_i = \lambda Y_i, \quad 1 \leq i \leq m - p - 1. \tag{2.3}$$

Then there exists $\underline{\xi}' \neq 0$ and $\underline{\lambda}$ such that:

$$a\left(\underline{\xi}'\right)Y_i = \underline{\lambda}Y_i, \quad 1 \leq i \leq m - p - 1$$

and that:

$$\underline{\lambda} = \lambda_1\left[a\left(\underline{\xi}'\right)\right] = \cdots = \lambda_{m-p-1}\left[a\left(\underline{\xi}'\right)\right].$$

Moreover, all $(\lambda, a(\xi'))$ satisfying (2.2) are of the type:

$$a\left(\xi'\right) = \alpha a\left(\underline{\xi}'\right), \quad \lambda = \alpha\underline{\lambda}.$$

Proof. It follows from previous lemma that there exists a mapping b such that $\lambda_1(b) = \cdots = \lambda_{m-p-1}(b) = 1$ and we may choose a basis of F such that b is diagonal. We denote by X_i the linearly independent eigenvectors corresponding to this eigenvalue as in preceding lemma. First, we prove Lemma 2.2 in the special case where:

$$Y_i = X_i, \quad 1 \leq i \leq m - p - 1,$$

by following the argument in [2]. $\qquad\qquad\Box$

If there exists $c = a(\chi')$ linearly independent from b such that:

$$cX_i = \mu X_i.$$

i)   when $\mu = 0$, by a similar argument to the one of Lemma 2.1, we are led to a contradiction.

ii) when $\mu \neq 0$, by replacing c by $c - \mu b$, we are also led to a contradiction.

Let now any of the Y_i, $1 \leq i \leq m - p - 1$ be linearly independent vectors. There exists a family $X_i(t)$, $1 \leq i \leq m - p - 1$ of linearly independent vectors depending continuously on t, $0 \leq t \leq 1$, such that:

$$X_i(0) = X_i, \qquad X_i(1) = Y_i, \qquad 1 \leq i \leq m - p - 1.$$

More precisely, there exists a continuous path in the connected space of invertible matrices of determinant $> 0 : t \mapsto U(t)$, such that: $U(0) = I, U(1)X_i = Y_i$, $1 \leq i \leq m - p - 1$, $U(t)X_i = X_i(t)$.

For any t, $0 \leq t \leq 1$, we consider the system in (λ, ξ'):

$$a\left(\xi'\right) X_i(t) = \lambda X_i(t), \qquad 1 \leq i \leq m - p - 1.$$

Let:

$$\tilde{a}_t\left(\xi'\right) = U^{-1}(t)a\left(\xi'\right) U(t),$$

the system becomes:

$$\tilde{a}_t\left(\xi'\right) X_i = \lambda X_i, \qquad 1 \leq i \leq m - p - 1.$$

$\tilde{a}_t$ is of reduced dimension larger or equal to: $m(m + 1)/2 - (p + 1)(p + 2)/2$ and is diagonalizable.

We also set $\lambda = -\xi_0$, $\tilde{a}_t = \Phi$; $\tilde{a}_{tj}^i = \Phi_j^i$, $1 \leq i \leq m$; and we will define $\mathcal{T}$, $\mathcal{I}_1$, $\mathcal{I}_2$ as in the preceding lemma. We then consider the system:

$$\left(\xi_0 I + \Phi\left(\xi'\right)\right) X_i = 0, \qquad 1 \leq i \leq m - p - 1, \tag{2.4}$$

i.e.,

$$\left(\xi_0 I + \Phi\left(\xi'\right)\right)_j^i = 0, \qquad 1 \leq i \leq m, \ 1 \leq j \leq m - p - 1.$$

The dimension of the vector space spanned by the forms of type $(\xi_0 I + \Phi(\xi'))_j^i$, $\Phi_j^i \in \mathcal{I}_1$ is at most:

$$m(m + 1)/2 - (p + 1)(p + 2)/2.$$

Choose $(\xi_0, \xi' \neq 0)$ so that we have:

$$\left(\xi_0 I + \Phi\left(\xi'\right)\right)_j^i = 0, \qquad 1 \leq j \leq i \leq m, \ 1 \leq j \leq m - p - 1,$$

(where we have set $\Phi(\xi')_1^1 = \psi(\xi')_1$). Then $\xi_0 = -\psi_1(\xi')$ is a characteristic root of order $m - p - 1$; the corresponding minors of order $p + 2$ vanish; moreover, the minors obtained with $p + 1$ last lines and $p + 1$ last columns cannot vanish (or else $\xi_0 = 0$ would be a root of order $m - p$ which contradicts the hypothesis (2.2)); we deduce from this as in the previous lemma that $\Phi_j^i \in \mathcal{T}$ depend linearly on Φ_j^i, such that:

$$1 \leq j \leq i \leq m, \qquad 1 \leq j \leq m - p - 1;$$

the system (2.4) reduces to:

$$\xi_0 I_j^i + \Phi_j^i\left(\xi'\right) = 0, \qquad 1 \leq j \leq i \leq m, \ 1 \leq j \leq m - p - 1.$$

Since $d \geq m(m+1)/2 - (p+1)(p+2)/2 + 1$, this system admits at least a non-trivial solution: rank of the system (2.4) is at least equal to $m(m+1)/2 - (p+1)(p+2)/2$ for $0 \leq t \leq 1$. Since for $t = 0$ we have seen that there was exactly one non-trivial solution defined up to a multiplicative constant, there exists $\varepsilon > 0$ such that the rank of the system (2.4) is exactly equal to $m(m+1)/2 - (p+1)(p+2)/2$ for $0 \leq t \leq \varepsilon$, by semi-continuity; then there is, up to a multiplicative constant, a pair $(\lambda(t), a[\xi'(t)])$, $a[\xi'(t)] \neq 0$, which verifies (2.4).

We continue as in [2]. We impose $\|a[\xi'(t)]\| = \|a(t)\| = 1$, for a standard norm; $a(t)$ can be chosen continuous, as far as the rank of the system remains constant, we have also: $\lambda_1[a(t)] = \lambda(t)$.

Assume there exists at least a t such that, for this t, we have two linearly independent solutions $(-\lambda(t), a(t))$ and note t_0 the smallest of these t. The continuity in t of coefficients of the linear system implies the existence of $b^* = a[\xi'(t_0)] \neq 0$ and λ^* such that:

$$b^* X_i(t_0) = \lambda^* X_i(t_0), \quad 1 \leq i \leq m - p - 1;$$

more:

$$\lambda^* = \lambda_1(b^*) = \cdots = \lambda_{m-p-1}(b^*) \lambda_{m-p}(b^*).$$

Let $c = a[\eta'(t_0)]$ linearly independent from b^* such that:

$$c X_i(t_0) = \mu X_i(t_0), \quad 1 \leq i \leq m - p - 1.$$

If $\mu = 0$, by considering $c(\alpha) = b^* + \alpha c$ as in Lemma 2.1, we are led to a contradiction.

If $\mu \neq 0$:

$$c(\alpha) X_i(t_0) = (\lambda^* + \alpha\mu) X_i(t_0), \quad 1 \leq i \leq m - p - 1.$$

For α^* small enough, we have:

$$\lambda_1[c(\alpha^*)] \neq 0, \quad \lambda_1[c(\alpha^*)] > \lambda_{m-p-1}[c(\alpha^*)].$$

Setting $c(\alpha^*) = b_1$ we have:

$$b_1 X_i(t_0) = \lambda_1(b_1) X_i(t_0), \quad 1 \leq i \leq m - p - 1.$$

Setting moreover $c_1 = c - \mu b_1/\lambda_1(b_1)$, we have

$$c_1 X_i(t_0) = 0, \quad 1 \leq i \leq m - p - 1.$$

Since b_1 and c_1 are linearly independent, using the same reasoning as in beginning of the lemma, we are led to a contradiction. The rank of the system is constant for all t, and the lemma is proved.

Lemma 2.3. *We assume that:*

$$d \geq m(m+1)/2 - (p+1)(p+2)/2 + 1, \quad 1 \leq p \leq m - 2.$$

If $m \geq 6$, we assume also that:

$$m(m+1)/2 - (p+1)(p+2)/2 - 1 > m(m-p-1). \qquad (2.5)$$

Moreover we assume that, for all $\xi' \neq 0$, we have

$$\lambda_1\left[a\left(\xi'\right)\right] > \lambda_{m-p}\left[a\left(\xi'\right)\right].$$

Then there exists $\xi' \neq 0$ such that $c = a(\xi')$ verifies:

$$\lambda_1(c) > \lambda_2(c) = \cdots = \lambda_{m-p}(c) > \lambda_m(c).$$

Proof. For $p = m-2$, the result is obvious. We next suppose that $1 \leq p \leq m-3$. As previously, there exists b such that:

$$\lambda_1(b) = \cdots = \lambda_{m-p-1}(b) = 1 > \lambda_{m-p}(b).$$

Choosing b as an element of basis of the matrix space, we have

$$a\left(\xi'\right) = \psi b + \Phi\left(\xi''\right),$$

where: $d[\Phi] \geq m(m+1)/2 - (p+1)(p+2)/2 - 1$.

Let us prove there exists a matrix $c = \Phi(\xi'') \neq 0$ such that:

$$cX_i = 0, \quad 2 \leq i \leq m - p - 1.$$

We have a system with $m(m-p-2)$ equations and at least $m(m+1)/2 - (p+1)(p+2)/2 - 1$ unknowns; if assumption (2.5) is satisfied, we have obviously a non-trivial solution c. If $\lambda_2(c) = \cdots \lambda_{m-p}(c) = 0$, the lemma is proved. If it is not satisfied, then by replacing possibly c by $-c$, we have:

$$\lambda_1(c) \geq \lambda_2(c) > 0.$$

We set: $c(\alpha) = b + \alpha c$; $c(\alpha)X_i = X_i, 2 \leq i \leq m - p - 1, \lambda_1(b) = 1$ is then an eigenvalue of $c(\alpha)$ of order $m - p - 2$ for all α. If α tends to 0, the eigenvalues of $c(\alpha)$ are close to these of b and we have:

$$\lambda_{m-p}\left[c(\alpha)\right] < 1.$$

If α tends to infinity, then $\lambda_1[c(\alpha)]$ and $\lambda_2[c(\alpha)]$ tend to infinity; then there exists at least $\alpha > 0$ such that: $\lambda_{m-p}[c(\alpha)] = 1$. Let γ be the smallest α such that $\lambda_{m-p}[c(\alpha)] = 1$; by continuity it is impossible that $\lambda_1[c(\alpha)] > \lambda_{m-p}[c(\alpha)]$. We then have:

$$\lambda_1\left[c(\alpha)\right] > \lambda_2\left[c(\alpha)\right] = \cdots = \lambda_{m-p}\left[c(\alpha)\right] > \lambda_m\left[c(\alpha)\right]$$

and the lemma is proved. $\qquad \square$

Remark. Let us set $m - p = q$ and express condition (2.5) using q; we then obtain:

$$q^2 - 3q + 4 - 2m < 0;$$

the discriminant of this polynomial is $\Delta = -7 + 8m > 0$; the inequality is satisfied if:

$$q < \left(3 + \sqrt{8m - 7}\right)/2$$

and, since $q \leq m - 1$, if:

$$2m - 5 < \sqrt{8m - 7} \text{ which implies } m^2 - 7m + 8 < 0,$$

i.e.,

$$m \leq 5.$$

Then for $m \leq 5$, condition 2.5 is useless.

For $m \geq 6$, a more precise study of the matrix $\Phi(\xi'')$ must be accomplished; it will be done in an ulterior publication.

Proof of the theorem. Let us assume that

$$\forall \xi' \neq 0, \quad \lambda_1\left[a\left(\xi'\right)\right] > \lambda_{m-p}\left[a\left(\xi'\right)\right].$$

Let c be the matrix obtained in Lemma 2.3: $\lambda_2(c) = \cdots = \lambda_{m-p}(c)$ is eigenvalue of multiplicity $m - p - 1$; let Y_i be $m - p - 1$ independent proper vectors corresponding to this eigenvalue; we have:

$$cY_i = \lambda_2(c)Y_i, \quad 2 \leq i \leq m - p.$$

On the other hand, by Lemma 2.2, there exists $c_1 \neq 0$ and $\underline{\lambda}$ such that:

$$c_1 Y_i = \underline{\lambda} Y_i$$

with:

$$\underline{\lambda} = \lambda_1(c_1) = \cdots \lambda_{m-p-1}(c_1)$$

and every solution of that system is of the type:

$$c_1' = \mu c_1, \quad \underline{\lambda}' = \mu \underline{\lambda};$$

then we have:

$$c = \mu c_1;$$

the largest eigenvalue of c is of multiplicity $m - p$, contrary to the assumption. Therefore there exists $\xi' \neq 0$ such that:

$$\lambda_1\left[a\left(\xi'\right)\right] = \cdots = \lambda_{m-p}\left[a\left(\xi'\right)\right]$$

and the theorem is proved.   □

Remark. The counterexample of [2] made with a symmetric therefore diagonalizable matrix a(ξ') applies: the estimation $d \geq m(m+1)/2 - (p+1)(p+2)/2 + 1$ is optimal.

References

[1]   Delquié, H. and Vaillant, J., Dimension réduite et valeurs propres multiples d'une matrice diagonalisable 4×4, *Bull. Sci. Math.* (2000), 1–13.

[2]   Friedland, S. and Loewy R., Subspaces of symmetric matrices containing matrices with a multiple first eigenvalue, *Pacific Journal of Mathematics (2)* **62** (1976).

[3]   Kasahara, K. and Yamaguti, M., Strongly hyperbolic systems of linear partial differential equations with constant coefficients, *Mem. Coll. Sci. Univ. Kyoto (A)* **33** (1960), 1–33.

[4]   Nishitani, T., Symmetrization of hyperbolic systems with real constant coefficients, *Annali Scuola Normale Superiore Pisa. (4)*, **21**(1) (1994).

[5]   Nishitani, T. and Vaillant J., Smoothly symmetrizable systems and the reduced dimensions, *Tsukuba Journal*, **25**(1) (2001), 165–177.

[6]   Oshime, Y., Canonical forms of 3×3 strongly hyperbolic systems with real constant coefficients, *J. Math. Kyoto Univ.* **31–4** (1991), 937–982.

[7]   Vaillant, J., Symétrisabilité des matrices localisées d'une matrice fortement hyperbolique en un point multiple, *Annali Scuola Normale Superiore - Pisa. Classe di Scienze. (4)*, **5**(2) (1978).

[8]   _____, Systèmes fortement hyperboliques et systèmes symétriques, *C.R.A.S. Acad. Sci. Paris*, (I), t. **328** (1999), 407–412.

[9]   _____, Symétrie des systèmes fortement hyperboliques 4×4 et dimension réduite, *Annali della Scuola Normale Superiore di Pisa, Ser. IV*, **29**(4) (2000), 839–890.

[10]   _____, Symétrie des opérateurs fortement hyperboliques 4×4 ayant un point triple caractéristique dans R^3, *Colloque Cattabriga, Annali di mat. di Ferrare, Ser. VII, Sc. Mat. Suppl.*, **45** (1999), 339–363.

LARGE TEMPS DE VIE DES SOLUTIONS D'UN PROBLÈME DE CAUCHY NON LINÉAIRE

DANIEL GOURDIN AND MUSTAPHA MECHAB

Contents

1 Notations et résultats 67
2 Esquise des preuves 68
    2.1   Les algèbres de Banach $\mathbf{G}_{T,\zeta}^{\omega,d}(\Omega_t)$ 69
    2.2   Préliminaires 70
    2.3 $\mathscr{L}$ est une contraction 71
    2.4   Preuve du théorème 1.1 73
    2.5   Preuve du théorème 1.2 74

M. de Gosson (ed.), Jean Leray '99 Conference Proceedings, 65-74.

Résumé. *On étudie un problème de Cauchy global non linéaire avec des données initiales ayant une vitesse de variation, dans l'espace des fonctions holomorphes en t et de classe de Gevrey par rapport aux autres variables sans hypothèse d'hyperbolicité.*

1. Notations et résultats

Soient $m, n \in \mathbb{N}^*$, $t \in \mathbb{C}$, $x = (x_1, \ldots, x_n) \in \mathbb{R}^n$, B une partie de

$$\{(j, \alpha) \in \mathbb{N} \times \mathbb{N}^n; \ j + |\alpha| \leq m, \ j < m \text{ et } \alpha \neq 0\}$$

où $|\alpha| = \sum_{i=1}^{n} \alpha_i$ est la longueur de α; r est le cardinal de B, $D^B u = (D^\sigma u)_{\sigma \in B}$, $D_t^{-1} u$ est la primitive de u relativement à t s'annulant pour $t = 0$, $d = \inf\{(m - j)/|\alpha|, \ (j, \alpha) \in B\}$, Ω est un voisinage ouvert de $0 \in \mathbb{R}^n$, tel que $\lambda x \in \Omega$ pour tout $x \in \Omega$ et $\lambda \in [0, 1]$. On note $\mathcal{U}_T = \{t \in \mathbb{C}; |t| \leq T\} = \mathcal{B}_\mathbb{C}(0, T)$ et $\Omega_T = \mathcal{U}_T \times \Omega$ $(T > 0)$, $\mathbf{G}^d(\Omega)$ la classe de Gevrey des fonctions $\varphi \in \mathscr{C}^\infty(\Omega, \mathbb{R})$ avec la propriété:

$$\exists C > 0, \ \forall \alpha \in \mathbb{N}^n, \ \forall x \in \Omega, \quad \left| D^\alpha \varphi(x) \right| \leq C^{|\alpha|+1} (\alpha!)^d$$

ce qui équivaut à l'existence de constantes $([\varphi]_0, [\varphi]_1) \in [\mathbb{R}_+^*]^2$ telles que

$$\forall \alpha \in \mathbb{N}^n, \ \forall x \in \Omega, \quad \left| D^\alpha \varphi(x) \right| \leq [\varphi]_0 \cdot ([\varphi]_1)^{|\alpha|} (\alpha!)^d$$

avec la possibilité de choisir $[\varphi]_0 = \sup_{x \in \Omega} |\varphi(x)|$.
$\mathbf{G}^{\omega,d}(\Omega_T)$ est l'algèbre des fonctions $u : \Omega_T \to \mathbb{C}$ admettant des dérivées continues $D_x^\alpha u$ (relativement à x) sur Ω_T $(\forall \alpha \in \mathbb{N}^n)$ holomorphes relativement à t, et telles que

$$\exists C > 0, \ \forall \alpha \in \mathbb{N}^n, \ \forall t \in \mathcal{U}_T, \ \forall x \in \Omega, \quad \left| D_x^\alpha u(t, x) \right| \leq C^{|\alpha|+1} (|\alpha|!)^d$$

Pour $\varphi_j \in \mathbf{G}^d(\Omega)$, $(j = 1, \ldots, m - 1)$ et $\varepsilon > 0$ petit, on considère le problème de Cauchy non linéaire

$$\begin{cases} D_t^m u(t, x) = f\left(t, D^B u(t, x)\right), \\ D_t^j u(0, x) = \varphi_j(\varepsilon \cdot x), \qquad \forall j = 0, \ldots, m - 1 \end{cases} \tag{1.1}$$

On supposera que $f(t, 0) = 0$ et on note $\varphi_j^\varepsilon(x) = \varphi_j(\varepsilon x)$.
On obtient les résultats suivants.

Théorème 1.1. *Pour toute fonction holomorphe f dans un voisinage de $\Omega_T \subset \mathbb{C} \times \mathbb{C}^r$ et $\varphi_j \in \mathbf{G}^d(\Omega)$ $(j = 0, \ldots, m - 1)$, il existe $\varepsilon_0 > 0$ tel que pour tout $\varepsilon < \varepsilon_0$, le problème de Cauchy (1.1) ait une solution unique $u_\varepsilon \in \mathbf{G}^{\omega,d}(\Omega_T)$.*

Si Ω est borné, la solution de (1.1) avec les données initiales nulles est stable au sens où: si $\varphi_j(0) = 0$ ($j = 0, \ldots, m - 1$), $u_\varepsilon \to_{\varepsilon \to 0} 0$ dans $\mathbf{G}^{\omega,d}(\Omega_T)$ pour la topologie usuelle limite projective lorsque $0 < T' < T$ de la famille $(\mathscr{E}'_T)_{T'}$ des espaces topologiques limites inductives

$$\mathscr{E}_{T'} = \bigcup_{\substack{\to \\ \zeta_k \to +\infty \\ 1 \leq k \leq n}} \mathbf{G}^{\omega,d}_{T',\zeta=(\zeta_1;\ldots,\zeta_n)}(\Omega_{T'})$$

Quand f ne dépend pas de t, on a:

Théorème 1.2. *Pour toute fonction holomorphe dans un voisinage de $0 \in \mathbb{C}^r$ et $\varphi_j \in \mathbf{G}^d(\Omega)$ ($j = 0, \ldots, m - 1$) et pour tout $\varepsilon > 0$, le problème de Cauchy (1.1) a une solution unique $u_\varepsilon \in \mathbf{G}^{\omega,d}(\Omega_{T_\varepsilon})$ avec l'estimation asymptotique suivante du temps de vie*

$$T_\varepsilon \geq \mu \left(\frac{1}{\varepsilon}\right)^s$$

avec $s = \inf\{|\delta|/(m - k), (k, \delta) \in B\}$

2. Esquise des preuves

On effectue les changements successifs suivants d'inconnues

$$v(t, x) = u(t, x) - \sum_{j=0}^{m-1} \frac{1}{j!} t^j \varphi_j^\varepsilon(x) \tag{2.1}$$

$$w = D_t^m v \tag{2.2}$$

$$\rho(t, x) = w(t, x) - f\left(0, \left(D_x^\alpha \varphi_k^\varepsilon\right)_{(k,\alpha)\in B}\right) = w(t, x) - \psi_\varepsilon(x) \tag{2.3}$$

Puisque $(k, \alpha) \in B \Rightarrow k < m$, alors

$$D^B u = D^B \left[D_t^{-m} \left(\rho + f\left(0, \left(D_x^\alpha \varphi_k^\varepsilon\right)_{(k,\alpha)\in B}\right)\right) + \sum_{j=0}^{m-1} \frac{1}{j!} t^j \varphi_j^\varepsilon(x) \right]$$

$$= D^B \left[D_t^{-m}\rho + \frac{t^m}{m!}\psi_\varepsilon(x) + \sum_{j=0}^{m-1} \frac{1}{j!} t^j \varphi_j^\varepsilon(x) \right]$$

et (1.1) est équivalent à l'équation

$$\rho(t, x) = f\left(t, D^B \left[D_t^{-m}\rho + \frac{t^m}{m!}\psi_\varepsilon(x) + \sum_{j=0}^{m-1} \frac{t^j}{j!} \varphi_j \right] (t, x)\right) - \psi_\varepsilon$$

avec $\rho(0, x) = 0$, $\forall x \in \Omega$.

On considère l'application $\mathscr{L}$ définie par

$$\mathscr{L}\rho(t,x) = f\left(t, D^B\left[D_t^{-m}\rho + \frac{t^m}{m!}\psi_\varepsilon(x) + \sum_{j=0}^{m-1}\frac{t^j}{j!}\varphi_j(x)\right](t,x)\right) - \psi_\varepsilon \quad (2.4)$$

Résoudre (1.1) revient à trouver les points fixes de $\mathscr{L}$.

Nous allons montrer que $\mathscr{L}$ est une contraction stricte dans les algèbres de Banach $\mathbf{G}_{T,\zeta}^{\omega,d}(\Omega_T)$ définies dans [9] et [3] avec $\xi = 1$.

2.1. LES ALGÈBRES DE BANACH $\mathbf{G}_{T,\zeta}^{\omega,d}(\Omega_T)$

On utilise une famille d'algèbres de Banach $\mathbf{G}_{T,\zeta}^{\omega,d}(\Omega_T)$ telle que $\bigcup_{\zeta \in (\mathbb{R}_+^*)^n} \mathbf{G}_{T,\zeta}^{\omega,d} = \mathbf{G}^{\omega,d}(\Omega_T)$ ([9] et [3]).

Rappelons brièvement la définition.

Soit $\theta(t) = \sum_{j=0}^{+\infty} t^j/(j+1)^2$ la fonction de Lax avec la propriété, pour une constante $K > 0$

$$\theta^2 \ll K\theta$$

suivant la définition suivante des séries majorantes, notée $\ll$:

$$v = \sum_\alpha v_\alpha x^\alpha \quad \text{et} \quad u = \sum_\alpha u_\alpha x^\alpha,$$

où $v_\alpha \in \mathbb{C}$ et $u_\alpha \in \mathbb{R}^+$ vérifient $v \ll u$ si et seulement si

$$|v_\alpha| \le u_\alpha \quad (\forall \alpha).$$

Ainsi, $\phi_T(t) = (1/K)\theta(t/T)$ satisfait

$$\phi_T^2 \ll \phi_T$$

Pour $\zeta = (\zeta_1, \ldots, \zeta_n) \in (\mathbb{R}_+^*)^n$, $|t| < T$ on considère

$$\Phi_{T,\zeta}^{\omega,d}(t,x) = \sum_{k\in\mathbb{N}} \frac{(\zeta \cdot x)^k}{k!}(k!)^{d-1} D^k \phi_T(t)$$

Définition. $u \in \mathbf{G}_{T,\zeta}^{\omega,d}(\Omega_T)$ si $u : \mathscr{U}_T \times \Omega = \Omega_T \to \mathbb{C}$ admet des dérivées de tout ordre en x, continues sur Ω_T et holomorphes par rapport à t telles que:

$$\exists C > 0, \ \forall \alpha \in \mathbb{N}^n, \ \forall x \in \Omega, \quad D_x^\alpha u(t,x) \ll C\zeta^\alpha (|\alpha|!)^{d-1} D^{|\alpha|}\phi_T(t). \ (2.5)$$

(i.e., $|D_t^\beta D_x^\alpha u(0,x)| \le C\zeta^\alpha(|\alpha|!)^{d-1} D^{|\alpha|+\beta}\phi_T(0)$).

On pourra noter cette relation

$$u \ll C \cdot \Phi_{T,\zeta}^{\omega,d} \quad (2.6)$$

Les fonctions $\Phi_{T,\zeta}^{\omega,d}$ vérifient la relation

$$\left(\Phi_{T,\zeta}^{\omega,d}\right)^2 \ll \Phi_{T,\zeta}^{\omega,d}$$

et $\mathbf{G}_{T,\zeta}^{\omega,d}(\Omega_T)$ avec la norme définie par $\|u\| = \inf\{C > 0; u \ll C\Phi_{T,\zeta}^{\omega,d}\}$ est une algèbre de Banach.

Proposition 2.1.   1. $\mathbf{G}_{T,\zeta}^{\omega,d}(\Omega_T) \subset \mathbf{G}^{\omega,d}(\Omega_T)$.

2. Soit $u \in \mathbf{G}^{\omega,d}(\Omega_T)$ tel que $u(0,\cdot) = 0$, $\zeta \in (\mathbb{R}_+^*)^n$ et $C > 0$, alors il existe $T_0 > 0$ tel que pour tout $T' \in]0, T_0[$, $u \in \mathbf{G}_{T',\zeta}^{\omega,d}(\Omega_{T'})$ et $\|u\| \le C$.

3. Si $u \in \mathbf{G}_{T,\zeta}^{\omega,d}(\Omega_T)$ satisfait $\|u\| < R$, alors $R/(R-u) \in \mathbf{G}_{T,\zeta}^{\omega,d}(\Omega_T)$ et

$$\left\|\frac{R}{R-u}\right\| \le \left(K + \frac{\|u\|}{R-\|u\|}\right).$$

Proposition 2.2. Pour tout $(k,\alpha) \in \mathbb{Z} \times \mathbb{N}^n$ avec $k + d|\alpha| \le 0$, il existe $C_{k,\alpha} > 0$ tel que:
$\forall T > 0$, $\forall u, u' \in \mathbf{G}_{T,\zeta}^{\omega,d}(\Omega_T)$,

$$\left\|D_t^k D_x^\alpha u\right\| \le C_{k,\alpha}\zeta^\alpha T^{-k-|\alpha|}\|u\|$$

$$\left\|D_x^k D_y^\alpha u - D_x^k D_y^\alpha u'\right\| \le C_{k,\alpha}\zeta^\alpha T^{-|k|-|\alpha|}\left\|u - u'\right\|$$

2.2. PRÉLIMINAIRES

Lemme 2.3. *Pour tout* $\alpha \in \mathbb{N}^n$, *il existe* $C_{|\alpha|} > 0$ *tel que pour tout* $\varphi \in \mathbf{G}^d(\Omega)$, $D_x^\alpha\varphi \in \mathbf{G}^d(\Omega)$ *et nous pouvons choisir:*

$$\left([D_x^\alpha\varphi]_0, [D_x^\alpha\varphi]_1\right) = \left(\left[\left(C_{|\alpha|}\right)^d [\varphi]_1\right]^{|\alpha|} [\varphi]_0, \left[\left(C_{|\alpha|}\right)^d\right]^{|\alpha|} \cdot [\varphi]_1\right)$$

Ce résultat est impliqué par $\lim_{|\beta| \to +\infty} \log(|\beta| + j)/(|\beta| + 1) = 0$, $\forall j \in \mathbb{N}$. Du lemme 2.3 on déduit:

Lemme 2.4. *Soit* $\varphi \in \mathbf{G}^d(\Omega)$, $T > 0$ *et* $\alpha \in \mathbb{N}^n$, *alors:*
pour tout $\zeta \ge ([(C_{|\alpha|})^d]^{|\alpha|}[\varphi]_1 T e^2, \dots, [(C_{|\alpha|})^d]^{|\alpha|}[\varphi]_1 T e^2)$, $D_x^\alpha\varphi \in \mathbf{G}_{T,\zeta}^{\omega,d}(\Omega_T)$ *et*

$$\left\|D_x^\alpha\varphi\right\| \le K \cdot [\varphi]_0 \left[\left(C_{|\alpha|}\right)^d \cdot [\varphi]_1\right]^{|\alpha|}. \tag{2.7}$$

et de la proposition 2.2 on obtient.

Lemme 2.5. *Soit* $k \in \mathbb{N}$, $\varphi \in \mathbf{G}^d(\Omega)$, *alors* $t^k \varphi \in \mathbf{G}^{\omega,d}_{T,\zeta}(\Omega_T)$ *pour tout* $T > 0$ *et on a:*
Pour tout $\zeta \geq \zeta^0 = (T \cdot e^2 \cdot [\varphi]_1, \dots, T \cdot e^2 \cdot [\varphi]_1)$,

$$\| t^k \varphi \| \leq K \cdot C_{k,0} \cdot T^k \cdot [\varphi]_0$$

De plus:

Lemme 2.6. *Il existe deux constantes* C_B, $C > 0$ *tel que pour* $\varphi_j \in \mathbf{G}^d(\Omega)$ ($j = 0, \dots, m-1$), $\forall T > 0$, $\forall \sigma = (k, \alpha) \in B$, $D^\sigma [\sum_{j=0}^{m-1} (t^j/j!)\varphi_j] \in \mathbf{G}^{\omega,d}_{T,\zeta^0}(\Omega_T)$ *et on a*

$$\left\| D^\sigma \left[\sum_{j=0}^{m-1} \frac{t^j}{j!} \varphi_j \right] \right\| \leq C \cdot \sum_{j=k}^{m-1} T^{j-k} C_{0,\varphi} \cdot \left[C_{1,\varphi} \right]^{|\alpha|} \tag{2.8}$$

avec $\zeta^0 = (C_B \cdot T \cdot C_{1,\varphi}, \dots, C_B \cdot T \cdot C_{1,\varphi})$, $C_{0,\varphi} = \sup_{j=0,\dots,m-1}[\varphi_j]_0$, $C_{1,\varphi} = \sup_{j=0,\dots,m-1}[\varphi_j]_1$.

2.3. $\mathscr{L}$ EST UNE CONTRACTION

Soit $\eta > 1$ et $T, R > 0$ tel que f soit holomorphe et bornée dans

$$\Delta = \Delta_{T,R,\eta} = \left\{ (t, z) \in \mathbb{C} \times \mathbb{C}^n, \, | \, |t| < \eta T \text{ et } |z_\sigma| < \eta R, \forall \sigma \in B \right\}$$

Par les inégalités de Cauchy on peut trouver une constante $C = C_{(f,T,R,\eta)} > 0$ telle que pour tout $\zeta \in (\mathbb{R}_+^*)^n$,

$$f(t, z) \ll C_{(f,T,R,\eta)} \frac{\eta R}{\eta R - 1 \cdot z} \Phi^{\omega,d}_{T,\zeta}(t, x) \tag{2.9}$$

Proposition 2.7. *Pour* $f(t, z)$ *fonction holomorphe dans un voisinage de* $\mathcal{U}_T \times \{0\} \subset \mathbb{C} \times \mathbb{C}^r$, *il existe deux constantes* $K_f > 0$ *et* $\varepsilon > 0$ *tel que pour toutes fonctions* $\varphi_j \in \mathbf{G}^d(\Omega)$, $j = 0, \dots, m-1$, *vérifiant* $C_{1,\varphi} < \varepsilon$ *on ait:*

$$\forall a \geq K_f, \quad \mathscr{L}(\mathscr{B}(0, a)) \subset \mathscr{B}(0, a)$$

où $\mathscr{B}(0, a)$ *est la boule fermée de centre 0 et de rayon* a *dans* $\mathbf{G}^{\omega,d}_{T,\zeta^0}(\Omega_T)$ *et* $\zeta^0 = (C_B \cdot T \cdot C_{1,\varphi}, \dots, C_B \cdot T \cdot C_{1,\varphi})$.

Soit $a > 0$ et $u \in \mathscr{B}(0, a) \subset \mathbf{G}^{\omega,d}_{T,\zeta^0}(\Omega_T)$. Notons (pour $\sigma \in B$, i.e., $k < m$ et $k + |\alpha| \leq m$)

$$z_\sigma = z_{(k,\alpha)} = D_t^k D_x^\alpha \left[D_t^{-m} u + \sum_{j=0}^{m-1} \frac{t^j}{j!} \varphi_j \right]$$

Pour simplifier on suppose $\psi_\varepsilon(x) = 0$.
Par la proposition 2.2, on obtient ($k - m + d|\alpha| \le 0$)

$$\left\| D_t^k D_x^\alpha \left[D_t^{-m} u \right] \right\| \le C_{k-m,\alpha} \zeta^\alpha T^{(m-k)-|\alpha|} \|u\| \tag{2.10}$$

Comme $f(t, 0) = 0$ on peut écrire

$$\mathscr{L}(u) = f(t, z) = \sum_{\sigma \in B} f_\sigma(t, z) \cdot z_\sigma \tag{2.11}$$

où $f_\sigma(t, z)$ sont des fonctions holomorphes dans $\Delta_{T,\eta,R}$.
Avec η plus petit, on peut supposer ces fonctions bornées dans $\Delta_{T,R,\eta}$, et d'après
(2.9) on obtient

$$\forall (t, z) \in \Delta_{T,\eta,R}, \quad f_\sigma(t, z) \ll C_{f_\sigma, T, R, \eta} \left(\frac{\eta R}{\eta R - 1 \cdot z} \right) \Phi_{T,\zeta}^{\omega,d}(t, x) \tag{2.12}$$

D'après les inégalités (2.8), (2.10) et l'expression de ζ nous obtenons

$$\|z_\sigma\| \le C \cdot \left[C_{1,\varphi} \right]^{|\alpha|} T^{m-k} \|u\| + C \cdot \sum_{j=k}^{m-1} T^{j-k} C_{0,\varphi} \left[C_{1,\varphi} \right]^{|\alpha|} \tag{2.13}$$

Comme $\alpha \ne 0$, pour a et T fixés et $C_{1,\varphi}$ petit on a $\|z\| \le \eta \cdot R$; grâce à la
proposition 2.1, d'après (2.12) nous obtenons

$$\|f_\sigma(t, z)\| \le C_{(f_\sigma, T, \eta, R)} \left(K + \mathscr{E} \left(C_{1,\varphi} \right) \right) \tag{2.14}$$

avec $\lim_{s \to 0} \mathscr{E}(s) = 0$. Comme $\mathbf{G}_{T,\zeta}^{\omega,d}(\Omega_T)$ est une algèbres de Banach, ces
inégalités utilisées dans (2.11) conduisent à

$$\|\mathscr{L}(u)\| \le \sum_{(k,\alpha) \in B} C_{(f_\sigma, T, \eta, R)} \left(K + \mathscr{E} \left(C_{1,\varphi} \right) \right) \tag{2.15}$$

$$\times \left[C_{k-m,\alpha} \cdot \left[C_{1,\varphi} \right]^{|\alpha|} T^{m-k} \|u\| + C \cdot \sum_{j=k}^{m-1} T^{j-k} C_{0,\varphi} \left[C_{1,\varphi} \right]^{|\alpha|} \right]$$

$$\le \sum_{\sigma \in B} C_{f_\sigma, T, \eta, R} \left(K + \mathscr{E} \left(C_{1,\varphi} \right) \right) \left[\mathscr{E} \left(C_{1,\varphi} \right) \|u\| + \mathscr{E} \left(C_{1,\varphi} \right) \right]$$

Pour $C_{1,\varphi}$ petit on peut écrire (2.15) de la façon suivante

$$\forall u \in \mathscr{B}(0, a) \subset \mathbf{G}_{T,\zeta}^{\omega,d}(\Omega_T), \quad \|\mathscr{L}(u)\| \le C_0 \|u\| + K_f \tag{2.16}$$

où $0 < C_0 < 1$ et $K_f > 0$; ainsi pour $C_0 = 1/2$ et $C_{1,\varphi}$ petit on a

$$\forall a \ge 2K_f, \quad \mathscr{L}(\mathscr{B}(0, a)) \subset \mathscr{B}(0, a) \tag{2.17}$$

et la proposition est démontrée.

Le résultat suivant a une démonstration analogue.

Proposition 2.8. Soit $f(t, z)$ une fonction holomorphe dans un voisinage de $\mathcal{U}_T \times \{0\} \subset \mathbb{C} \times \mathbb{C}^r$, alors il existe $\varepsilon > 0$ tel que pour tout $a > 0$ et toutes fonctions $\varphi_j \in \mathbf{G}^d(\Omega)$, $j = 0, \ldots, m-1$, avec $C_{1,\varphi} < \varepsilon$ on peut trouver une constante $C \in]0, 1[$ telle que

$$\forall u, u' \in \mathcal{B}(0, a) \subset \mathbf{G}^{\omega,d}_{T,\zeta^0}(\Omega_T), \qquad \left\| \mathcal{L}u - \mathcal{L}u' \right\| \leq C \left\| u - u' \right\|$$

2.4. PREUVE DU THÉORÈME 1.1

On utilise le lemme suivant:

Lemme 2.9. Soit $\varphi \in \mathbf{G}^d(\Omega)$, alors pour tout $\varepsilon \in [0, 1]$, la fonction $x \to \varphi^\varepsilon(x) = \varphi(\varepsilon \cdot x)$ appartient à $\mathbf{G}^d(\Omega)$ et on peut choisir

$$\left([\varphi^\varepsilon]_0, [\varphi^\varepsilon]_1 \right) = \left([\varphi]_0, \varepsilon[\varphi]_1 \right).$$

Par le lemme 2.9 on a $C_{1,\varphi^\varepsilon} = \varepsilon C_{1,\varphi} \to_{\varepsilon \to 0} 0$.
Soit T et $R > 0$ et $f(t, z)$ une fonction holomorphe dans un voisinage de $\mathcal{U}_T \times \{0\}$.

Par les propositions 2.7 et 2.8 et le lemme 2.9, on voit que pour $\varepsilon > 0$ petit, il existe ζ tel que $\mathcal{L}$ soit une contraction d'une boule $\mathcal{B}(0, a) \subset \mathbf{G}^{\omega,d}_{T,\zeta}(\Omega_T)$; $\mathcal{L}$ a donc un point fixe unique. Par la proposition 2.1 on peut affirmer que ce point fixe est dans $\mathbf{G}^{\omega,d}(\Omega_T)$ ainsi le problème (1.1) admet une solution dans $\mathbf{G}^{\omega,d}(\Omega_T)$.

Pour l'unicité, on utilise la seconde partie de la proposition 2.1.

Preuve de la stabilité. En utilisant les inégalités (2.15), la proposition 2.2 et les lemmes 2.5 et 2.9 on a:

$$\|u_\varepsilon\| \leq \sum_{(k,\alpha) \in B} C \cdot \left(K + \mathcal{E}\left(\varepsilon C_{1,\varphi} \right) \right)$$

$$\cdot \varepsilon^{|\alpha|} C_{1,\varphi}{}^{|\alpha|} \left\{ 2K_f + C_{0,\varphi^\varepsilon} \right\} + C \cdot C_{0,\varphi^\varepsilon} \qquad (2.18)$$

dans $\mathbf{G}^{\omega,d}_{T,\zeta_\varepsilon}(\Omega_T)$, avec u_ε solution du problème (1.1) et $\zeta_\varepsilon = (C_B T C_{1,\varphi^\varepsilon}, \ldots, C_B T C_{1,\varphi^\varepsilon})$. $\varphi_j(0) = 0$ implique $\varphi_j^\varepsilon(x) = \lim_{\varepsilon \to 0} [\varphi_j^\varepsilon]_0 = 0$.
Ainsi le deuxième membre χ_ε de l'inégalité (2.18) admet 0 pour limite quand $\varepsilon \to 0$.

De plus (2.18) peut être exprimée avec les fonctions majorantes et on a

$$\forall \alpha \in \mathbb{N}^n, \ \forall x \in \Omega, \quad D_x^\alpha u_\varepsilon(t, x) \ll \chi_\varepsilon \cdot (\zeta_0)^\alpha \left(|\alpha|! \right)^{d-1} D^{|\alpha|} \phi_{T'}(t) \quad (2.19)$$

pour $T' \in]0, T[$, d'où la convergence de u_ε vers zéro, lorsque ε tend vers zéro, dans $\mathbf{G}^{\omega,d}(\mathcal{U}_T \times \Omega)$, pour la topologie limite projective lorsque $0 < T' < T$

de la famille $(\mathscr{E}_{\mathscr{T}'})_{T'}$ des espaces topologiques

$$\mathscr{E}_{T'} = \bigcup_{\substack{\to \\ \zeta_k \to +\infty \\ 1 \le k \le n}} \mathbf{G}_{T',\zeta=(\zeta_1;\ldots,\zeta_n)}^{\omega,d} (\Omega_{T'})$$

limites inductives lorsque $\zeta \to +\infty$ ($\zeta_k \to +\infty, \forall k = 1, \ldots, n$), des $\mathbf{G}_{T',\zeta}^{\omega,d}(\Omega_{T'})$, ce qui termine la preuve du théorème 1.1.

2.5. PREUVE DU THÉORÈME 1.2

En utilisant les majorations de $\|\mathscr{L}u\|$ et $\|\mathscr{L}u - \mathscr{L}u'\|$ on obtient

$$C_f \left[K + \mathscr{E}\left(\varepsilon C_{1,\varphi}\right) \right] \sum_{(k,\alpha)\in B} C_{k-m,\alpha} \cdot \varepsilon^{|\alpha|} \left(C_{1,\varphi}\right)^{|\alpha|} T_\varepsilon^{m-k} < 1 \qquad (2.20)$$

si il existe une constante $C > 0$ qui ne dépend pas de x tel que

$$\forall (k,\alpha) \in B, \qquad T_\varepsilon^{m-k}\varepsilon^{|\alpha|} \le \frac{1}{2C}$$

où $C = C_f[K + \mathscr{E}(\varepsilon C_{1,\varphi})] \sum_{(k,\alpha)\in B} C_{k-m,\alpha} \cdot (C_{1,\varphi})^{|\alpha|}$.
Ainsi (2.20) est satisfait si

$$T_\varepsilon = \inf_{(k,\alpha)\in B} \left(\frac{1}{2C}\right)^{k-m} \left(\frac{1}{\varepsilon}\right)^s$$

Références

[1]   Arosio, A. and Spagnolo, S., *J. Math. Pures et Appl.* **65** (1984), 263–305.
[2]   Bony, J. and Schapira, P., *Invent. Math.* **17** (1972), 95–105.
[3]   Gourdin, D. and Mechab, M., *J. Math. Pures Appl.* **75** (1996), 569–593.
[4]   _____, *Comptes Rendus Acad. Sci. (Paris), T.* **328** (1999), Série I, 485–488.
[5]   Leray, J., Cours de Princeton (1953), (paragraphe 73).
[6]   Leray, J. and Ohya Y., *Colloque de Liege*, CBRM, (1964), 105–144.
[7]   _____, *Math. Ann.* **170** (1967), 167–205.
[8]   Leray, J. and Waelbroeck, L., *Colloque de Liege*, CBRM, (1964), 145–152.
[9]   Wagschal, C. *J. Math. Pure et Appl.* **58** (1979), 309–337.

UNE REMARQUE SUR UN PROLONGEMENT ANALYTIQUE DE LA SOLUTION DU PROBLÈME DE CAUCHY

YÛSAKU HAMADA

Contents

1 Introduction et résultats 77
2 Équation différentielle aux dérivées partielles du premier ordre 78
3 Fonction modulaire et son équation différentielle ordinaire 79
4 Équation différentielle aux dérivées partielles du premier ordre particulière 80
5 La preuve des propositions 80

M. de Gosson (ed.), Jean Leray '99 Conference Proceedings, 75-82.

1. Introduction et résultats

J. Leray [11] et L. Gårding, T. Kotake et J. Leray [4] ont étudié les singularités
et un prolongement analytique de la solution du problème de Cauchy dans le
domaine complexe.

[6], [8], [13] et [15] ont étudié le cas de l'opérateur différentiel à coefficients
des fonctions entières.

On considère $a(x, D)$ un opérateur différentiel d'ordre m, à coefficients
de fonctions entières sur $\mathbf{C}^{n+1}$. $[x = (x_0, x') \in \mathbf{C}^{n+1}, x' = (x_1, x''), x'' = (x_2, \ldots, x_n), D = (D_0, \ldots, D_n), D_i = \partial/\partial x_i]$. Sa partie principale est notée
$g(x, D)$ et supposons que $g(x; 1, 0, \ldots, 0) = 1$.

Soit S l'hyperplan $x_0 = 0$ donc non caractéristique pour g.

Étudions le problème de Cauchy

$$a(x, D)u(x) = v(x), \qquad D_0^h u\left(0, x'\right) = w_h\left(x'\right), \quad 0 \le h \le m - 1, \quad (1.1)$$

où $v(x)$, $w_h(x')$ sont des fonctions entières sur $\mathbf{C}^{n+1}$.

D'après le théorème de Cauchy–Kowalewski, il existe une unique solution
holomorphe au voisinage de S dans $\mathbf{C}^{n+1}$. Nous étudions un prolongement
analytique de cette solution. En général, des diverses phénomènes y se passent.
Dans [8], en appliquant les résultats de L. Bieberbach et de P. Fatou, nous avons
construit un exemple tel que le domaine d'holomorphie de la solution admet un
point extérieur dans $\mathbf{C}^{n+1}$.

Dans cet exposé, nous donnons une remarque sur le domaine d'holomorphie
de la solution du problème de Cauchy pour l'opérateur différentiel de partie
principale à coefficients polynomiaux.

Rappelons d'abord les définitions d'un domaine de Riemann (Ω, π), c'est-
à-dire, un domaine étalé au-dessus de $\mathbf{C}^n$ et d'un point frontière de Ω: Ω est un
espace topologique connexe, séparé, π est une application continue de Ω dans
$\mathbf{C}^n$, qui est localement un homéomorphisme, et le point $\underline{p} = \pi(p), p \in \Omega$ est
appelé la projection de p. (Voir [2], [5] et [12]).

Soient p un point frontière de Ω et (A) la condition suivante:

Condition (A). *Il existe une constante $\delta_0 > 0$ telle que pour tout polydisque*
$\underline{V}_\delta (0 < \delta \le \delta_0)$ *de centre $\underline{p}$ et de rayon δ dans $\mathbf{C}^n$, la projection $\pi(V_\delta)$ de
la composante connexe V_δ de $\pi^{-1}(\underline{V}_\delta)$, ayant p comme un point frontière,
a l'extérieur non vide dans $\underline{V}_\delta$.* Si un domaine étalé a au moins un point
frontière satisfaisant la condition (A), nous disons que ce domaine possède
la propriété (E).

Nous avons alors

Proposition 1.1. *Le domaine d'holomorphie de la solution du problème de
Cauchy suivant pour l'opérateur différentiel à coefficients polynomiaux possède*

la propriété (E). $[x = (x_0, \ldots, x_4)]$.

$$\left[D_0 - x_1^2 A_1 \left(x'' \right) D_1 + \sum_{i=2}^{4} A_i \left(x'' \right) D_i \right] D_0 U_1(x) = 0, \tag{1.2}$$

$$U_1 \left(0, x' \right) = 0, \quad D_0 U_1 \left(0, x' \right) = x_1,$$

où

$$A_1 \left(x'' \right) = x_2^2 (1 - x_2)^2 x_3, \qquad A_2 \left(x'' \right) = x_2^2 (1 - x_2)^2 x_3^2,$$

$$A_3 \left(x'' \right) = x_2^2 (1 - x_2)^2 x_3 x_4,$$

$$A_4 \left(x'' \right) = \frac{3}{2} x_2^2 (1 - x_2)^2 x_4^2 - \frac{1}{2} \left(1 - x_2 + x_2^2 \right) x_3^4.$$

Nous avons:

Proposition 1.2. *Le domaine d'holomorphie de la solution du problème de Cauchy suivant possède la propriété* (E).

$$\left[D_0 - x_1^2 A_1 \left(x'' \right) D_1 + \sum_{i=2}^{4} A_i \left(x'' \right) D_i + x_1 D_5 \right] U_2 \left(x, x_5 \right) = 0, \tag{1.3}$$

$$U_2 \left(0, x', x_5 \right) = x_5.$$

En effet, $U_2(x, x_5) = x_5 - U_1(x)$.

Effectuons le changement de variable $X_1 = 1/x_1$ et récrivons X_1 par x_1, alors le problème (1.3) se transforme au problème suivant.

$$\left[D_0 + \sum_{i=1}^{4} A_i \left(x'' \right) D_i + \frac{1}{x_1} D_5 \right] U_3 \left(x, x_5 \right) = 0, \quad U_3 \left(0, x', x_5 \right) = x_5. \tag{1.4}$$

On a donc $U_2(x, x_5) = U_3(x_0, 1/x_1, x'', x_5)$.

Nous avons:

Proposition 1.3. *Le domaine d'holomorphie de la solution du problème de Cauchy* (1.4) *possède la propriété* (E).

2. Équation différentielle aux dérivées partielles du premier ordre

Soit $L = D_0 + M$, $M = \sum_{i=1}^{n} a_i(x') D_i$, où $a_i(x')$ sont des fonctions entiers, et considérons le problème de Cauchy

$$\left[L + \frac{1}{x_1} D_{n+1} \right] U \left(x, x_{n+1} \right) = 0, \quad U \left(0, x', x_{n+1} \right) = x_{n+1}.$$

On a $U(x, x_{n+1}) = x_{n+1} - u(x)$, $u(x)$ étant la solution du problème: $Lu(x) = 1/x_1, u(0, x') = 0$. Donc en écrivant par $\varphi(x)$ la solution du problème: $L\varphi(x) = 0$, $\varphi(0, x') = x_1$, on a

$$u(x) = \int_0^{x_0} \frac{1}{\varphi(\sigma, x')}\, d\sigma.$$

Supposons qu'il existe un point $x''^{(0)}$ voisin de 0 tel que $a_1(0, x''^{(0)}) \neq 0$, alors il existe une unique fonction holomorphe $\delta(x')$ au voisinage de $(0, x''^{(0)})$ telle que $\varphi(\delta(x'), x') = 0$, $\delta(0, x'') = 0$. Soit $x^{(0)} = (x_0^{(0)}, x_1^{(0)}, x''^{(0)})$ $[x_0^{(0)} = \delta(x'^{(0)}), x_1^{(0)} \neq 0]$ un point voisin de 0. Au voisinage de $x'^{(0)}$, on a $D_0\varphi(\delta(x'), x') \neq 0$.

Posons $\Gamma(x') = 1/D_0\varphi(\delta(x'), x')$.

Prenons x' voisin de $x'^{(0)}$, et écrivons par $\gamma(x')$ un chemin fermé d'origine et d'extrémité $(0, x')$, dépendant continûment de x', qui fait un tour autour du point $(\delta(x'), x')$ dans la droite complexe. Prolongeons analytiquement $u(x)$ le long de $\gamma(x')$, alors le germe u au point d'extrémité est égal à son germe au point d'origine augmenté du germe $2\pi i\Gamma(x')$.

On a:

Lemme 2.1. $\Gamma(x')$ *est la solution du problème* $M\Gamma(x') = 0$, $\Gamma(0, x'') = -1/a_1(0, x'')$.

3. Fonction modulaire et son équation différentielle ordinaire

Soit $w = \lambda(z)$ la fontion modulaire. Son domaine d'holomorphie est $\{z; \Im z > 0\}$ et $\lambda'(z) \neq 0$. Sa fonction inverse $\nu(w)$ est holomorphe au voisinage d'un point de $\mathbf{C} \setminus \{0, 1\}$ et elle se prolonge analytiquement sur le revêtement universel $\mathfrak{R}[\mathbf{C} \setminus \{0, 1\}]$ de $\mathbf{C} \setminus \{0, 1\}$. Elle donne une représentation conforme de $\mathfrak{R}[\mathbf{C} \setminus \{0, 1\}]$ sur $\{z; \Im z > 0\}$.

$W = \lambda((at + b)/(ct + d))$, a, b, c, d étant des constantes, $ad - bc = 1$, satisfait

$\{W; t\} = -R(W)(dW/dt)^2$, où $\{W; t\}$ est la dérivée de Schwarz:

$$\{W; t\} = \frac{d}{dt}\left(\frac{d^2W}{dt^2} \bigg/ \frac{dW}{dt}\right) - \frac{1}{2}\left(\frac{d^2W}{dt^2} \bigg/ \frac{dW}{dt}\right)^2,$$

$R(W) = (1 - W + W^2)/2W^2(1 - W)^2$. [10]

Les $w_1 = W$, $w_2 = dW/dt$, $w_3 = d^2W/dt^2$ satisfont

$$\frac{dw_1}{dt} = w_2, \qquad \frac{dw_2}{dt} = w_3, \qquad \frac{dw_3}{dt} = \frac{3w_3^2}{2w_2} - \frac{(1 - w_1 + w_1^2)\,w_2^3}{2w_1^2\,(1 - w_1)^2}. \tag{3.1}$$

Lemme 3.1. *Par une transformation $z = (at + b)/(ct + d)$, (C) un cercle ou une droite dans le plan de t se transforme en l'axe réel dans le plan de z. Pour $\Im(a\bar{c}) \neq 0$, (C) est le cercle*

$$\left| t - \frac{\bar{a}d - b\bar{c}}{a\bar{c} - \bar{a}c} \right| = \left| \frac{ad - bc}{a\bar{c} - \bar{a}c} \right|,$$

et pour $\Im(a\bar{c}) = 0$, (C) est la droite $\Im[(a\bar{d} - \bar{b}c)t + b\bar{d}] = 0$.

Supposons $\Im(b\bar{d}) > 0$. Alors, le demi-plan supérieur se transforme respectivement au domaine suivant.

1. *Si $\Im(a\bar{c}) > 0$, l'extérieur de (C) contenant $t = 0$.*

2. *Si $\Im(a\bar{c}) < 0$, l'intérieur de (C) contenant $t = 0$.*

3. *Si $\Im(a\bar{c}) = 0$, le demi-plan $\Im[(a\bar{d} - \bar{b}c)t + b\bar{d}] > 0$ contenant $t = 0$.*

4. Équation différentielle aux dérivées partielles du premier ordre particulière

Soit $\Phi(x)$ la solution du problème $[D_0 + \sum_{i=1}^{4} A_i(x'')D_i]\Phi(x) = 0$ avec $\Phi(0, x') = x_1$. La solution de (4) s'écrit sous la forme:

$$U_3(x, x_5) = x_5 - U_4(x) = x_5 - \int_0^{x_0} \frac{1}{\Phi(\sigma, x')} \, d\sigma.$$

Prenons un point x''^0 voisin de 0 tel que $A_1(x''^0) \neq 0$. Il existe une unique fonction holomorphe $\delta(x')$ au voisinage de $(0, x''^0)$ satisfaisant $\Phi(\delta(x'), x') = 0$ et $\delta(0, x'') = 0$. Prolongeons analytiquement $U_4(x)$ le long de $\gamma(x')$, alors elle augmente de $2\pi i\Theta(x')$, où $\Theta(x') = 1/D_0\Phi(\delta(x'), x')$. Vu le lemme 2.1, $\Theta(x')$ est la solution de l'équation:

$$\left[\sum_{i=1}^{4} A_i(x'') D_i \right] \Theta(x') = 0, \qquad \Theta(0, x'') = -1/A_1(x'').$$

5. La preuve des propositions

D'après la section 3, $\Theta(x')$ s'écrit au voisinage de $(0, x''^0)$ comme suit: $\Theta(x') = -1/y_2(x')^2(1 - y_2(x'))^2 y_3(x')$, où

$$y_2(x') = \lambda \left(\frac{p(x'') x_1 + q(x'')}{r(x'') x_1 + s(x'')} \right),$$

$$y_3 \left(x' \right) = \lambda' \left(\frac{p \left(x'' \right) x_1 + q \left(x'' \right)}{r \left(x'' \right) x_1 + s \left(x'' \right)} \right)$$

$$\times 4\lambda' \left(v \left(x_2 \right) \right)^3 x_3^3 \times \left(2\lambda' \left(v \left(x_2 \right) \right)^2 x_3 + r \left(x'' \right) x_1 \right)^{-2},$$

$$p \left(x'' \right) = v \left(x_2 \right) r \left(x'' \right) - 2\lambda' \left(v \left(x_2 \right) \right) x_3^2, \qquad q \left(x'' \right) = v \left(x_2 \right) s \left(x'' \right),$$

$$r \left(x'' \right) = - \left\{ \lambda'' \left(v \left(x_2 \right) \right) x_3^2 - \lambda' \left(v \left(x_2 \right) \right)^2 x_4 \right\}, \qquad s \left(x'' \right) = 2\lambda' \left(v \left(x_2 \right) \right)^2 x_3.$$

Définissons les domaines suivants.

$$\mathscr{A}_+ = \left\{ \tilde{x}'' = \left(\tilde{x}_2, x_3, x_4 \right) \in \mathscr{R} \left[\mathbf{C} \setminus \{0, 1\} \right] \times \mathbf{C}^2, \, \Im \left(p \left(\tilde{x}'' \right) \overline{r \left(\tilde{x}'' \right)} \right) > 0 \right\},$$

$$\mathscr{A}_- = \left\{ \tilde{x}'' = \left(\tilde{x}_2, x_3, x_4 \right) \in \mathscr{R} \left[\mathbf{C} \setminus \{0, 1\} \right] \times \mathbf{C}^2, \, \Im \left(p \left(\tilde{x}'' \right) \overline{r \left(\tilde{x}'' \right)} \right) < 0 \right\},$$

$$\mathscr{A}_0 = \left\{ \tilde{x}'' = \left(\tilde{x}_2, x_3, x_4 \right) \in \mathscr{R} \left[\mathbf{C} \setminus \{0, 1\} \right] \times \mathbf{C}^2, \, \Im \left(p \left(\tilde{x}'' \right) \overline{r \left(\tilde{x}'' \right)} \right) = 0 \right\},$$

et

$$\mathscr{D}_+ = \left\{ \tilde{x}' = \left(x_1, \tilde{x}'' \right) \in \mathbf{C} \times \mathscr{A}_+, \, \mid x_1 - P(\tilde{x}'') \mid > R(\tilde{x}'') \right\},$$

$$\mathscr{D}_- = \left\{ \tilde{x}' = \left(x_1, \tilde{x}'' \right) \in \mathbf{C} \times \mathscr{A}_-, \, \mid x_1 - P(\tilde{x}'') \mid < R(\tilde{x}'') \right\},$$

$$\mathscr{D}_0 = \left\{ \tilde{x}' = \left(x_1, \tilde{x}'' \right) \in \mathbf{C} \times \mathscr{A}_0, \right.$$

$$\left. \Im \left[\left\{ p(\tilde{x}'')\overline{s(\tilde{x}'')} - \overline{q(\tilde{x}'')}r(\tilde{x}'') \right\} x_1 + q(\tilde{x}'')\overline{s(\tilde{x}'')} \right] > 0 \right\},$$

où

$$P \left(\tilde{x}'' \right) = \left\{ \overline{p \left(\tilde{x}'' \right)} s \left(\tilde{x}'' \right) - q \left(\tilde{x}'' \right) \overline{r \left(\tilde{x}'' \right)} \right\} \Big/ \left\{ p \left(\tilde{x}'' \right) \overline{r \left(\tilde{x}'' \right)} - \overline{p \left(\tilde{x}'' \right)} r \left(\tilde{x}'' \right) \right\},$$

$$R \left(\tilde{x}'' \right) = \left| p \left(\tilde{x}'' \right) s \left(\tilde{x}'' \right) - q \left(\tilde{x}'' \right) r \left(\tilde{x}'' \right) \right| \Big/ \left| p \left(\tilde{x}'' \right) \overline{r \left(\tilde{x}'' \right)} - \overline{p \left(\tilde{x}'' \right)} r \left(\tilde{x}'' \right) \right|,$$

$\mathscr{D} = \mathscr{D}_+ \cup \mathscr{D}_- \cup \mathscr{D}_0$ est un domaine étalé au-dessus de $\mathbf{C}^4$ possédant la propriété (E).

Lemme 5.1. *Le domaine d'holomorphie de* $\Theta(x')$ *est* $\mathscr{D} \setminus \{x_3 = 0\}$, *donc il possède la propriété* (E).

La $U_4(x)$ est holomorphe au voisinage de $S \setminus \{x_1 = 0\}$ dans $\mathbf{C}^5$, et en se prolongeant analytiquement le long de $\gamma(x')$, elle augmente de $2\pi i \Theta(x')$. Le lemme 5.1 preuve donc la proposition 1.3.

On trouvera les démonstrations détaillées de ces résultats dans [9].

Références

[1]   Bieberbach, L., *S. B. Preuss, Acad. Wiss.* (1933), 476–479.

[2] Cartan, H., *Théorie Élémentaire des Fonctions Analytiques d'une ou Plusieurs Variables Complexes*, Hermann, Paris, (1961).

[3] Fatou, P., *C. R. Acad. Sc. Paris* **175** (1922), 862–865; 1030–1033.

[4] Gårding, L., Kotake, T., and Leray, J., *Bull. Soc. Math. France* **92** (1964), 263–361.

[5] Gunning, R. C., *Introduction to holomorphic functions of several variables*, Vol. 1, Wadsworth & Brooks/Cole (1991).

[6] Hamada, Y., Leray, J., and Takeuchi, A., *J. Math. Pures et Appl.* **64** (1985), 257–319.

[7] Hamada, Y., *Recent developments in hyperbolic equations, Pitman research notes in mathematics series*, Vol. 183, 82–95, Longman 1988.

[8] ———, *Tôhoku Math. J.* **50** (1998), 133–138.

[9] ———, *à paraître à Tôhoku Math. J.*

[10] Hille, E., *Ordinary differential equations in the complex domain*, John Wiley 1976.

[11] Leray, J., *Bull. Soc. Math. France* **85** (1957), 389–429.

[12] Nishino, T., *Theory of functions of several complex variables*, (en japonais), Univ. of Tokyo Press (1996).

[13] Persson, J., *Ark. Mat.* **9** (1971), 171–180.

[14] Picard, E., *C. R. Acad. Sc. Paris* **139** (1904), 5–9.

[15] Pongérard, P. and Wagschal, C., *J. Math. Pures et Appl.* **75** (1996), 409–418.

CONORMALITY AND LAGRANGIAN PROPERTIES ALONG DIFFRACTIVE RAYS

PASCAL LAUBIN

Contents

1 The geometry in the complex domain 85
   1.1   FBI TRANSFORM 85
   1.2   ISOTROPIC AND INVOLUTIVE SUBMANIFOLDS 87
   1.3   LAGRANGIAN SUBMANIFOLDS 88
2 Second wave front set 89
3 Bilagrangian geometry 90
   3.1   PHASE FUNCTIONS 90
   3.2   DISTRIBUTION CLASS 93
4 An application 95

M. de Gosson (ed.), Jean Leray '99 Conference Proceedings, 83-96.
© *2003 Kluwer Academic Publishers.*

Representation of micro-distributions or microfunctions into the complex domain is an important tool in linear PDE. The purpose of this paper is to present some ways to perform complex canonical maps from the cotangent bundle of $\mathbb{R}^n$ to a complex manifold and to show how they can be used to obtain lagrangian properties of solutions to some classical PDE's. We refind the classical point of view by considering lagrangian structure at the 2-microlocal level.

1. The geometry in the complex domain

Our purpose is to prepare the definition of the phase functions used to characterize the bilagrangian distributions in the formalism of the Fourier-Bros-Iagolnitzer transform. In the microlocal case, we closely follow [7] and collect some material from [10], see also [15].

As usual, we identify

- $\mathbb{C}^n$ with $\mathbb{R}^n \times \mathbb{R}^n$ and write $z = x + iy$,

- $\zeta \in T_z^* \mathbb{C}^n$ with $(\zeta_1, \ldots, \zeta_n) \in \mathbb{C}^n$ using $\zeta(h) = \sum_j \zeta_j h_j$,

- $T_z^* \mathbb{C}^n$ with $T_{(x,y)}^* \mathbb{R}^{2n}$ by mapping the $\mathbb{C}$-linear form $\zeta \in T_z^* \mathbb{C}^n$ to the $\mathbb{R}$-linear form $h \mapsto -\Im \zeta(h)$.

This map is symplectic if $T^* \mathbb{R}^{2n}$ is endowed with the usual canonical 2-form and $T^* \mathbb{C}^n$ with the 2-form $-\Im \sigma$ defined below.

It follows that if f is a holomorphic function, $\partial f \in T_z^* \mathbb{C}^n$ is identified with $d(-\Im f) \in T_{(x,y)}^* \mathbb{R}^{2n}$ since $d(-\Im f) = -\Im(df) = -\Im(\partial f)$.

In the same way, if φ is a real function then $d\varphi \in T_{(x,y)}^* \mathbb{R}^{2n}$ is identified with $(2/i)D_z\varphi \in \mathbb{C}^n$.

All the constructions described in this section are local even this is not stated explicitly.

1.1. FBI TRANSFORM

Writing $z = x + iy$ and $\zeta = \xi + i\eta$, the canonical 2-form on $T^* \mathbb{C}^n$ is

$$\sigma = \sum_j d\zeta_j \wedge dz_j.$$

Its real and imaginary parts

$$\Re\sigma = \sum_j \left(d\xi_j \wedge dx_j - d\eta_j \wedge dy_j \right), \qquad \Im\sigma = \sum_j \left(d\eta_j \wedge dx_j + d\xi_j \wedge dy_j \right)$$

are symplectic forms on $\mathbb{R}^{2n}$.

Let φ be a real C_1 function defined in a neighborhood of $z_0 \in \mathbb{C}^n$ and

$$\Lambda_\varphi = \left\{ \left(z, \frac{2}{i} D_z \varphi(z) \right) : z \in \mathbb{C}^n \right\}.$$

This manifold is $\Im$-lagrangian since it is identified with

$$\{ (z, d\varphi(z)) : z \in \mathbb{C}^n \} \subset T^* \mathbb{R}^{2n}.$$

If j_φ denotes the immersion $z \mapsto (z, (2/i) D_z \varphi(z))$ then

$$j_\varphi^*(\Re\sigma) = j_\varphi^*(\sigma) = j_\varphi^*(d(\zeta\, dz)) = d\left(\frac{2}{i} \partial\varphi \right) = \frac{2}{i} \bar{\partial}\partial\varphi.$$

It follows that, if $\bar{\partial}\partial\varphi$ is non degenerate, j_φ is a symplectic map from $(\mathbb{C}^n, (2/i)\bar{\partial}\partial\varphi)$ onto $(\Lambda_\varphi, \Re\sigma)$. Its inverse is the projection.

Let us remark that

$$\frac{2}{i} \bar{\partial}\partial\varphi(z)(u, v) = \frac{2}{i} \sum_{j,k} D_{z_j} D_{\bar{z}_k} \varphi(z) \left(\bar{u}_k v_j - u_j \bar{v}_k \right)$$

is real for every $u, v \in \mathbb{C}^n$.

The tangent space to Λ_φ at the point $j_\varphi(z)$ is given by

$$\left\{ \left(h, \frac{2}{i} \left(\partial_z^2 \varphi(z) \cdot h + \partial_{\bar{z}} \partial_z \varphi(z) \cdot \bar{h} \right) \right) : h \in \mathbb{C}^n \right\}.$$

This shows that if $\bar{\partial}\partial\varphi$ is non degenerate then Λ_φ is a totally real submanifold and that its complexification is $T^* \mathbb{C}^n$. We have the following result.

Theorem 1.1. *Let φ be a strictly plurisubharmonic function near $z_0 \in \mathbb{C}^n$ and $\chi : \dot{T}^* \mathbb{R}^n \to \Lambda_\varphi$ a canonical transform defined near (y_0, η_0) such that*

$$\chi(y_0, \eta_0) = \left(z_0, \frac{2}{i} D_z \varphi(z_0) \right).$$

Theorem 1.2. *Here Λ_φ is endowed with the 2-form $\Re\sigma$. There is a unique holomorphic function g near (z_0, y_0), such that*

- *the complexification of χ is*

$$\chi^{\mathbb{C}} : T^* \mathbb{C}^n \longrightarrow T^* \mathbb{C}^n : (y, -D_y g(z, y)) \longmapsto (z, D_z g(z, y)),$$

- $ig(z_0, y_0) = \varphi(z_0), \; -D_y g(z_0, y_0) = \eta_0,$

- *the function $y \mapsto -\Im g(z, y)$ has a non degenerate critical point $y(z)$
  with signature $(0, n)$ and critical value $\varphi(z)$. Moreover, we have*

$$\left(y(z), -D_y g\left(z, y(z) \right) \right) = \chi^{-1}\left(z, \frac{2}{i} D_z \varphi(z) \right).$$

For example, if

$$\chi : (x, \xi) \longmapsto (x - i\xi, \xi), \qquad \varphi(z) = \frac{1}{2}|\Im z|^2,$$

then

$$g(z, y) = \frac{i}{2}(z - y)^2.$$

The FBI transform associated to φ, χ near the points (y_0, η_0), z_0 is

$$T_\chi u(z, \lambda) = \int e^{i\lambda g(z,y)} a(z, y, \lambda) u(y) \, dy,$$

where a is a classical symbol.

Assume $\varphi(z) = (1/2)|\Im z|^2$. The function

$$f(z) = \int_0^{+\infty} e^{-\lambda/2} T_\chi u(z, \lambda) \, d\lambda$$

is holomorphic in $|\Im z|^2 < 1$ and extends near a point where the equality holds if and only if $\rho(z) \notin WF_a u$. This a local complex canonical map.

1.2. ISOTROPIC AND INVOLUTIVE SUBMANIFOLDS

The following result links the isotropic submanifolds of $(\mathbb{C}^n, (2/i)\bar\partial\partial\varphi)$ to the complex structure.

Proposition 1.3. Let φ be an analytic strictly plurisubharmonic function near $z_0 \in \mathbb{C}^n$. If Γ is an isotropic submanifold of $(\mathbb{C}^n, (2/i)\bar\partial\partial\varphi)$ then Γ is totally real and there is a unique pluriharmonic function h on $\Gamma^{\mathbb{C}}$ such that

$$\varphi_{|\Gamma^{\mathbb{C}}} - h \geq 0 \qquad \text{and} \qquad (\varphi - h)_{|\Gamma} = 0.$$

Moreover, there is $C > 0$ such that

$$\varphi_{|\Gamma^{\mathbb{C}}} - h \geq C \, d(z, \Gamma)^2.$$

Let us now consider an involutive submanifold V of $(\mathbb{C}^n, \omega = (2/i)\bar\partial\partial\varphi)$. If $z \in V$ then

$$T_z V \oplus i \, (T_z V)^\omega = \mathbb{C}^n.$$

Indeed, if $u \in T_z V$ and $iu \in (T_z V)^\omega$ we have

$$0 = \frac{2}{i} \bar{\partial} \partial \varphi(u, iu) = 2 \sum_{j,k} D_{z_j} D_{\bar{z}_k} \varphi(z) u_j \bar{u}_k$$

hence $u = 0$.

This shows that the union of the complexifications of the bicharacteristic leaves of V can be locally identified with $\mathbb{C}^n$. It follows from the Proposition 1.3 that there is a unique real analytic function φ_V equal to φ on V, pluriharmonic on the complexification of the bicharacteristic leaves of V and such that $\varphi_V \leq \varphi$.

Proposition 1.4. Let φ be an analytic strictly plurisubharmonic function near $z_0 \in \mathbb{C}^n$. If V is an involutive submanifold of $(\mathbb{C}^n, (2/i)\bar{\partial}\partial\varphi)$ then

$$\Lambda_{\varphi_V} = \left\{ \left(z, \frac{2}{i} D_z \varphi_V(z) \right) : z \in \mathbb{C}^n \right\}$$

is the union of the complexification of the bicharacteristic leaves of $j_\varphi(V)$.

Remark that j_Γ is holomorphic since it has a holomorphic inverse.

1.3. LAGRANGIAN SUBMANIFOLDS

For a lagrangian submanifold, we can use a holomorphic function.

Proposition 1.5. Let Λ be a lagrangian submanifold of $\dot{T}^*\mathbb{R}^n$, h be a phase function of Λ near ρ_0 and χ be a local canonical map from $\dot{T}^*\mathbb{R}^n$ to Λ_φ mapping ρ_0 to z_0. If g the FBI phase defined in Theorem 1.1 and

$$\phi_\Lambda(z) = \mathrm{cv}_{(x,\theta)}\big(g(z, x) + h(x, \theta)\big)$$

then $\varphi_\Lambda = -\Im\phi_\Lambda$. The critical points are given by

$$(x, \theta) = j_{\mathbb{C}}^{-1} \circ \chi_{\mathbb{C}}^{-1}\big(z, D_z\phi_\Lambda(z)\big).$$

Here j is the immersion $(x, \theta) \mapsto (x, h'_x)$ and $j_{\mathbb{C}}$ is its complexification.

We have

$$\chi^{\mathbb{C}}\left(\Lambda^{\mathbb{C}}\right) = \{(z, D_z\phi_\Lambda(z)) : z \in \mathbb{C}^n\}$$

and

$$\varphi_\Lambda(z) \leq \varphi(z).$$

The equality holds if and only if $(z, (2/i)D_z\varphi(z)) \in \chi(\Lambda)$.

In this formalism, the lagrangian distributions are defined in the following way.

Definition. Let u be a distribution in an open subset Ω of $\mathbb{R}^n$, Λ a lagrangian submanifold of $\dot{T}^*\Omega$. With the notations of Proposition 1.5, u is said lagrangian at ρ_0 if, in a neighborhood of z_0, we have

$$(T_\chi u)(z, \lambda) = e^{i\lambda\phi_\Lambda(z)} b(z, \lambda),$$

where b is a classical analytic symbol.

This is equivalent to the fact that u can be written $u = u_1 + u_2$ where h is a phase function for Λ, $\rho_0 = j_h(x_0, \theta_0)$ not in the singular spectrum of u_2 and

$$u_1(x) = \int_\Gamma e^{ih(x,\theta)} a(x, \theta)\, d\theta.$$

Here Γ is a conic neighborhood of θ_0 and a is a classical analytic symbol near (x_0, θ_0).

2. Second wave front set

We first need some basic facts concerning the second wave front set. The proofs can be found in [7] and [5].

Let V be an involutive submanifold of $T^*\Omega$ with codimension k. Write $z = (z', z'') \in \mathbb{C}^k \times \mathbb{C}^{n-k}$. An FBI phase function g near (z_0, y_0), $\Im z_0' = 0$, is adapted to V if $\rho(z) \in V$ when $\Im z' = 0$. Let $(y_0, \eta_0) = \rho_0 = \rho(z_0)$.

Define

$$T^{(2)} u(z, \mu, \lambda) = \int_{|y' - \Re z_0'| < r} e^{-\lambda\mu^2 (z' - y')^2 / (2(1-\mu^2))} T u\left(y', z'', \lambda\right) dy'.$$

If z_1 is near z_0, $\Im z_1' = 0$ and $\sigma_1 \in \mathbb{R}^k \setminus \{0\}$ then

$$\tau(z_1, \sigma_1') = \partial_s \rho\left(z_1' - i s \sigma_1', z_1''\right)_{|s=0}$$

defines an element of

$$T_V\left(T^*\mathbb{R}^n\right) \sim \bigcup_F T^*F,$$

where F runs over the bicharacteristic leaves of V.

Definition. $\mathrm{WF}^{(2)}_{a,V}(u)$ is the subset of $T_V(T^*\mathbb{R}^n)$ defined by $\tau(z_1, \sigma_1') \notin \mathrm{WF}^{(2)}_{a,V}(u)$ if

$$\left|(T^{(2)} u)(z, \mu, \lambda)\right| \leq e^{(\lambda\mu^2/2)|\Im z'|^2 + (\lambda/2)|\Im z''|^2 - \epsilon\lambda\mu^2}$$

when

$$\left|z' - (z_1' - i\sigma_1')\right| < r, \qquad \left|z'' - z_1''\right| < r,$$
$$0 < \mu < \mu_0, \qquad \lambda > f(\mu).$$

Let $\pi_V : T_V(T^*\mathbb{R}^n) \to V$ be the projection. We have

$$\mathrm{supp}_{a,V}^{(2)}(u) = \pi_V\left(\mathrm{WF}_{a,V}^{(2)}(u)\right) \subset V \cap \mathrm{WF}_a u.$$

Moreover $\rho(z_1) \notin \mathrm{supp}_{a,V}^{(2)}(u)$ iff

$$|(Tu)(z,\lambda)| \leq e^{1/2|\Im z''|^2 + \epsilon\lambda}$$

if $|z - z_1| < r$ and $\lambda > \lambda_\epsilon$. The following result is the basic relation between $\mathrm{WF}_a u$ and $\mathrm{WF}_{a,V}^{(2)}(u)$.

Proposition 2.1. If ω is an open connected subset of a bicharacteristic leaf of V and $\omega \cap \mathrm{supp}_{a,V}^{(2)}(u) = \emptyset$ then

$$\omega \subset \mathrm{WF}_a u \qquad \text{or} \qquad \omega \cap \mathrm{WF}_a u = \emptyset.$$

3. Bilagrangian geometry

Of course, in the definition of the lagrangian structure at the 2-microlocal level, we want

- an intrinsic definition independant of phases and amplitude,

- that the second wave front lies on a lagrangian manifold,

- the invariance through canonical transformations.

3.1. PHASE FUNCTIONS

Here we collect some material from [6]. A conic lagrangian submanifold of a manifold X can be written $\Lambda_\varphi = \{(x, \varphi_x') : \varphi_\theta' = 0\}$ where φ is 1-homogeous, $(\varphi_x', \varphi_\theta') \neq 0$ and $\mathrm{rk}(\varphi_{\theta x}'', \varphi_{\theta\theta}'')$ is maximal.

If Λ is a conic lagrangian submanifold of $\dot{T}^*X$ then $\dot{T}^*\Lambda \sim \dot{T}_\Lambda \dot{T}^*X$. Here, there are two homogeneities: one inherited from Λ and another one generated by the cotangent bundle.

A lagrangian submanifold of $\dot{T}^*\Lambda$ is said *conic bilagrangian* if it is conic for both homogeneities.

Let X be a C^∞ manifold and Γ_0 an open subset of $X \times \mathbb{R}^N \setminus \{0\} \times \mathbb{R}^M \setminus \{0\}$ such that $(x, \theta, \eta) \in \Gamma_0$ and $s, t > 0$ imply $(x, t\theta, st\eta) \in \Gamma_0$. Such an open set is called a *profile*. An open subset Γ of $X \times \mathbb{R}^N \setminus \{0\} \times \mathbb{R}^M \setminus \{0\}$ is said *biconic with profile* Γ_0 if

- $(x, \theta, \eta) \in \Gamma$ and $t > 0$ imply $(x, t\theta, t\eta) \in \Gamma$,

- for each compact subset K of Γ_0, there is $\epsilon > 0$ such that $(x, \theta, s\eta) \in \Gamma$ if $(x, \theta, \eta) \in K$ and $0 < s < \epsilon$.

Let $p, q \in \mathbb{R}$ and $r \in \mathbb{N}_0$. A C^∞ function $f : \Gamma \to \mathbb{R}^m$ is said *bihomogeneous of degree* $(p, q; r)$ if

- $f(x, t\theta, t\eta) = t^p f(x, \theta, \eta)$ if $(x, \theta, \eta) \in \Gamma, t > 0$,

- for every $(x_0, \theta_0, \eta_0) \in \Gamma_0$, there is an open neighborhood V of (x_0, θ_0, η_0) and a C^∞ function F in $V \times \,] - \epsilon, \epsilon[$ satisfying

$$f(x, \theta, s\eta) = s^q F\left(x, \theta, \eta, s^{1/r}\right)$$

if $(x, \theta, \eta, s) \in V \times \,]0, \epsilon[$.

A function $f : \Gamma \to \mathbb{R}^m$ is said *bihomogeneous of degree* $(p, q; r)$ if $f(x, t\theta, t\eta) = t^p f(x, \theta, \eta)$ and $f(x, \theta, s\eta) = s^q F(x, \theta, \eta, s^{1/r})$.

Definition. Let

- Λ be a conic lagrangian submanifold of $\dot{T}^*X$,

- φ be an analytic real valued function which is homogeneous of degree 1 in Γ_1,

- ψ be an analytic real valued function which is bihomogeneous of degree $(1, 1; r)$ in Γ

and

$$C_{\varphi, \psi} = \left\{ (x, \theta, \eta) \in \Gamma_0 : \varphi'_\theta(x, \theta) = 0, \, \psi'_{1,\eta}(x, \theta, \eta) = 0 \right\}.$$

The pair (φ, ψ) is a local 2-phase function of Λ (with regularity r) if

- φ is a local phase function that parameterizes Λ,

- at each point of $C_{\varphi, \psi}$, the vector $(\psi'_{1,x}, \psi'_{1,\theta})$ is different from 0 and

$$\mathrm{rk} \begin{pmatrix} \psi''_{1,\eta x} & \psi''_{1,\eta \theta} & \psi''_{1,\eta \eta} \\ \varphi''_{\theta x} & \varphi''_{\theta \theta} & 0 \end{pmatrix} = N + M.$$

In this situation, we obtain easily the following facts.

a) The map $(\rho, \eta) \mapsto \psi_1(j_\varphi^{-1}(\rho), \eta)$ is a local phase function of Λ.

b) The map

$$j_{\varphi,\psi} : C_{\varphi,\psi} \longrightarrow \dot{T}^*\Lambda : (x, \theta, \eta) \longmapsto \left((x, \varphi_x'), j_{\varphi*}\left((\psi_{1,x}', \psi_{1,\theta}')_{|TC_\varphi} \right) \right).$$

is a lagrangian immersion. We denote its image by $\Lambda_{\varphi,\psi}$.

The global geometric setting can be introduced in the following way. If Y is a submanifold of X, the *blowup* of X along Y is

$$\widehat{X}_Y = (X \setminus Y) \cup \dot{T}_Y X.$$

Definition. A pair (Λ_0, Λ_1) is a 2-microlocal pair of lagrangian submanifolds of $\dot{T}^*X$ if

- Λ_0 is a conic lagrangian submanifolds of $\dot{T}^*X$, $\Lambda_1 \subset (\dot{T}^*X)_{\Lambda_0}^\wedge$,

- $\Lambda_1 \cap (\dot{T}^*X \setminus \Lambda_0)$ is a conic lagrangian submanifold of $\dot{T}^*X$,

- for each $(\rho, h) \in \Lambda_1 \cap \dot{T}_{\Lambda_0}T^*X$, there is a 2-phase function (φ, ψ) such that

$$\Lambda_0 \cap \pi(V) = \Lambda_\varphi \qquad \text{and} \qquad \Lambda_1 \cap V = \Lambda_{\varphi+\psi} \cup \Lambda_{\varphi,\psi}$$

near (ρ, h).

In this situation, we say that the 2-phase function (φ, ψ) defines (Λ_0, Λ_1). Let $T_{\Lambda_0}\Lambda_1 = \Lambda_1 \cap \dot{T}_{\Lambda_0}(T^*X)$. This is a conic bilagrangian submanifold of $\dot{T}^*\Lambda_0$.

Example. In $\dot{T}^*\mathbb{R}^n$, consider

$$\varphi(x, \xi) = x \cdot \xi, \qquad \psi(x, \xi, \eta') = \frac{\eta' \cdot \xi'}{\xi_n} - H(\eta', \xi_n),$$

where $\xi = (\xi', \xi_n)$ and H is bihomogeneous of degree $(1, 1; r)$. We have

$$\Lambda_\varphi = \{(0, \xi) : \xi_n \neq 0\}$$

and

$$\Lambda_{\varphi+\psi} = \left\{ \left(\left(-\frac{\eta'}{\xi_n}, \frac{\eta' \cdot H_{\eta'}'}{\xi_n} + H_{\xi_n}' \right), \left(\xi_n H_{\eta'}', \xi_n \right) \right) : \xi_n \neq 0 \right\}.$$

If $H(\eta', \xi_n) = \eta_1^3/\eta_2^2$ in $\mathbb{R}^3$, the projection of $T_{\Lambda_\varphi}\Lambda_{\varphi+\psi}$ on Λ_φ is the cusp

$$\left\{ (0, \xi) : \left(\frac{\xi_1}{3} \right)^3 = \left(\frac{\xi_2}{2} \right)^2 \xi_3 : \xi_3 \neq 0 \right\}.$$

The following properties show that we have met the requirements at the beginning of this section.

a) The property of being a pair of lagrangian submanifolds is preserved by an homogeneous canonical transformation.

b) Two 2-phase functions (φ, ψ) and $(\widetilde{\varphi}, \widetilde{\psi})$ are said *equivalent* if there is a C^∞ diffeomorphism $\Gamma \to \widetilde{\Gamma} : (x, \theta, \eta) \mapsto (x, f(x, \theta, \eta), g(x, \theta, \eta))$ such that

- $\varphi(x, f(x, \theta, \eta)) + \psi(x, f(x, \theta, \eta), g(x, \theta, \eta)) = \widetilde{\varphi}(x, \theta) + \widetilde{\psi}(x, \theta, \eta)$,

- f is strictly bihomogeneous of degree $(1, 0; r)$ and g is bihomogeneous of degree $(1, 1; r)$,

- $D_\theta f_0$ and $D_\eta g_1$ are invertible in Γ_0.

These phases define the same 2-microlocal pair (Λ_0, Λ_1).

c) If Δ is a diagonal real invertible matrix, the pair of phases

$$\varphi(x, \theta) = \widetilde{\varphi}\left(x, \theta''\right) + \frac{\langle \Delta\theta', \theta' \rangle}{2 |\theta''|}, \qquad \psi\left(x, \theta'', \eta\right) = \widetilde{\psi}\left(x, \theta'', \eta\right)$$

defines the same lagrangian submanifolds as $\widetilde{\varphi}$ and $\widetilde{\psi}$. In the same way,

$$\varphi(x, \theta) = \widetilde{\varphi}(x, \theta), \qquad \psi(x, \theta, \eta) = \widetilde{\psi}\left(x, \theta, \eta''\right) + \frac{\langle \Delta\eta', \eta' \rangle}{2 |\eta''|}$$

defines the same lagrangian submanifolds as $\widetilde{\varphi}$ and $\widetilde{\psi}$.

d) If two 2-phase functions define the same 2-microlocal pair (Λ_0, Λ_1) they can be reduced locally to the same one by some iterations of the process b) and c).

3.2. DISTRIBUTION CLASS

Assume that (Λ_0, Λ_1) is a pair of lagrangian submanifolds. We first consider the C_∞ case. Consider the class of distributions of the form

$$I_{\varphi, \psi, a}(x) = \int \int e^{i\varphi(x, \theta) + \psi(x, \theta, \eta)} a(x, \theta, \eta) \, d\theta d\eta,$$

where φ, ψ is a 2-phase function for (Λ_0, Λ_1) and the symbol $a \in S^{m, p}(\Omega, \mathbb{R}^N, \mathbb{R}^M)$ satisfies

$$\left| D_x^\alpha D_\theta^\beta D_\eta^\gamma a(x, \theta, \eta) \right| \leq C_{\alpha, \beta, \gamma} \, (1 + |\theta| + |\eta|)^{m - |\beta|} \, (1 + |\eta|)^{p - |\gamma|}.$$

The space $I^{m,p}(X, \Lambda_0, \Lambda_1)$ is sum of $I^m(X, \Lambda_0)$, $I^{m+p}(X, \Lambda_1 \cap T^*X)$ and of distributions $I_{\varphi,\psi,a}$ where (φ, ψ) is a 2-phase function of (Λ_0, Λ_1) and

$$a \in S^{m+(n-2N)/4, p-M/2}\left(\Omega, \mathbb{R}^N, \mathbb{R}^M\right).$$

It can be shown that this space in invariant by the conjugaison with a Fourier integral operator. Moreover, we have the following result.

Theorem 3.1. *If* $u \in I^{m,p}(X, \Lambda_0, \Lambda_1)$ *then*

$$\mathrm{WF}(u) \subset \Lambda_0 \cup \Lambda_1$$

and

$$\mathrm{WF}_\Lambda^{(2)}(u) \subset T_{\Lambda_0}\Lambda_1.$$

In the analytic case, assume that

- (Λ_0, Λ_1) is a pair of lagrangian submanifolds,

- (h, ψ) is a 2-phase function for this pair,

- h is analytic and ψ is an analytic function of $(x, \theta, \sigma^{1/2})$ satisfying

$$\psi(x, \theta, \sigma) = \psi_1(x, \theta)\sigma + \psi_{3/2}(x, \theta)\sigma^{3/2} + \psi_2(x, \theta)\sigma^2 + \mathbb{O}\left(\sigma^{5/2}\right). \quad (3.1)$$

If g is an FBI phase function, we then have

$$\phi(z, \sigma) = \mathrm{cv}_{(x,\theta)}\left(g(z, x) + h(x, \theta) + \psi(x, \theta, \sigma)\right)$$

$$= \Phi_{\Lambda_0}(z) + \Phi_1(z)\sigma + \Phi_{3/2}(z)\sigma^{3/2} + \Phi_2(z)\sigma^2 + \mathbb{O}\left(\sigma^{5/2}\right).$$

Here Φ_1 and $\Phi_{3/2}$ are real on $\pi \circ \chi(\Lambda_0)$, $\Phi_1(z_0) = 0$, $D_z\Phi_1(z_0) \neq 0$ and $\Im\Phi_2(z_0) > 0$.

A distribution u is said *analytic bilagrangian* at ρ_0 with respect to (Λ_0, Λ_1) if, in a neighborhood of z_0, we have

$$(T_\chi u)(z, \lambda) = \int_0^\delta e^{i\phi(z,\sigma)}a(z, \sigma, \lambda)\, d\sigma.$$

The symbol a is assumed to be holomorphic in

$$\{(z, \sigma) \in \mathbb{C}^n \times \mathbb{C} : |z - z_0| < \epsilon, |\Im\sigma| < c\Re\sigma\}$$

and bounded by $C\lambda^m$.

For example, if

$$\Lambda_0 = \left\{\left((0, x_n), (\xi', 0)\right)\right\}, \qquad \Lambda_1 = \left\{\left((0, 0), (\xi', \xi_n)\right)\right\}$$

and $g(z, y) = i(z - y)^2/2$, we have

$$\Phi_{\Lambda_0}(z) = \frac{i}{2} z'^2, \qquad \Phi_{\Lambda_1}(z) = \frac{i}{2} z^2$$

and

$$\phi(z, \sigma) = \frac{iz'^2}{2} + \sigma z_n + \frac{i\sigma^2}{2}.$$

4. An application

Let M be a real manifold with boundary and P a second order differential operator with smooth coefficients and real principal symbol p. We assume that p is of real principal type and not characteristic on the boundary. Let us consider the classical Dirichlet problem

$$Pu = 0 \quad \text{in } M, \qquad u_{|\partial M} = \delta_{x_0},$$

where δ_{x_0} is the Dirac mass at some point of the boundary. If the equation of the boundary is $f = 0$ with $f > 0$ in M, the diffractive region is defined by

$$\mathcal{G}_+ = \left\{ \rho \in \dot{T}^*\partial M : p(\rho) = 0, \{p, f\} = 0, \frac{\{p, \{p, f\}\}_\rho}{\{\{p, f\}, f\}_\rho} > 0 \right\}$$

and corresponds to rays tangent to the boundary. The principal symbol p defines a boundary hamiltonian H_r on $\mathcal{G}_+$, see for example [2].

Denote by ν a non-vanishing normal vector field to M. Two lagrangian manifolds are naturally involved here:

- the flowout Λ_0 of $\{\rho \in T^*_{x_0}\partial M : p(\rho) = 0\}$ through H_r on the boundary and followed by H_p in M,

- the flowout Λ_1 of $\{\rho \in T^*_{x_0}M : p(\rho) = 0, \rho(\nu) \neq 0\}$ through H_p.

The manifolds Λ_0, Λ_1 define a 2-microlocal pair of lagrangian submanifolds [6]. Moreover, the phase functions satisfy the assumption (3.1).

It is known that $\mathrm{WF}(u)$ is included in the closure of Λ_1 but u is not lagrangian on Λ_1. We have the following lagrangian description at the 2-microlocal level.

Theorem 4.1. *The solution u of the previous boundary value problem belongs to*

$$\left(I^{(n/4)-1,(3/4)} \left(M, \Lambda_0, \Lambda_1 \cup T_{\Lambda_0}\Lambda_1 \right) \right) + I^{n/4-1/2}_{2/3} \left(M, \Lambda_0 \right).$$

Here $I_\rho^m(X, \Lambda_0)$ is the set of lagrangian distributions with symbol in $S_\rho^m = S_{\rho,1-\rho}^m$. This describes the behavior of the solution at the transition from the shadow to the illuminated region.

In the analytic category, we have the following result.

Theorem 4.2. *At the transition between the shadow and the illunimated region, the solution u of the boundary value problem can be written $u_1 + u_2$ where u_1 is analytic bilagrangian and*

$$|u_2(z, \lambda)| \leq C_\epsilon e^{\lambda(\varphi_{\Lambda_0}(z) + Cd(z, \pi \circ \chi(\Lambda_0))^3) + \epsilon\lambda}$$

for every $\epsilon > 0$.

References

[1] Delort, J.-M., Deuxième microlocalisation simultanée et front d'onde de produits, *Ann. scient. Ec. Norm. Sup.* **23** (1990), 257–310.

[2] Hörmander, L., *The analysis of linear partial differential operators I-IV*, Springer–Verlag 1983–85.

[3] Friedlander, F.G. and Melrose, R.B., The wave front set of the solution of a simple initial-boundary value problem with glancing rays II, *Math. Proc. Camb. Phil. Soc.* **87** (1977), 97–120.

[4] Lafitte, O., The kernel of the Neumann operator for a strictly diffractive analytic problem, *Comm. in Part. Diff. Eq.* **20** (1995), 419–483.

[5] Laubin, P., Etude 2-microlocale de la diffraction, *Bull. Soc. Roy. Sc. Liège* **4** (1987), 295–416.

[6] Laubin, P. and Willems, B., Distributions associated to a 2-microlocal pair of lagrangian manifolds, *Comm. in Part. Diff. Eq.* **19** (1994), 1581–1610.

[7] Lebeau, G., Deuxième microlocalisation sur les sous-variétés isotropes, *Ann. Inst. Fourier, Grenoble* **35** (1985), 145–216.

[8] Lebeau, G., Régularité Gevrey 3 pour la diffraction, *Comm. in Part. Diff. Eq. (15)* **9**, 1984, 1437–1494.

[9] Lebeau, G., *Propagation des singularités Gevrey pour le problème de Dirichlet*, Advances in microlocal analysis, Nato ASI, C168 1986, 203–223.

[10] Lebeau, G., Scattering frequencies and Gevrey 3-singularités, *Invent. math.* **90** (1987), 77–114.

[11] Melrose, R.B. and Sjöstrand J., Singularities in boundary value problem I, *Comm. Pure Appl. Math.* **31** (1978), 593–617.

[12] Melrose, R.B., Local Fourier-Airy integral operators, *Duke Math. J.* **42** (1975), 583–604.

[13] Melrose, R.B., Transformation of boundary value problems, *Acta Math. J.* **147** (1981), 149–236.

[14] Sjöstrand, J., Propagation of analytic singularities for second order Dirichlet problems, I-III, *Comm. Part. Diff. Eq.* **5**(1) (1980), 41–94; **5**(2) (1980), 187–207; **6**(5) (1981), 499–567.

[15] Sjöstrand, J., Singularités analytiques microlocales, *Astérisque* **95** (1982), 1–166.

CARACTÉRISATION DES OPÉRATEURS DIFFÉRENTIELS HYPERBOLIQUES

YUJIRO OHYA

Contents

1 Introduction 99
2 Résolution dans la classe de Gevrey 101
    2.1 Le départ 101
    2.2 Collaboration avec J. Leray 103
    2.3 La résolution finale dans la classe de Gevrey 103
3 Résolution dans la classe C^∞ 104
    3.1 Multiplicité constante 104
    3.2 Multiplicité variable 105

M. de Gosson (ed.), Jean Leray '99 Conference Proceedings, 97-107.

1. Introduction

Historique.

A la fin du dix-neuvième siècle, S. Kovalesky [5] avait montré l'existence de solution locale pour le problème de Cauchy dans la classe de fonctions analytiques, J. Hadamard a proposé la nécéssité du théorème d'existence de solution globale dans la classe de fonctions indéfiniment dérivables compte tenu du point de vue physique J. Schauder [19] l'a rélisé en employant l'inégalité énergétique [2], [18] a généralisé ce type de traitement et développé la théorie moderne des équations aux dérivées partielles du type hyperbolique. J. Leray [8] a remarqué et corrigé une erreur dans l'article de Petrowsky.

Vers la fin de 1960, le problème de Cauchy pour les équations strictement hyperboliques a été complètement résolu dans la classe de fonctions indéfiniment dérivables.

Ici, hyperbolicité stricte signifie que le polynôme caractéristique a des racines réelles et distinctes.

D'un autre côté P.D. Lax [7] (sous l'hypothèse pour les racines simples) et S. Mizohata [10] (sous l'hypothèse même pour les racines multiples) ont découvert qu'il nous faut supposer des racines réelles pour le problème de Cauchy bien posé.

Alors, pour étendre la notion d'hyperbolicité, il suffit que l'on discute le problème de Cauchy pour les équations différentielles aux racines caractéristiques réelles multiples—sous cette seule hypothèse.

Notation. Soit $P(t, x, D_t, D_x)$ un opérateur différentiel du type kovalevskien:

$$P(t, x, D_t; D_x) = D_t^m + \sum_{\substack{j+|v|\le m \\ j<m}} a_{jv}(t, x) D_t^j D_x^v$$

avec $D_t = -i\partial_t$ et $D_x = (D_{x_1}, \ldots, D_{x_\ell})$, $D_{x_j} = -i\partial_{x_j}$.

On considère le problème de Cauchy

$$\begin{cases} P(u(t, x)) = f(t, x) & \text{sur } \Omega = [0, \infty[\times\mathbb{R}^\ell, \\ D_t^j u(0, x) = \varphi_j(x) & (0 \le j \le m - 1), \end{cases} \tag{1.1}$$

et on note

$$P_m(t, x, D_t, D_x) = D_t^m + \sum_{\substack{j+|v|=m \\ j<m}} a_{jv}(t, x) D_t^j D_x^v,$$

$$P_k(t, x, D_t, D_x) = \sum_{j+|v|=k} a_{jv}(t, x) D_t^j D_x^v (0 \le k \le m - 1).$$

On définit le polynôme caractéristique de P:

$$p_m(t, x, \tau, \xi) = \tau^m + \sum_{j+|\nu|=m} a_{j\nu}(t, x)\tau^j \xi^\nu.$$

Il se factorise en

$$p_m(t, x, \tau, \xi) = \prod_{i=1}^{k} (\tau - \lambda_i(t, x; \xi))^{\nu_i} \tag{1.2}$$

où $\sum_{i=1}^{k} \nu_i = m$ et $\lambda_i(t, x; \xi)$ est réelle pour $(t, x) \in \Omega$, $\xi \in \mathbb{R}^\ell$.

Dans la suite, on emploie le calcul élémentaire de la théorie des opérateurs pseudo-différentiels (voir par exemple L. Nirenberg [12]).

Définition. On définit la classe de symboles $S_{\rho,\delta}^m(\Omega \times \mathbb{R}^\ell)$ ($0 \leq \delta \leq \rho \leq 1$) des fonctions $a \in C^\infty(\Omega \times (\mathbb{R}^\ell \backslash 0))$ telles que pour tout compact $K \subset \mathbb{R}^\ell$ et tous multi-indices α, β il existe $C_{K,\alpha,\beta} \geq 0$ tel que

$$\sup_{(t,x)\in[0,\infty[} \left| D_\xi^\alpha D_x^\beta a(t, x; \xi) \right| \leq C_{K,\alpha,\beta}(1 + |\xi|)^{m-\rho|\alpha|+\delta|\beta|}.$$

Nous n'emploierons en fait que le cas $\rho = 1$, $\delta = 0$, auquel cas l'on note $S_{1,0}^m(\Omega \times \mathbb{R}^\ell) = S^m(\Omega \times \mathbb{R}^\ell)$. On définit aussi l'opérateur pseudo-différentiel $A(t, x; D)$ associé au symbole a par

$$A(t, x; D)u(t, x) = (2\pi)^{-\ell/2} \int e^{i\langle x,\xi\rangle} a(t, x; \xi)\hat{u}(t, \xi)d\xi \tag{1.3}$$

pour $u \in C^\infty([0, \infty], \mathscr{S}(\mathbb{R}^\ell))$; $\hat{u}$ est la transformée de Fourier de u en x. On emploie les notations suivantes:

$$\|f(t)\|_\sigma^2 = \int \left(1 + |\xi|^2\right)^\sigma \left|\hat{f}(t, \xi)\right|^2 d\xi \tag{1.4}$$

$$\left\|D^n f(t)\right\|_\sigma^2 = \sup_{j+|\nu|\leq n} \left\|D_t^j D_x^\nu f(t)\right\|_\sigma^2 \tag{1.5}$$

pour $f(t, x) \in L^2([0, \infty], \mathscr{S}(\mathbb{R}^\ell))$. Si on note $\langle\xi\rangle = (1 + |\xi|^2)^{1/2}$, alors (1.3) s'écrit encore

$$\|f(t)\|_\sigma^2 = \int \left|\langle\xi\rangle^\sigma \hat{f}(t, \xi)\right|^2 d\xi \tag{1.6}$$

qui équivaut à

$$\sum_{|\alpha|\leq\sigma} \left|D_x^\alpha f(t, x)\right|^2 dx$$

vu la formule de Parseval–Plancherel.

2. Résolution dans la classe de Gevrey

2.1. LE DÉPART

L'auteur a essayé la résolution dans cette classe pour le cas le plus simple que voici:

$$p_m(t, x, \tau, \xi) = (\tau - a(t, x; \xi))^m \qquad (2.1)$$

En supposant $\lambda_i(t, x; \xi) = a(t, x; \xi) \in S^1(\Omega \times \mathbb{R}^\ell)$ dans (1.2) pour tout i uniformément en $(t, x, \xi) \in \Omega \times \mathbb{R}^\ell$.

On associe $A(t, x, D)$ par (1.3), et on considère $(D_t - A(t, x, D))^m$ à la place de $P(t, x, D)$. C'est à dire

$$(D_t - A)^m [u(t, x)] = \left\{ (D_t - A)^m - P \right\} [u(t, x)] + f(t, x)$$

que l'on écrit pour simplifier

$$L[u(t, x)] = Q[u(t, x)] + f(t, x). \qquad (2.2)$$

remarquons que si L est strictement hyperbolique, alors on aura

$$\left\| D^{n+m-1} u(t) \right\|_\sigma \leq \gamma_0 \left\| D^{n+m-1} u(0) \right\|_\sigma + \int_0^t \left\| D^n f(\tau) \right\|_\sigma d\tau \qquad (2.3)$$

pour tout n, puisque l'ordre de $L = m$. Mais, sous l'hypothèse (2.1), on n'a que

$$\left\| D^n u(t) \right\|_\sigma \leq \gamma_0 \left\| D^n u(0) \right\|_\sigma + \int_0^t \left\| D^n f(\tau) \right\|_\sigma d\tau \qquad (2.4)$$

même pour $m \geq 2$.

Pour corriger cette situation, il y a deux possiblilités, dont la première est de construire une suite de solution $\{u_N(t, x)\}$ telle que

$$\begin{cases} L[u_0(t, x)] = f(t, x) \\ D_t^j u_0(0, x) = 0 \end{cases} \qquad j = 0, 1 \cdots m - 1 \qquad (2.5)$$

$$\begin{cases} L[u_N(t, x)] = Q[u_{N-1}(t, x)] \\ D_t^j u_N(0, x) = 0 \end{cases} \qquad j = 0, 1 \cdots m - 1 \qquad (2.6)$$

où l'on a amené d'abord $\varphi_i(x) \equiv 0$ dans (1.1) et l'autre est de montrer la convergence de $\sum_{N=0}^{\infty} u_N(t, x)$, où il nous a fallu employer l'espace de fonctions de classe de Gevrey.

Définition. (Classes de Gevrey). On définit $\gamma^s(\mathbb{R}^\ell)$ comme l'ensemble des $f \in C^\infty(\mathbb{R}^\ell)$ tels que pour tout compact K et tout α, il existe A et C tels que

$$\sup_{x \in K} \left| D^\alpha f(x) \right| \leq C A^{|\alpha|} (|\alpha|!)^s,$$

et on définit $\gamma_0^s(\mathbb{R}^\ell) = \gamma^s(\mathbb{R}^\ell) \cap C_0^\infty(\mathbb{R}^\ell)$.

Pour donner une première motivation au procédé, on va d'abord citer le lemme suivant

Lemme 2.1 (Ohya [14]). *Pour le problème de Cauchy*

$$[D_t - A(t, x, D)] u(t, x) = f(t, x)$$

avec $u(0, x) = 0$, si l'on suppose que pour tous n, s

$$\left\| D^n f(t) \right\|_\sigma \leq C e^{\gamma_0 t} K(t)^{n+\sigma} A^{n+\sigma} (n+\sigma)!^s \qquad (2.7)$$

$$\sup_{|\xi|=1} \left| D_\xi^\alpha a(t, x; \xi) \right| \leq \gamma_1 \left(\frac{A}{\rho} \right)^{|\alpha|-1} (|\alpha|-1)!^s \qquad (2.8)$$

où γ_0, γ_1, CA et ρ (≥ 4) sont des constantes, et $K(t) = (1 + \gamma_t)^{2\gamma_t}$, alors on a

$$\left\| D^n u(t) \right\|_\sigma \leq C e^{\gamma_0 t} K(t)^{n+\sigma} A^{n+\sigma} (n+\sigma)!^s t.$$

Compte tenu de ce lemme, on obtient

$$\left\| D^n u_0(t) \right\|_\sigma \leq 2^m C e^{\gamma_0 t} K(t)^{n+\sigma} A^{n+\sigma} (n+\sigma)!^s \frac{t^m}{m!}$$

et

$$\left\| D^n u_N(t) \right\|_\sigma \leq 2^m C \left(2^m B \right)^N e^{\gamma_0 t} K(t)^{n+\sigma+N(m-1)}$$
$$\times A^{n+\sigma+N(m-1)} (n + N(m-1))! \frac{t^{(N+1)m}}{(N+1)m!}$$

successivement. Finalement on obtient

$$\left\| D^n u(t) \right\|_\sigma \leq \sum_{n=0}^{N} \left\| D^n u_N(t) \right\|_\sigma$$

$$\precsim C A^{n+\sigma} (n+\sigma)!^s \sum_{n=0}^{\infty} \frac{\left(\text{const.} t^m (k(t) A)^{m-1} \right)^N}{N(m - (m-1)s)!}$$

où '$\precsim$' signifie 'partie essentielle'. Maintenant il est clair que si $1 < s < m/(m-1)$, alors pour tout t on a convergence; si $s = m/(m-1)$ alors on a convergence localement par rapport à t.

Théorème 2.2 (Ohya [14]). *On suppose que les coefficients*

$$a_{j\nu}(t, x), f(t, x) \text{ et } \{\varphi_j(x)\}_0^{m-1}$$

appartiennent à $\gamma^s(\mathbb{R}^\ell)$ par rapport à x, et $a_{j\nu}(t, x)$ sont bornés dans Ω. Alors il existe une solution $u(t, x) \in C^m([0, \infty[, \gamma^s(\mathbb{R}^\ell))$ pour $1 < s < m/(m-1)$.

2.2. COLLABORATION AVEC J. LERAY

L'auteur à proposé J. Leray l'article du numéro 1, avant publication, qui a été déposé auprès de S. Mizohata pour publication au *J. Math. Soc. Japan.*

La première intention de J. Leray fut d'étendre ce résultat aux systèmes.

La collaboration a commencé à l'automne 1963 au bureau du Collège de France, ou chez lui, à Robinson.

Il eut beaucoup d'idées nouvelles, dont l'une était de majorer les solutions pour des séries formelles, et l'autre l'emploi du théorème de Cauchy–Kovalevska pour assurer la convergence de ces séries. Voici le résultat:

Théorème 2.3 (Leray–Ohya [9]). *Le problème de Cauchy*

$$\begin{cases} P[u(t, x)] = f(t, x) \in \gamma^s(\Omega) \\ D_t^j u(0, x) = \varphi_j(x) \in \gamma^s(\mathbb{R}^\ell) \\ j = 0, \ldots, m-1 \end{cases} \quad (2.9)$$

a une, et une seule, solution dans $C^m([0, \infty[, \gamma^s(\mathbb{R}^\ell))$ pour $1 < s < p/q$, si l'on suppose que

$$L = A_1(t, x, D_t, D) \cdots A_p(t, x, D_t, D) \quad (2.10)$$

où les $A_i(t, x, D_t, D)$ sont des opérateurs pseudo-différentiels strictement hyperboliques par rapport à x à multiplicité constante (coefficients dans $\gamma^s(\mathbb{R}^\ell)$) et

$$\text{ordre}(P - L) \le m - p + q \quad (q \le p - 1) \quad (2.11)$$

Remarque. L'ordre de $P - L$ est au plus $m - p + q$ $(q \le p - 1)$.

2.3. LA RÉSOLUTION FINALE DANS LA CLASSE DE GEVREY

M. D. Brohnstein [1] réussi à la construction de paramétrix pour le problème de Cauchy en 1976. En nous appuyant sur sa méthode, nous avons obtenu

Théorème 2.4 (Ohya–Tarama [17], Nishitani [13]). *Soit r la multiplicité la plus haute pour que si $j + |\nu| = m$ (resp. $j + |\nu| = m$) $a_{j\nu}(t, x)$ appartienne à $C^k([0, \infty[, \gamma^s(\mathbb{R}^\ell))$ (resp. $C^0([0, \infty[, \gamma^s(\mathbb{R}^\ell)))$ bornés dans Ω, et que $f(t, x) \in C^0([0, \infty[, \gamma^s(\mathbb{R}^\ell))$, $\varphi_j(x) \in \gamma^s(\mathbb{R}^\ell)$ pour $j = 0, \dots, m-1$, alors le problème de Cauchy (1.1) est bien posé dans $C^m([0, \infty[, \gamma^s(\mathbb{R}^\ell))$ pour $1 < s < \min(1 + k/r, r/(r-1))$ où $0 < k \le 2$.*

3. Résolution dans la classe C^∞

Compte tenu du Théorème 2.3, on remarque que l'on pourra obtenir la résolution dans C^∞ si l'on peut découvrir le cas $q = 0$ par le choix moralement très convenable de $A_i(t, x, D_t, D)$. Pour cette direction, il y a l'article à deux variables de A. Lax [6] en 1956, et l'extension de M. Yamaguti [20] à plusieurs variables. Mais leur résultat est trop artificiel pour le problème de Cauchy original.

Alors on a trouvé un article de E.E. Levi en 1909 dans un coin des bagages transportés du Japon. Cet article a été distribué au Séminaire Mizohata en 1963. Cet article de E.E. Levi a donné le traitement extrêmement fondamental de la résolution du problème de Cauchy à deux variables dans C^∞.

3.1. MULTIPLICITÉ CONSTANTE

Commençons par nous limiter au cas des caractéristiques doubles à multiplicité constante:

$$p_m(t, x; \tau, \xi) = \prod_{i=1}^{s} (\tau - \lambda_i(t, x; \xi)) \prod_{j=s+1}^{m-s} \left(\tau - \mu_j(t, x; \xi)\right). \qquad (3.1)$$

Théorème 3.1 (Mizohata–Ohya [11]). *Sous l'hypothèse (3.1), pour que le problème de Cauchy (1.1) soit bien posé dans la classe C^∞, il faut et il suffit que l'on ait*

$$\left[p_{m-1} + \frac{1}{2}\left(\frac{\partial p'_m}{\partial \tau}\frac{\partial \lambda_j}{\partial t} + \sum_{\alpha=1}^{\ell}\frac{\partial p'_m}{\partial \xi_\alpha}\frac{\partial \lambda_j}{\partial x_\alpha}\right)\right](t, x; \lambda_j(t, x, \xi), \xi) = 0 \quad (3.2)$$

pour tout $(t, x, \xi) \in \Omega \times \mathbb{R}^\ell$, $1 \le j \le s$ et $p'_m = \partial p_m / \partial \tau$.

L'hypothèse (3.2) a été nommée la condition de E.E. Levi, puisque cette condition a été énoncée précisément dans l'article de E.E. Levi dans le cas de deux variables.

Remarque. Il faut ajouter une remarque très importante. La condition (3.2) s'écrit encore

$$\left[p_{m-1} - \frac{1}{2} \sum_{\alpha=0}^{\ell} \frac{\partial}{\partial \xi_\alpha} D_{x_\alpha} p_m \right] (\lambda_j) \equiv 0 \qquad (3.3)$$

x_0 étant t et ξ_0 étant τ. En effet, en supposant $p_m = r^2 q$ où

$$r = \prod_{i=1}^{s} (\tau - \lambda_i) \quad \text{et} \quad q = \prod_{j=s+1}^{m-s} \left(\tau - \mu_j(t, x; \xi) \right)$$

on obtient

$$\left[\frac{1}{2} \sum_{\alpha=0}^{\ell} \frac{\partial^2}{\partial \xi_\alpha \partial x_\alpha} p_m \right] (\lambda_i) = \left[\sum_{\alpha=0}^{\ell} \frac{\partial r}{\partial \xi_\alpha} \frac{\partial r}{\partial x_\alpha} q \right] (\lambda_i)$$

d'où la dénomination de (3.3) de 'partie sous-principale'. Il nous reste à discuter le cas à multiplicité variable.

3.2. MULTIPLICITÉ VARIABLE

Il y a l'article de Ivrii–Petkov [4] où est introduit une notion nouvelle (du point de vue de la condition nécessaire), la matrice fondamentale. Citons ici un résultat bien dans l'esprit de cet article: supposons

$$p_m = \prod_{i=1}^{m-s} (\tau - \lambda_i(t, x; \xi)) \prod_{j=1}^{s} \left(\tau - \mu_j(t, x; \xi) \right)$$

et définissons

$$\prod_{i=1}^{m-s} (\tau - \lambda_i(t, x; \xi)) = q_{m-s}(t, x; \tau, \xi) \qquad (3.4)$$

$$\prod_{i=1}^{s} (\tau - \mu_i(t, x; \xi)) = r_s(t, x; \tau, \xi); \qquad (3.5)$$

$q_{m-s}(t, x; \tau, \xi)$ et $r_s(t, x; \tau, \xi)$ sont des polynômes strictement hyperboliques. Pour de mêmes indices i on permet à $\lambda_i(t, x; \xi)$ et $\mu_i(t, x; \xi)$ de coincider; sinon P est un opérateur strictement hyperbolique, donc le problème de Cauchy pour P est résolu dans la classe C^∞.

Théorème 3.2 (Ohya, [15]). *Si l'on suppose que*

$$\frac{L_{m-1}(t, x; \mu_i(t, x; \xi), \xi)}{\mu_i(t, x; \xi) - \lambda_i(t, x; \xi)} \tag{3.6}$$

appartient à $C^\infty(\Omega \times \mathbb{R}^\ell \backslash 0)$ pour tout i ($1 \leq i \leq s$), alors le problème de Cauchy (1.1) est bien posé dans la classe C^∞,

$$L_{m-1}(t, x; \tau, \xi) = p_{m-1} - \sum_{\alpha=0}^{\ell} \frac{\partial q_{m-s}}{\partial \xi_\alpha} \frac{\partial r_s}{\partial x_\alpha}. \tag{3.7}$$

En particulier si $(\mu_i - \lambda_i)/t \in C^\infty(\Omega \times \mathbb{R}^\ell \backslash 0)$, alors (3.6) est remplacé par

$$\frac{L_{m-1}(t, x; \mu_i(t, x; \xi), \xi)}{\frac{\mu_i - \lambda_i}{t}} \in C^\infty\left(\Omega \times \mathbb{R}^\ell \backslash 0\right) \tag{3.8}$$

pour tout i ($1 \leq i \leq s$).

Conclusion. L'auteur a voulu au début la caractérisation complète des opérateurs différentiels hyperboliques. Concernant le problème de Cauchy dans les classes de Geverey, le résultat est complet, compte tenu de la preuve de la non-unicité d'après J. Leray, avec celle de E. de Giorgi. Mais la résolution dans la classe C^∞ n'est pas encore complète, puisque l'on ne réussit pas encore d'avoir un résultat analogue au Théorème 3.1 dans le cas de la multiplicité variable.

Reférénces

[1] Bronshtein, M.D., The parametrix of the Cauchy problem for hyperbolic operators with characteristics of variable multiplicity. *Fumk. Anal, i Prilozen*, **19** (1976), 83–84.

[2] Friedrichs, K.O. and Lewy, H., Über die eindeutigkeit und das Abhängigkeitsgebiet des Lösungen beim Anfangswert Problem lineare hyperbolisher. *Differentialgleichungen. Math. Ann*, **98** (1928), 192–204.

[3] Hadamard, J., Les fonctions de classe supérieure dans l'équation de Volterra. *J. Anlyse Math.* (1951).

[4] Ivrii, V. and Ya–Petkov, V.M., Necessary conditions for the Cauchy problem for non-strictly hyperbolic equations to be well posed. *Russ. Math. Survey.* **29** (1974), 1–70.

[5] Kovalevska, S., Zur Theorie der partiellen Differentialgleichungen. *J. reine. und ang. Math.* **80** (1875), 1–32.

[6] Lax, A., On Cauchy's problem for partial differential equations with multiple characteristics. *Comm. Pure Appl. Math.* **9** (1956), 135–165.

[7] Lax, P.D., Asymptotic solutions of oscillatory initial problem. *Duke Math. J.* **24** (1957), 627–646.

[8]   Leray, J., *Hyperbolic differential equations*. Inst. Adv. Study. Princeton (1952).

[9]   Leray, J. and Ohya, Y., Systèmes linéaires, hyperboliques non strictes. *Colloque de Liège.* **10** (1964), 105–152.

[10] Mizohata, S., Some remarks on the Cauchy problem. *J. Math. Kyoto Univ.* **1** (1961), 109–127.

[11] Mizohata, S. and Ohya, Y., Sur la condition de E.E. Levi concernant des équations hyperboliques. *Publ. RIMS, Kyoto.* **4** (1971), 63–104.

[12] Nirenberg, L., Pseudo-differential operators in global analysis. *Proc. Sympo. Pure Math.* **16** (1970), 149–167.

[13] Nishitani, T., Sur les équations hyperboliques à coefficients en t et de classe de Gevrey en x. *Bull. Soc. Math. France.* **107** (1983), 113–138.

[14] Ohya, Y., Le problème de Cauchy pour les équations hyperboliques à caractéristique multiple. *J. Math. Soc. Japan.* **16** (1964), 268–286.

[15] _____, Le problème de Cauchy à caractéristiques multiples. *Ann.Scuola Norm.Sup.* **IV**, (1977) 757–805.

[16] _____, *Le problème de Cauchy pour les équations hyperboliques à caractéristiques multiples*. Cours à Univ. Pierre et Marie Curie (1979–80).

[17] Ohya, Y. and Tarama, S., Le problème de Cauchy à caractéristique multiple dans la classe de Gevrey. *HERT, Taniguchi Symp*, (1984) 273–306.

[18] Petrowsky, I.-G., Über das Cauchysche Problem für ein Systeme lineare partieller Differentialgleichungen im Gebiete der nichtanalytischen Funktionen. *Bull. Uni Moscou.* **1** (1938), 1–72.

[19] Schauder, J., Das Anfangswertproblem einer quasilinearen hyperbolischen Differentialgleichung zweiter Ordung in beliebiger Anzahl unabhängigen Veränderlichen. *Fund. Math.* **24** (1935), 213–246.

[20] Yamaguti, M., Le problème de Cauchy et les opérateurs d'intégrale singulière. *Mem. Coll. Kyoto Univ.* **32** (1959), 121–151.

A DEPENDENCE DOMAIN FOR A CLASS OF MICRO-DIFFERENTIAL EQUATIONS WITH INVOLUTIVE DOUBLE CHARACTERISTICS

Y. OKADA AND N. TOSE

Contents

1   Statement of the main theorem 111
2   Theorem in the model case 112

M. de Gosson (ed.), Jean Leray '99 Conference Proceedings, 109-114.
© *2003 Kluwer Academic Publishers.*

1. Statement of the main theorem

Let M be a real analytic manifold with a complex neighborhood X. Let P be a microdifferential operator defined in a neighborhood U in T^*X of $\dot{q} \in T^*_M X \setminus M$. We assume that the characteristic variety of P satisfies

$$\text{Char}(P) \subset \{q \in U; \ p_1(q) \cdot p_2(q) = 0\}$$

with homogeneous holomorphic functions p_1 and p_2 on U. We assume that

$$p_1 \text{ and } p_2 \text{ are real valued on } T^*_M X, \tag{1}$$
$$dp_1 \wedge dp_2 \wedge \omega_X(q) \neq 0 \quad \text{if } p_1(q) = p_2(q) = 0, \tag{2}$$
$$\{p_1, p_2\}(q) = 0 \quad \text{if } p_1(q) = p_2(q) = 0. \tag{3}$$

Here ω_X is the canonical 1-form of T^*X, and $\{\cdot, \cdot\}$ the Poisson bracket on T^*X.

In this situation, we can define a regular involutive submanifold V by

$$V = \{q \in U; \ p_1(q) = p_2(q) = 0\}.$$

We assume, for simplicity, that $\dot{q} \in V$. Moreover Γ denotes the canonical leaf of V passing through $\dot{q}$.

A set K in Γ is called, in this article, a Γ-rectangle if there exists an injective real analytic map

$$\Phi : [0, 1] \times [0, 1] \longrightarrow \Gamma$$

with the following three properties.

- $\Phi([0, 1] \times [0, 1]) = K$

- $\Phi(\cdot, t)$ is an integral curve of the Hamiltonian vector field H_{p_1} for any fixed $t \in [0, 1]$.

- $\Phi(s, \cdot)$ is an integral curve of the Hamiltonian vector field H_{p_2} for any fixed $s \in [0, 1]$.

We give, in this situation:

Theorem 1.1. *There exists an open neighborhood U_0 of $\dot{q}$ in Γ with the property that for any Γ-rectangle K contained in U_0 with the four vertices q_0, q_1, q_2, and q_3 and for any microfuntion solution u to $Pu = 0$ on K,*

$$q_1, q_2, q_3 \notin \text{supp}(u) \implies q_0 \notin \text{supp}(u).$$

This theorem can be deduced from the model case given in the next section, where several remarks are also given.

2. Theorem in the model case

Let M be an open subset of $\mathbb{R}^n$ with a complex neighborhood X in $\mathbb{C}^n$ ($n \geq 3$). We take a coordinate system of M (resp. X) as $x = (x_1, \ldots, x_n)$ (resp. $z = (z_1, \ldots, z_n)$). Then $(x; \sqrt{-1}\xi \cdot dx)$ (resp. $(z; \zeta \cdot dz)$) denotes a point in $T_M^* X$ (resp. $T^* X$) with $\xi = (\xi_1, \ldots, \xi_n)$ (resp. $\zeta = (\zeta_1, \ldots, \zeta_n)$).

We take a point $q_0 = (0; \sqrt{-1}\, dx_n) \in T_M^* X$. Let P be a microdifferential operator defined in a neighborhood of q_0 whose principal symbol is of the form

$$\zeta_1^{m_1} \zeta_2^{m_2}$$

with $m_1, m_2 \geq 1$. We define an involutive manifold V of $T_M^* X$ by

$$V = \left\{ \left(x; \sqrt{-1}\xi \cdot dx \right); \xi_1 = \xi_2 = 0 \right\}$$

and denote by Γ the leaf of V passing through the point q_0. We take a rectangle K on Γ defined by

$$K = \left\{ \left(x_1, x_2, x'' = 0; \sqrt{-1}\, dx_n \right); 0 \leq x_1 \leq t_1, \, 0 \leq x_2 \leq t_2 \right\}.$$

The vertices of K are denoted by

$$q_0, q_1 = \left(t_1, 0, 0; \sqrt{-1}\, dx_n \right), \qquad q_2 = \left(0, t_2, x'' = 0; \sqrt{-1}\, dx_n \right),$$

$$q_3 = \left(t_1, t_2, x'' = 0; \sqrt{-1}\, dx_n \right).$$

Here $x'' = (x_3, \ldots, x_n)$. Then we have

Theorem 2.1. *Let u be a microfunction defined in a neighborhood of K. We assume that u satisfies*

$$Pu = 0$$

and that the three points q_1, q_2, q_3 are not in supp(u):

$$q_1, q_2, q_3 \notin \text{supp}(u).$$

Then

$$q_0 \notin \text{supp}(u).$$

Remark 2.1. *The phenomenon in the above theorem was first observed by Y. Okada [4] for C^∞ wavefront set of microdistribution solutions. His result concerns the case $m_1 = m_2 = 1$ under a Levi condition on the lower order term of P. He employed a microlocal version of the Goursat problem in the complex domain.*

Remark 2.2. *It is inevitable to assume the condition $q_3 \notin \mathrm{supp}(u)$ in Theorem 2.1 For example, we define a hyperfunction*

$$u\,(x_1, x_2, x_3) = (Y\,(x_1) - Y\,(x_2)) \cdot \delta\,(x_3)\,,$$

which is a solution to $D_1 D_2 u = 0$. Then we have, for a positive constant $t > 0$,

$$q_1 = \left(t, -t, 0;\, \sqrt{-1}\,dx_3\right), \qquad q_2 = \left(-t, t, 0;\, \sqrt{-1}\,dx_3\right) \notin \mathrm{supp}(u),$$

but

$$q_0 = \left(t, t, 0;\, \sqrt{-1}\,dx_3\right), \qquad q_2 = \left(-t, -t, 0;\, \sqrt{-1}\,dx_3\right) \in \mathrm{supp}(u).$$

To give an implication of Theorem 1.1, we recall a result obtained by N. Tose [7].

Theorem 2.2. *Let u be a microfunction solution to $Pu = 0$ on an open subset U of Γ. Then there exist a family $\{b_\lambda^{(1)}\}_{\lambda \in \Lambda_1}$ of integral curves on Γ of $\partial/\partial x_1$ and another family $\{b_\lambda^{(2)}\}_{\lambda \in \Lambda_2}$ of integral curves on Γ of $\partial/\partial x_2$ which satisfy the property that $\mathrm{supp}(u)$ has unique continuation property on the set*

$$\Omega = U \setminus \left(\bigcup_{\lambda \in \Lambda_1} b_\lambda^{(1)} \cup \bigcup_{\lambda \in \Lambda_2} b_\lambda^{(2)} \right).$$

More precisely, if a point $q \in \Omega$ is not in $\mathrm{supp}(u)$, then the connected component of Ω containing q is disjoint with $\mathrm{supp}(u)$.

In the situation of Theorem 2.2, we take a point

$$\dot{q} = \left(s_1, s_2, x'' = 0;\, \sqrt{-1}\,dx_n\right) \in \Gamma.$$

We assume that, for a neighborhood U_1 of $\dot{q}$, the only one integral curve $b_{\lambda_1}^{(1)}$ of $\partial/\partial x_1$ and the only one $b_{\lambda_2}^{(2)}$ of $\partial/\partial x_2$ pass U_1. We assume, for simplicity, that the both two curves pass $\dot{q}$:

$$\dot{q} \in b_{\lambda_j}^{(j)}, \qquad (j = 1, 2).$$

We assume that

$$\text{supp}(u) \cap U_1 \cap \left\{ \left(x_1, x_2, x'' = 0, \sqrt{-1}\, dx_n \right); \ x_1 < s_1, \ x_2 > s_2 \right\} = \emptyset$$

and that

$$\text{supp}(u) \cap U_1 \cap \left\{ \left(x_1, x_2, x'' = 0, \sqrt{-1}\, dx_n \right); \ x_1 > s_1, \ x_2 < s_2 \right\} = \emptyset.$$

In this situation, if a point

$$\dot{q}' \in \left\{ \left(x_1, x_2, x'' = 0, \sqrt{-1}\, dx_n \right); \ x_1 < s_1, \ x_2 < s_2 \right\}$$

does not belong to $\text{supp}(u)$, then it follows from Theorem 2.1 that

$$\text{supp}(u) \cap \left\{ \left(x_1, x_2, x'' = 0, \sqrt{-1}\, dx_n \right); \ x_1 > s_1, \ x_2 > s_2 \right\} = \emptyset.$$

References

[1]   Kashiwara, M. and Laurent,Y., *Théorè me d'annulation et deuxième microlocalisation*, Prépublication d'Orsay 1983.

[2]   Kataoka, K., Okada, Y. and Tose, N., *Decomposition of second microlocal singularities,* Microlocal Geometry (edited by T. Monteiro-Fernandes and P. Schapira) 1990.

[3]   Kashiwara, M. and Schapira, P., *Sheaves on manifolds*, Grundlehren der Math., Springer 1990.

[4]   Okada, Y., *Differential singularities of solutions of microdifferential equations with double characteristics*, preprint.

[5]   Okada, Y. and Tose, N., FBI transformation and second microlcoalization, *J. de math. pures et appl.* **70**(4) (1991), 427–455.

[6]   Tose, N., On a class of microdifferential equations with involutive characteristics—as an application of second microlocalization, *J. Fac. Sci., Univ. of Tokyo, Sect. IA, Math.* **33** (1986), 619–634.

[7]   ———, On a class of 2-microhyperbolic systems, *J. de Math. pures et appl.* **67** (1988), 1–15.

RAMIFICATION NON ABÉLIENNE

CLAUDE WAGSCHAL

Contents

1 Introduction.. 117
2 Un problème de Cauchy intégro-différentiel......................... 119

M. de Gosson (ed.), Jean Leray '99 Conference Proceedings, 115-121.

Cet exposé a pour objet de présenter la résolution du problème de Cauchy ramifié lorsque la ramification du second membre autour des hypersurfaces caratéristiques est quelconque. Ce problème a été traité par E. Leichtnam [4]. On propose ici une présentation unifiée des principaux théorèmes relatifs au problème de Cauchy ramifié; les démonstrations détaillées ont déjà été publiées dans un travail effectué en collaboration avec P. Pongérard [7].

1. Introduction

Les coordonnées d'un point x de $\mathbf{C}^{n+1}$ seront notées $(x_0, \ldots, x_n)$; on note D_j l'opérateur de dérivation par rapport à la variable complexe x_j et $D^\alpha = D_0^{\alpha_0} \circ \cdots \circ D_n^{\alpha_n}$ si $\alpha = (\alpha_0, \ldots, \alpha_n) \in \mathbf{N}^{n+1}$ est un multi-indice de dérivation.

Si E est un espace de Banach complexe, on note $E\{x\}$ l'espace vectoriel des séries entières convergentes. Si X est une variété analytique complexe, on note $\mathcal{H}(X; E)$ l'espace vectoriel des fonctions holomorphes $f : X \to E$ et, si X est connexe, $\mathcal{R}(X)$ son revêtement universel.

Un opérateur différentiel linéaire d'ordre $\leq m$ et à coefficients holomorphes au voisinage de l'origine de $\mathbf{C}^{n+1}$ s'écrira

$$a(x, D) = \sum_{|\alpha| \leq m} a_\alpha(x) D^\alpha$$

où les coefficients a_α sont supposés appartenir à l'espace $\mathbf{C}\{x\}$; le symbole principal ou polynôme caractéristique de $a(x, D)$, en tant qu'opérateur d'ordre m, est défini par

$$g(x, \xi) = \sum_{|\alpha| = m} a_\alpha(x)\xi^\alpha, \quad \xi^\alpha = \xi_0^{\alpha_0} \times \cdots \times \xi_n^{\alpha_n};$$

il s'agit d'un polynôme homogène de degré m à coefficients dans l'espace $\mathbf{C}\{x\}$. On suppose l'hyperplan $S : x_0 = 0$ non caractéristique à l'origine. Ceci signifie que $g(0; 1, 0, \ldots, 0) \neq 0$, c'est-à-dire $a_{(m,0,\ldots,0)}(0) \neq 0$; par division par la fonction $a_{(m,0,\ldots,0)}$, on peut supposer que le coefficient de D_0^m est égal à 1. Nous noterons Ω_0 un voisinage ouvert de l'origine de $\mathbf{C}^{n+1}$ tel que toutes les fonctions a_α soient définies et holomorphes sur Ω_0.

Soit a un point de $S \cap \Omega_0$, on se donne une fonction holomorphe v au voisinage de ce point et des fonctions holomorphes w_h, $0 \leq h < m$, dans un voisinage de a relativement à S. On considère alors le problème de Cauchy

$$\begin{cases} a(x, D)u(x) = v(x), \\ D_0^h u(x) = w_h(x') \quad \text{pour } x_0 = 0, \, 0 \leq h < m, \end{cases} \tag{1}$$

où $x' = (x_1, \ldots, x_n)$. D'après le théorème de Cauchy–Kowalevski, ce problème admet une unique solution u holomorphe au voisinage de a. On se propose de déterminer, sous certaines hypothèses, les singularités de u connaissant celles des données v et w_h. Nous allons d'abord expliciter ces hypothèses.

En ce qui concerne l'opérateur $a(x, D)$, nous supposerons qu'il s'agit d'un opérateur à caractéristiques multiples de multiplicité constante: ceci signifie qu'il existe des fonctions $(x, \xi') \mapsto \lambda_i(x, \xi')$, $1 \leq i \leq d$, holomorphes au voisinage du point $\overline{x} = 0$, $\overline{\xi}' = (1, 0, \ldots, 0)$ et des entiers $m_i \geq 1$ telles que

$$g(x, \xi) = \prod_{i=1}^{d} \left(\xi_0 - \lambda_i \left(x, \xi' \right) \right)^{m_i} \text{ pour } \left(x, \xi' \right) \text{ voisin de } \left(\overline{x}, \overline{\xi}' \right) \quad (2)$$

et en posant $\lambda_i = \lambda_i(\overline{x}, \overline{\xi}')$,

$$\lambda_i \neq \lambda_j \quad \text{si } i \neq j. \quad (3)$$

On peut alors résoudre le problème de Cauchy non-linéaire du premier ordre

$$\begin{cases} D_0 k_i(x) = \lambda_i \left(x, D'k_i(x) \right), \\ k_i(x) = x_1 \end{cases} \quad \text{pour } x_0 = 0, \quad (4)$$

où $D'k_i(x) = (D_1 k_i(x), \ldots, D_n k_i(x))$. Ce problème admet une unique solution holomorphe au voisinage de l'origine. On notera que $Dk_i(0) = (\lambda_i, 1, 0, \ldots, 0)$; on peut donc supposer les fonctions k_i définies et holomorphes sur Ω_0 et $Dk_i(x) \neq 0$ pour $x \in \Omega_0$. Ceci permet de définir des hypersurfaces $K_i = \{x \in \Omega_0; k_i(x) = 0\}$; si T désigne l'hyperplan $x_0 = x_1 = 0$ de S, ces hypersurfaces vérifient $K_i \cap S = \Omega_0 \cap T$: ce sont donc les hypersurfaces caractéristiques issues de T.

Indiquons ensuite les hypothèses sur les données v et w_h. Étant donné un voisinage ouvert connexe $\Omega \subset \Omega_0$ de l'origine de $\mathbf{C}^{n+1}$ tel que $\Omega \cap S$ soit connexe et un point $a \in \Omega \cap S - T$, nous supposerons que les fonctions w_h, pour $0 \leq h < m$, sont des germes au point a relativement à S qui se prolongent en des fonctions holomorphes sur $\mathcal{R}(\Omega \cap S - T)$ (on notera que $\Omega \cap S - T$ est un ouvert connexe de S): nous dirons que les données de Cauchy sont ramifiées autour de T. Quant à v, nous supposerons que v est un germe au point a se prolongeant en une fonction holomorphe sur $\mathcal{R}(\Omega - \bigcup_{i=1}^{d} K_i)$: nous dirons que v est ramifié autour de $\bigcup_{i=1}^{d} K_i$. D'après le théorème de Cauchy–Kowalevski, le problème de Cauchy (1) définit un germe u au point a et on se propose de démontrer que ce germe se ramifie autour de la réunion des caractéristiques K_i.

Énonçons précisément le théorème dû à E. Leichtnam [4].

Théorème 1.1. *Soit $\Omega \subset \Omega_0$ un voisinage ouvert connexe de l'origine de $\mathbf{C}^{n+1}$ tel que $\Omega \cap S$ soit connexe, il existe un voisinage ouvert connexe $\Omega' \subset \Omega$ de* *l'origine de $\mathbf{C}^{n+1}$ tel que: soient $a \in \Omega' \cap S - T$, $(w_h)_{0 \le h < m}$ et v des germes au point a se prolongeant en des fonctions holomorphes sur $\mathcal{R}(\Omega \cap S - T)$ et $\mathcal{R}(\Omega - \bigcup_{i=1}^{d} K_i)$ respectivement, alors le germe au point a, solution du problème de Cauchy (1), se prolonge en une fonction holomorphe sur $\mathcal{R}(\Omega' - \bigcup_{i=1}^{d} K_i)$.*

La démonstation de ce théorème repose sur la construction de solutions ramifiées autour de chacune des hypersurfaces K_i. Cette construction (voir [7]) se réduit à la résolution d'un problème intégro-différentiel.

2. Un problème de Cauchy intégro-différentiel

Outre l'espace $\mathbf{C}^{n+1}$ de la variable x, on considère l'espace $\mathbf{C}^2$ dont la variable est notée $t = (t_0, t_1)$. Les dérivations en t sont notées D_{t_0} et D_{t_1}. On définit les primitives en t de la façon suivante. On se donne un point $b = (b_0, b_1) \in \mathbf{C}^2$, un ouvert Ω de $\mathbf{C}^{n+1}$ et on considère une fonction holomorphe $u : (t, x) \mapsto u(t, x)$ au voisinage de $\{b\} \times \Omega$ et à valeurs dans un espace de Banach complexe F. On pose

$$D_{t_0}^{-1} u(t, x) = \int_{b_0}^{t_0} u(\tau, t_1, x) \, d\tau \qquad (5)$$

et

$$D_{t_1}^{-1} u(t, x) = \int_{b_1}^{t_1} u(t_0, \tau, x) \, d\tau \qquad (6)$$

pour t voisin de b et $x \in \Omega$, où la première intégrale s'effectue sur le segment joignant b_0 et t_0 et la seconde sur le segment joignant b_1 et t_1. Ceci permet de définir des opérateurs $D_{t_i}^{l_i}$ pour tout $l_i \in \mathbf{Z}$ et, pour tout $l = (l_0, l_1) \in \mathbf{Z}^2$; on pose $D_t^l = D_{t_0}^{l_0} \circ D_{t_1}^{l_1}$: on obtient ainsi des fonctions $D_t^l u$ holomorphes au voisinage de $\{b\} \times \Omega$.

On considère alors le problème de Cauchy intégro-différentiel

$$\begin{cases} \left(D_0^m - A_m(x, D)\right) u(t, x) = \displaystyle\sum_{l \in \mathcal{L}} A_l^m(x, D) D_t^{-l} u(t, x) + w_m(t, x), \\ \left(D_0^h - A_h(x, D)\right) u(t, x) = \displaystyle\sum_{l \in \mathcal{L}} A_l^h(x, D) D_t^{-l} u(t, x) + w_h(t, x) \\ \text{pour } x_0 = 0, \ 0 \le h < m. \end{cases} \qquad (7)$$

Les hypothèses sont les suivantes. On se donne deux espaces de Banach complexes E, F et une application bilinéaire continue $(a, u) \mapsto au$ de $E \times F$ dans F de norme ≤ 1. Les opérateurs $A_h(x, D)$ et $A_l^h(x, D)$ sont des opérateurs différentiels linéaires dont les coefficients appartiennent à l'espace $\mathcal{H}(\Omega_1; E)$

où Ω_1 est un voisinage ouvert de l'origine de $\mathbf{C}^{n+1}$. L'ensemble $\mathscr{L}$ est une partie finie de $\mathbf{Z}^2$ telle que

$$\mathscr{L} \subset \left\{ l = (l_0, l_1) \in \mathbf{Z}^2; l \neq (0, 0) \text{ et } |l| = l_0 + l_1 \geq 0 \text{ lorsque } l_0 < 0 \right\}. \quad (8)$$

On suppose enfin que les ordres des opérateurs vérifient pour tout $0 \leq h \leq m$

$$\text{ordre } A_h \leq h, \qquad \text{ordre}_{x_0} A_h < h, \qquad\qquad\qquad (9)$$

$$\text{ordre } A_l^h \leq \begin{cases} h + |l| - 1 & \text{si } l_1 < 0, \\ h + |l| & \text{si } l_1 \geq 0. \end{cases} \qquad\qquad (10)$$

On se donne en outre un ouvert connexe $O = O_0 \times O_1$ de $\mathbf{C}^2$ et on choisit un point $b = (b_0, b_1) \in O$. On a alors le

Théorème 2.1. *Soit Ω un voisinage ouvert de l'origine de $\mathbf{C}^{n+1}$, il existe un voisinage ouvert $\Omega' \subset \Omega \cap \Omega_1$ de l'origine de $\mathbf{C}^{n+1}$ et un réel $\delta > 0$ tels que: soit $c_0 > 0$ et soient $w_h : \mathscr{R}(O) \times \Omega \to F$ des fonctions holomorphes telles que, pour tout compact $K \subset \mathscr{R}(O) \times \Omega$, il existe une constante $c_K \geq 0$ telle que*

$$\left\| D_t^p w_h(t, x) \right\| \leq c_0^{p_0} c_K^{p_1+1} p! \text{ pour tout } p$$

$$= (p_0, p_1) \in \mathbf{N}^2 \text{ et tout } (t, x) \in K, \qquad (11)$$

alors si le diamètre δ_0 de O_0 est $\leq \delta$ et si le diamètre δ_1 de O_1 est $\leq \delta \min(c_0^{-1}, \delta)$, le problème (7) admet une unique solution holomorphe $u : \mathscr{R}(O) \times \Omega' \to F$.

Ce théorème contient le théorème de [1] concernant le problème de Cauchy ramifié lorsque v est de la forme $v = \sum_i v_i$ où v_i est ramifié autour de K_i. Il contient également le théorème d'Hamada–Takeuchi [2] concernant des données de Cauchy ramifiées dans un domaine. En effet, supposons que $(l_0, l_1) \in \mathscr{L}$ implique $l_0 = 0$ et que les fonctions w_h ne dépendent que de t_1 et de x: il en est alors de même pour la solution u. En notant t la variable t_1, u est donc la solution du problème (7) où $\mathscr{L}$ est une partie finie de $\mathbf{Z}^*$ et où les opérateurs A_h vérifient (9) et les opérateurs A_l^h

$$\text{ordre } A_l^h \leq \begin{cases} h + l - 1 & \text{si } l < 0, \\ h + l & \text{si } l > 0. \end{cases} \qquad\qquad (12)$$

La propriété (11) étant vérifiée avec $c_0 = 1$, par exemple, on obtient ainsi la

Proposition. *Soit Ω un voisinage ouvert de l'origine de $\mathbf{C}^{n+1}$, il existe un voisinage ouvert $\Omega' \subset \Omega \cap \Omega_1$ de l'origine de $\mathbf{C}^{n+1}$ et un réel $\delta > 0$ tels que: soient O un ouvert connexe de $\mathbf{C}$ de diamètre $\leq \delta$, $w_h : \mathscr{R}(O) \times \Omega \to F$, $0 \leq h \leq m$, des fonctions holomorphes, alors le problème (7) admet une unique solution holomorphe $u : \mathscr{R}(O) \times \Omega' \to F$.*

Le théorème 2.1 permet enfin de construire des solutions ramifiées autour de l'une des hypersurfaces K_i et, plus précisément, on a la

Proposition. *Il existe un voisinage ouvert V de $0 \in \mathbf{C}^{n+1}$, $r > 0$ et, pour tout $\alpha > 0$, une fonction holomorphe sur le revêtement universel de*

$$\{x \in V; \alpha < |x_0| \ et \ 0 < |k_i(x)| < r|x_0|\}$$

tel que $a(x, D)u(x) = v(x)$.

En utilisant ce théorème, des techniques de prolongement analytique permettent alors d'établir le théorème 1.1.

Références

[1]   Hamada, Y., Leray, J. et Wagschal, C., Systèmes d'équations aux dérivées partielles à caractéristiques multiples: problème de Cauchy ramifié; hyperbolicité partielle, *J. Math. Pures Appl.* **55** (1976), 297–352.

[2]   Hamada, Y. et Takeuchi, A., Le domaine d'existence et le prolongement analytique des solutions des problèmes de Goursat et de Cauchy à données singulières, *C. R. Acad. Sci. Paris* **295** (1982), 377–380.

[3]   _____, *Le domaine d'existence et le prolongement analytique des solutions des problèmes de Goursat et de Cauchy à données singulières*, Taniguchi Symp. HERT Kataka (1984), 51–62.

[4]   Leichtnam, E., Le problème de Cauchy ramifié, *Ann. scient. Éc. Norm. Sup.*, t. **23** (1990), 369–443.

[5]   Pongérard, P., *Problème de Cauchy holomorphe*, Thèse, Université Paul Sabatier, Toulouse, 1996.

[6]   Pongérard, P. et Wagschal, C., Problème de Cauchy dans des espaces de fonctions entières, *J. Math. Pures Appl.* **75** (1996), 409–418.

[7]   _____, Ramification non abélienne, *J. Math. Pures Appl.* **77** (1998), 51–88.

[8]   Wagschal, C., Diverses formulations du problème de Cauchy pour un système d'équations aux dérivées partielles, *J. Math. Pures Appl.* **53** (1974) 51–70.

[9]   _____, Une généralisation du problème de Goursat pour des systèmes d'équations intégro-différentielles holomorphes ou partiellement holomorphes, *J. Math. Pures Appl.* **53** (1974), 99–132.

[10] _____, Sur le problème de Cauchy ramifié, *J. Math. Pures Appl.* **53** (1974), 147–164.

[11] _____, Le problème de Goursat non linéaire, *J. Math. Pures Appl.* **58** (1979), 309–337.

[12] _____, Problème de Cauchy ramifié à caractéristiques multiples holomorphes de multiplicité variable, *J. Math. Pures Appl.* **62** (1983), 99–127.

PART II

SYMPLECTIC MECHANICS AND GEOMETRY

EXTENSION DU CALCUL DIFFÉRENTIEL ET APPLICATION À LA THÉORIE DES GROUPES DE LIE EN DIMENSION INFINIE

PIERRE DAZORD

Contents

1   Introduction 127
2   Rappels: Notion de difféologie 128
3   Revêtements. Revêtement universel 131
4   Espace tangent d'un espace difféologique 133
5   Exemples d'espaces tangents difféologiques 135
6   Sous-espaces tangents. Groupes de Lie 138
        6.1   Notion de groupe de Lie 138
7   Appendice. Variétés non compactes et champs complets 139

M. de Gosson (ed.), Jean Leray '99 Conference Proceedings, 125-141.

Résumé. *L'efficacité de l'analyse fonctionnelle a fait oublier l'origine géométrique de nombreux problèmes. C'est à un retour aux sources, en quelque sorte, que l'on s'attelle ici, grâce aux «difféologies» de J.M. Souriau. On est ainsi amené à construire le revêtement universel et l'espace tangent d'un espace difféologique, ce qui conduit dans la dernière partie à une rapide esquisse de la théorie des groupes de Lie. Dans un appendice on prouve enfin un résultat qui ne semble pas connu: tout champ de vecteurs sur une variété C^∞ paracompacte, de dimension finie, est somme d'au plus deux champs complets.*

Abstract. *The great efficiency of functional analysis has led to forget the geometrical origin of many problems. In this paper, we go back to the origins, with the help of J.M. Souriau's diffeologies, and construct, in this general setting, universal coverings and tangent bundles. Then we show, rapidly, the importance of this new tool to treat the theory of Lie groups in infinite dimension. In an appendix we prove a result which seems to be unknown: on a finite dimensional C^∞ paracompact manifold, any vector field is the sum of at most two complete vector fields.*

1. Introduction

En 1976, l'Université de Lyon eut l'honneur d'accueillir les premiers cours décentralisés du Collège de France: pendant 15 jours alternèrent des conférences de Jean Leray consacrées au Calcul Lagrangien et d'André Lichnérowicz consacrées aux produits-étoiles. J'avais été l'élève d'André Lichnérowicz et c'est à cette occasion que je fis la connaissance de Jean Leray, avec qui je restais en relation plusieurs années au cours desquelles je répondais à quelques questions posées dans son cours: dans la situation la plus générale—et sans aucune condition de transversalité—je donnai une construction de cet «indice des chemins» dont parlait Maslov, qui généralisait l'indice de Morse des géodésiques, et qui jouait un rôle central dans le calcul Lagrangien. Ces résultats furent exposés dans plusieurs articles et notamment dans «Invariants homotopiques attachés aux fibrés symplectiques» (*Ann. Inst. Fourier. Grenoble. 29(2), 1979, 25–78*) et «Sur la géométrie des sous-fibrés et des feuilletages Lagrangiens» (*Ann. Scient. E.N.S., 4e série, 13, 1981, 465–480*). Cet indice des chemins permit à G. Patissier et moi-même d'établir la condition de quantification asymptotique

$$\frac{1}{2\pi h}[\sigma] - \frac{1}{2}C_1(M) \in H^2(M, \mathbb{Z})$$

conjecturée par de nombreux physiciens.

Je suis donc particulièrement reconnaissant aux organisateurs, M. de Gosson et J. Vaillant, de me donner l'occasion de rendre hommage à Jean Leray.

Dans le prolongement de l'intérêt que, sous l'influence d'André Lichnérow-icz, j'ai porté aux aspects géométriques des problèmes issus de la Physique mathématique, l'exposé qui suit est consacré à la géométrie des Groupes de Lie de dimension infinie. A la différence de nombreux auteurs (Birkhoff, Milnor, Omori, Maeda,...), ce n'est pas à l'analyse fonctionnelle mais à la géométrie différentielle que je fais appel pour aborder ces problèmes, renouant ainsi avec une tradition remontant, au moins, à Sophus Lie.

C'est ainsi que les quatre premiers paragraphes sont consacrés d'une part à un bref rappel de la théorie des difféologies de J.M. Souriau [11] (a), [11] (b) et de son école, P. Donato [4] et P. Iglesias [6], et d'autre part à des constructions nouvelles: revêtements galoisiens et revêtement universel, utilisant la notion de chemin difféologique par morceaux, (§2), fibré tangent (§3). En particulier tout «champ de vecteurs tangent» est une dérivation de l'algèbre des fonctions dif-féologiques réelles, la réciproque étant vraie pour une variété C^∞ de dimension finie. Après les exemples de fibrés tangents au §4, le dernier paragraphe est consacré à une rapide esquisse de la théorie des groupes de Lie en dimension infinie développée en [2]a, [2]b et [2]c.

Enfin, un appendice est consacré à deux résultats (utilisés au §4) qui ne semblent pas connus et dont le plus spectaculaire est que sur toute variété para-compacte, tout champ de vecteurs est somme d'au plus deux champs complets.

Cet article a bénéficié de l'ambiance studieuse et féconde du colloque Leray. Je remercie également H. Maillot pour d'utiles conversations et Mme J. Lefranc qui a assuré la frappe du manuscrit.

Nota bene. On a limité la bibliographie à ce qui est strictement nécessaire aux développements nouveaux de cet article. Pour les théories (groupoïdes de Lie, algébroïdes de Lie, etc.) auxquelles il est fait allusion, on se reportera aux bibliographies des articles cités notamment [2]a,b,c.

2. Rappels: Notion de difféologie

Une *difféologie* sur un ensemble non vide M [4], [6] et [11] est la donnée d'un ensemble $\mathscr{D}_M = \bigcup_{p \in \mathbb{N}} \mathscr{D}(M, p)$ ($\mathscr{D}_M$ est l'ensemble des *plaques*, $\mathscr{D}(M, p)$ l'ensemble des *p-plaques*) où $\mathscr{D}(M, p)$ est un sous-ensemble non vide de l'ensemble des applications d'ouverts non vides de $\mathbb{R}^p$ dans M, verifiant les axiomes suivants:

AD 1 Tout point de M appartient à l'image d'au moins une plaque de M.

AD 2 Si $f : U \to M$ est une *p*-plaque et $V \subset U$ un ouvert non vide, $f|_V$ est une *p*-plaque.

AD 3 Si $(U_i)_{i \in I}$ est un recouvrement par des ouverts non vides de l'ouvert U, $f : U \to M$ est une *p*-plaque si et seulement si pour tout $i \in I$, $f_i = f|_{U_i} \in \mathscr{D}(M, p)$.

AD 4 Si $f : U \to M$ est une p-plaque et si φ est une application C^∞ d'un ouvert V non vide de $\mathbb{R}^q$ dans U, $f \circ \varphi \in \mathcal{D}(M, q)$.

Un *espace difféologique* est un couple $(M, \mathcal{D}_M)$. Il est *connexe* si pour tout couple $(x, y) \in M^2$, il existe une 1-plaque g de source un intervalle I telle que $g(I)$ contienne x et y. Un espace difféologique est *discret* si les seules plaques à sources connexes sont les constantes.

Les difféologies sur un même espace M sont ordonnées par inclusion: $\mathcal{D}_M^1$ est plus fine que $\mathcal{D}_M^2$ si $\mathcal{D}_M^1 \subset \mathcal{D}_M^2$. Il existe une *difféologie sur M plus fine* que toutes les difféologies, la *difféologie discrète* dont les plaques de sources connexes sont les constantes et une *difféologie sur M moins fine* que toutes les difféologies la *difféologie grossière* dont les plaques sont n'importe quelles applications d'ouverts d'espaces numériques dans M. Plus généralement si $(\mathcal{D}_i)_{i \in I}$ est une famille de difféologie sur M, il existe une borne inférieure et une borne supérieure de $(\mathcal{D}_i)_{i \in I}$.

Une *application difféologique* $f : M \to N$ où M et N sont difféologiques est une application telle que $f \circ \mathcal{D}_M \subset \mathcal{D}_N$. On note $\mathcal{D}(N, M)$ l'ensemble des applications difféologiques de M dans N.

Si $(M_i, \mathcal{D}_{M_i})$ est une famille d'espaces difféologiques, on appelle *espace difféologique produit*, $\prod_{i \in I} M_i$ muni de la borne supérieure des difféologies rendant chaque projection $p_i : \prod_{i \in I} M_i \to M_i$ difféologique.

Les espaces et applications difféologiques constituent une catégorie: la *catégorie difféologique*, qui est une catégorie à produit quelconque.

Toute variété C^∞ de dimension finie M est canoniquement munie d'une difféologie dont les plaques sont les applications C^∞ d'ouverts d'espaces numériques dans M. La classe des variétés C^∞ de dimension finie (resp. et séparés) et des applications C^∞ constitue une sous-catégorie pleine de la catégorie difféologique.

Plus généralement, rappelons que M est une *variété C^∞ de Sataké* de dimension n [10] si M est un espace topologique (séparé ou non) possédant un recouvrement ouvert $(U_i)_{i \in I}$ (appelé *atlas*) tel que

(i) Pour tout i, il existe un ouvert V_i de $\mathbb{R}^n$ et un groupe fini G_i de difféomorphismes C^∞ de V_i tel que U_i soit le quotient de V_i par G_i.

(ii) Si $U_i \cap U_j \neq \emptyset$ et si l'on note $\pi_i : V_i \to U_i$ l'application quotient, l'application identique de $U_i \cap U_j$ se relève en une application C^∞ de $V_{ij} = \pi_i^{-1}(U_i \cap U_j)$ dans $V_{ji} = \pi_j^{-1}(U_j \cap U_i)$ ψ_{ji} et une application C^∞ de $\pi_j^{-1}(U_j \cap U_i)$ dans $\pi_i^{-1}(U_i \cap U_j)$, ψ_{ij}, telles que $\psi_{ji} \circ \psi_{ij} \in G_j|_{V_{ji}}$ et $\psi_{ij} \circ \psi_{ji} \in G_i|_{V_{ij}}$.

(iii) Si $U_i \cap U_j \cap U_k \neq \emptyset$, avec des notations évidentes $\widehat{\psi}_{ij} \circ \widehat{\psi}_{jk} \circ \widehat{\psi}_{ki} \in G_i$ où $\widehat{\psi}_{ij}$ désigne l'adéquate restriction de ψ_{ij}, etc.

Si M et N sont deux variétés de Sataké d'atlas respectifs (U_i^M), (U_j^N), une application $f : M \to N$ est un *morphisme de Sataké* si pour tout $x_0 \in M$ il existe un voisinage W_{x_0} de x_0 contenu dans un U_i^M et un ouvert U_j^N tels que $f(W_{x_0}) \subset U_j^N$, et un relèvement C^∞ de $f|_{W_{x_0}}$ de $\pi_i^{-1} W_{x_0}$ dans V_j^N.

Par exemple $[0, 1]$ (ou plus généralement tout segment) est une variété de Sataké, $U_1 = [0, 1[$, $U_2 =]0, 1]$, $V_1 =]-1, 1[$, $V_2 =]0, 2[$, G_1 (resp., G_2) est le groupe engendré par la symétrie autour de 0 (resp., 1).

Les variétés et morphismes de Sataké forment une *sous-catégorie pleine de la catégorie difféologique*: si M est de Sataké et (U_i^M) un atlas, $f : W \to M$ est une p-plaque s'il existe un recouvrement de W par des ouverts non vides W_k tel que pour tout k il exite i tel que $f(W_k) \subset U_i^M$ et une application C^∞ $g_k : W_k \to V_i^M$ telle que $\pi_i \circ g_k = f|_{W_k}$.

Si M et N sont deux espaces difféologiques, on munit $\mathscr{D}(N, M)$ de la *difféologie fonctionnelle* dont les plaques $f : U \to \mathscr{D}(N, M)$ sont les applications telles que l'application $\tilde{f} : U \times M \to N$ définie par $\tilde{f}(u, x) = f(u)(x)$ est difféologique quand on munit $U \times M$ de la difféologie produit.

Si $(M, \mathscr{D}_M)$ est un espace difféologique et f une application de M dans l'ensemble N, l'ensemble des difféologies sur N pour lesquelles f est difféologique possède une borne inférieure appelée *difféologie image* de $\mathscr{D}_M$ par f. Si $f(M) = N$ on dit que f est une *surmersion*. En particulier pour toute relation d'équivalence $\mathscr{D}$ dans M, sur $M \mid \mathscr{R}$, la difféologie image par $\pi : M \to M \mid \mathscr{R}$ s'appelle la *difféologie quotient*.

Si $(N, \mathscr{D}_N)$ est un espace difféologique et $f : M \to N$ une application de l'ensemble M dans N, la difféologie sur M borne supérieure des difféologies rendant f difféologique s'appelle la difféologie *image réciproque de $\mathscr{D}_N$ par f*.

Si A est une partie de N, la difféologie image réciproque par l'inclusion s'appelle la difféologie induite $\mathscr{D}_A$ et on dit que $(A, \mathscr{D}_A)$ est un *sous-espace difféologique* de $(N, \mathscr{D}_N)$.

$f : (M, \mathscr{D}_M) \to (N, \mathscr{D}_N)$ est une *immersion* si $\mathscr{D}_M$ est l'image réciproque de $\mathscr{D}_N$ par f.

Si $\mathscr{S}$ est une structure algébrique sur M, $(M, \mathscr{D}_M, \mathscr{S})$ est un espace $\mathscr{S}$-*difféologique* si les lois de $\mathscr{S}$ sont difféologiques.

Si $(M_i, \mathscr{D}_{M_i})$ $(i = 1, 2, 3)$ sont trois espaces difféologiques, la composition

$$\mathscr{D}(M_3, M_2) \times \mathscr{D}(M_2, M_1) \longrightarrow \mathscr{D}(M_3, M_1)$$

est difféologique. Il en résulte que si $(M, \mathscr{D}_M)$ est difféologique et si l'on note $\mathscr{D}(M)$ l'espace des bijections de M difféologiques ainsi que leurs inverses, $\mathscr{D}(M)$ muni de la difféologie induite par $\mathscr{D}(M, M)$ est un *groupe difféologique*.

Un *fibré difféologique* $\pi : E \to M$ est un triple formé de deux espaces difféologiques $(E, \mathscr{D}_M)$ et $(M, \mathscr{D}_M)$, et d'une surmersion π. Si de plus pour tout x, $E_x = \pi^{-1}(x)$ est un espace vectoriel difféologique (avec la difféologie induite), on dit que $\pi : E \to M$ est un fibré vectoriel. Un morphisme de fibrés difféologiques (resp., vectoriels) $\pi_i : E_i \to M_i$ est un morphisme fibré qui est une application difféologique (resp., linéaire sur les fibrés). Les fibrés et les morphismes fibrés difféologiques (resp., et vectoriels) forment une sous-catégorie de la catégorie difféologique. Si $\pi : E \to M$ est un fibré vectoriel difféologique, $\mathscr{A}(E)$ espace des sections difféologiques muni de la difféologie induite par $\mathscr{D}(E, M)$ est un espace vectoriel difféologique. Si $(E, [\,,\,], \rho)$ est un algébroïde de Lie [9] et [2]a, $\mathscr{A}(E)$ est une algèbre de Lie difféologique. De même $\mathscr{A}_c(E) = \{ s \in \mathscr{A}(E) \mid \rho \circ s$ est à support compact$\}$ est une algèbre de Lie difféologique munie de la difféologie à contrôle compact [2]a: $g \in \mathscr{D}(\mathscr{A}_c(E), p)$ si $g \in \mathscr{D}(\mathscr{A}(E), p)$ et si $g : U \to \mathscr{A}_c(E)$ pour tout $u_0 \in U$ il existe un voisinage ouvert V de u_0 dans U et un compact K de M tel que pour tout $u \in V$, $\rho \circ g(u)$ a son support dans K.

3. Revêtements. Revêtement universel

Soit $(M, \mathscr{D}_M)$ un espace difféologique. Une application $\gamma : [0, 1] \to M$ est un *chemin difféologique par morceaux* s'il existe une subdivision fini $0 = t_0 < t_1 < \cdots < t_k = 1$ de $[0, 1]$ telle que pour tout $i \in \{0, k - 1\}$,

$$\gamma_i = \gamma|_{[t_i, t_{i+1}]} \in \mathscr{D}(M, [t_i, t_{i+1}])$$

où $[t_i, t_{i+1}]$ est muni de sa difféologie d'espace de Sataké. On munit l'espace $\mathscr{C}_b(M)$ des chemins difféologiques par morceaux de M de la difféologie dont les plaques $g : U \to \mathscr{C}_b(M)$ sont les applications telles que pour tout $u_0 \in U$ il existe une partition finie $0 = t_0 < t_1 < \cdots < t_k = 1$ de $[0, 1]$ et un voisinage V de u_0 dans U tel que

$$\widetilde{g}_i : [t_i, t_{i+1}] \times V \longrightarrow M$$
$$(t, u) \longrightarrow \widetilde{g}_i(t, u) = g(u)(t)$$

est difféologique pour tout $i \in \{0, k - 1\}$.

L'inclusion d'intervalles $[a_1, b_1] \subset [a_2, b_2]$ où $a_2 \leq a_1 < b_1 \leq b_2$ étant difféologique, on peut toujours remplacer dans les définitions précédentes une subdivision par subdivision plus fine.

Ceci permet de montrer que, dans $\mathscr{C}_b(M)$, la relation $\gamma_0 \sim \gamma_1$ (*homotopie difféologique à extrêmités fixées*) définie ci-dessous est une relation d'équivalence:

$\gamma_0 \sim \gamma_1$ si et seulement si $\exists H : [0, 1]^2 \to M$, $(t, \tau) \to H_\tau(t) = H(t, \tau)$ telle que

(i) $H_0 = \gamma_0$; $H_1 = \gamma_1$;

(ii) $H_\tau(0) = \gamma_0(0) = \gamma_1(0)$; $H_\tau(1) = \gamma_0(1) = \gamma_1(1)$;

(iii) il existe deux subdivisions finies $0 = t_0 < t_1 < \cdots < t_k = 1$ et $0 = \tau_0 < \tau_1 < \cdots < \tau_\ell = 1$ telles que $H_{ij} = H|_{[t_i, t_{i+1}] \times [\tau_j, \tau_{j+1}]}$ soit pour tout $(i, j) \in \{0, k-1\} \times \{0, \ell-1\}$ difféologique de $[t_i, t_{i+1}] \times [\tau_j, \tau_{j+1}]$ dans M.

Appelons *groupoïde difféologique* un groupoïde ensembliste [2]a, $\Gamma \underset{\beta}{\overset{\alpha}{\rightrightarrows}} \Gamma_0$, $\Gamma_0 \subset \Gamma$ tel que Γ est muni d'une difféologie, Γ_0 de la difféologie induite (Γ_0 est l'espace des unités) et:

(i) α et β sont des surmersions de Γ dans Γ_0;

(ii) $x \to x^{-1}$ est un isomorphisme difféologique;

(iii) si l'on note $\Gamma_2 = \{(x, y) \mid \alpha(x) = \beta(y)\}$ l'espace des couples composables muni de la difféologie induite par $\Gamma \times \Gamma$, le produit

$$\Gamma_2 \longrightarrow \Gamma(x, y) \longrightarrow x \cdot y$$

est difféologique.

Avec cette définition $\sqcap_1 M = \mathscr{C}_b(M)/\sim$ est un *groupoïde difféologique*, dont l'espace des unités est l'espace des chemins constants identifié à M, $\alpha : \sqcap_1 M \to M$ (source) est le quotient de $\gamma \to \gamma(0)$ et $\beta : \sqcap_1 M \to M$ (but) est le quotient de $\gamma \to \gamma(1)$.

$\sqcap_1 M$ s'appelle le *groupoïde fondamental* de M.

Un fibré difféologique $\pi : E \to M$ est un *fibré G-principal* où G est un groupe difféologique si E est un fibré (ensembliste) G-principal et si:

(i) l'action à droite $E \times G \to G$ est difféologique;

(ii) pour tout $x \in E$ et tout $z \in E_x$, la bijection

$$G \longrightarrow E_x : g \longrightarrow zg$$

est un isomorphisme difféologique.

Un *revêtement galoisien* $\pi : \widehat{M} \to M$ de M difféologique connexe est un fibré principal difféologique à groupe structural discret tel que, si $\widehat{M} \boxtimes \mathscr{C}_b(M)$ désigne l'espace des couples $(\widehat{x}, \gamma)$ vérifiant $\pi(\widehat{x}) = \gamma(0)$, il existe une application difféologique ψ (relèvement des chemins)

$$\psi : \widetilde{M} \boxtimes \mathscr{C}_b(M) \longrightarrow \mathscr{C}_b(\widehat{M})$$
$$(\widehat{x}, \gamma) \longrightarrow \widetilde{\gamma}$$

telle que $\pi \circ \widehat{\gamma} = \gamma$.

Le groupe structural étant discret, tout chemin $\widehat{\gamma}_1$ tel que $\pi \circ \widehat{\gamma}_1 = \gamma$ s'écrit $\widehat{\gamma}_1 = \psi(\widehat{\gamma}_1(0), \gamma)$.

Théorème 3.1. *Les fibrés principaux $\alpha^{-1}(x_0) \to M$ et $\beta^{-1}(x_0) \to M$ sont tous isomorphes et sont des revêtements galoisiens de M.*

Si $\widehat{M} \xrightarrow{\pi} M$ est un revêtement galoisien il existe une surmersion $\chi : \alpha^{-1}(x_0) \to \widehat{M}$ qui est un morphisme de fibrés principaux.

Ceci conduit à appeler *revêtement universel* de M, l'un quelconque des fibrés $\alpha^{-1}(x_0)$ ou $\beta^{-1}(x_0)$.

4. Espace tangent d'un espace difféologique

Si $(M, \mathscr{D}_M)$ est un espace difféologique, on note $\mathscr{D}_0(M, 1)$ l'espace des 1-plaques de M définies sur un voisinage ouvert de 0. On munit $\mathscr{D}_0(M, 1)$ de la difféologie dont les p-plaques

$$\gamma : U \longrightarrow \mathscr{D}_0(M, 1)$$

sont les applications telles qu'il existe un voisinage ouvert Ω de $U \times 0$ dans $U \times$ et une $(p+1)$-plaque $\widetilde{\gamma}$ de $M : \widetilde{\gamma} : \Omega \to M$ définie par $\gamma(u)(t) = \widetilde{\gamma}(u, t)$ et telle que, pour tout u, $\gamma(u)$ soit la 1-plaque de source $p^{-1}(u)$ où $p : \Omega \to U$ est la première projection.

On introduit dans $\mathscr{D}_0(M, 1)$ la relation d'équivalence: $\gamma_1 \sim \gamma_2$ si et seulement si pour tout application difféologie $f : M \to \mathbb{R}$, $j_0^1 f \circ \gamma_1 = j_0^1 f \circ \gamma_2$ où $j_0^1 f \circ \gamma_i$ désigne le 1-jet en 0 de l'application C^∞ $f \circ \gamma_i$.

L'espace tangent à M est $TM = \mathscr{D}_0(M, 1)/ \sim$, c'est un espace fibré difféologique noté $\pi : TM \to M$ dont la fibre en x est le quotient de $\mathscr{D}_{x,0}(M, 1)$, où $\mathscr{D}_{x,0}(M, 1)$ désigne le sous-espace difféologique des 1-plaques $\gamma \in \mathscr{D}_0(M, 1)$ telles que $\gamma(0) = x$.

C'est un espace fibré $\mathbb{R}$-homogène c-á-d, muni d'une action à droite par $\mathbb{R}$ qui est le quotient de l'action à droite de $\mathbb{R}$ dans $\mathscr{D}_0(M, 1)$ définie par

$$\gamma : I \longrightarrow M, (\gamma, \tau) \longrightarrow \tau \cdot \gamma : t \longrightarrow \gamma(\tau t),$$
$$t \in \tau^{-1} I \quad \text{si } \tau \neq 0, \quad t \in \mathbb{R} \quad \text{si } \tau = 0.$$

Pour tout $x \in M$, on note $T_x M$ l'espace tangent en x et O_x l'image dans $T_x M$ des 1-plaques constantes au voisinage de 0. Pour tout $X \in T_x M$, $O \cdot X = O_x$.

Notation. Pour tout chemin $\gamma \in \mathscr{D}_0(M, 1)$ on notera $j_0^1 \gamma$ sa classe (jet du premier ordre de γ en 0) et, si $\gamma(0) = x$, $(d\gamma/dt)(0)$ (vitesse de γ en 0) la classe de γ dans $\mathscr{D}_{x,0}(M, 1)$. Cette notation étend la notation en usage dans la théorie des variétés. (cf infra).

On appelle champ de «vitesses» difféologiques, toute section difféologique de $\pi : TM \to M$, et on note $\mathscr{X}(M)$ leur ensemble.

Pour tout application difféologique $\varphi : M \to N$ l'application $\mathcal{D}_0(M, 1) \to \mathcal{D}_0(N, 1)$ définie par $\gamma \to \varphi \circ \gamma$ définit par passage au quotient une application difféologique fibrée $\mathbb{R}$-homogène $T\varphi : TM \to TN$; on notera $T_x\varphi$ la restriction à T_xM à valeurs dans $T_{\varphi(x)}N$, où T est un foncteur covariant de la catégorie difféologique dans la catégorie des fibrés difféologiques $\mathbb{R}$-homogènes.

En général TM n'est pas un fibré vectoriel, comme le montre l'exemple suivant: $M = D_1 \cup D_2$ où D_1 et D_2 sont deux droites distinctes concourantes en 0. $T_0M = D_1 \cup D_2$. Par contre si G est un groupe difféologique $TG \simeq G \times T_eG$ où T_eG espace tangent en l'élément neutre est un espace vectoriel difféologique.

Si V est un espace vectoriel difféologique, TV est isomorphe à $V \times T_0V$. De plus, l'application $v \to [t \to tv]$ induit une application linéaire de V dans T_0V.

On dira qu'un espace vectoriel difféologique V *est plein* si cette application $V \to T_0V$ est un isomorphisme difféologique.

En général, on ne peut rien spécifier sur cette application. Par exemple si on munit V de la difféologie (d'espace vectoriel) grossière, l'espace tangent est réduit à 0. Par contre si V' désigne le dual difféologique de V c-à-d, l'espace vectoriel difféologique des applications linéaires difféologiques de V dans V' est plein.

Remarque. Si $f \in \mathcal{D}(\mathbb{R}, M)$ et $X \in T_xM$, X est la classe d'un chemin γ et

$$X \cdot f = \frac{d}{dt}f \circ \gamma(t)|_{t=0} = Tf\left(\frac{d\gamma}{dt}(0)\right)$$

par définition ne dépend que de X. Si $X \in \mathcal{X}(M)$ $f \to X \cdot f$ est une application de l'algèbre $\mathcal{D}(\mathbb{R}, M)$ dans elle-même qui est une dérivation difféologique et $\mathcal{X}(M)$ est un sous-ensemble difféologique de Der $\mathcal{D}(\mathbb{R}, M)$.

Il en résulte que si M est une variété C^∞, de dimension n, l'espace tangent construit est l'espace tangent ordinaire, avec sa difféologie canonique de fibré vectoriel C^∞. La définition générale d'espace tangent étend donc la définition usuelle pour les variétés C^∞ de dimension finie. Cependant si M est difféologique, $\mathcal{X}(M)$ est un sous-espace difféologique de Der $\mathcal{D}(\mathbb{R}, M)$ ce qui interdit de définir dans $\mathcal{X}(M)$ le crochet en toute généralité.

De même si M est de Sataké, TM est son espace tangent comme variété de Sataké. Ainsi si $M = [0, 1]$, $T_0M = 0 = T_1M$, et pour $x \in]0, 1[$, $T_xM = \mathbb{R}$.

Si $\pi : \widehat{M} \to M$ est un revêtement galoisien, $T\pi$ induit un isomorphisme, pour tout $\widehat{x} \in \widehat{M}$, de $T_{\widehat{x}}\widehat{M}$ sur T_xM où $x = \pi(\widehat{x})$. En particulier si G est un groupe difféologique, $\widehat{G}$ son revêtement universel, $T_e\widehat{G} = T_eG$ et $T\widehat{G} = \widehat{G} \times T_eG$.

5. Exemples d'espaces tangents difféologiques

Soit $\pi : E \to N$ un fibré vectoriel C^∞ de rang fini sur une variété C^∞ de dimension n. Soit $\mathcal{A}(E)$ l'espace des sections C^∞ de E muni de la difféologie induite par $\mathcal{D}(E, N)$; pour tout espace difféologique M, $\mathcal{D}(\mathcal{A}(E), M)$ est un espace vectoriel difféologique.

Théorème 5.1. $\mathcal{D}(\mathcal{A}(E), M)$ *est plein.*
Autrement dit $T\mathcal{D}(\mathcal{A}(E), M) \simeq \mathcal{D}(\mathcal{A}(E), M) \times \mathcal{D}(\mathcal{A}(E), M)$

Démonstration. Soit $\mathscr{E}_{x,y}$ l'application (difféologique) de $\mathcal{D}(\mathcal{A}(E), M)$ dans l'espace vectoriel de dimension finie E_y définie par

$$s \longrightarrow s(x)(y), \quad x \in M, \ y \in N.$$

On définit une application linéaire difféologique

$$\mathscr{E} : T_0\mathcal{D}(\mathcal{A}(E), M) \longrightarrow \mathcal{D}(\mathcal{A}(E), M)$$
$$X \longrightarrow \left(x \longrightarrow \left(y \longrightarrow T\mathscr{E}_{x,y}X \right) \right)$$

$\mathscr{E}$ est surjective car si $Y \in \mathcal{D}(\mathcal{A}(E), M)$ et si X est la classe de $t \to tY$, $\mathscr{E}X = Y$. D'autre part si $\mathscr{E}X = 0$, où X est la classe de $t \to s(t)$,

$$\frac{d}{dt}s(t)(x, y)|_{t=0} = 0.$$

On pose:

$$\varepsilon(t, x, y) = \int_0^1 s_t'(t\tau, x, y)d\tau.$$

Alors $s(t)(x, y) = t\varepsilon(t, x, y)$, et on définit $\varepsilon(t) \in \mathcal{D}(\mathcal{A}(E), M)$ par $\varepsilon(t)(x)(y) = \varepsilon(t, x, y)$. $(t \to \varepsilon(t)) \in \mathcal{D}_0(\mathcal{D}(\mathcal{A}(E), M), 1)$ car si $u \to g(u)$ est une plaque de M, $\varepsilon(t, g(u), y)$ est C^∞ par rapport aux 3 variables (t, u, y). Donc $s(t) = t\varepsilon(t)$ et, $\mathcal{D}(\mathcal{A}(E), M)$ étant un espace vectoriel difféologique, $(t, \tau) \to t\varepsilon(\tau)$ est difféologique, et pour tout $f \in \mathcal{D}(\mathcal{D}(\mathcal{A}(E), M))$, $(t, \tau) \to f(t\varepsilon(\tau))$ est C^∞ en t et τ. Donc

$$\frac{d}{dt}f(t\varepsilon(t))|_{t=0} = \frac{d}{dt}f(t\varepsilon(0))|_{t=0} + \frac{d}{d\tau}f(0 \cdot \varepsilon(\tau))|_{\tau=0}$$

Donc

$$\frac{d}{dt}f(s(t))|_{t=0} = 0.$$

Autrement dit $X = 0$, $\mathscr{E}$ est donc injective, ce qui achève la démonstration.

Le résultat précédent s'applique en particulier si M ou N est réduit à un point.

Corollaire. $\mathscr{A}(E)$ est plein.

Corollaire. $\mathscr{D}(\mathbb{R}^n, M)$ est plein.

Si $(E, [\,,], \rho)$ est un algébroïde de Lie, $\mathscr{A}_c(E)$ l'espace des sections à contrôle compact, on a le resultat suivant.

Théorème 5.2. $\mathscr{A}_c(E)$ *est plein.*

La seule chose nouvelle dans la preuve est la vérification que la condition de contrôle compact entraîne que $\mathscr{E}$ envoie $T_0\mathscr{A}_c(E)$ dans $\mathscr{A}_c(E)$ et que l'application $t \to \varepsilon(t)$ construite est à valeurs dans $\mathscr{D}(\mathscr{A}_c(E), M)$.

Soit $\Gamma \underset{\beta}{\overset{\alpha}{\rightrightarrows}} \Gamma_0$ un groupoïde de Lie [1] et [5]. Soit $\widehat{G}_\Gamma$ le groupe des bissections de Γ [1] et [5], c'est à dire des sous-variétés S de Γ telles que $\alpha|_S$ et $\beta|_S$ soient des difféomorphismes sur Γ_0. Les lois dans $\widehat{G}_\Gamma$ sont induites par les lois de Γ étendues aux parties de Γ. On note $S \to S^\beta$ l'application de $\widehat{G}_\Gamma$ dans $C_\beta^\infty(\Gamma_0, \Gamma)$ espace des sections de $\beta : \Gamma \to \Gamma_0$ définie par $S^\beta = (\beta|_S)^{-1}$. On note $S \to \phi_S^r$ l'antireprésentation de $\widehat{G}_\Gamma$ dans Difféo Γ, groupe des difféomorphismes de Γ, définie par $\phi_S^r(x) = x \cdot S = x \cdot S^\beta(\alpha(x))$.

On munit $\widehat{G}_\Gamma$ de la difféologie image réciproque de celle de $C_\beta^\infty(\Gamma_0, \Gamma)$ par l'application $S \to S^\beta$, qui est, sur $\widehat{G}_\Gamma$, une difféologie de groupe. On désigne par G_Γ le sous-ensemble de $\widehat{G}_\Gamma$ des S tels que $x_0 \to \alpha \circ S^\beta(x_0)$ soit à support compact. G_Γ est un sous-groupe de $\widehat{G}_\Gamma$ qu'on munit de la difféologie de groupe "à contrôle compact" dont les plaques $g : U \to G_\Gamma$ sont les plaques de $\widehat{G}_\Gamma$ telles que pour tout $u_0 \in U$, il existe un voisinage ouvert V de u_0 dans U et un compact K de Γ_0 tel que pour tout $u \in V$, $x_0 \to \alpha \circ g^\beta(u)(x_0)$ ait son support dans K. [2]a. C'est une difféologie sur G_Γ *plus fine* que la difféologie induite par $\widehat{G}_\Gamma$. Elle est induite sur G_Γ par la difféologie «à contrôle compact» de $C_{\beta c}^\infty(\Gamma_0, \Gamma)$ espace des sections de β à support compact.

Si $\Gamma = \Gamma_0 \times \Gamma_0 \rightrightarrows \Gamma_0$ où α est la première projection et β la seconde et où $(x, y)(y, z) = (x, z)$ et $(x, y)^{-1} = (y, x)$ sont les lois de groupoïde (on dit que Γ est le *groupoïde grossier de* Γ_0), $\widehat{G}_\Gamma$ est le groupe des difféomorphismes de Γ_0 identifiés à leurs graphes.

Soit $\mathscr{E}_{x_0} : \widehat{G}_\Gamma \to \beta^{-1}(x_0)$ l'application $S \to S^\beta(x_0)$. On note $\mathscr{E}$ l'application de $T_{\Gamma_0}\widehat{G}_\Gamma$ dans $(\operatorname{Ker} T\beta|_{\Gamma_0} \to \Gamma_0)$ définie par $\widehat{X} \to (x_0 \to T\mathscr{E}_{x_0}((\widehat{X})))$. $\operatorname{Ker} T\beta|_{\Gamma_0} \to \Gamma_0$ s'identifie canoniquement à l'algébroïde de Lie $\underline{\Gamma}$ de Γ. $\mathscr{E}$ est donc un morphisme vectoriel difféologique de $T_{\Gamma_0}\widehat{G}_\Gamma$ dans $\mathscr{A}(\underline{\Gamma})$. On vérifie aisément que la même construction appliquée à $T_{\Gamma_0}G_\Gamma$ fournit un morphisme difféologique vectoriel de $T_{\Gamma_0}G_\Gamma$ dans $\mathscr{A}_c(\underline{\Gamma})$

Théorème 5.3. (i) $\mathscr{E} : T_{\Gamma_0}\widehat{G}_\Gamma \to \mathscr{A}(\underline{\Gamma})$ *est surjectif*
(ii) $\mathscr{E} : T_{\Gamma_0}G_\Gamma \to \mathscr{A}_c(\underline{\Gamma})$ *est un isomorphisme.*

Démonstration. (i) Soit $X \in \mathcal{A}(\Gamma)$ et $Y = \rho \circ X$ l'image de X par l'ancre ρ. Γ_0 étant paracompacte et connexe il existe (cf. Appendice Théorème A.1.) un recouvrement ouvert de Γ_0 par deux ouverts W_1 et W_2 dont les composantes connexes W_{ij} sont relativement compactes et [Théorème A.2] si (φ_i) est une partition de l'unité subordonnée au recouvrement W_i, les champs $Y_i = \varphi_i \circ Y$ sont complets et $Y = Y_1 + Y_2$. Il résulte d'un théorème de Lichnérowicz [8]a, que les champs invariants à gauche $Y_i^{\ell} = \alpha^* \varphi_i \cdot X^{\ell}$ sur Γ sont complets où X^{ℓ} désigne le champ invariant à gauche sur Γ canoniquement associé à $X \in \mathcal{A}(\Gamma)$. On définit

$$S(t) = \varphi_t^{Y_1^{\ell}} \circ \varphi_t^{Y_2^{\ell}} (\Gamma_0)$$

où φ_t^Z désigne le flot du champ Z.

$t \to S(t) \in \mathcal{D}(\widehat{G}_\Gamma, 1)$ et, par construction même, si $\widehat{X}$ désigne la classe dans $T_{\Gamma_0} \widehat{G}_\Gamma$ de $t \to S(t)$, $\mathcal{E}(\widehat{X}) = X$ ce qui prouve la surjectivité de $\mathcal{E}$.

(ii) Si $X \in \mathcal{A}_c(\Gamma)$, X est complet et on note $\exp t X = \varphi_t^{X^{\ell}}(\Gamma_0)$. $X \to (t \to \exp t X)$ est un morphisme difféologique de $\mathcal{A}_c(F)$ dans $\mathcal{D}_0(G_\Gamma, 1)$ qui définit par passage au quotient un inverse à droite de $\mathcal{E} : T_{\Gamma_0} G_\Gamma \to \mathcal{A}_c(\Gamma)$. Il suffit donc de prouver que $\mathcal{E}$ est injectif pour achever la preuve de (ii).

Si $\mathcal{E}(\widehat{X}) = 0$, $\widehat{X}$ est la classe d'une 1-plaque $t \to S(t)$, $S(0) = \Gamma_0$, telle que pour tout $x_0 \in \Gamma_0$, $(d/dt) S^{\beta}(t)(x_0)|_{t=0} = 0$, $t \to S(t)$ étant une 1-plaque de G_Γ il existe $\eta > 0$ et un compact K de Γ_0 tel que si $|t| < \eta$, $x_0 \notin K$, $S^{\beta}(t)(x_0) = x_0$.

Or Γ étant un groupoïde de Lie, il existe un voisinage $\mathcal{U}$ de Γ_0 dans Γ qui est séparé (bien qu'en général Γ ne le soit pas), et dans lequel Γ_0 est fermé [1]. Ceci entraîne l'existence d'un voisinage tubulaire de Γ_0 dans $\mathcal{U}$, $U(\Gamma_0)$ [7] et on peut imposer que $U(\Gamma_0)$ soit un sous-fibré de $\beta : \Gamma_0 \to \Gamma$ et qu'il existe un plongement χ de $U(\Gamma_0)$ dans $\underline{\Gamma} \to \Gamma_0$ qui est un morphisme de fibrés et se réduit à l'identité sur Γ_0.

Comme $S^{\beta}(t)(x_0) = x_0$ si $|t| < \eta$ et $x_0 \notin K$, il existe $\eta_1, 0 < \eta_1 < \eta$ tel que si $|t| < \eta_1$, pour tout $x_0 \in \Gamma_0$, $S^{\beta}(t)(x_0) \in U(\Gamma_0)$. On pose $s(t) = \chi \circ S^{\beta}(t)$.

Par construction $s \in \mathcal{D}(\mathcal{A}_c(F), 1)$ et $(d/dt) s(t)(x_0)|_{t=0} = 0$. La formule de Taylor reste intégrale et permet de construire une application C^{∞}, $\varepsilon :] - \eta_1, \eta_1[\times \Gamma_0 \to \Gamma$ telle que $\varepsilon(0)(x_0) \equiv 0$, $s(t)(x_0) \equiv t \varepsilon(t)(x_0)$ où l'on pose $\varepsilon(t)(x_0) = \varepsilon(t, x_0)$.

ε est donc une 1-plaque de $\mathcal{A}_c(\Gamma)$ et $\varepsilon(0) = 0$ implique qu'il existe η_2, $0 < \eta_2 < \inf(\eta_1, 1)$ tel que $\varepsilon(t)(x_0) \in \chi(U(\Gamma_0))$ pour tout t, $|t| < \eta_2$. On considère l'application C^{∞} sur $] - \eta_2, \eta_2[^2$ définie par $S^{\beta}(t, \tau) = \chi^{-1} \circ (t \varepsilon(\tau))$; $S^{\beta}(t, t) = S^{\beta}(t)$ et pour tout x_0, par construction, $\beta \circ S^{\beta}(t, \tau)(x_0) = x_0$. D'autre part $\varphi_{S(t,\tau)}^r(x_0) = \alpha \circ S^{\beta}(t, \tau)(x_0)$ vaut l'identité si $x_0 \notin K$ et également pour tout x_0 si $t = 0 = \tau$. Ceci entraîne l'existence d'un η_3, $0 < \eta_3 < \eta_2$ tel que si $(t, \tau) \in] - \eta_3, \eta_3[^2$, $\varphi_{S(t,\tau)}^r$ est un C^{∞}-difféomorphisme de Γ_0, ce qui implique que $S^\beta(t, \tau)(\Gamma_0)$ est une bissection. $(t, \tau) \to S^\beta(t, \tau)(\Gamma_0) = S(t, \tau)$ est donc une 2-plaque de G_Γ.

On peut alors comme précédemment calculer $(d/dt)f(S(t))|_{t=0}$ en calculant successivement $(d/dt)f(S(t, 0))|_{t=0}$ et $(d/d\tau)f(S(0, \tau))|_{\tau=0}$ ce qui assure que $(d/dt)f(S(t))|_{t=0} = 0$, autrement dit que $(dS/dt)(0) = 0$, et donc que $\mathscr{E}$ est injective, ce qui achève la démonstration.

Remarque. Si on munit $C_{\beta,c}^\infty(\Gamma_0, \Gamma)$ de sa difféologie à contrôle compact, G_Γ est un sous-espace difféologique de $C_{\beta,c}^\infty(\Gamma_0, \Gamma)$ et en raisonnant comme précédemment on peut montrer que $T_{\Gamma_0}C_{\beta,c}^\infty(\Gamma_0, \Gamma) = \mathscr{A}_c(\underline{\Gamma})$.

En particulier $T_{\Gamma_0}C_{\beta,c}^\infty(\Gamma_0, \Gamma) = T_{\Gamma_0}G_\Gamma$. Ceci généralise la propriété bien connue: pour un espace de Banach E, de dimension finie ou non, $T_e \, GL(E) = \mathscr{G}L(E) = T_e \mathscr{G}L(E)$, ce qui dans ce cas est une conséquence de ce que $GL(E)$ est ouvert dans $\mathscr{G}L(E)$.

6. Sous-espaces tangents. Groupes de Lie

Soit $(M, \mathscr{D}_M)$ un espace difféologique et $N \subset M$ un *sous-espace difféologique* de M. On note i l'inclusion. $Ti : TN \to TM$ est un morphisme difféologique -homogène relevant l'identité.

Définition. $T^r N = Ti(TN)$ s'appelle le *sous-espace tangent de N* ou *espace tangent réduit*.

C'est un fibré sur N, homogène, dont la difféologie est la difféologie de fibré vectoriel induite par TM qui coïncide avec la difféologie image de TN par Ti.

Exemples. (1) Si N est une sous-variété de la variété $C^\infty M$ de dimension finie, TN et $T^r N$ coïncident.

(2) Si V est un sous-espace vectoriel difféologique de $\mathscr{A}_c(E)$, où E est un algébroïde de Lie, TV et $T^r V$ coïncident.

6.1. NOTION DE GROUPE DE LIE

Soit $G \subset G_\Gamma$ un sous-groupe difféologique de G_Γ. TG est le fibré trivial $G \times T_{\Gamma_0}G$. $T^r G$ fibré sous-tangent s'écrit $G \times \underline{G}$ où $\underline{G}$ est l'image de $T_{\Gamma_0}G$ dans $T_{\Gamma_0}G_\Gamma = \mathscr{A}_c(\underline{\Gamma})$.

Définition. [2]a $G \subset G_\Gamma$ est un groupe de Lie si:

1. $\underline{G}$ est plein;

2. Pour tout $X \in \underline{G}$, $t \to \exp tX \in \mathscr{D}(G, 1)$.

La condition (1) implique que $\underline{G}$ est une sous-algèbre de Lie de $\mathcal{A}_c(\underline{\Gamma})$, appelée *algèbre de Lie de G*.

La condition (2) fournit une section du morphisme surjectif $T_{\Gamma_0}G \to \underline{G}$. Si on note $T_{\Gamma_0}^{\infty}G = \{\widehat{X} \in T_{\Gamma_0}G \mid Ti\widehat{X} = 0\}$ où $i : G \hookrightarrow G_{\Gamma}$

$$T_{\Gamma_0}G = \underline{G} \oplus T_{\Gamma_0}^{\infty}G.$$

Extension de la définition ci-dessus Tout revêtement d'un groupe de Lie est encore appelé groupe de Lie. Si $G_1 \to G$ est un revêtement, $T_eG_1 \equiv T_eG$. En particulier, avec des notations évidentes $\underline{G_1} \equiv \underline{G}$. G_1 et G ont la même algèbre de Lie.

Remarque. Si Γ_0 est réduit à un point $\widehat{G_{\Gamma}} = G_{\Gamma} = \Gamma$, ce sera un *groupe de Lie* au sens usuel et il résulte d'un théorème de Yamabé [13] que les conditions (1) et (2) sont toujours vérifiées. La théorie proposée se réduit donc, en dimension finie, à la théorie usuelle.

Afin de rester dans les limites de cet article, on se contente de donner quelques résultats concernant les groupes de Lie.

1. Si $(\Gamma_0, \wedge_0)$ est une *variété de Poisson* [8]b, qui est l'espace des unités d'un groupoïde symplectique [1] et [2]a, $(\Gamma, \sigma) \rightrightarrows (\Gamma_0, \wedge_0)$, G_{Γ}^{σ} sous-groupe de G_{Γ} des bissections qui sont des sous-variétés Lagrangiennes de (Γ, σ) est un groupe de Lie d'algèbre de Lie $Z\Omega_c^1(\Gamma_0, \wedge_0)$ algèbre des 1-formes fermées sur Γ_0 à contrôle compact.

Si $G_{\Gamma}^{\sigma,ex}$ désigne le sous-groupe de G_{Γ}^{σ} des extrémités d'isotopies hamiltoniennes [2]a, $G_{\Gamma}^{\sigma,ex}$ est un groupe de Lie d'algèbre de Lie $B\Omega_c^1(\Gamma_0, \wedge_0)$ sous-algèbre de Lie de $Z\Omega_c^1(\Gamma_0, \wedge_0)$ formée des 1-formes exactes [2]c.

2. Si $(\Gamma_0, E_0, \wedge_0)$ est une variété de Jacobi au sens de Lichnérowicz [8]c qui est l'espace des unités d'un groupoïde de contact (Γ, θ), G_{Γ}^{θ} espace des bissections legendriennes est un groupe de Lie d'algèbre de Lie $C^{\infty}(\Gamma_0,)$ muni de la structure d'algèbre de Lie associée à la struture de Jacobi [2]b.

Les deux exemples précédents (variétés de Poisson et de Jacobi) contiennent (pratiquement) toutes les situations correspondant aux algèbres de Kirillov de rang 1. L'étude comparée des groupes de Lie associés à une variété de Poisson $(\Gamma_0, \wedge_0)$ et à cette même variété considérée comme variété de Jacobi $(\Gamma_0, E_0 = 0, \wedge_0)$ [2]b est en relation étroite avec la préquantification géométrique telle que présentée, dans cette même conférence, par G. Tuynman dans le cas où $(\Gamma_0, \wedge_0) = (\Gamma_0, \sigma_0)$ est une variété symplectique.

7. Appendice. Variétés non compactes et champs complets

Les résultats qui suivent ne semblent pas être connus. On note $\mathcal{X}(M)$ l'algèbre des champs de vecteurs C^{∞} de M.

Théorème A.1. *Si M est une variété C^∞ de dimension finie paracompacte **et non compacte**, il existe un recouvrement de M par deux ouverts W_1 et W_2 dont les composantes connexes (W_{1i}) et (W_{2j}) sont relativement compactes.*

Démonstration. Si $M = \mathbb{R}^n$, on pose:

$$\bullet\, U_1 = \bigcup_{p=1}^{+\infty} U_{1p} \qquad \text{où} \qquad U_{11} = \{x \mid \|x\| < 1\}$$

et pour $p \geq 2$,

$$U_{1p} = \{x \mid p - 1 < \|x\| < p - 2\}.$$

$$\bullet\, U_2 = \bigcup_{p=1}^{+\infty} U_{2p} \qquad \text{où} \qquad U_{2p} = \left\{x \mid p - \frac{1}{4} < \|x\| < p + \frac{1}{4}\right\}$$

pour $p \geq 1$.

$W_i = U_i$, $i = 1, 2$ répond à la question pour $\mathbb{R}^n$.

Si M est une variété connexe et paracompacte de dimension n, le théorème de Whitney [12] assure l'existence d'un plongement propre $\chi : M \to \mathbb{R}^{2n+1}$.

(U_i) étant le recouvrement précédemment construit sur $\mathbb{R}^{2n+1}$, on pose $W_i = \chi^{-1} U_i$. Si $W_{i\alpha}$ est une composante connexe de W_i, il existe j tel que $\chi(W_{i\alpha} \subset U_{ij})$ et donc $W_{i\alpha} \subset \chi^{-1}(\overline{U_{ij}})$ qui est compact ce qui achève la démonstration dans le cas connexe. Pour le cas général, on procède sur chaque composante connexe de M.

Théorème A.2. *Les notations étant les mêmes que les précédentes soit $(\varphi_i)_{i=1,2}$ une partition de l'unité C^∞ subordonnée à (W_i). Soit $E_i = \varphi_i \cdot \mathscr{X}(M)$. Les champs de vecteurs appartenant à E_i sont complets et*

$$\mathscr{X}(M) = E_1 + E_2.$$

Commentaire. Tout champ de vecteurs sur une variété compacte est complet et, sur une variété non compacte, somme de deux (au plus) champs complets.

Démonstration. La seule chose à prouver est que si $X \in \mathscr{X}(M)$, $X_i = \varphi_i \cdot X$ est complet. Soit donc $x \in M$. Si $X_i(x) = 0$, le flot de X_i sur x est défini sur $\mathbb{R}$. Si $X_i(x) \neq 0$ soit $]t^-(x), t^+(x)[$ l'intervalle de définition du flot. Pour tout $t \in]t^-(x), t^+(x)[$, $X_i(\varphi_t^{X_i}(x)) \neq 0$. Donc $\varphi_i(\varphi_t^{X_i}(x)) \neq 0$. Autrement dit $\varphi_t^{X_i}(x) \in F_i$ pou tout $t \in]t^-(x), t^+(x)[$, où F_i est le support de φ_i.

Lemme A.3. $F_{ij} = F_i \cap W_{ij}$ *est compact.*

Preuve du lemme. Soit $z \in \overline{F_{ij}}$. $z \in F_i \subset W_i$ donc il existe k tel que $z \in W_{ik}$ et $W_{ik} \cap F_{ij} \neq \emptyset$. Donc $k = j$, z appartient donc à $F_i \cap W_{ij} = F_{ij}$. F_{ij} est donc fermé et compact.

Fin de la démonstration. Si $x \in F_i$ il existe j tel que $x \in F_{ij}$. Comme $\varphi_t^{X_i}]t^-(x), t^+(x)[$ est connexe, $\varphi_t^{X_i}(x) \in W_{ij}$ et donc à $W_{ij} \cap F_i = F_{ij}$ mais F_{ij} est compact ce qui entraîne que $t^-(x) = -\infty$, $t(x) = +\infty$ et prouve que X_i est compact.

Références

[1]   Coste, A., Dazord, P., and Weinstein, A., *Groupoïdes symplectiques*, Publ. Dept. Math. Lyon 1987, 2/A, 1–62.

[2]   Dazord, P., a) Lie groups and algebras in infinite dimension, *Contemporary Mathematics* **179** (1994), 17–44.

[3]   b) Sur l'intégration des algèbres de Lie locales et la préquantification, *Bull. Sci. Math.* **121** (1997), 423–462.

c) Groupes et Algèbres de Lie de dimension infinie d'un point de vue géométrique in *Analysis on infinite-dimensional Lie groups and Algebra*, World Scientific, Publishing co. 1998, 47–66.

[4]   Donato, P., *Revêtements d'orbite difféologiques*, Travaux en cours, **25**, Hermann ed., (1987), 11–24.

[5]   Ehresmann, C., *Oeuvres complètes*, Tome 1, Amiens, 1984.

[6]   Iglésias, P., *Connexions et Difféologies*, Travaux en cours, **25**, Hermann ed., (1987), 61–78.

[7]   Lang, S., *IntroductionauxVariétésdifférentiables*, Dunod, ed. 1966.

[8]   Lichnérowicz, A., a) *Théorie globale des Connexions et des groupes d'holonomie*. Cremonese, ed., (1955).

b) Les variétés de Poisson et leurs algèbres de Lie associées, *J. Differential Geometry* **12** (1977), 253–300.

c) Les variétés de Jacobi et leurs algèbres de Lie associées, *J. Math. Pures et appli.* **51** (1978), 453–488.

[9]   Pradines, J., Théorie de Lie pour les groupoïdes infinitésimaux, *C. R. Acad. Sc. Paris* **264** (1967), 245–248.

[10] Sataké, I., The Gauss-Bonnet formula for V-manifolds, *J. Math. Society of Japan* **9** (1957), 464–492.

[11] Souriau, J.-M., a) *Groupes différentiels et Physique Mathématique*, Travaux en cours, **6**, Hermann éd., (1984), 73–120.

b) *Un algorithme générateur de stuctures quantiques*, Astérisque hors série, S.M.F., 1985, 341–400.

[12] Whitney, H., Geometric integration theory, Princeton University Press 1957.

[13] Yamabé, H., On an armise connected subgroup of a Lie Group, *Osaka Math. J. (1)* **2** (1950), 13–14.

THE COHOMOLOGICAL MEANING OF MASLOV'S LAGRANGIAN PATH INTERSECTION INDEX

M. DE GOSSON AND S. DE GOSSON

To Jean Leray...with a vengeance!

Contents

1	Introduction	145
2	Notations and terminology	146
3	Signature and composition form	147
	3.1 The signature function	147
	3.2 The composition form	148
4	Leray indices	150
	4.1 The Leray index on $\Lambda^2_\infty(n)$	150
	4.2 Practical computation of μ	152
	4.3 Generalizations	153
	4.4 The Leray index on $\mathrm{Sp}_\infty(n)$	154
5	Path intersection indices	156
	5.1 Lagrangian paths	156
	5.2 Symplectic paths	158
6	The Robbin–Salamon theory	158
	6.1 The index $\mu_{RS}(\Lambda, \ell)$	158
	6.2 The index $\mu_{RS}(\Psi)$	159
7	In cauda venenum...	160

M. de Gosson (ed.), Jean Leray '99 Conference Proceedings, 143-162.
© 2003 Kluwer Academic Publishers.

Abstract. *We study the relation between the complete Maslov index defined by Leray and the author, and the Lagrangian path intersection index defined by Robbin and Salamon, and used by McDuff and Salamon in their study of symplectic topology.*

1. Introduction

Leray and one of us (MdG) were in 1989 exchanging letters about the interest of defining a Maslov index for Lagrangian paths with non-transversal endpoints. After all, argued Leray, it was the jumps of that index at the caustics that were relevant for the study of caustics in Lagrangian Analysis, and not its point-wise values. Leray came with the following argument, intended to show that the definition of such a complete index was perhaps a merely academic exercise:

‹‹*l'indice de Kronecker le plus simple est le nombre de points d'un ensemble fini contenu dans un domaine, dont le bord ne contient aucun de ces points (condition de transversalité). Son plus bel exemple est le fameux théorème de Cauchy: soit une fonction f holomorphe sur un domaine D de C et sur son bord; alors le nombre m de zéros de f contenus dans D est*

(1)

$$m = \frac{1}{2\pi i} \int_{\partial D} \frac{df}{f},$$

si D ne contient aucun zéro de f.

Je ne crois pas qu'il y ait interêt a supprimer cette condition de transversalité, à définir

$$m = (\textit{nombre de zéros} \in D) + \frac{1}{2}(\textit{nombre de zéros} \in \partial D),$$

à employer dans (1) la valeur principale de l'intégrale, à constater que (1) vaut encore, à proclamer alors qu'on à fait mieux que Cauchy!

Ce n'est là qu'une parabole et je ne suis pas prophète, hélàs!›› (‹‹hélàs›› was deleted by Leray, and replaced by: *heureusement!*)

It turns out that the theory of the Maslov index, the study of which was initiated in [18], has had since our exchange of letters unexpected applications in several fields of pure mathematics, particularly in functional analysis since Floer established

'Maslov index = spectral flow'

type of theorems (see the review paper by Cappell *et al.* [3] for a comprehensive and interesting study of other related notions of Maslov indices. Dazord [4] gives a construction on general symplectic bundles). It also seems that the Maslov index could play an essential role in the theory of Schubert manifolds.

In [20, section 2] (1993), Robbin and Salamon define a half-integer valued function μ of pairs of paths of Lagrangian subspaces of the standard symplectic space $(\mathbb{R}^{2n}, \omega)$. They then use (*ibidem*, section 4) the properties of that function (which they call *Maslov index*) to define a function of paths of symplectic matrices; that function is also called *Maslov index* by the authors; some of their constructions are also exposed in the treatise [17] of McDuff and Salamon. The Robbin–Salamon theory has recently been extended in a highly non-trivial way to the infinite-dimensional case by Booss–Bavnbek and Furutani ([1] and [2]), with applications to the study of the 'Fredholm Lagrangian' which plays an important role in many questions of global analysis.

Our purpose is to compare the Robbin Salamon–McDuff indices [17], [19] and [20], defined in 1993, with those defined by Leray [13] and [14] in 1978, and whose definition in the non-transversal case is due to one of us [6] and [7] in 1990.

2. Notations and terminology

The vector space $\mathbb{R}^{2n} \equiv \mathbb{R}^n \oplus (\mathbb{R}^n)^*$ is equipped with the symplectic form defined by

$$\omega\left(z, z'\right) = \langle p, x'\rangle - \langle p', x\rangle$$

if $z = (x, p)$, $z' = (x', p')$. We denote by $\mathrm{Lag}(n)$ (resp. $\mathrm{Sp}(n)$) the Lagrangian Grassmannian (resp. the symplectic group) of the symplectic space $(\mathbb{R}^{2n}, \omega)$: $\ell \in \mathrm{Lag}(n)$ if and only if ℓ is a n-dimensional linear subspace on which the symplectic form vanishes identically, and $s \in \mathrm{Sp}(n)$ if s is a linear automorphism of $\mathbb{R}^{2n}$ preserving the symplectic form, that is $\omega(sz, sz') = \omega(z, z')$ for all vectors z and z' in $\mathbb{R}^{2n}$. We will use the notations $X = \mathbb{R}^n \times 0$ and $X^* = 0 \times \mathbb{R}^n$ to denote the 'horizontal' and 'vertical' Lagrangian planes, respectively.

We will use the following standard notations and terminology from algebraic topology. Let F be a set, k an integer ≥ 0, $(G, +)$ an Abelian group. A G-valued k-cochain on F is a mapping $c : F^{k+1} \to G$. Such a cochain c is called a k-cocycle if $\partial c = 0$, where ∂ is the coboundary operator: it is the mapping which to every k-cochain c associates the $(k + 1)$-cochain ∂c defined by the formula

$$\partial c\left(x_0, \ldots, x_{k+1}\right) = \sum_{j=0}^{k+1} (-1)^j c\left(x_0, \ldots, \hat{x}_j, \ldots, x_{k+1}\right),$$

where the cap $\hat{\ }$ suppresses the term it covers. The coboundary operator satisfies $\partial^2 c = 0$ for every cochain c. A k-cochain c is called a k-coboundary if there exists a $(k - 1)$-cochain m such that $c = \partial m$; a k-coboundary is evidently a k-cocycle.

3. Signature and composition form

3.1. THE SIGNATURE FUNCTION

We begin by recalling some standard results about the notion of signature $\sigma(\ell, \ell', \ell'')$ of a triple of Lagrangian planes (ℓ, ℓ', ℓ'') (see [5], [15] and [16]). It is defined as follows: consider the quadratic form

$$R\left(z, z', z''\right) = \omega\left(z, z'\right) + \omega\left(z', z''\right) + \omega\left(z'', z\right)$$

on $\ell \oplus \ell' \oplus \ell''$; then $\sigma(\ell, \ell', \ell'')$ is, by definition, the signature of that quadratic form: $\sigma(\ell, \ell', \ell'')$ is difference $\sigma_+ - \sigma_-$ between the number of > 0 and < 0 eigenvalues of R, and one proves (see [15]) that

$$-n \le \sigma\left(\ell, \ell', \ell''\right) \le n.$$

The signature is an antisymmetric and $\mathrm{Sp}(n)$-invariant function such that

$$\sigma\left(\ell, \ell', \ell''\right) \equiv n + \partial \dim\left(\ell, \ell', \ell''\right) \quad \text{mod } 2, \tag{3.1}$$

where $\partial \dim$ is the coboundary of the 1-chain

$$\dim\left(\ell, \ell'\right) = \dim \ell \cap \ell'.$$

If $\ell \cap \ell'' = 0$, then $\sigma(\ell, \ell', \ell'')$ is simply the signature (in the sense above) of the quadratic form

$$R'\left(z'\right) = \omega\left(z', P_{\ell'',\ell}z'\right) = \omega\left(P_{\ell,\ell'}z', z'\right) \tag{3.2}$$

on ℓ' where $P_{\ell'',\ell}$ (resp. $P_{\ell,\ell'}$) is the projection operator on ℓ'' along ℓ (resp. the projection operator ℓ along ℓ'').

The signature has the two following essential properties:

Theorem 3.1. *(1)* σ *is a* $\mathbb{Z}$*-valued 2-cocycle:* $\partial\sigma = 0$*, that is*

$$\sigma(\ell', \ell'', \ell''') - \sigma\left(\ell, \ell'', \ell'''\right) + \sigma\left(\ell, \ell', \ell'''\right) - \sigma\left(\ell, \ell', \ell''\right) = 0 \tag{3.3}$$

for all 4-tuples $(\ell, \ell', \ell'', \ell''')$ *of Lagrangian planes; (2)* $\sigma(\ell, \ell', \ell'')$ *remains constant when the Lagrangian planes* ℓ, ℓ', ℓ'' *move continuously in such a way that the dimensions of the intersections* $\ell \cap \ell'$, $\ell \cap \ell''$ *and* $\ell'' \cap \ell$ *remain constant.*

Proof. See [15] and [16]. $\square$

An immediate consequence of Theorem 3.1(1) is:

Corollary 3.2. For every Lagrangian plane ℓ the mapping $\sigma_\ell : (\mathrm{Sp}(n))^2 \to \mathbb{Z}$ defined by the formula

$$\sigma_\ell\left(s, s'\right) = \sigma\left(\ell, s\ell, ss'\ell\right) \tag{3.4}$$

has the following property:

$$ss's'' = I \implies \sigma_\ell\left(ss', s''\right) - \sigma_\ell\left(s', ss''\right) + \sigma_\ell\left(s, s'\right) = 0. \tag{3.5}$$

The mapping σ_ℓ is the *group cocycle* associated with σ and ℓ.

3.2. THE COMPOSITION FORM

We next proceed to compare Robbin and Salamon's *composition form* ([20], section 5) with the signature defined above. Identifying every $s \in \mathrm{Sp}(n)$ with its matrix in the canonical symplectic basis of $\mathbb{R}^{2n}$ we have

$$\begin{pmatrix} A & B \\ C & D \end{pmatrix} \in \mathrm{Sp}(n) \iff \begin{cases} A^T C = C^T A, \\ B^T D = D^T B, \\ A^T D - C^T B = I, \end{cases} \tag{3.6}$$

where A^T is the transpose of A, etc. Of course A, B, C, D are here real $n \times n$ matrices and I the identity. Denoting by $\mathrm{Sp}_0(n)$ the subset of $\mathrm{Sp}(n)$ consisting of all symplectic matrices

$$\begin{pmatrix} A & B \\ C & D \end{pmatrix}, \quad \det(B) \neq 0,$$

the composition form of Robbin–Salamon is the mapping

$$Q : \left(\mathrm{Sp}_0(n)\right)^2 \longrightarrow G\ell(n, \mathbb{R})$$

defined by

$$Q\left(s, s'\right) = B^{-1}A + D'\left(B'\right)^{-1} = B^{-1}B''\left(B'\right)^{-1} \tag{3.7}$$

for

$$s = \begin{pmatrix} A & B \\ C & D \end{pmatrix}, \quad \text{and} \quad s' = \begin{pmatrix} A' & B' \\ C' & D' \end{pmatrix} \tag{3.8}$$

and

$$s'' = ss' = \begin{pmatrix} A'' & B'' \\ C'' & D'' \end{pmatrix}.$$

Robbin–Salamon composition form is related to the group cocycle associated to the signature and the vertical plane:

Proposition 3.3. We have

$$\mathrm{sign}\, Q\left(s, s'\right) = \sigma_{X*}\left(s, s'\right) \tag{3.9}$$

for all $s, s' \in \mathrm{Sp}(n)$ such that $sX^* \cap X^* = s'X^* \cap X^* = 0$.

Proof. Let us first prove (3.9) when $s'X^* = X$. Using successively the antisymmetry and the $Sp(n)$-invariance of σ we have

$$\sigma\left(X^*, sX^*, ss'X^*\right) = -\sigma\left(X^*, s^{-1}X^*, X\right) \tag{3.10}$$

so that by (3.2)

$$\sigma\left(X^*, sX^*, ss'X^*\right) = \text{sign}\left(B^{-1}A\right). \tag{3.11}$$

On the other hand, the condition $s'X^* = X$ implies that s' has the form

$$s' = \begin{pmatrix} A' & B' \\ C' & 0 \end{pmatrix}$$

and hence $Q(s, s') = B^{-1}A$ in view of (3.7), which proves (3.9) in the case $s'X^* = X$. The general case is easily reduced to the former, recalling that the symplectic group acts transitively on all pairs of transverse Lagrangian planes (see [8] and [13]). Since

$$s'X^* \cap X^* = X \cap X^* = 0$$

we can thus find $h \in Sp(n)$ such that

$$\left(X^*, s'X^*\right) = h\left(X^*, X\right)$$

that is

$$s'X^* = hX, \quad \text{and} \quad hX^* = X^*$$

from which follows, using again the antisymmetry and $Sp(n)$-invariance of σ:

$$\sigma\left(X^*, sX^*, ss'X^*\right) = \sigma\left(X^*, shX^*, shX\right)$$
$$= -\sigma\left(X^*, (sh)^{-1}X^*, X\right)$$

which is (3.10) with s replaced by sh. Changing s' into $h^{-1}s'$ (and hence leaving ss' unchanged) we are thus led back to the first case. Since $hX^* = X^*$, h must be of the type

$$h = \begin{pmatrix} L & 0 \\ P & (L^{-1})^T \end{pmatrix}, \quad \det(L) \neq 0, \ P = P^T$$

in view of (3.6). Writing again s in block-matrix form (3.8) we have

$$sh = \begin{pmatrix} * & B\left(L^T\right)^{-1} \\ * & * \end{pmatrix}, \quad h^{-1}s' = \begin{pmatrix} * & L^{-1}B' \\ * & * \end{pmatrix}$$

hence, by formula (3.7):

$$\sigma\left(X^*, sX^*, ss'X^*\right) = \text{sign}\left(L^T B^{-1} B'' B'^{-1} L\right)$$
$$= \text{sign}\left(B^{-1} B'' B'^{-1}\right),$$

which proves (3.9) in the general case.   □

Notice that formula (3.9) *defines* the composition form in the general case.

Corollary 3.4. The composition form Q satisfies

$$\text{sign } Q\left(ss', s''\right) - \text{sign } Q\left(s', s''\right) + \text{sign } Q\left(s, s'\right) = 0 \qquad (3.12)$$

when s, s', s'' are in $\text{Sp}_0(n)$ and $ss's'' = I$.

Proof. Formula (3.12) is just a restatement of formula (3.5) in Corollary 3.2.   □

Notice that the identity (3.12) does not immediately follow from the definition (3.7) of Robbin–Salamon.

4. Leray indices

4.1. THE LERAY INDEX ON $\Lambda_\infty^2(n)$

We recall that the fundamental group $\pi_1(\text{Lag}(n))$ is isomorphic to $(\mathbb{Z}, +)$ so that $\text{Lag}(n)$ has coverings of all orders $q = 1, 2, \ldots, +\infty$; $\text{Lag}_\infty(n)$ is the universal covering of $\text{Lag}(n)$. Following previous work of [11], V.P. Maslov introduced in [18] an index describing the jumps of the phase of the asymptotic solutions to certain partial differential equations depending on a small parameter (for instance Planck's constant, or the inverse of the velocity of light) undergoes on the caustics. Arnold clarified Maslov's definition by using the properties of the Kronecker intersection index from algebraic topology. In [13, Chapter I, §§2.4–2.5], Leray extends Arnold's construction by defining an integer-valued function m on pairs $(\ell_\infty, \ell'_\infty)$ of $\text{Lag}_\infty(n)$ such that $\ell \cap \ell' = 0$ (ℓ and ℓ' the projections on $\text{Lag}(n)$ of ℓ_∞ and ℓ'_∞), and he calls that function "Maslov index on $\Lambda_\infty^2(n)$". Leray makes use of the properties of Kronecker's chain intersection index [12] to define m, and then shows (*ibidem*, Theorem 5.10, p. 46) that m is locally constant on its domain, and that it satisfies the essential algebraic relation

$$m\left(\ell_\infty, \ell'_\infty\right) - m\left(\ell_\infty, \ell''_\infty\right) + m\left(\ell'_\infty, \ell''_\infty\right) = \text{Inert}\left(\ell, \ell', \ell''\right) \qquad (4.1)$$

for all Lagrangian planes ℓ, ℓ', ℓ'' such that

$$\ell \cap \ell' = \ell' \cap \ell'' = \ell'' \cap \ell = 0. \qquad (4.2)$$

Inert(ℓ, ℓ', ℓ'') is the *index of inertia* of the triple (ℓ, ℓ', ℓ''): it is the common index of inertia of the three quadratic forms

$$z \longrightarrow \omega\left(z, z'\right), \qquad z' \longrightarrow \omega\left(z', z''\right), \qquad z'' \longrightarrow \omega\left(z'', z\right),$$

where $(z, z', z'') \in \ell \times \ell' \times \ell''$ is chosen such that $z + z' + z'' = 0$ ([13, Ch.I, §2.4]). Using (3.2) it is easy to show that

$$\sigma\left(\ell, \ell', \ell''\right) = 2\operatorname{Inert}\left(\ell, \ell', \ell''\right) - n \quad (\bmod\ 2) \tag{4.3}$$

(see [7]). Set now, as we did in [7]:

$$\mu\left(\ell'_\infty, \ell''_\infty\right) = 2m\left(\ell'_\infty, \ell''_\infty\right) - n;$$

then, in view of (4.1)

$$\mu\left(\ell_\infty, \ell'_\infty\right) - \mu\left(\ell_\infty, \ell''_\infty\right) + \mu\left(\ell'_\infty, \ell''_\infty\right) = \sigma\left(\ell, \ell', \ell''\right) \tag{4.4}$$

for all $\ell_\infty, \ell'_\infty, \ell''_\infty$ satisfying the transversality assumption (4.2). Since $\sigma(\ell, \ell', \ell'')$ is defined even in the non-transversal case, we can *define* $\mu(\ell_\infty, \ell'_\infty)$ for arbitrary $\ell_\infty, \ell'_\infty$ by the formula

$$\mu\left(\ell_\infty, \ell'_\infty\right) = \mu\left(\ell_\infty, \ell''_\infty\right) - \mu\left(\ell'_\infty, \ell''_\infty\right) + \sigma\left(\ell, \ell', \ell''\right), \tag{4.5}$$

where ℓ''_∞ is chosen transversal to both ℓ_∞ and ℓ'_∞. One then checks, using the cocycle property (3.3) of the signature, that $\mu(\ell_\infty, \ell'_\infty)$ is well defined, i.e., that it is independent of the choice of ℓ''_∞. Moreover, formula (4.4) will hold in the general case as well. We will call the function μ just defined the *Leray index* on $\Lambda^2_\infty(n)$.

Theorem 4.1. *(1) The Leray index is the only function $\mu : \Lambda^2_\infty(n) \to \mathbb{Z}$ having the two following properties:*

(i) the coboundary of μ is the signature; by this we mean:

$$\mu\left(\ell_\infty, \ell'_\infty\right) - \mu\left(\ell_\infty, \ell''_\infty\right) + \mu\left(\ell'_\infty, \ell''_\infty\right) = \sigma\left(\ell, \ell', \ell''\right) \tag{4.6}$$

for all $\ell_\infty, \ell'_\infty, \ell''_\infty$ in $\mathrm{Lag}_\infty(n)$;

(ii) the mapping $\Lambda^2_\infty(n) \times \mathrm{Lag}(n) \to \mathbb{Z}$:

$$\left(\ell_\infty, \ell'_\infty, \ell''\right) \longrightarrow \mu\left(\ell_\infty, \ell'_\infty\right) - \sigma\left(\ell, \ell', \ell''\right) \tag{4.7}$$

is locally constant when $\ell \cap \ell'' = \ell' \cap \ell'' = 0$.

(2) The Maslov index μ moreover has the following three properties

$$\mu\left(\beta^r \ell_\infty, \beta^{r'} \ell'_\infty\right) = \mu\left(\ell_\infty, \ell'_\infty\right) + 2\left(r - r'\right) \tag{4.8}$$

for all integers r, r'; β is the generator of $\pi_1(\mathrm{Lag}_\infty(n)) \equiv (\mathbb{Z}, +)$ whose natural image in $\mathbb{Z}$ is $+1$;

$$\mu\left(s_\infty \ell_\infty, s_\infty \ell'_\infty\right) = \mu\left(\ell_\infty, \ell'_\infty\right) \tag{4.9}$$

for all $s_\infty \in \mathrm{Sp}_\infty(n)$;

$$\mu\left(\ell_\infty, \ell'_\infty\right) + \mu\left(\ell'_\infty, \ell_\infty\right) = 0 \tag{4.10}$$

(and hence in particular $\mu(\ell_\infty, \ell_\infty) = 0$).

Remark. If one relaxes the topological condition (4.7), then there are infinitely many functions μ satisfying (4.6). Take for instance an arbitrary Lagrangian plane ℓ_0 and set $\mu_0(\ell_\infty, \ell'_\infty) = \sigma(\ell, \ell', \ell_0)$. Formula (4.6) then holds for μ_0 in view of the cocycle property of the signature, but μ_0 does not satisfy the topological property (ii) because $\sigma(\ell, \ell', \ell_0)$ has jumps when ℓ or ℓ' cross ℓ_0.

Defining the 2-cocycle $\pi^*\sigma$ on $\mathrm{Lag}_\infty(n)$ by:

$$\pi^*\sigma\left(\ell_\infty, \ell'_\infty, \ell''_\infty\right) = \sigma\left(\ell, \ell', \ell''\right), \tag{4.11}$$

where the pull-back $\pi^*\sigma$ is defined by

$$\left(\ell, \ell', \ell''\right) = \left(\pi\left(\ell_\infty\right), \pi\left(\ell'_\infty\right), \pi\left(\ell''_\infty\right)\right)$$

(π the projection $\mathrm{Lag}_\infty(n) \rightarrow \mathrm{Lag}(n)$), one immediately sees that property (4.1) can be rewritten in cohomological form as

$$\pi^*\sigma = \partial\mu. \tag{4.12}$$

4.2. PRACTICAL COMPUTATION OF μ

Leray's original definition of μ made use of chain intersection theory; μ is thus a purely topological object. In practice one can calculate the Leray index as follows. In [21] Souriau defines a mapping

$$w : \mathrm{Lag}(n) \longrightarrow U(n)$$

by the formula

$$w(\ell) = u\bar{u}^* = u\left(u^T\right) \quad \text{if } \ell = u\left(X^*\right)$$

and shows that this mapping is one-to-one. This allows him to identify $\mathrm{Lag}(n)$ with the manifold $W(n)$ of all symmetric unitary matrices:

$$\mathrm{Lag}(n) \equiv W(n) = \left\{ w \in U(n), w = w^T \right\}.$$

The universal covering $\mathrm{Lag}_\infty(n)$ of $\mathrm{Lag}(n)$ is then identified with the subset

$$W_\infty(n, \mathbb{C}) = \left\{ (w, \alpha) : w \in W(n, \mathbb{C}), \det(w) = e^{i\alpha} \right\} \qquad (4.13)$$

of $U(n, \mathbb{C}) \times \mathbb{C}$, the covering projection being simply the mapping $(w, \theta) \to w$. Suppose now that $\ell_\infty \equiv (w, \alpha)$ and $\ell'_\infty \equiv (w', \alpha')$ are such that $\ell \cap \ell' = 0$; then the Leray index $\mu(\ell_\infty, \ell'_\infty)$ is given by (see again [21]):

$$\mu\left(\ell_\infty, \ell'_\infty\right) = \frac{1}{\pi} \left(\alpha - \alpha' + i \operatorname{Tr Log} \left(-w \left(w'\right)^{-1} \right) \right). \qquad (4.14)$$

To calculate $\mu(\ell_\infty, \ell'_\infty)$ when the Lagrangian planes no longer are transversal, it suffices to choose ℓ'' such that $\ell \cap \ell'' = \ell' \cap \ell'' = 0$ and to apply (4.5) together with (3.2) (see [8] for details).

4.3. GENERALIZATIONS

More generally, we will call *Leray index on* $\mathrm{Lag}_\infty(n)$ any mapping

$$\mu : \left(\mathrm{Lag}_\infty(n) \right)^2 \longrightarrow \mathbb{Z}$$

having the two following properties:

(1) the coboundary of μ, viewed as a 1-cochain, descends to a $\mathrm{Sp}(n)$-invariant cocycle on $\mathrm{Lag}(n)$:

$$\partial\mu\left(\ell_\infty, \ell'_\infty, \ell''_\infty\right) = f\left(\ell, \ell', \ell''\right) \qquad (4.15)$$

and

$$f\left(s\ell, s\ell', s\ell''\right) = f\left(\ell, \ell', \ell''\right) \quad \text{for all } s \in \mathrm{Sp}(n) \qquad (4.16)$$

(2) m is locally constant on each of the subsets

$$\left\{ (\ell_\infty, \ell'_\infty) : \dim \left(\ell \cap \ell' \right) = k \right\} \quad (0 \leq k \leq n) \qquad (4.17)$$

of $(\mathrm{Lag}_\infty(n))^2$.

We note that f is a 2-cocycle on $\mathrm{Lag}(n)$: $\partial f = 0$, and that it is locally constant on each of the sets

$$\{(\ell, \ell', \ell'') : \dim\left(\ell \cap \ell'\right) = k, \dim\left(\ell' \cap \ell''\right) = k', \dim\left(\ell'' \cap \ell\right) = k''\} \quad (4.18)$$

$(0 \le k, k', k'' \le n)$. Moreover, given such a 2-cocycle f, there exists at most one Leray index m satisfying (4.15), because of the following lemma (see [7]):

Lemma 4.2. *A function* $v : (\mathrm{Lag}(n))^3 \to \mathbb{Z}$ *which is locally constant on any of the sets (4.17) and such that* $\partial v = 0$ *is identically zero.*

We also have:

Lemma 4.3. *Suppose* m *is a real function defined on all the pairs* $(\ell_\infty, \ell'_\infty)$ *such that* $\ell \cap \ell' = 0$, *and such that (4.15) holds for some 2-cocycle* f *on* $\mathrm{Lag}(n)$. *Then, the formula*

$$m\left(\ell_\infty, \ell'_\infty\right) = m\left(\ell_\infty, \ell''_\infty\right) - m\left(\ell'_\infty, \ell''_\infty\right) + f\left(\ell, \ell', \ell''\right), \quad (4.19)$$

where ℓ''_∞ *is chosen such that*

$$\ell \cap \ell'' = \ell' \cap \ell'' = 0$$

defines unambiguously $m(\ell_\infty, \ell'_\infty)$ *for all* $(\ell_\infty, \ell'_\infty) \in (\mathrm{Lag}_\infty(n))^2$.

Proof. It is sufficient to verify that $m(\ell_\infty, \ell'_\infty)$ is independent of the choice of ℓ''_∞, but this is an immediate consequence of the cocycle property of f. □

4.4. THE LERAY INDEX ON $\mathrm{Sp}_\infty(n)$

Let us denote by α is the generator of $\pi_1(\mathrm{Sp}(n)) \equiv (\mathbb{Z}, +)$ whose natural image in $\mathbb{Z}$ is $+1$. For ℓ an arbitrary Lagrangian plane the Leray index μ_ℓ on $\mathrm{Sp}_\infty(n)$ is defined by

$$\mu_\ell\left(s_\infty\right) = \mu\left(s_\infty \ell_\infty, \ell_\infty\right), \quad s_\infty \in \mathrm{Sp}_\infty(n), \quad (4.20)$$

where ℓ_∞ is any element of $\mathrm{Lag}_\infty(n)$ with projection ℓ. That $\mu_\ell(s_\infty)$ only depends on ℓ (thus motivating the notation μ_ℓ) follows from properties (4.8), (4.9) of μ: if $\ell'_\infty = \beta^r \ell_\infty$ for some integer r, then

$$\mu\left(s_\infty \ell'_\infty, \ell'_\infty\right) = \mu\left(s_\infty\left(\beta^r \ell_\infty\right), \beta^r \ell_\infty\right)$$

$$= \mu\left(s_\infty\left(\beta^r \ell_\infty\right), \ell_\infty\right) - 2r$$

$$= \mu\left(\beta^r \ell_\infty\right), s_\infty^{-1} \ell_\infty - 2r$$

$$= \mu\left(\ell_\infty, s_\infty^{-1} \ell_\infty\right)$$

$$= \mu\left(s_\infty \ell'_\infty, \ell'_\infty\right).$$

In ([7]) we proved the following result:

Proposition 4.4. For every $\ell \in \text{Lag}(n)$ there exists a unique mapping

$$\mu_\ell : \text{Sp}_\infty(n) \longrightarrow \mathbb{Z}$$

having the two following properties:

(i) for all $s_\infty s'_\infty \in \text{Sp}_\infty(n)$

$$\mu_\ell \left(s_\infty s'_\infty \right) = \mu_\ell \left(s_\infty \right) + \mu_\ell \left(s'_\infty \right) + \sigma_\ell \left(s, s' \right);$$

(ii) the mapping $(s_\infty, \ell, \ell'') \to \mu_\ell(s_\infty) - \sigma(s\ell, \ell, \ell'')$ is locally constant on the subset $\{(s_\infty, \ell, \ell'); s\ell \cap \ell'' = \ell \cap \ell'' = 0\}$ of $\text{Sp}_\infty(n) \times \text{Lag}(n) \times \text{Lag}(n)$.

In particular μ_ℓ is locally constant on $\{s_\infty, ; s\ell \cap \ell = 0\} \subset \text{Sp}_\infty(n)$.

Proposition 4.5. The mapping μ_ℓ has moreover the three properties

$$\mu_\ell \left(\alpha s_\infty \right) = \mu_\ell \left(s_\infty \right) + 4$$

$$\mu_\ell \left(s_\infty \right) = \mu_\ell \left(s_\infty^{-1} \right) = 0 \tag{4.21}$$

$$\mu_\ell \left(s_\infty \right) \equiv n - \dim(s\ell, \ell), \quad \mod 2.$$

The proof of this result is straightforward, using Theorem 4.1. Let $\text{Sp}_{0,\infty}(n)$ be the subset of $\text{Sp}_\infty(n)$ consisting of all $s_\infty \in \text{Sp}_\infty(n)$ with projection $s \in \text{Sp}_0(n)$. Leray ([13], Ch. I, §2.7) shows, by a direct construction involving the Maslov index on $\text{Lag}_\infty(n)$, the existence and uniqueness of a function

$$m : \text{Sp}_{0,\infty}(n) \longrightarrow \mathbb{Z}$$

which he calls the *Maslov index*. That function m is characterized by the two following properties (*ibidem*, Theorem 7.1, p. 52):

$$m \text{ is locally constant on its domain} \tag{4.22}$$

and if $s_\infty s'_\infty s''_\infty = I_\infty$ (the identity in $\text{Sp}_\infty(n)$) are such that $s, s', s'' \in \text{Sp}_0(n)$ then

$$m \left(s_\infty \right) - m \left(s_\infty^{'-1} \right) + m \left(s''_\infty \right) = \text{Inert} \left(s\ell_0, \ell_0, s''^{-1}\ell_0 \right), \tag{4.23}$$

where ℓ_0 is again the vertical Lagrangian plane X^*; replacing s''_∞ by its inverse $(s''_\infty)^{-1}$ and using the relation

$$m \left(s_\infty \right) + m \left(s_\infty^{-1} \right) = n \tag{4.24}$$

([13], p. 52) formula (4.23) is equivalent to

$$m \left(s_\infty s'_\infty \right) = m \left(s_\infty \right) + m \left(s'_\infty \right) - \text{Inert} \left(ss' \ell_0, s \ell_0, \ell_0 \right) \tag{4.25}$$

for all s_∞, s'_∞ in $\text{Sp}_0(n)$.

5. Path intersection indices

5.1. LAGRANGIAN PATHS

Let us introduce the following conventions and notations. We identify the universal covering $\text{Lag}_\infty(n)$ with the set of all homotopy classes ℓ_∞ of continuous path in $\text{Lag}(n)$ starting at a given point $\ell_0 \in \text{Lag}(n)$ and finishing at ℓ. The projection $\pi : \text{Lag}_\infty(n) \to \text{Lag}(n)$ is then the mapping which to ℓ_∞ associates the endpoint ℓ. For a path $\Lambda : [a, b] \to \text{Lag}(n)$ with *arbitrary* endpoints ℓ_a and ℓ_b we denote by:

(1) $\ell_{a,\infty}$ the homotopy class of an arbitrary path with origin ℓ_0 and endpoint ℓ_a: the projection of $\ell_{a,\infty}$ is thus ℓ_a;

(2) $\ell_{b,\infty}$ the homotopy class of the path obtaining by juxtaposition of a representant of $\ell_{a,\infty}$ and of the path Λ: the projection of $\ell_{b,\infty}$ is thus ℓ_b.

(3) $\ell_{0,\infty}$ the homotopy class of an arbitrary loop through ℓ_0 in $\text{Lag}(n)$.

Definition. Let Λ be a Lagrangian path; the integer

$$\mu_{\text{Lag}}(\Lambda, \ell) = \mu \left(\ell_{b,\infty}, \ell_{0,\infty} \right) - \mu \left(\ell_{a,\infty}, \ell_{0,\infty} \right) \tag{5.1}$$

is called the intersection index of Λ and ℓ_0.

In view of (4.6) we can rewrite definition (5.1) as:

$$\mu_{\text{Lag}} \left(\Lambda, \ell_0 \right) = -\mu_{\text{Lag}} \left(\ell_{a,\infty}, \ell_{b,\infty} \right) + \sigma \left(\ell_a, \ell_b, \ell_0 \right). \tag{5.2}$$

The intersection index $\mu_{\text{Lag}}(\Lambda, \ell_0)$ only depends on Λ and ℓ_0, and *not* on how we choose $\ell_{a,\infty}, \ell_{0,\infty}$. This is because if we replace $\ell_{a,\infty}$ by another element $\beta^r \ell_{a,\infty}$ of $\text{Lag}_\infty(n)$, then $\ell_{b,\infty}$ will be replaced by $\beta^r \ell_{b,\infty}$ and $\mu_{\text{Lag}}(\Lambda, \ell_0)$ will keep the same value by property (4.8) of the Maslov index. The same argument shows that $\mu_{\text{Lag}}(\Lambda, \ell_0)$ does not depend on the choice of $\ell_{0,\infty}$ (which is anyway clear from (5.2)!).

Theorem 5.1. *The index $\mu_{\text{Lag}}(\cdot, \ell_0)$ has the following properties: (1) homotopy invariance: two Lagrangian paths Λ and Λ' with same endpoints are homotopic*

if and only if they have same intersection indices: $\mu_{\text{Lag}}(\Lambda, \ell_0) = \mu_{\text{Lag}}(\Lambda', \ell_0)$;
(2) additivity: for $a < c < b$ we have

$$\mu_{\text{Lag}}(\Lambda, \ell_0) = \mu_{\text{Lag}}\left(\Lambda|_{[a,c]}, \ell_0\right) + \mu_{\text{Lag}}\left(\Lambda|_{[c,b]}, \ell_0\right); \tag{5.3}$$

(3) Sp(n)-invariance: for every $s \in \text{Sp}(n)$

$$\mu_{\text{Lag}}(s\Lambda, s\ell_0) = \mu_{\text{Lag}}(\Lambda, \ell_0) \tag{5.4}$$

(4) zero in strata: $\mu_{\text{Lag}}(\Lambda, \ell_0) = 0$ *if there exists an integer k $(0 \leq k \leq n)$ such that* $\dim(\Lambda(t) \cap \ell_0) = k$ *for $a \leq t \leq b$.*

Proof. (1): If Λ and Λ' are homotopic with fixed endpoints we can associate to them the same elements $\ell_{a,\infty}$ and $\ell_{b,\infty}$ of $\text{Lag}_\infty(n)$, and hence $\mu_{\text{Lag}}(\Lambda, \ell_0) = \mu_{\text{Lag}}(\Lambda', \ell_0)$ by (5.1), (5.2). Suppose conversely that Λ and Λ' are two Lagrangian paths with same endpoints, and denote by $\ell_{b,\infty}$, $\ell'_{b,\infty}$ the corresponding elements of $\text{Lag}_\infty(n)$. Since these elements have some projection $\Lambda(b)$ on $\text{Lag}(n)$ we have $\ell'_{b,\infty} = \beta^r \ell_{b,\infty}$ for $\beta^r \in \pi_1(\text{Lag}(n))$, and hence, by (4.8):

$$\mu_{\text{Lag}}(\Lambda', \ell_0) = \mu_{\text{Lag}}(\Lambda, \ell_0) + r.$$

If $\mu_{\text{Lag}}(\Lambda', \ell_0) = \mu_{\text{Lag}}(\Lambda, \ell_0)$ then $r = 0$ and $\ell'_{b,\infty} = \ell_{b,\infty}$ which implies that Λ and Λ' are homotopic. (2): This property immediately follows from (5.1). (3): It is an immediate consequence of (5.2) using (4.9). (4). It follows from property 2 in Theorem 3.1 of σ, and property (4.10) of the Leray index. $\square$

Particularly interesting is the following:

Proposition 5.2. Define, for $a \leq t \leq b$, $\Lambda(t) = \{(x, A(t)x) : x \in X\}$ where $t \mapsto A(t)$ is a continuous family of symmetric $n \times n$ matrices. Then

$$\mu_{RS}(\Lambda, X) = \text{sign}\, A(b) - \text{sign}\, A(a). \tag{5.5}$$

Proof. We have $\Lambda(t) \cap X^* = 0$ for $a \leq t \leq b$ and hence, by...:

$$\mu\left(\ell_{b,\infty}, X_\infty^*\right) - \mu\left(\ell_{a,\infty}, X_\infty^*\right) = 0.$$

It follows, using definition (5.1) and (4.6) that

$$\begin{aligned}
\mu_{\text{Lag}}(\Lambda, X) &= \mu\left(\ell_{b,\infty}, X\right) - \mu\left(\ell_{a,\infty}, X\right) \\
&= \mu\left(\ell_{b,\infty}, X\right) - \mu\left(\ell_{b,\infty}, X_\infty^*\right) \\
&\quad - \left(\mu\left(\ell_{a,\infty}, X\right) - \mu\left(\ell_{a,\infty}, X_\infty^*\right)\right) \\
&= \sigma\left(\ell_b, X^*, X\right) - \sigma\left(\ell_a, X^*, X\right)
\end{aligned}$$

hence (5.5), using (3.2). $\square$

5.2. SYMPLECTIC PATHS

The Maslov index and its properties allow us to define an intersection index for paths of symplectic matrices. An element of $Sp_\infty(n)$ will be viewed as a homotopy class of smooth path $[a, b] \to Sp(n)$ joining the identity $I \in Sp(n)$ to $s \in Sp(n)$; the projection on $Sp(n)$ of such a homotopy class s_∞ is s. The group structure is defined by juxtaposition of paths.

Definition. Let $\psi : [a, b] \to Sp(n)$ be a continuous path; the intersection index $\mu_{Sp}(\psi)$ is the integer

$$\mu_{Sp}(\psi) = \mu_{Lag}\left(\psi X^*, X^*\right). \tag{5.6}$$

Suppose the path ψ starts at the identity element I of $Sp(n)$: $\psi(a) = I$. Then $\mu_{Sp}(\psi)$ is just the index μ_{X^*}.

Proposition 5.3. The intersection index $\mu_{Sp}(\cdot)$ for symplectic paths has the following properties: (1) homotopy invariance: two paths $\psi, \psi' : [a, b] \to Sp(n)$ with same endpoints are homotopic if and only if $\mu_{Sp}(\psi) = \mu_{Sp}(\psi')$. (2) Additivity: for $a < c < b$ we have

$$\mu_{Sp}(\psi) = \mu_{Sp}\left(\psi|_{[a,c]}\right) + \mu_{Sp}\left(\psi|_{[c,b]}\right) \tag{5.7}$$

(3) zero in strata: $\mu_{Sp}(\psi) = 0$ if there exists an integer k $(0 \le k \le n)$ such that $\dim(\psi(t)X^* \cap X^*) = k$.

These properties are of course straightforward consequences of the properties of the Lagrangian path intersection index listed in Theorem 5.1.

6. The Robbin–Salamon theory

6.1. THE INDEX $\mu_{RS}(\Lambda, \ell)$

In [20], section 2, Robbin and Salamon associate to every smooth path $\Lambda :$ $[a, b] \to Lag(n)$ and every $\ell_0 \in Lag(n)$ a half-integer $\mu_{RS}(\Lambda, \ell_0)$. Their definition makes use of the signature of a 'crossing form' which is expressed in terms of an intersection number associated with ℓ_0, and the Lagrangian path Λ. The authors then show, in the same section of their article, that their index has the following properties:

RS1 *Homotopy invariance*: If Λ and Λ' are homotopic then $\mu_{RS}(\Lambda, \ell) = \mu_{RS}(\Lambda', \ell)$ for all $\ell \in Lag(n)$.

RS2 *Additivity:* $\mu_{RS}(\Lambda + \Lambda', \ell) = \mu_{RS}(\Lambda, \ell) + \mu_{RS}(\Lambda', \ell)$.

RS3 *Symplectic invariance*: if $s \in \mathrm{Sp}(n)$, then $\mu_{RS}(s\Lambda, s\ell) = \mu_{RS}(\Lambda, \ell)$.

RS4 *Zero in strata*: if Λ is defined on $[a, b]$ and if there exists an integer k $(0 \leq k \leq n)$ such that $\dim \Lambda(t) \cap \ell = k$ for $a \leq t \leq b$ then $\mu_{RS}(\Lambda, \ell) = 0$.

RS4 *Normalization*: For every integer, we have $\mu_{RS}(\beta^k, \ell) = k/2$.

RS5 *Localization*: Define, for $a \leq t \leq b$, $\Lambda(t) = \{(x, A(t)x) : x \in X\}$ where $t \mapsto A(t)$ is a continuous family of symmetric $n \times n$ matrices. Then

$$\mu_{RS}(\Lambda, \ell) = \frac{1}{2} \left(\mathrm{sign}\, A(b) - \mathrm{sign}\, A(a)\right). \tag{6.1}$$

Formula (6.1) in Proposition 5.2 suggests that there might be a simple relationship between the Lagrangian path intersection indices $\mu(\Lambda, \ell_0)$ and $\mu_{RS}(\Lambda, \ell_0)$. In fact, we have proved in [9] that

Theorem 6.1. *([9], Th. 3(a)). For all Lagrangian paths Λ and all $\ell_0 \in \mathrm{Lag}(n)$ we have*

$$\mu(\Lambda, \ell_0) = 2\mu_{RS}(\Lambda, \ell_0). \tag{6.2}$$

6.2. THE INDEX $\mu_{RS}(\Psi)$

Robbin and Salamon define in [20], section 4, a half-integer valued function calculated on paths ψ of symplectic matrices. That function, which they denote by $\mu(\psi)$, but which we will denoted by $\mu_{RS}(\psi)$, is defined in terms of their index of Lagrangian paths. Robbin and Salamon prove that the function $\mu_{RS}(\psi)$ has four essential properties. Two of these properties ('homotopy' and 'composition') are sufficient for us to prove the identity of $2\mu_{RS}(\psi)$ with the index μ_{X^*}. These two properties are the following:

(1) *Homotopy* ([20], Theorem 4.1, p. 837, l. 14): two symplectic paths ψ and ψ' with the same end points are homotopic if and only if $\mu_{RS}(\psi) = \mu_{RS}(\psi')$.

   That property implies that the restriction of μ_{RS} to all paths starting at $I \in \mathrm{Sp}(n)$ is a mapping

$$\mu_{RS} : \mathrm{Sp}_\infty(n) \longrightarrow \mathbb{R}.$$

(2) *Composition (ibidem, Theorem 5.1, p. 842, l. 3): the restriction of $\mu_{RS}(\psi)$ to all $s_\infty \in \mathrm{Sp}_\infty(n)$ with projection $s \in \mathrm{Sp}_\infty(n)$ such that $sX^* \cap X^* = 0$ is the unique locally constant map such that*

$$\mu_{RS}(s_\infty s_\infty') = \mu_{RS}(s_\infty) + \mu_{RS}(s_\infty') + \frac{1}{2}\,\mathrm{sign}\, Q(s, s') \tag{6.3}$$

(s, $s\prime$ the projections of $s_\infty s'_\infty$ respectively). In view of formula (3.9) in Corollary 3.4 this can be rewritten as

$$\mu_{RS}\left(s_\infty s'_\infty\right) = \mu_{RS}\left(s_\infty\right) + \mu_{RS}\left(s'_\infty\right) + \frac{1}{2}\sigma_{X^*}\left(s, s'\right). \tag{6.4}$$

The following relation between μ_{RS} and μ_{X^*} immediately follows:

Proposition 6.2. For all $s_\infty \in \mathrm{Sp}_\infty(n)$, with projection $s \in \mathrm{Sp}_0(n)$ we have

$$2\mu_{RS}\left(s_\infty\right) = \mu_{X^*}\left(s_\infty\right) \tag{6.5}$$

and hence

$$2\mu_{RS}(\psi) = \mu_{\mathrm{Sp}}(\psi) \tag{6.6}$$

for all loops in $\mathrm{Sp}(n)$ through the identity.

Proof. We first notice that the Leray index μ_{X^*} is the only function $\mathrm{Sp}_\infty(n) \to \mathbb{Z}$ which is locally constant on $\{s_\infty; sX^* \cap X^* = 0\}$ and such that

$$\mu_{X^*}\left(s_\infty s'_\infty\right) = \mu_{X^*}\left(s_\infty\right) + \mu_{X^*}\left(s'_\infty\right) + \sigma_{X^*}\left(s, s'\right). \tag{6.7}$$

Suppose in fact that μ'_{X^*} is another such mapping, and set $\upsilon = \mu_{X^*} - \mu'_{X^*}$. Then υ is such that

$$\upsilon\left(s_\infty s'_\infty\right) = \upsilon\left(s_\infty\right) + \upsilon\left(s'_\infty\right)$$

for all s_∞, s'_∞, and is locally constant on the set $\{s_\infty; s\ell \cap \ell = 0\}$. Now every $s''_\infty \in \mathrm{Sp}_\infty(n)$ is, the product of two elements s_∞, s'_∞ whose projections s, $s\prime$ are such that $s\ell \cap \ell = s'\ell \cap \ell = 0$ (this is an immediate consequence of the fact that the action of $\mathrm{Sp}(n)$ on pairs of transverse Lagrangian planes is transitive), hence υ is locally constant on the connected group $\mathrm{Sp}_\infty(n)$, and hence equal to 0. The function $2\mu_{RS}(s_\infty)$ is locally constant on $\{s_\infty : sX^* \cap X^* = 0\}$ and satisfies (6.7) in view of (6.4); the proposition follows. □

It turns out that the indices $2\mu_{RS}(\psi)$ and $\mu_{\mathrm{Sp}}(\psi)$ are actually equal for all symplectic paths:

Theorem 6.3. *For all symplectic paths* $\psi : [a, b] \to \mathrm{Sp}(n)$ *following identity holds:*

$$2\mu_{RS}(\psi) = \mu_{\mathrm{Sp}}(\psi). \tag{6.8}$$

Proof. Apply Theorem 6.1. □

7. In cauda venenum...

In the introduction to their article [20] Robbin and Salamon claim that the superiority of their definitions lies in their being able to compute intersection indices for symplectic and Lagrangian paths with arbitrary endpoints. The Leray index μ, whose complete definition goes back to 1989, allows us to reach the same goals in a much simpler way; it is moreover clear from Leray's original construction [13] that the definition of μ is topological (whereas differentiable paths are used in [20]). Moreover their composition form is immediately deduced, in a more general framework, from the properties of the signature of a triple of Lagrangian planes, whose definition is to be found in [5] and [16].

The alleged superiority of the constructions of McDuff, Robbin, and Salamon is thus *illusory*.

References

[1] Booss–Bavnbek, B. and Furutani, K., The Maslov index: a functional analytical definition and the spectral flow formula, *Tokyo J. math. (1)* **21** (1998).

[2] _____, Symplectic functional analysis and spectral invariants, geometric aspects of PDEs, *Cont. Math.* **242**, Amer. Math. Soc. (1999).

[3] Cappell, S.E., Lee, R., and Miller, E.Y., On the Maslov index, *Comm. Pure and Appl. Math.* **XLVII** (1994).

[4] Dazord, P., Invariants homotopiques attachés aux fibrés symplectiques, *Ann. Inst. Fourier, Grenoble (2)* **29** (1979), 25–78.

[5] Demazure, M., *Classe de Maslov II*, Exposé no. 10, Séminaire sur le fibré cotangent, Orsay (1975–76).

[6] Gosson, M. de., 1) La définition de l'indice de Maslov sans hypothèse de transversalité, *C. R. Acad. Sci., Paris* **309** série I (1990).
2) La relation entre Sp_∞, revêtement universel du groupe symplectique Sp et $Sp \times Z$, *C. R. Acad. Sci., Paris* **310** série I (1990).

[7] _____, The structure of q-symplectic geometry, *J. Math. Pures Appl.* **71** (1992), 429–453.

[8] _____, *Maslov classes, metaplectic representation and Lagrangian quantization*, Research Notes in Mathematics, vol. 95, Wiley–VCH, Berlin 1997.

[9] _____, *Lagrangian path intersections and the Leray index, geometry and topology*, Aarhus, Contemporary Mathematics, vol. 258, 2000, 177–184.

[10] Gosson, S. de., *Master's Thesis*, Kalmar, 1998.

[11] Keller, J.B., Corrected Bohr–Sommerfeld quantum conditions for nonseparable systems, *Ann. of Physics* **4** (1958), 180–188.

[12] Lefschetz, S., *Algebraic topology*, Amer. Math. Soc. Colloq. Publications, **27**, (1942).

[13] Leray, J., *Lagrangian analysis*, MIT Press, Cambridge, Mass., London 1981; Analyse Lagrangienne RCP 25, Strasbourg, Collège de France (1976–1977).

[14]   ——, The meaning of Maslov's asymptotic method the need of Planck's constant in mathematics, *Bull. of the Amer. Math. Soc.*, Symposium on the Mathematical Heritage of Henri Poincaré (1980).

[15]   Libermann, P. and Marle, C.-M., *Symplectic geometry and analytical mechanics*, D. Reidel Publishing Company 1990.

[16]   Lion, G. and Vergne, M., *The Weil representation, Maslov index and theta series*, Progress in mathematics, vol. 6, Birkhäuser 1980.

[17]   McDuff, D. and Salamon, D., *Introduction to symplectic topology*, Oxford science Publications 1998.

[18]   Maslov, V.P., *Théorie des Perturbations et Méthodes Asymptotiques*, Dunod, Paris 1972 [original Russian edition 1965].

[19]   Robbin, J. and Salamon, D.A., *The Spectral flow and the Maslov index*.

[20]   ——, The Maslov index for paths, *Topology* **32** (1993).

[21]   Souriau, J.M., *Construction explicite de l'indice de Maslov, Group theoretical methods in physics*, Lecture Notes in Physics, vol. 50, Springer–Verlag 1975, 17–148.

A KÄHLER STRUCTURE ON THE PUNCTURED COTANGENT BUNDLE OF THE CAYLEY PROJECTIVE PLANE

KENRO FURUTANI

Dedicated to the memory of Jean Leray

Contents

1 Introduction 165
2 Cayley projective plane 166
3 A Kähler structure 168
4 Freudenthal product and determinant 170
5 Proof of the theorem 172
6 An embedding into $M(8, \mathbb{C})$ 179

M. de Gosson (ed.), Jean Leray '99 Conference Proceedings, 163-182.
© 2003 Kluwer Academic Publishers.

Abstract. *We construct a Kähler structure on the punctured cotangent bundle of the Cayley projective plane whose Kähler form coincides with the natural symplectic form on the cotangent bundle and we show that the geodesic flow action is holomorphic and is expressed in a quite explicit form. We also give an embedding of the punctured cotangent bundle of the Cayley projective plane into the space of 8×8 complex matrices.*

1. Introduction

In the paper [8] a Kähler structure on the punctured cotangent bundle $T_0^* S^n = T^* S^n \setminus S^n$ of the sphere S^n is constructed through the mapping τ_S

$$\tau_S : T_0^* S^n \longrightarrow \mathbb{C}^{n+1},$$
$$(x, y) \longmapsto \|y\|x + \sqrt{-1}y. \tag{1.1}$$

It is shown that the natural symplectic form $\omega = \omega_S$ on the cotangent bundle coincides with the Kähler form $\sqrt{-2}\,\bar{\partial}\partial\|z\|$ (see also [10]). Moreover the geodesic flow action is holomorphic.

In the paper [2] we constructed a Kähler structure on the punctured cotangent bundle of complex and quaternion projective spaces with similar properties as for the sphere cases (see also [5]). This Kähler structure is just a positive complex polarization on the cotangent bundle as a symplectic manifold and is applied to construct a quantization operator by the method of pairing of polarizations. The operator quantizes geodesic flows of such manifolds. In other words such an operator gives a correspondence between the geodesic flow and the one-parameter group of Fourier integral operators generated by the square root of the Laplacian ([3] and [9]).

In this paper we construct a Kähler structure on the punctured cotangent bundle of the Cayley projective plane whose Kähler form coincides with the natural symplectic form on the cotangent bundle and is invariant under the action of the geodesic flow (Theorem 3.1).

In general it will not be easy to find such Riemannian manifolds whose (punctured) cotangent bundle has a Kähler structure where the symplectic form coincides with the Kähler form and is invariant under the action of the geodesic flow (see [11] and [12]). Such a Kähler structure for complex and quaternion projective spaces is constructed by making use of the Hopf fibration and the map above for the sphere. Although the Cayley projective plane has no fiber bundle like the Hopf fiber bundle, we prove here that a map similar to the cases of complex and quaternion projective spaces gives an embedding of the punctured cotangent bundle of the Cayley projective plane into the space of a complexified exceptional Jordan algebra. It is well known that the Cayley projective plane is one of the compact symmetric spaces of rank one and that the exceptional Lie group F_4 acts on it two-point homogeneously. Some properties which we prove in Theorem 3.1 could be shown quite easily if we used this property of minimal rank for symmetric spaces, but we prove our main theorem through elementary calculus in a Jordan algebra where the Cayley projective plane is realized as a subset consisting of primitive idempotents. Further we give an embedding of this image in the complexified exceptional Jordan algebra into the space of 8×8 complex matrices by composing with a map given by [13].

In §2 we describe the Cayley projective plane as a subset consisting of primitive idempotents in an exceptional Jordan algebra. In §3 we state our main theorem and a corollary. In §4 we recall some basic facts about the Jordan algebra including, so called, the Freudenthal product and the determinant on the Jodran algebra. In §4 we prove our main theorem and in §6 we describe an embedding of the punctured cotangent bundle of the Cayley projective plane into the space of 8×8 complex matrices.

2. Cayley projective plane

In this section we describe the Cayley projective plane as a subset in the exceptional Jordan algebra over the real number field $\mathbb{R}$ (see [1] and [7]).

Let $\mathbb{H}$ be the quaternion number field, that is, $\mathbb{H}$ is an algebra over $\mathbb{R}$ generated by $\{e_i\}_{i=0}^3$ with the relations:

$$
\begin{aligned}
e_0 e_i &= e_i e_0 \quad (i = 0, 1, 2, 3), \\
e_i^2 &= -e_0 \quad (i = 1, 2, 3), \\
e_i e_j &= -e_j e_i = e_k, \quad \text{mod } 3.
\end{aligned}
\tag{2.1}
$$

The Cayley number field $\mathbb{O}$ is a division algebra over $\mathbb{R}$ generated by $\{e_i\}_{i=0}^7$ with $e_i e_j$ given by the table

	e_0	e_1	e_2	e_3	e_4	e_5	e_6	e_7
e_0	e_0	e_1	e_2	e_3	e_4	e_5	e_6	e_7
e_1	e_1	$-e_0$	e_3	$-e_2$	e_5	$-e_4$	$-e_7$	e_6
e_2	e_2	$-e_3$	$-e_0$	e_1	e_6	e_7	$-e_4$	$-e_5$
e_3	e_3	e_2	$-e_1$	$-e_0$	e_7	$-e_6$	e_5	$-e_4$
e_4	e_4	$-e_5$	$-e_6$	$-e_7$	$-e_0$	e_1	e_2	e_3
e_5	e_5	e_4	$-e_7$	e_6	$-e_1$	$-e_0$	$-e_3$	e_2
e_6	e_6	e_7	e_4	$-e_5$	$-e_2$	e_3	$-e_0$	$-e_1$
e_7	e_7	$-e_6$	e_5	e_4	$-e_3$	$-e_2$	e_1	$-e_0$

$$
\tag{2.2}
$$

In particular,

$$\mathbf{e}_1\mathbf{e}_4 = \mathbf{e}_5, \qquad \mathbf{e}_2\mathbf{e}_4 = \mathbf{e}_6, \qquad \mathbf{e}_3\mathbf{e}_4 = \mathbf{e}_7. \tag{2.3}$$

Hence $\mathbb{O}$ is identified with

$$\mathbb{H} \oplus \mathbb{H}\mathbf{e}_4 \tag{2.4}$$

and the multiplication between $x = a + b\mathbf{e}_4$ and $y = h + k\mathbf{e}_4 \in \mathbb{H} \oplus \mathbb{H}\mathbf{e}_4$ is given by

$$x \cdot y = ah - \theta(k)b + \{ka + b\theta(h)\}\,\mathbf{e}_4, \tag{2.5}$$

where $h = \sum_{i=0}^{3} h_i\mathbf{e}_i\,(h_i \in \mathbb{R})$ and $\theta(h) = h_0\mathbf{e}_0 - h_1\mathbf{e}_1 - h_2\mathbf{e}_2 - h_3\mathbf{e}_3$, and so on. We assume that the basis $\{\mathbf{e}_i\}_{i=0}^{7}$ are *orthonormal* and we will sometimes omit $\mathbf{e}_0$ (= identity element) and identify $\mathbb{R} = \mathbb{R}\mathbf{e}_0 \subset \mathbb{H} \subset \mathbb{O}$.

For $h = \sum_{i=0}^{7} h_i\mathbf{e}_i \in \mathbb{O}$, we denote

$$\theta(h) = h_0\mathbf{e}_0 - \sum_{i=1}^{7} h_i\mathbf{e}_i, \tag{2.6}$$

as for $h \in \mathbb{H}$.

Let $M(3, \mathbb{O})$ be the space of 3×3 matrices with entries in $\mathbb{O}$. By identifying $M(3, \mathbb{O}) \cong M(3, \mathbb{R}) \otimes_{\mathbb{R}} \mathbb{O}$, we denote for $X \in M(3, \mathbb{O})$

$$\theta(X) = X_0 \otimes \mathbf{e}_0 - \sum_{i=1}^{7} X_i \otimes \mathbf{e}_i,$$
$$^{t}X = \sum_{i=1}^{7} {}^{t}X_i \otimes \mathbf{e}_i \tag{2.7}$$

and

$$\mathrm{tr}\, X = \sum_{i=0}^{7} (\mathrm{tr}\, X_i)\,\mathbf{e}_i, \tag{2.8}$$

where $X = \sum_{i=0}^{7} X_i \otimes \mathbf{e}_i$, $X_i \in M(3, \mathbb{R})$.

Now consider the subspace $\mathfrak{J}$ in $M(3, \mathbb{O})$ defined by

$$\mathfrak{J} = \left\{ X \in M(3, \mathbb{O}) \mid \theta\left({}^{t}X\right) = X \right\}, \tag{2.9}$$

then $\dim_{\mathbb{R}} \mathfrak{J} = 27$ and any $X \in \mathfrak{J}$ has the form

$$X = \begin{pmatrix} \xi_1\mathbf{e}_0 & x_3 & \theta(x_2) \\ \theta(x_3) & \xi_2\mathbf{e}_0 & x_1 \\ x_2 & \theta(x_1) & \xi_3\mathbf{e}_0 \end{pmatrix},$$

where $\xi_i \in \mathbb{R}$, $x_i \in \mathbb{O}$.

The space $\mathfrak{J}$ is called an exceptional Jordan algebra with the Jordan product

$$X \circ Y = \frac{1}{2}(XY + YX) \in \mathfrak{J}, \qquad (2.10)$$

$X, Y \in \mathfrak{J}$.

$\mathfrak{J}$ has an inner product given by

$$\operatorname{tr}(X \circ Y) = (X, Y)\mathbf{e}_0. \qquad (2.11)$$

In fact, for $X \in \mathfrak{J}$, we have $\operatorname{tr} X \in \mathbb{R}\mathbf{e}_0$, and

$$(X, Y) = \sum_{i=1}^{3} (\xi_i \eta_i + 2 \langle x_i, y_i \rangle), \qquad (2.12)$$

where

$$X = \begin{pmatrix} \xi_1 \mathbf{e}_0 & x_3 & \theta(x_2) \\ \theta(x_3) & \xi_2 \mathbf{e}_0 & x_1 \\ x_2 & \theta(x_1) & \xi_3 \mathbf{e}_0 \end{pmatrix},$$

$$Y = \begin{pmatrix} \eta_1 \mathbf{e}_0 & y_3 & \theta(y_2) \\ \theta(y_3) & \eta_2 \mathbf{e}_0 & y_1 \\ y_2 & \theta(y_1) & \eta_3 \mathbf{e}_0 \end{pmatrix},$$

and $\langle x_i, y_i \rangle$ denotes the inner product on $\mathbb{O}$.

Now the Cayley projective plane $P^2\mathbb{O}$ is defined as

Definition. $P^2\mathbb{O} = \{X \in \mathfrak{J} \mid X \circ X = X, \operatorname{tr} X = 1\}$.

The exceptional Lie group F_4 is defined as a group of algebra automorphisms of $\mathfrak{J}$, and acts on $P^2\mathbb{O}$ two-point homogeneously. So we have $P^2\mathbb{O} \cong F_4/\operatorname{Spin}(9)$, where $\operatorname{Spin}(9)$ is realized as a subgroup of F_4 consisting of those elements which leave $\begin{pmatrix} 1 & 0 & 0 \\ 0 & 0 & 0 \\ 0 & 0 & 0 \end{pmatrix} \in \mathfrak{J}$ invariant.

3. A Kähler structure

In this section we describe a Kähler structure on the punctured cotangent bundle $T_0^* P^2\mathbb{O} = T^* P^2\mathbb{O} \backslash P^2\mathbb{O}$ and state our main Theorem 3.1.

The tangent bundle $T P^2\mathbb{O}$ is identified with the subset in $\mathfrak{J} \times \mathfrak{J}$ as

$$T P^2\mathbb{O} = \left\{ (X, Y) \in \mathfrak{J} \times \mathfrak{J} \mid X \circ X = X, \operatorname{tr} X = 1, X \circ Y = \frac{1}{2}Y \right\}. \quad (3.1)$$

We introduce the Riemannian metric on $P^2\mathbb{O}$ such that

$$(Y_1, Y_2)_{\mathbb{P}} = \frac{1}{2}\operatorname{tr}(Y_1 \circ Y_2) = \frac{1}{2}(Y_1, Y_2), \tag{3.2}$$

where $(X, Y_1), (X, Y_2) \in TP^2\mathbb{O}$.

Then from the inclusions $\mathbb{C} \subset \mathbb{H} \subset \mathbb{O}$, the complex projective plane $P^2\mathbb{C}$ and the quaternion projective plane $P^2\mathbb{H}$ are embedded isometrically into $P^2\mathbb{O}$ as totally geodesic submanifolds.

In the following we identify the tangent bundle $TP^2\mathbb{O}$ and the cotangent bundle $T^*P^2\mathbb{O}$ through the metric above. Under this identification the symplectic form $\omega_{\mathbb{O}}$ on $T^*P^2\mathbb{O}$ is given by

$$\omega_{\mathbb{O}} = -\frac{1}{2}(dX, dY), \tag{3.3}$$

where we should interpret the inner product (dX, dY) as a two-form in such a way that

$$-(dX, dY) = \sum_{i=1}^{3} d\eta_i \wedge d\xi_i + 2\sum_{i=1}^{3}\sum_{\alpha=0}^{7} dy_\alpha^i \wedge dx_\alpha^i \tag{3.4}$$

restricted to $TP^2\mathbb{O}$, that is, we notice that $\sum \eta_i \xi_i$ is replaced by $\sum d\eta_i \wedge d\xi_i$ and so on in the definition of the inner product on $\mathfrak{J}$

We can extend $\theta : \mathfrak{J} \to \mathfrak{J}, {}^t : \mathfrak{J} \to \mathfrak{J}, \operatorname{tr} : \mathfrak{J} \to \mathbb{O}$, and the inner product $(\cdot, \cdot)$ to the complexification $\mathfrak{J} \otimes_{\mathbb{R}} \mathbb{C} = \mathfrak{J}^{\mathbb{C}}$ in a natural way. So the Hermite inner product $\langle \cdot, \cdot \rangle$ on $\mathfrak{J}^{\mathbb{C}}$ is given by

$$\langle X, Y \rangle = (X, \overline{Y}), \tag{3.5}$$

where $\overline{X} = \sum_{\alpha=0}^{7} \overline{X}_\alpha \otimes e_\alpha$, $X_\alpha \in M(3, \mathbb{C})$ and $\overline{X}_\alpha$ is the complex conjugate of X_α. The norm of these elements in $\mathfrak{J}$ and $\mathfrak{J}^{\mathbb{C}}$ is always written as $\|\cdot\|$, and we write the norm of the tangent vector $Y \in T_X(P^2\mathbb{O})$ by $\|Y\|_{\mathbb{P}}$.

Now consider the map $\tau_{\mathbb{O}} : T_0^* P^2\mathbb{O}(\cong T_0 P^2\mathbb{O}) \to \mathfrak{J}^{\mathbb{C}}$ defined by

$$\begin{aligned}
\tau_{\mathbb{O}}(X, Y) &= \left(\|Y\|^2 X - Y \circ Y\right) \otimes 1 + \frac{1}{\sqrt{2}}\|Y\|Y \otimes \sqrt{-1} \\
&= \left(2\|Y\|_{\mathbb{P}}^2 X - Y \circ Y\right) \otimes 1 + \|Y\|_{\mathbb{P}}Y \otimes \sqrt{-1}, \tag{3.6}
\end{aligned}$$

then we have:

Theorem 3.1. *The map $\tau_{\mathbb{O}}$ gives an isomorphism between $T_0^* P^2\mathbb{O}$ and $\mathbb{E} = \{A \in \mathfrak{J}^{\mathbb{C}} \mid A \circ A = 0, A \neq 0\}$. Moreover,*

$$\tau_{\mathbb{O}}^*\left(\sqrt{-1}\,\overline{\partial}\partial\|A\|^{1/2}\right) = \frac{1}{\sqrt{2}}\omega_{\mathbb{O}}. \tag{3.7}$$

The two-form $\sqrt{-2}\,\bar{\partial}\partial\|A\|^{1/2}$ is itself a Kähler form on $\mathfrak{J}^{\mathbb{C}}\backslash\{0\}$, so that we can regard $\mathfrak{J}^{\mathbb{C}}\backslash\{0\}$ is a symplectic manifold. On this symplectic manifold the flow $\{\phi_t\}_{t\in\mathbb{R}}$ defined by

$$\phi_t : A \longmapsto \phi_t(A) = e^{-2\sqrt{-1}t} \cdot A$$

is a Hamilton flow. The Hamiltonian of this flow is given by the function $f : A \mapsto (1/\sqrt{2})\|A\|^{1/2}$. Since $\mathbb{E}$ is holomorphic and the flow $\{\phi_t\}$ leaves $\mathbb{E}$ invariant, the Hamiltonian of this flow on $\mathbb{E}$ is just the restriction of f to $\mathbb{E}$, that is, the Hamiltonian is the square root of the metric function. So the flow $\{\phi_t\}$ is the bicharacteristic flow of the square root of the Laplacian on $P^2\mathbb{O}$. Especially the flow restricted to the unit sphere $= \{(X, Y) \in TP^2\mathbb{O} : \|Y\|_{\mathbb{P}} = 1\}$ coincides with the geodesic flow. So we have

Corollary. The geodesics $\gamma(t)$ on $P^2\mathbb{O}$ through a point X with the direction Y ($\|Y\|_{\mathbb{P}} = 1$ and $X \circ Y = (1/2)Y$) is given by

$$\gamma(t) = \cos 2t \cdot \left(X - \frac{1}{2}Y \circ Y\right) + \frac{1}{2}\sin 2t \cdot Y + \frac{1}{2}Y \circ Y. \qquad (3.8)$$

4. Freudenthal product and determinant

In this section we recall several formulas in the Jordan algebra $\mathfrak{J}$ for later use (see [7]).

Let $X, Y \in \mathfrak{J}$, then the 'Freudenthal product' $X \times Y \in \mathfrak{J}$ is defined by the formula

$$X \times Y = \frac{1}{2}\{2X \circ Y - (\text{tr } X)Y - (\text{tr } Y)X + (\text{tr } X \cdot \text{tr } Y - \text{tr}(X \circ Y)) E\}, \quad (4.1)$$

where $E = \begin{pmatrix} 1 & 0 & 0 \\ 0 & 1 & 0 \\ 0 & 0 & 1 \end{pmatrix}$ and the determinant, 'det X', for $X \in \mathfrak{J}$ is defined by

$$\det X = \frac{1}{3}\text{tr}(X \circ (X \times X)). \qquad (4.2)$$

Then we have

Proposition.   (i) $(X \circ Y, Z) = (X, Y \circ Z), \forall X, Y, Z \in \mathfrak{J}$

(ii) $X \circ (X \times X) = \det X \cdot E$ *(Cayley–Hamilton)*

(iii) $(X \times X) \times (X \times X) = \det X \cdot X$.

As we explained above

$$F_4 = \{g \in GL(\mathfrak{J}) \mid g(X \circ Y) = g(X) \circ g(Y), \forall X, Y \in \mathfrak{J}\}. \tag{4.3}$$

Now F_4 is also given in the following ways:

$$
\begin{aligned}
F_4 &= \{g \in GL(\mathfrak{J}) \mid \text{for any } X, Y \in \mathfrak{J}\} \\
&= \{g \in GL(\mathfrak{J}) \mid \text{for any } X, Y \in \mathfrak{J}\} \\
&= \{g \in GL(\mathfrak{J}) \mid \text{for any } X, Y \in \mathfrak{J}\}.
\end{aligned}
$$

We also have for $g \in F_4$

$$\text{tr } gX = \text{tr } X. \tag{4.4}$$

The 'Freudenthal product' and 'det' on $\mathfrak{J}$ are extended naturally to the complexification $\mathfrak{J}^{\mathbb{C}}$, and we denote them with the same notations. Then the complexification of F_4 is defined in the same way:

Definition. The complex simple Lie group $F_4^{\mathbb{C}}$ is

$$
\begin{aligned}
F_4^{\mathbb{C}} &= \left\{g \in GL\left(\mathfrak{J}^{\mathbb{C}}\right) \mid g(X \circ Y) = g(X) \circ g(Y)\right\} \\
&= \left\{g \in GL\left(\mathfrak{J}^{\mathbb{C}}\right) \mid \det(gX) = \det X, (g(X), g(Y)) = (X, Y)\right\} \\
&= \left\{g \in GL\left(\mathfrak{J}^{\mathbb{C}}\right) \mid g(X \times Y) = g(X) \times g(Y)\right\}.
\end{aligned}
$$

The two-point homogeneity of F_4 on $P^2\mathbb{O}$ is equivalent to:

Proposition. Let $S(P^2\mathbb{O}) = \{(X, Y) \mid (X, Y) \in TP^2\mathbb{O} \subset \mathfrak{J} \times \mathfrak{J}, \|Y\| = 1\}$, then F_4 acts on $S(P^2\mathbb{O})$ transitively and we have

$$S\left(P^2\mathbb{O}\right) = F_4 / \text{Spin}(7), \tag{4.5}$$

where the stationary subgroup at the point $\left(\begin{pmatrix} 1 & 0 & 0 \\ 0 & 0 & 0 \\ 0 & 0 & 0 \end{pmatrix}, \begin{pmatrix} 0 & 1/\sqrt{2} & 0 \\ 1/\sqrt{2} & 0 & 0 \\ 0 & 0 & 0 \end{pmatrix}\right) \in S(P^2\mathbb{O})$ is identified with $\text{Spin}(7)$.

Remark. The representation of $F_4^{\mathbb{C}}$ on $\mathfrak{J}_0^{\mathbb{C}}$ ($=\{A \in \mathfrak{J}^{\mathbb{C}} \mid \text{tr } A = 0\}$) is irreducible, and the subspace $\mathbb{E}$ is the orbit of the highest weight vector. According to the theorem by Lichtenstein ([6]), then such an orbit is characterized as the null set of a certain system of quadric equations. Thus our equation $A \circ A = 0$ (also this is equivalent to $A \times A = 0$ and $\text{tr } A = 0$) is nothing else than an example of this theorem in [6], however we need the map τ_0 to put a Kähler structure on $T_0^* P^2\mathbb{O}$.

5. Proof of the theorem

We give a proof of Theorem 3.1 in a series of lemmas and proposition.
First we prove

Lemma 5.1. *Let* $(X, Y) \in TP^2\mathbb{O}$, *then*

(i) $\det X = 0$,

(ii) $\operatorname{tr} Y = 0$,

(iii) $\det Y = 0$.

Proof. Since

$$X \times X = \frac{1}{2} \left\{ 2X \circ X - 2 \operatorname{tr} X \cdot X + \left((\operatorname{tr} X)^2 - (X, X) \right) E \right\} = 0,$$

we have

$$\det X = 0$$

by the definition of the determinant

$$X \circ (X \times X) = \det X \cdot E = 0.$$

From the equality

$$(X \circ Y, Z) = (X, Y \circ Z), \quad (X, Y, Z \in \mathfrak{J})$$

we have

$$(X \circ Y, X) = \frac{1}{2}(Y, X) = (Y, X \circ X) = (Y, X).$$

Hence $\operatorname{tr} Y = (1/2) \operatorname{tr}(X \circ Y) = (X, Y) = 0$.
Since $\operatorname{tr} Y = 0$, the Freudenthal product $Y \times Y$ is expressed as

$$Y \times Y = Y \circ Y - \frac{1}{2} \|Y\|^2 E,$$

and we have

$$Y \circ (Y \times Y) = Y \circ (Y \circ Y) - \frac{1}{2} \|Y\|^2 Y = \det Y \cdot E. \tag{5.1}$$

Then by taking the trace we have

$$\operatorname{tr}(Y \circ (Y \circ Y)) = 3 \det Y.$$

Now

$$\text{tr}(Y \circ (Y \circ Y)) = \left(Y, Y \times Y + \frac{1}{2}\|Y\|^2 E\right)$$

$$= 2\left(X \circ Y, Y \times Y + \frac{1}{2}\|Y\|^2 E\right)$$

$$= 2\left(X, Y \circ \left(Y \times Y + \frac{1}{2}\|Y\|^2 E\right)\right)$$

$$= 2(X, Y \circ (Y \times Y)) + \left(X, \|Y\|^2 Y\right)$$

$$= 2(X, \det Y \cdot E)$$

$$= 2 \det Y.$$

Hence we have

$$\det Y = 0. \tag{5.2}$$

$\square$

Lemma 5.2. *For* $(X, Y) \in T_0 P^2 \mathbb{O}$, $\tau_0(X, Y) \in \mathbb{E}$, *that is,* $\tau_0(X, Y) \circ \tau_0(X, Y) = 0$.

Proof. Let $X, Y \in T_0 P^2$, then

$$\tau_0(X, Y) \circ \tau_0(X, Y) = \left(\left(\|Y\|^2 X - Y \circ Y\right)^2 - \frac{1}{2}\|Y\|^2 Y \circ Y\right) \otimes 1$$

$$+ \frac{2}{\sqrt{2}}\|Y\|Y \circ \left(\|Y\|^2 X - Y \circ Y\right) \otimes \sqrt{-1}$$

$$= \left(\|Y\|^4 X - 2\|Y\|^2 X \circ (Y \circ Y)\right.$$

$$+ (Y \circ Y) \circ (Y \circ Y) - \frac{1}{2}\|Y\|^2 Y \circ Y\left.\right) \otimes 1$$

$$+ \sqrt{2}\left(\frac{1}{2}\|Y\|^3 Y - \|Y\|Y \circ (Y \circ Y)\right) \otimes \sqrt{-1}.$$

Here we notice the following formulas: let $(X, Y) \in T(P^2\mathbb{O})$, then:

(i) $Y \circ (Y \circ Y) = (1/2)\|Y\|^2 Y$;

(ii) $(X + Y) \times (X + Y) = (X - Y) \times (X - Y) = Y \circ Y - (1/2)\|Y\|^2 E$;

(iii) $\det(X \pm Y) = 0$;

(iv) $X \circ (Y \circ Y) = (1/2)\|Y\|^2 X$.

(i) is obtained by the Cayley–Hamilton

$$Y \circ (Y \times Y) = \det Y \cdot E = 0.$$

and (ii) is easily shown. (iii) and (iv) are proved by first calculating

$$(X \pm Y) \circ ((X \pm Y) \times (X \pm Y)) = (X \pm Y) \circ \left(Y^2 - \frac{1}{2} \|Y\|^2 E \right)$$

$$= X \circ (Y \circ Y) - \frac{1}{2} \|Y\|^2 X$$

$$= \det(X \pm Y) \cdot E,$$

and by taking trace of both sides, we know $\det(X \pm Y) = 0$. Hence we have also (iv).

Next, from the formula $(Y \times Y) \times (Y \times Y) = (\det Y)Y = 0$ we have

$$0 = \left(Y \circ Y - \frac{1}{2} \|Y\|^2 E \right) \times \left(Y \circ Y - \frac{1}{2} \|Y\|^2 E \right)$$

and so we have

$$\|Y \circ Y\|^2 = \frac{1}{2} \|Y\|^4 \tag{5.3}$$

and

$$(Y \circ Y) \circ (Y \circ Y) = \frac{1}{2} \|Y\|^2 Y \circ Y. \tag{5.4}$$

Finally by making use of these formulas we can prove

$$\tau_0(X, Y) \circ \tau_0(X, Y)$$

$$= \left(\|Y\|^4 X - 2\|Y\|^2 \cdot \frac{1}{2} \|Y\|^2 X + \frac{1}{2} \|Y\|^2 Y \circ Y - \frac{1}{2} \|Y\|^2 Y \circ Y \otimes 1 \right)$$

$$+ \sqrt{2} \left(\|Y\|^3 \cdot \frac{1}{2} Y - \|Y\| \cdot \frac{1}{2} \|Y\|^2 Y \right) \otimes \sqrt{-1}$$

$$= 0. \qquad \qquad \qquad \qquad \qquad \qquad \qquad \qquad \square$$

Let a map $\sigma : \mathbb{E} \to \mathfrak{J} \times \mathfrak{J}$ be

$$\sigma : A \longmapsto (X, Y), \tag{5.5}$$

where X and Y are given by the following formulas:

$$X = \frac{1}{2} \frac{1}{\|A\|} \left(A + \overline{A} \right) + \frac{A \circ \overline{A}}{\|A\|^2}, \tag{5.6}$$

$$Y = -\frac{\sqrt{-1}}{\sqrt{2}} \|A\|^{-1/2} \left(A - \overline{A} \right). \tag{5.7}$$

Proposition. Let X, Y be defined as above for $A \in \mathbb{E}$, then;

(i) $X \circ X = X,\ {}^t\theta(X) = X,\ \operatorname{tr} X = 1$

(ii) $Y \circ X = (1/2)Y,\ \operatorname{tr} Y = 0.$

Proof. By the definition of the Freudenthal product we have

$$
\begin{aligned}
A \times A &= \frac{1}{2}\left\{2A \circ A - 2\operatorname{tr} A \cdot A + \left((\operatorname{tr} A)^2 - (A, A)\right) E\right\}\\
&= \frac{1}{2}\left\{-2\operatorname{tr} A \cdot A + (\operatorname{tr} A)^2 E\right\},
\end{aligned}
$$

and

$$
\begin{aligned}
A \circ (A \times A) &= \det A \cdot E\\
&= A \circ \left(-\operatorname{tr} A \cdot A + \frac{1}{2}(\operatorname{tr} A)^2 E\right)\\
&= \frac{1}{2}(\operatorname{tr} A)^2 A.
\end{aligned}
$$

So we have

$$
\det A \cdot A = \frac{1}{2}(\operatorname{tr} A)^2 A \circ A = 0 \tag{5.8}
$$

and then we have

$$
\det A = 0, \qquad \operatorname{tr} A = 0. \tag{5.9}
$$

It follows easily that ${}^t\theta(X) = X$ because of ${}^t\theta(A) = A$ and ${}^t\theta(\overline{A}) = \overline{A}$.
Put $A = a \otimes 1 + b \otimes \sqrt{-1} \in \mathbb{E}$, where $a, b \in \mathfrak{J}$. Now we can assume $\|A\| = 1$, because of the homogeneity of the map σ. Then we have $a \circ a = b \circ b$, $a \circ b = 0$, $\|a\|^2 = \|b\|^2 = 1/2$, and $\operatorname{tr} a = \operatorname{tr} b = 0$. Also we show

$$
\det a = \det b = 0, \qquad \det(a \pm b) = 0. \tag{5.10}
$$

The last equalities are proved by the following argument: from the equalities $a \times a = a \circ a - (1/4)E$ and $b \times b = b \circ b - (1/4)E$, we have

$$
a \circ (a \times a) = a \circ (a \circ a) - \frac{1}{4}a = \det a \cdot E,
$$

$$
b \circ (b \times b) = b \circ (b \circ b) - \frac{1}{4}b = \det b \cdot E,
$$

$$
(a \pm b) \times (a \pm b) = 2a \circ a - \frac{1}{2}E = 2b \circ b - \frac{1}{2}E.
$$

Then

$$(a \pm b) \circ ((a \pm b) \times (a \pm b)) = (a \pm b) \circ \left(2a \circ a - \frac{1}{2}E\right)$$

$$= (a \pm b) \circ \left(2b \circ b - \frac{1}{2}E\right)$$

$$= 2a \circ \left(a \circ a - \frac{1}{4}E\right) + 2b \circ \left(b \circ b - \frac{1}{4}E\right)$$

$$= \det(a \pm b)E.$$

Hence we have

$$2(\det a \pm \det b) = \det(a \pm b).$$

On the other hand

$$((a \pm b) \times (a \pm b)) \times ((a \pm b) \times (a \pm b)) = \det(a \pm b)(a \pm b) \quad (5.11)$$

is given by

$$((a \pm b) \times (a \pm b)) \times ((a \pm b) \times (a \pm b))$$

$$= \left(2a \circ a - \frac{1}{2}E\right) \times \left(2a \circ a - \frac{1}{2}E\right).$$

Note that the last equality shows that (5.11) does not depend on the sign. Hence

$$\det(a + b) \cdot (a + b) = \det(a - b) \cdot (a - b).$$

So we have

$$\det(a + b) = \det(a - b),$$

since $\mathrm{tr}(a \circ a) = 1/2$ and

$$\det a = \det b = 0. \quad (5.12)$$

Then these finally imply

$$\det(a \pm b) = 0. \quad (5.13)$$

By making use of these formulas we prove $X \circ X = X$. Again we may assume $\|A\| = 1$, then X is written as

$$X = a \otimes 1 + \left(a \otimes 1 + b \otimes \sqrt{-1}\right) \circ \left(a \otimes 1 - b \otimes \sqrt{-1}\right)$$

$$= a + 2a \circ a.$$

Hence

$$
\begin{aligned}
X \circ X &= (a + 2a \circ a) \circ (a + 2a \circ a) \\
&= a \circ a + 4a \circ (a \circ a) + 4(a \circ a) \circ (a \circ a) \\
&= a \circ a + 4 \cdot \frac{a}{4} + a \circ a = a + 2a \circ a = X,
\end{aligned}
$$

where we have used the equality

$$
(a \circ a) \circ (a \circ a) = \frac{1}{4} a \circ a.
$$

Also we have $\operatorname{tr} X = 1 = 2 \operatorname{tr}(a \circ a)$.

Next we show

$$
X \circ Y = \frac{1}{2} Y.
$$

Since

$$
Y = \frac{\sqrt{-1}}{\sqrt{2}} \left(a \otimes 1 + b \otimes \sqrt{-1} - \left(a \otimes 1 - b \otimes \sqrt{-1} \right) \right),
$$

$$
\begin{aligned}
X \circ Y &= -\sqrt{2}(a + 2a \circ a) \circ b \\
&= -\sqrt{2} \left(a \circ b + 2b \circ (a \circ a) \right) \\
&= -2\sqrt{2} b \circ \left(a \times a + \frac{1}{4} E \right) \\
&= -2\sqrt{2} b \circ \left(b \times b + \frac{1}{4} E \right) \\
&= -2\sqrt{2} \left(2 \det b \cdot E + \frac{1}{2} b \right) \\
&= \frac{1}{2} \cdot \left(-2\sqrt{2} b \right) = \frac{1}{2} Y.
\end{aligned}
$$

From these we have proved that τ_0 is a bijection between $T_0^* P^2 \mathbb{O}$ in $\mathfrak{J} \times \mathfrak{J}$ and $\mathbb{E}$ in $M(3, \mathbb{O}) \otimes_{\mathbb{R}} \mathbb{C}$ and that σ is the inverse map.

Next we prove $\tau_0^* (\sqrt{-2} \, \partial \bar\partial \sqrt{\|A\|}) = \omega_0$, that is, the Kähler form $\sqrt{-2} \, \partial \bar\partial \sqrt{\|A\|}$ coincides with the symplectic form ω_0 on $T_0^* P^2 \mathbb{O}$.

First we have

$$
\begin{aligned}
\tau_0^* \left(\sqrt{-1} \, \bar\partial \partial \|A\|^{1/2} \right) &= \tau_0^* \left(\sqrt{-1} \, \bar\partial \partial \, (A, \bar A)^{1/4} \right) \\
&= \frac{\sqrt{-1}}{4} d \left(\tau_0^* (A, \bar A)^{-3/4} (dA, \bar A) \right) \\
&= \frac{\sqrt{-1}}{4} d \left(\|Y\|^{-3} (d\tau_0^* A, \tau_0^* (\bar A)) \right).
\end{aligned}
$$

Here we should consider $A \in \mathfrak{J}^{\mathbb{C}}$ to be the section

$$A : \mathbb{E} \longrightarrow \mathbb{E} \times \mathfrak{J}^{\mathbb{C}}, \quad A \longmapsto (A, A) \tag{5.14}$$

of the trivial bundle $\mathbb{E} \times \mathfrak{J}^{\mathbb{C}}$ on $\mathbb{E}$, and dA the section of $\mathfrak{J}^{\mathbb{C}} \otimes T^{*}\mathbb{E}$. Note that the inner product $(\cdot, \cdot)$ defines the pairing $\mathfrak{J} \times \mathfrak{J} \otimes T^{*}\mathbb{E} \to T^{*}\mathbb{E}$. In the calculations below we will use this pairing with the same notation $(\cdot, \cdot)$. Also we note $\|A\|^{2} = \|Y\|^{4}$ under the mapping τ_{0}.

We write

$$\tau_{0}^{*}(A) = \left(\|Y\|^{2}X - Y \circ Y \right) \otimes 1 + \frac{1}{\sqrt{2}}\|Y\|Y \otimes \sqrt{-1} \tag{5.15}$$

$$= a \otimes 1 + b \otimes \sqrt{-1}, \tag{5.16}$$

where $a = a(X, Y), b = b(X, Y)$. Then

$$\tau_{0}^{*}\left(dA, \overline{A} \right) = \left(d\tau_{0}^{*}A, \tau_{0}^{*}\overline{A} \right)$$

$$= \left(da \otimes 1 + d\left(b \otimes \sqrt{-1} \right), a \otimes 1 - b \otimes \sqrt{-1} \right)$$

$$= (da, a) + (db, b) + ((db, a) - (da, b)) \sqrt{-1}.$$

Now from

$$(a, a) = \frac{1}{2}\|Y\|^{4} = (b, b)$$

we have

$$(da, a) = (Y, Y)(dY, Y)$$
$$(db, b) = (Y, Y)(dY, Y).$$

So the real part of

$$\tau_{0}^{*}\left((dA, \overline{A}) \right)$$

is a closed form, since $d((Y, Y)(dY, Y)) = 2(dY, Y) \wedge (dY, Y) = 0$. From

$$(a, b) = \left(\|Y\|^{2}X - Y \circ Y, \frac{1}{\sqrt{2}}\|Y\|Y \right) = 0,$$

$$(db, a) - (b, da) = 2(db, a)$$

$$= 2\frac{1}{\sqrt{2}}\left(Y \otimes \frac{(dY, Y)}{\|Y\|} + \|Y\|Y, \|Y\|^{2}X - Y \circ Y \right)$$

$$= \frac{2}{\sqrt{2}}\left\{ \|Y\|^{3}(dY, X) - \|Y\|(dY, Y \circ Y) \right\}.$$

Hence

$$\frac{(\sqrt{-1})^2}{4} \cdot \frac{2}{\sqrt{2}} \left[d \left\{ \|Y\|^{-3} \cdot \left(\|Y\|^3 (dY, X) - \|Y\|(dY, Y \circ Y) \right) \right\} \right]$$

$$= \frac{-1}{2\sqrt{2}} \left\{ d(dY, X) - d \left(\frac{(dY, Y \circ Y)}{\|Y\|^2} \right) \right\}.$$

By the equality $(X, Y \circ Z) = (X \circ Y, Z)$ we have

$$(dY, Y \circ Y) = (Y, dY \circ Y)$$

and by $(Y, Y \circ Y) = 0$, we have

$$(dY, Y \circ Y) = 0.$$

Hence we finally proved

$$\tau_0^* \left(\sqrt{-1} \, \bar{\partial} \partial \sqrt{\|A\|} \right) = \frac{1}{2\sqrt{2}} (dY, dX) = \frac{1}{\sqrt{2}} \omega_0. \tag{5.17}$$

$\square$

Remark. The map τ_0 commutes with the actions of F_4 on $T_0 P^2 \mathbb{O}$ and on $\mathfrak{J}^{\mathbb{C}}$ (as a subgroup in $F_4^{\mathbb{C}}$). Of course all the elements in F_4 preserve the symplectic form ω_0. Then it will be true that the subgroup of $F_4^{\mathbb{C}}$ consisting of those elements that preserve the symplectic form ω_0 is compact. Hence it will coincide with F_4.

6. An embedding into $M(8, \mathbb{C})$

In this section we describe an embedding of the space $\mathbb{E} \subset \mathfrak{J}^{\mathbb{C}}$ into the space of 8×8 complex matrices $M(8, \mathbb{C})$.

By identifying $\mathbb{O} \cong \mathbb{H} \oplus \mathbb{H} e_4$, we define $\gamma : \mathbb{O} \to \mathbb{O}$ by

$$\gamma(h + k e_4) = h - k e_4. \tag{6.1}$$

The map γ is naturally extended to the complexification $\mathbb{O} \otimes \mathbb{C}$, where we regard $\mathbb{O} \otimes \mathbb{C} \cong \mathbb{H} \otimes \mathbb{C} \oplus \mathbb{H} \otimes \mathbb{C} e_4$. It is easily verified that γ is an algebra isomorphism of $\mathbb{O}$ (and of $\mathbb{O} \otimes \mathbb{C}$), that is, $\gamma \in G_2$ ($\cong$ the group of algebra isomorphisms of $\mathbb{O}$), $\gamma^2 = \mathrm{Id}$, and $\theta \circ \gamma = \gamma \circ \theta$.

Let $X \in \mathfrak{J}^{\mathbb{C}}$ be

$$X = \begin{pmatrix} \xi_1 & x_3 & \theta(x_2) \\ \theta(x_3) & \xi_2 & x_1 \\ x_2 & \theta(x_1) & \xi_3 \end{pmatrix},$$

$\xi_i \in \mathbb{C}$, $x_i \in \mathbb{O} \otimes \mathbb{C}$, $x_i = \sum_{\alpha=0}^{7} x_i^{\alpha} \mathbf{e}_{\alpha} (x_i^{\alpha} \in \mathbb{C})$, and we denote by

$$\gamma(X) = \begin{pmatrix} \xi_1 & \gamma(x_3) & \theta\,(\gamma(x_2)) \\ \theta\,(\gamma(x_3)) & \xi_2 & \gamma\,(x_1) \\ \gamma(x_2) & \theta\,(\gamma(x_1)) & \xi_3 \end{pmatrix}, \tag{6.2}$$

then

$$\gamma : \mathfrak{J}^{\mathbb{C}} \longrightarrow \mathfrak{J}^{\mathbb{C}} \tag{6.3}$$

is also an algebra isomorphism.

We decompose elements $X \in \mathfrak{J}^{\mathbb{C}}$ as

$$\begin{aligned}
X &= \begin{pmatrix} \xi_1 & x_3 & \theta(x_2) \\ \theta(x_3) & \xi_2 & x_1 \\ x_2 & \theta(x_1) & \xi_3 \end{pmatrix} \\
&= \begin{pmatrix} \xi_1 & m_3 & \theta(m_2) \\ \theta(m_3) & \xi_2 & m_1 \\ m_2 & \theta(m_1) & \xi_3 \end{pmatrix} + \begin{pmatrix} 0 & a_3 & -a_2 \\ -a_3 & 0 & a_1 \\ a_2 & -a_1 & 0 \end{pmatrix} \mathbf{e}_4 \\
&= M + A\mathbf{e}_4,
\end{aligned}$$

where $x_i = m_i + a_i \mathbf{e}_4 (m_i, a_i \in \mathbb{H} \otimes \mathbb{C})$. Then we can regard

$$\mathfrak{J} \otimes \mathbb{C} = \mathfrak{J}(3, \mathbb{H}) \otimes \mathbb{C} \oplus \mathbb{H}^3 \otimes \mathbb{C}.$$

Here we denote by $\mathfrak{J}(3, \mathbb{H})$ the Jordan algebra of 3×3 Hermitian matrices with entries in $\mathbb{H}$.

Now we define a map g following [13]:

$$g : \mathfrak{J} \otimes \mathbb{C} \longrightarrow \mathfrak{J}(4, \mathbb{H})_0 \otimes \mathbb{C}, \tag{6.4}$$

$$X = M + A\mathbf{e}_4 \longmapsto \begin{pmatrix} \frac{1}{2} \operatorname{tr} M & ia_1 & ia_2 & ia_3 \\ i\theta(a_1) & & & \\ i\theta(a_2) & & M - \frac{1}{2} \operatorname{tr} M \cdot E & \\ i\theta(a_3) & & & \end{pmatrix},$$

where we denote by $\mathfrak{J}(4, \mathbb{H})_0$ the subspace in the Jordan algebra $\mathfrak{J}(4, \mathbb{H})$ consisting of elements whose trace is zero.

The map g satisfies:

(i) $g(X) \circ g(Y) = g(\gamma(X \times Y)) + (1/4)(\gamma(X), Y) \cdot E$;

(ii) $(g(X), g(Y)) = (\gamma(X), Y) = \operatorname{tr}(\gamma(X) \circ Y) = \operatorname{tr}(g(X) \circ g(Y))$;

We remark that $\mathrm{tr}(A \circ B)$ for $A, B \in \mathfrak{J}(4, \mathbb{H})$ defines a Euclidean inner product on $\mathfrak{J}(4, \mathbb{H})$. We extend it to the complexification $\mathfrak{J}(4, \mathbb{H}) \otimes \mathbb{C}$.

If $A \in \mathfrak{J}^{\mathbb{C}}$, $A \circ A = 0$, then from (i) above we have at once

$$g(A) \circ g(A) = \frac{1}{2}(\gamma(A), A) \cdot E. \tag{6.5}$$

Let $\rho : \mathbb{H} \otimes \mathbb{C} \to M(2, \mathbb{C})$ be the isomorphism given by

$$\rho\left(\sum_{i=0}^{3} z_i e_i\right) \longmapsto \begin{pmatrix} z_0 + z_1 i & z_2 + z_3 i \\ -z_2 + z_3 i & z_0 - z_1 i \end{pmatrix}, \tag{6.6}$$

and we denote with the same notation ρ the map

$$\rho : \mathfrak{J}(4, \mathbb{H}) \otimes \mathbb{C} \longrightarrow M(8, \mathbb{C}), \qquad (h_{ij}) \longmapsto (\rho(h_{ij})), \tag{6.7}$$

where $h_{ij} \in \mathbb{H} \otimes \mathbb{C}$.

Proposition.

$$\rho(\mathfrak{J}(4, \mathbb{H})) = \{A \in M(8, \mathbb{C}) \mid \mathbb{J}A = {}^t A \mathbb{J}\}, \tag{6.8}$$

where

$$\mathbb{J} = \begin{pmatrix} J & 0 & 0 & 0 \\ 0 & J & 0 & 0 \\ 0 & 0 & J & 0 \\ 0 & 0 & 0 & J \end{pmatrix}, \quad J = \begin{pmatrix} 0 & 1 \\ -1 & 0 \end{pmatrix}.$$

Let $A \in \mathfrak{J} \otimes \mathbb{C}$ and $A \circ A = 0$, then as we know $\rho(g(A))^2$ is a scalar matrix and $\mathrm{tr}\, \rho(g(A)) = 0$. Conversely we have

Proposition. Let $A \in M(8, \mathbb{C})$, $A \neq 0$, $\mathbb{J}A = {}^t A \mathbb{J}$, and $A^2 = \lambda E$, $\lambda \in \mathbb{C}$, $\mathrm{tr}\, A = 0$, then there exists an element $X \in \mathfrak{J}^{\mathbb{C}}$ such that

$$X \circ X = 0, \qquad \rho(g(X)) = A$$

under the condition that A is of the form

$$A = \begin{pmatrix} 0 & 0 & & & \\ 0 & 0 & & & \\ & & \xi_1 & 0 & & \\ & & 0 & \xi_1 & & \\ \hline & & & & \xi_2 & 0 & \\ & & & & 0 & \xi_2 & \\ \hline & & & & & & \xi_3 & 0 \\ & & & & & & 0 & \xi_3 \end{pmatrix}, \qquad \sum_{i=1}^{3} \xi_i = 0. \tag{6.9}$$

Remark. The canonical line bundle of this complex structure on $T_0^* P^2 \mathbb{C}$ will be holomorphically trivial and this realization of $T_0^* P^2 \mathbb{C}$ in the space $M(8, \mathbb{C})$ given in the above proposition will be useful to construct an explicit holomorphic trivialization of the canonical line bundle.

References

[1]   Besse, A.L., *Manifolds all of whose Geodesics are closed*, Springer–Verlag 1978.

[2]   Furutani, K. and Tanaka, R.A., Kähler structure on the punctured cotangent bundle of complex and quaternion projective spaces and its application to geometric quantization I, *J. Math. Kyoto Univ.* **34–4** (1994), 57–69.

[3]   Furutani, K. and Yoshizawa, S., A Kähler structure on the punctured cotangent bundle of complex and quaternion projective spaces and its application to geometric quantization II, *Japanese J. Math.* **21** (1995), 355–392.

[4]   Ii, K., On a Bargmann-type transform and a Hilbert space of holomorphic functions, *Tôhoku Math. J. (1)* **38** (1986), 57–69.

[5]   Ii, K. and Morikawa, T., Kähler structures on tangent bundle of Riemannian manifolds of constant positive curvature, *Bull. Yamagata Univ. Natur. Sci. (3)* **14** (1999), 141–154.

[6]   Lichtenstein, W., A system of quadrics describing the orbit of the highest weight vector, *Proceedings of Amer. Math. Soc.*, vol. 84, 1982, no. 4, 605–608.

[7]   Murakami, S., *Exceptional simple Lie groups and related topics in recent differential geometry, in differential geometry and topology*, Springer Lecture Notes 1369, Proceedings, at Tianjir 1986–1987.

[8]   Rawnsley, J.H., Coherent states and Kähler manifolds, *Quart. J. Math. Oxford* **28** (1977), 403–415.

[9]   _____, A non-unitary pairing of polarization for the Kepler problem, *Trans. Amer. Math. Soc.* **250** (1979), 167–180.

[10] Souriau, J.M., Sur la variété de Kepler, *Symposia Math.* **14** (1974), 343–360.

[11] Szőke, R., Complex structures on the tangent bundle of Riemannian manifolds, *Math. Ann.* **291** (1991), 409–428.

[12] _____, Adapted complex structures and geometric Quantization, *Nagoya J. Math.* **154** (1999), 171–183.

[13] Yokota, I., Realizations of involutive automorphisms σ and G^σ of exceptional linear Lie groups G. Part I, $G = G_2$, F_4 and E_6, *Tsukuba J. Math. (1)* **14** (1990), 185–223.

ON MECHANICAL SYSTEMS WITH A LIE GROUP AS CONFIGURATION SPACE

CHARLES-MICHEL MARLE

*À la mémoire du Professeur Jean Leray, en témoignage
d'admiration et de respect*

Contents

1 Introduction 185
2 The right and left actions of a Lie group on its cotangent bundle 187
3 Taking into account a magnetic field 191
4 Reduction of one of the actions $\widehat{L}$ or $\widehat{R}$ to a subgroup 194
5 The rigid body with a fixed point 196
6 The motion of an ideal incompressible fluid 197
7 The Korteweg–de Vries equation as an Euler equation 201

M. de Gosson (ed.), Jean Leray '99 Conference Proceedings, 183-203.

1. Introduction

Several physically important mechanical systems have a configuration space which may be identified with a Lie group. Let us give some examples.

Example 1.1. Let us consider a massive point particle moving in the physical space E, of dimension 3. The configuration space is the affine space E. Once a particular point O of E has been chosen as origin (that point being looked at as a reference configuration of the system) we can consider E as a vector space, therefore as a Lie group with the addition of vectors as composition law.
More generally, let us consider a system of n massive point particles moving in the physical 3-dimensional space E. The configuration space is now $E \times \cdots \times E$, product of n copies of E, one for each particle. We allow two distinct particles to occupy the same position in space at a given time (otherwise we should take as configuration space an open subset of $E \times \cdots \times E$). As above, once a particular point $O = (O_1, \ldots, O_n)$ of $E \times \cdots \times E$ is chosen, the configuration space of the system can be considered as a vector space, therefore as a Lie group.

Example 1.2. Let us consider a massive rigid body moving in the Euclidean 3-dimensional space E. We assume that the body has at least three distinct points not on the same straight line. Let S_0 be a particular position of the body in E, which will be considered as a reference configuration of the system. Then for any other configuration, i.e., for any other position S of the rigid body in E, there exists a unique element g of the group $\mathscr{E}(E)$ of Euclidean displacements of E which maps S_0 onto S. Therefore, once a reference configuration is chosen the configuration space of the system can be identified with the Lie group $\mathscr{E}(E)$.

Example 1.3. Let us particularize Example 1.2 a little by assuming that our rigid body has a point which remains fixed in space, and that it rotates around that point. We take that point as origin. As above, once a reference configuration is chosen the configuration space of the system can be identified with the subgroup of elements $g \in \mathscr{E}(E)$ such that $g(0) = 0$, i.e., with the linear group $SO(E, 0)$ (the space E, with 0 as origin, being now considered as an Euclidean vector space).

Example 1.4. Consider an ideal, incompressible fluid which fills a vessel V fixed in the physical space E, and flows in that vessel. We assume that V is a compact, connected part of E, bounded by a smooth surface ∂V. To describe the configuration space of the system, we introduce the set of all fluid particles; it is an abstract set $\mathscr{V}$, with a smooth three-dimensional manifold structure with boundary, and a volume 3-form. A configuration of the system is a diffeomorphism $\varphi : \mathscr{V} \to V$ such that the pullback of the natural (meaning determined by the Euclidean structure) volume 3-form v of V is equal to the 3-form given of $\mathscr{V}$. Let $\varphi_0 : \mathscr{V} \to V$ be a reference configuration of the system.

For any other configuration $\varphi : \mathcal{V} \to V$, there exists a unique element g of the group $\mathrm{Diff}(V, v)$ of volume-preserving diffeomorphisms of V such that $\varphi = g \circ \varphi_0$. Conversely, for any $g \in \mathrm{Diff}(V, v)$, $g \circ \varphi_0$ is a configuration of the system. Therefore the configuration space of the system can be identified with $\mathrm{Diff}(V, v)$, which is an infinite-dimensional Lie group.

Example 1.5. As we shall see later, the well known Korteweg–de Vries equation on the circle S^1 can be considered as the Euler equation of a mechanical system whose configuration space is an infinite-dimensional Lie group, the Virasoro–Bott group.

Among these systems, the most remarkable are those whose phase space is a non-Abelian Lie group (Examples 1.2 to 1.5). These systems share remarkable properties which derive from the fact that a Lie group G acts on itself by two distinct actions: the left and right translations. These two actions can be canonically lifted into two actions,

$$\widehat{L} : G \times T^*G \longrightarrow T^*G, \quad \text{and} \quad \widehat{R} : T^*G \times G \longrightarrow T^*G$$

of the Lie group G on its cotangent bundle T^*G. These two actions are Hamiltonian with respect to the canonical symplectic structure of the cotangent bundle T^*G, and have Ad*-equivariant momentum maps $J_L : T^*G \to \mathcal{G}^*$ and $J_R : T^*G \to \mathcal{G}^*$. Moreover, the Hamiltonian $H : T^*G \to \mathbb{R}$ of the system is invariant, in some cases under the action $\widehat{L}$, in other cases under the other action $\widehat{R}$, and still in other cases under the restriction of $\widehat{L}$ (or of $\widehat{R}$) to a subgroup of G. As a consequence, the Hamiltonian vector field X_H on T^*G can be projected

(i)   by J_R into $\mathcal{G}^*$ when H is $\widehat{L}$-invariant,

(ii) by J_L into $\mathcal{G}^*$ when H is $\widehat{R}$-invariant,

(iii) by the momentum map of an action of a new Lie group G_1, which is a semi-direct product of G with an Abelian group, when H is invariant under the restriction of $\widehat{L}$ (or of $\widehat{R}$) to a suitable subgroup of G.

That projected vector field defines a differential equation (on $\mathcal{G}^*$ in cases (i), and (ii), and on $\mathcal{G}_1^*$ in case (iii)), called the Euler equation. That equation is Hamiltonian with respect to the Lie–Poisson structure of $\mathcal{G}^*$ (or of $\mathcal{G}_1^*$).

These remarkable properties were discovered by Euler [3] around 1765 for Examples 1.3 and 1.4 (the motion of a rigid body around a fixed point, and the motion of an ideal, incompressible fluid). They were expressed with the modern concepts of Lie groups and Lie algebras, and generalized, by V. Arnol'd [1] in 1966. The construction of the group G_1 and of its action on T^*G, needed when the Hamiltonian H is invariant by the restriction of $\widehat{L}$ (or of $\widehat{R}$) to a subgroup of G only, appears in the works of Iacob and Sternberg [5], Marsden, Ratiu and Weinstein [7], Guillemin and Sternberg [4]. The recent book by Arnol'd and Khesin [2] fully develops applications of Lie groups to Hydrodynamics.

In section 2 we discuss the actions $\widehat{L}$ and $\widehat{R}$ of a Lie group G on its cotangent bundle T^*G, and we obtain the expressions and properties of their momentum maps J_L and J_R.

In section 3 we present a slight generalization, useful for mechanical systems involving a magnetic field. The phase space of such systems is the cotangent bundle to their configuration manifold equipped with a symplectic form which differs from the canonical symplectic form of a cotangent bundle: it is the sum of the canonical symplectic form and of the pull-back of a closed 2-form on the configuration space.

In section 4 we discuss the construction of the Lie group G_1 and of its action on T^*G, when one of the two actions $\widehat{L}$ or $\widehat{R}$ is restricted to a subgroup of G.

Finally in sections 5 to 7 we develop the examples briefly sketched above.

2. The right and left actions of a Lie group on its cotangent bundle

Let G be a Lie group. We will denote by $\mathcal{G}$ its Lie algebra, and by $\mathcal{G}^*$ the dual of $\mathcal{G}$. For any $g \in G$, we denote by $L_g : G \to G$ the left translation $L_g(h) = gh$, and by $R_g : G \to G$ the right translation $R_g(h) = hg$. We denote by $TL_g : TG \to TG$ and $TR_g : TG \to TG$ the canonical lifts to the tangent bundle TG of L_g and R_g, respectively. We define the maps $\widehat{L}_g : T^*G \to T^*G$ and $\widehat{R}_g : T^*G \to T^*G$, as the transpose of $TL_{g^{-1}}$ and $TR_{g^{-1}}$, respectively:

$$\widehat{L}_g = {}^t(TL_{g^{-1}}), \qquad \widehat{R}_g = {}^t(TR_{g^{-1}}).$$

We recall that the Liouville 1-form α on the cotangent bundle T^*G is given by the formula, where $w \in T(T^*G)$,

$$\langle \alpha(p_{T^*G}(w)), w \rangle = \langle p_{T^*G}(w), Tq_G(w) \rangle,$$

where $p_{T^*G} : T(T^*G) \to T^*G$ and $q_G : T^*G \to G$ are the canonical projections.

We denote by $\Omega = d\alpha$ the canonical symplectic form on T^*G.

Theorem 2.1. *The maps*

$$\widehat{L} : G \times T^*G \longrightarrow T^*G, \qquad \widehat{L}(g, \xi) = \widehat{L}_g \xi,$$
$$\widehat{R} : T^*G \times G \longrightarrow T^*G, \qquad \widehat{R}(\xi, g) = \widehat{R}_g \xi,$$

are two commuting Hamiltonian actions of the Lie group G on the symplectic manifold (T^*G, Ω), respectively, on the left and on the right. They admit as

Ad^*-*equivariant momentum maps, respectively, the maps* $J_L : T^*G \to \mathscr{G}^*$ *and* $J_R : T^*G \to \mathscr{G}^*$,

$$J_L(\xi) = \widehat{R}_{(q_G(\xi))^{-1}}\xi, \qquad J_R(\xi) = \widehat{L}_{(q_G(\xi))^{-1}}\xi.$$

Proof. The actions L and R of G on itself by left and right translations satisfy, for all g and $h \in G$,

$$L_g \circ L_h = L_{gh}, \qquad R_g \circ R_h = R_{hg}, \qquad L_g \circ R_h = R_h \circ L_g.$$

Therefore L is an action on the left and R an action on the right, and these two actions commute. Since $\widehat{L}$ and $\widehat{R}$ are the canonical lifts to the cotangent bundle T^*G of the actions L and G, respectively, they satisfy, for all g and $h \in G$,

$$\widehat{L}_g \circ \widehat{L}_h = \widehat{L}_{gh}, \qquad \widehat{R}_g \circ \widehat{R}_h = \widehat{R}_{hg}, \qquad \widehat{L}_g \circ \widehat{R}_h = \widehat{R}_h \circ \widehat{L}_g.$$

This means that $\widehat{L}$ is an action on the left, $\widehat{R}$ an action on the right, and that these two actions commute. For any $g \in G$, $\widehat{L}_g$ is the canonical lift to the cotangent bundle T^*G of the diffeomorphism $L_g : G \to G$; therefore, it satisfies

$$\left(\widehat{L}_g\right)^* \alpha = \alpha, \qquad \left(\widehat{L}_g\right)^* \Omega = \Omega.$$

This shows that $\widehat{L}$ is a symplectic action of G on (T^*G, Ω). For the same reason, $\widehat{R}$ is a symplectic action of G on (T^*G, Ω). For every $X \in \mathscr{G}$, we denote by $X_{T^*G}^L$ and $X_{T^*G}^R$ the vector fields on T^*G defined by

$$X_{T^*G}^L(\xi) = \frac{d}{dt}\left(\widehat{L}_{\exp(tX)}\xi\right)\big|_{t=0}, \qquad X_{T^*G}^R(\xi) = \frac{d}{dt}\left(\widehat{R}_{\exp(tX)}\xi\right)\big|_{t=0}.$$

They are called the fundamental vector fields on T^*G associated to the element X of the Lie algebra $\mathscr{G}$, for the actions $\widehat{L}$ and $\widehat{R}$, respectively. In order to prove that the action L is Hamiltonian and admits as a momentum map the map $J_L : T^*G \to \mathscr{G}^*$, we must check that for any $X \in \mathscr{G}$, the fundamental vector field $X_{T^*G}^L$ is Hamiltonian and admits as Hamiltonian the function $X \circ J_L$ (we consider here X as a linear form on the dual $\mathscr{G}^*$ of the Lie algebra $\mathscr{G}$). In other words, we must check that

$$i\left(X_{T^*G}^L\right)\Omega = -d\left(X \circ J_L\right).$$

Since the action $\widehat{L}$ leaves invariant the Liouville 1-form α, we have

$$\mathscr{L}\left(X_{T^*G}^L\right)\alpha = i\left(X_{T^*G}^L\right)d\alpha + di\left(X_{T^*G}^L\right)\alpha = 0,$$

therefore, since $\Omega = d\alpha$,

$$i\left(X_{T^*G}^L\right)\Omega = -di\left(X_{T^*G}^L\right)\alpha.$$

According to the definition of the Liouville 1-form α, we have, for any $\xi \in T^*G$,

$$i\left(X_{T^*G}^L\right)\alpha(\xi) = \left\langle \xi, Tq_G\left(X_{T^*G}^L(\xi)\right)\right\rangle.$$

But since $\widehat{L}$ is the canonical lift to T^*G of the action of G on itself by left translations, we have

$$Tq_G\left(X_{T^*G}^L(\xi)\right) = \frac{d}{dt}\left(L_{\exp(tX)}q_G(\xi)\right)|_{t=0} = TR_{q_G(\xi)}X.$$

Therefore, since $J_L(\xi) = \widehat{R}_{q_G(\xi)^{-1}}(\xi)$,

$$\left\langle \xi, Tq_G\left(X_{T^*G}^L(\xi)\right)\right\rangle = \left\langle \widehat{R}_{q_G(\xi)^{-1}}\xi, X\right\rangle = X \circ J_L(\xi).$$

So we obtain

$$i\left(X_{T^*G}^L\right)\Omega = -d\left(X \circ J_L\right).$$

A similar calculation shows that

$$i\left(X_{T^*G}^R\right)\Omega = -d\left(X \circ J_R\right).$$

We have proved that the actions $\widehat{L}$ and $\widehat{R}$ are Hamiltonian and admit as momentum maps the maps J_L and J_R, respectively.

Finally, its very expression shows that J_L is equivariant for the $\widehat{L}$-action of G on T^*G and the Ad^*-action of G on $\mathcal{G}^*$. Similarly, J_R is equivariant for the $\widehat{R}$-action of G on T^*G and the Ad^*-action of G on $\mathcal{G}^*$. The action of G on T^*G being tacitly assumed to be $\widehat{L}$ when we consider J_L and $\widehat{R}$ when we consider J_R, we will say, in short, that both J_L and J_R are Ad^*-equivariant.   □

Remark. The momentum maps J_L and J_R are submersions. For any $\xi \in T^*G$, we observe that $J_L(\xi)$ is the unique point where the orbit of ξ under the action $\widehat{R}$ meets $\mathcal{G}^*$, considered as the cotangent space to G at the unit element. Similarly $J_R(\xi)$ is the unique point where the orbit of ξ under the action $\widehat{L}$ meets $\mathcal{G}^*$. Therefore the orbits of the $\widehat{L}$-action are the level sets of J_R, and the orbits of the $\widehat{R}$-action are the level sets of J_L.

Theorem 2.2. *Let n be the dimension of the Lie group G. For each point $\xi \in T^*G$, the tangent spaces at ξ to the $\widehat{L}$-orbit and to the $\widehat{R}$-orbit of that point are two n-dimensional vector subspaces of $T_\xi(T^*G)$, symplectically orthogonal to each other. The momentum map $J_L : T^*G \to \mathcal{G}^*$ is a Poisson map when $\mathcal{G}^*$ is equipped with the Lie–Poisson bracket, called the minus Lie–Poisson bracket, given by the formula (where f and g are two smooth functions on $\mathcal{G}^*$ and η a point in $\mathcal{G}^*$),*

$$\{f, g\}_-(\eta) = -\langle \eta, [df(\eta), dg(\eta)]\rangle.$$

Similarly, the momentum map $J_R : T^*G \to \mathcal{G}^*$ *is a Poisson map when* $\mathcal{G}^*$ *is equipped with the Lie–Poisson bracket, called the plus Poisson bracket, opposite of the previous one,*

$$\{f, g\}_+(\eta) = \langle \eta, [df(\eta), dg(\eta)] \rangle .$$

Application to a G-invariant Hamiltonian system on T^*G

Let $H : T^*G \to \mathbb{R}$ be a smooth function, and X_H the associated Hamiltonian vector field on T^*G, such that

$$i(X_H)\,\Omega = -dH.$$

Let us assume that H is invariant under the action $\widehat{L}$ of G on T^*G. Of course, similar results would hold, *mutatis mutandis*, if H were invariant under the action $\widehat{R}$. Since the $\widehat{L}$-orbits are the level sets of J_R, there exists a unique smooth function $\widehat{H} : \mathcal{G}^* \to \mathbb{R}$ such that

$$H = \widehat{H} \circ J_R.$$

Since J_R is a Poisson map (when $\mathcal{G}^*$ is equipped with the plus Lie–Poisson bracket), the Hamiltonian vector field X_H on T^*G projects, under the map J_R, onto $\mathcal{G}^*$, and its projection is the Hamiltonian vector field $X_{\widehat{H}}$ associated with the Hamiltonian $\widehat{H}$ on the Poisson manifold $(\mathcal{G}^*, \{\,,\,\}_+)$. The differential equation on $\mathcal{G}^*$,

$$\frac{d\varphi(t)}{dt} = X_{\widehat{H}}(\varphi(t)),$$

defined by the vector field $X_{\widehat{H}}$, is called the *Euler equation.*

Remark. Under the same assumption as above, Noether's theorem shows that the momentum map J_L is constant on each integral curve of X_H. In order to make easier the determination of these integral curves, one may use the well known Marsden–Weinstein reduction procedure [8]. Let η be an element of $\mathcal{G}^*$. To determine the integral curves of X_H on which the (constant) value taken by J_L is η, we first consider $J_L^{-1}(\eta)$, which contains all these integral curves. Since J_L is a submersion, $J_L^{-1}(\eta)$ is a submanifold of T^*G. Let G_η be the stabilizer of η (for the coadjoint action of G on $\mathcal{G}^*$). Since J_L is Ad*-equivariant, the restriction of the $\widehat{L}$-action to the subgroup G_η leaves $J_L^{-1}(\eta)$ invariant. The quotient manifold $J_L^{-1}(\eta)/G_\eta$, that means the set of orbits of the $\widehat{L}$-action of G_η on $J^{-1}(\eta)$, has a reduced symplectic structure. Moreover, there exists a unique smooth function $\widetilde{H} : J_L^{-1}(\eta)/G_\eta \to \mathbb{R}$ such that

$$H|_{J_L^{-1}(\eta)} = \widetilde{H} \circ \pi,$$

where $\pi : J_L^{-1}(\eta) \to J_L^{-1}(\eta)/G_\eta$ is the canonical projection. That projection maps the integral curves of X_H contained in $J_L^{-1}(\eta)$ onto the integral curves of the Hamiltonian vector field $X_{\tilde{H}}$, on the reduced symplectic manifold $J_L^{-1}(\eta)/G_\eta$.

Remark. So we see that two different procedures can be used in order to make easier the determination of the integral curves of X_H:

— the use of the Poisson map $J_R : T^*G \to \mathcal{G}$, which maps these curves onto the integral curves of the Euler equation on the Poisson manifold $\mathcal{G}^*$,

— the use of a level set $J_L^{-1}(\eta)$ of the momentum map J_L and of the canonical projection $\pi : J_L^{-1}(\eta) \to J_L^{-1}(\eta)/G_\eta$, which maps the integral curves of X_H contained in $J_L^{-1}(\eta)$ onto the integral curves of the Hamiltonian vector field $X_{\tilde{H}}$, on the reduced symplectic manifold $J_L^{-1}(\eta)/G_\eta$.

These two procedures are essentially equivalent: the connected components of the reduced symplectic manifolds of Marsden and Weinstein are symplectomorphic to the symplectic leaves of the Poisson manifold $(\mathcal{G}^*, \{\,,\,\}_+)$, which are exactly the connected components of the coadjoint orbits. However, the use of the Euler equation has the advantage of giving at once, by a single operation, the same result as that given by looking at all the reduced symplectic manifolds for all the values of η.

3. Taking into account a magnetic field

For mechanical systems made of material bodies moving in an electromagnetic field, one has to replace the canonical symplectic form Ω of the cotangent bundle T^*G by another symplectic form, sum of the canonical form Ω and of the pullback of a closed 2-form on the configuration space G (see, for example, the book by J.-M. Souriau [9]). This leads us to the following generalization.

As in section 2, G is a Lie group, $\mathcal{G}$ its Lie algebra and $\mathcal{G}^*$ the dual space of $\mathcal{G}$. Let $\theta : G \to \mathcal{G}$ be a smooth map such that, for all g and $h \in G$,

$$\theta(gh) = \mathrm{Ad}_g^* \, \theta(h) + \theta(g), \tag{*}$$

and such that its differential at the unit element, $\Theta = T_e\theta$, is skew-symmetric: for all X and $Y \in \mathcal{G}$ (identified with T_eG),

$$\langle T_e\theta(X), Y \rangle = -\langle T_e\theta(Y), X \rangle.$$

We set

$$\Theta(X, Y) = \langle T_e\theta(X), Y \rangle.$$

We can consider Θ as a bilinear skew-symmetric 2-form on the Lie algebra $\mathcal{G}$, or as a left invariant differential 2-form on the Lie group G. As a consequence of $(*)$, Θ is closed:

$$d\Theta = 0.$$

We now consider the 2-form on T^*G:

$$\Omega_\theta = d\alpha + q_G^*\Theta,$$

where α is the Liouville 1-form on T^*G. That 2-form is closed and non-degenerate, *i.e.*, symplectic. Since Θ is left invariant, Ω_θ is invariant under the $\widehat{L}$ action defined in section 2: for every $g \in G$, we have

$$\widehat{L}_g^* \Omega_\theta = \Omega_\theta.$$

However, Ω_θ is not invariant under the other action, $\widehat{R}$, defined in section 2. We define a new map $\widehat{R}^\theta : T^*G \times G \to T^*G$ by setting, for all $g \in G$, $\xi \in T^*G$,

$$\widehat{R}^\theta(\xi, g) = \widehat{R}_g^\theta(\xi) = \widehat{R}_g \xi + \widehat{L}_{q_G(\xi)g}\theta\left(g^{-1}\right).$$

An easy calculation shows that $\widehat{R}^\theta$ is an action of G on T^*G on the right which commutes with the action $\widehat{L}$. In other words it satisfies, for all g and $h \in G$,

$$\widehat{R}_g^\theta \circ \widehat{R}_h^\theta = \widehat{R}_{hg}^\theta, \qquad \widehat{L}_g \circ \widehat{R}_h^\theta = \widehat{R}_h^\theta \circ \widehat{L}_g.$$

Moreover, the action $\widehat{R}^\theta$ leaves invariant the symplectic 2-form Ω_θ: for all $g \in G$,

$$\left(\widehat{R}_g^\theta\right)^* \Omega_\theta = \Omega_\theta.$$

The following theorem generalizes Theorem 2.1 of section 2.

Theorem 3.1. *The maps*

$$\widehat{L} : G \times T^*G \longrightarrow T^*G, \qquad \widehat{L}(g, \xi) = \widehat{L}_g\xi,$$
$$\widehat{R}^\theta : T^*G \times G \longrightarrow T^*G, \qquad \widehat{R}(\xi, g) = \widehat{R}_g\xi,$$

are two commuting Hamiltonian actions of the Lie group G on the symplectic manifold (T^*G, Ω_θ), respectively, on the left and on the right. They admit as momentum map, respectively, the maps $J_L^\theta : T^*G \to \mathcal{G}^*$ and $J_R : T^*G \to \mathcal{G}^*$,

$$J_L^\theta(\xi) = \widehat{R}_{q_G(\xi)^{-1}}\xi + \theta\left(q_G(\xi)\right), \qquad J_R(\xi) = \widehat{L}_{q_G(\xi)^{-1}}\xi.$$

The momentum map J_L^θ is equivariant with respect to the action $\widehat{L}$ of G on T^*G and the affine action of G on $\mathcal{G}^*$:

$$(g, \eta) \longmapsto \mathrm{Ad}_g^* \eta + \theta(g),$$

while the momentum map J_R is equivariant with respect to the action $\widehat{R}^\theta$ of G on T^*G on the right and the affine action of G on $\mathcal{G}^*$ on the right:

$$(\eta, g) \longmapsto \mathrm{Ad}^*_{g^{-1}} \eta + \theta\left(g^{-1}\right).$$

Remarks

Let us set, for all $g \in G$, $\eta \in \mathcal{G}^*$,

$$a_\theta(g, \eta) = \mathrm{Ad}^*_g \eta + \theta(g).$$

Then $a_\theta : G \times \mathcal{G}^* \to \mathcal{G}^*$ is the action of G on $\mathcal{G}^*$ on the left for which J^θ_L is equivariant. The action of G on $\mathcal{G}^*$ on the right for which J_R is equivariant is the same action a_θ, but with g replaced by g^{-1}, in order to have an action on the right; in other words, it is the action on the right

$$(\eta, g) \longmapsto a_\theta\left(g^{-1}, \eta\right).$$

In [6] the two actions of G on $\mathcal{G}^*$ for which J^θ_L and J_R are equivariant were not so simply related, because at that time we made different sign conventions, leading to the replacement of J^θ_L by its opposite. The conventions made here now seem to us more natural.

As in section 2, the orbits of $\widehat{L}$ are the level sets of J_R, and the orbits of $\widehat{R}^\theta$ are the level sets of J^θ_L. As a consequence, we have the following theorem, which generalizes Theorem 2.2 of section 2.

Theorem 3.2. *Let n be the dimension of the Lie group G. For each point $\xi \in T^*G$, the tangent spaces at ξ to the $\widehat{L}$-orbit and to the $\widehat{R}^\theta$-orbit of that point are two n-dimensional vector subspaces of $T_\xi(T^*G)$, symplectically orthogonal to each other (with respect to the symplectic form $\Omega_\theta(\xi)$). The momentum map $J^\theta_L : T^*G \to \mathcal{G}^*$ is a Poisson map when T^*G is equipped with the symplectic 2-form Ω_θ and $\mathcal{G}^*$ with the modified Lie–Poisson bracket (called the θ-modified minus Lie–Poisson bracket) given by the formula (where f and g are two smooth functions on $\mathcal{G}^*$ and η a point in $\mathcal{G}^*$),*

$$\{f, g\}^\theta_-(\eta) = -\langle \eta, [df(\eta), dg(\eta)]\rangle + \Theta(df(\eta), dg(\eta)).$$

Similarly, the momentum map $J_R : T^*G \to \mathcal{G}^*$ is a Poisson map when T^*G is equipped with the symplectic 2-form Ω_θ and $\mathcal{G}^*$ with the modified Lie–Poisson bracket, opposite of the previous one (called the θ-modified plus Lie–Poisson bracket),

$$\{f, g\}^\theta_+(\eta) = \langle \eta, [df(\eta), dg(\eta)]\rangle - \Theta(df(\eta), dg(\eta)).$$

4. Reduction of one of the actions $\widehat{L}$ or $\widehat{R}$ to a subgroup

Several mechanical systems with a Lie group G as configuration space have a Hamiltonian $H : T^*G \to G$ which is invariant under the restriction of the action $\widehat{L}$ to a subgroup of G, rather than under the full action $\widehat{L}$ of G on T^*G. For example, the motion of a heavy rigid body with a fixed point (Example 1.3 of section 1) has the Lie group $SO(E, 0)$ as configuration space; when the body is submitted to the gravitational force, its Hamiltonian is no more invariant under the full action $\widehat{L}$, but only under the restriction of that action to the one-dimensional subgroup of $SO(E, 0)$ made by the rotations around the vertical axis through the fixed point of the body.

A very remarkable property allows us to obtain, even in the latter case, an Euler equation (at least when the subgroup of G for which the restricted $\widehat{L}$-action leaves invariant the Hamiltonian is the stabilizer of a point for a linear representation of G in a finite-dimensional vector space). Let us describe more fully that property.

As in the preceding sections, G is a Lie group, $\mathcal{G}$ its Lie algebra and $\mathcal{G}^*$ the dual space of $\mathcal{G}$. The cotangent bundle T^*G will be equipped with its canonical symplectic 2-form $\Omega = d\alpha$, as in section 2. (Of course, it is possible to extend the results when T^*G is equipped with a modified symplectic 2-form Ω_θ, as in section 3.)

Let $\rho : G \to GL(E)$ be a linear representation of G in a finite-dimensional vector space E. We denote by $\rho^* : G \to GL(E^*)$ the contragredient representation of G in the dual space E^* of E. We recall that for each $x \in E$, $\zeta \in E^*$ and $g \in G$, we have

$$\left\langle \rho_g^*\zeta, x \right\rangle = \left\langle \zeta, \rho_{g^{-1}}x \right\rangle.$$

Let η be a point in E^*, and let

$$G_\eta = \left\{ g \in G \mid \rho_g^*\eta = \eta \right\}$$

be the stabilizer of η; it is a closed subgroup of G.

Let $G_1 = G \times_\rho E$ be the semi-direct product of G with E, for the linear representation ρ. We recall that the product in G_1 is given by the formula

$$(g, x)(h, y) = \left(gh, x + \rho_g y \right),$$

where g and $h \in G$, x and $y \in E$.

With every $x \in E$, we associate the function f_x^η on G

$$f_x^\eta(g) = \left\langle \eta, \rho_g(x) \right\rangle.$$

Of course, the function f_x^η depends not only on $x \in E$, but also on $\eta \in E^*$; however, the parameter $\eta \in E^*$ will be considered as fixed.

We define an action ψ^η of E on T^*G by setting, for all $x \in E$ and $\xi \in T^*G$,

$$\psi^\eta_x(\xi) = \xi + df^\eta_x(q_G(\xi)) \,.$$

Finally, by composition of ψ^η with $\widehat{R}$, we obtain an action Φ_R of $G_1 = G \times_\rho E$ on T^*G, on the right, given by the formula

$$\Phi_{R(g,x)}(\xi) = \psi^\eta_x \circ \widehat{R}_g(\xi),$$

where $(g, x) \in G \times_\rho E$ and $\xi \in T^*G$.

The main result of this section is the following theorem.

Theorem 4.1. *With the above assumptions and notations, the restriction to G_η of the action $\widehat{L}$ and the action Φ_R are two commuting, Hamiltonian actions on the symplectic manifold (T^*G, Ω), respectively, of the Lie group G_η on the left, and of the Lie group $G_1 = G \times_\rho E$ on the right. Their momentum maps J^η_L and J^η_R take their values, respectively, in the dual $\mathscr{G}^*_\eta$ of the Lie algebra of G_η and in the dual $\mathscr{G}^* \times E^*$ of the Lie algebra $\mathscr{G} \times E$ of G_1. For each point $\xi \in T^*G$, the tangent spaces at ξ to the $\widehat{L}$-orbit of G_η and to the Φ_R-orbit of G_1 are two vector subspaces of the symplectic vector space $(T_\xi(T^*G), \Omega(\xi))$, symplectically orthogonal to each other. When T^*G is equipped with its canonical symplectic structure, $\mathscr{G}^*_\eta$ with its minus Lie–Poisson bracket and $\mathscr{G}^* \times E^*$ with its plus Lie–Poisson bracket, the momentum maps $J^\eta_L : T^*G \to \mathscr{G}^*_\eta$ and $J^\eta_R : T^*G \to \mathscr{G}^* \times E^*$ are Ad^*-equivariant Poisson maps. Moreover, the orbits of the action $\widehat{L}$ of G_η are the level sets of J^η_R, and the orbits of the action Φ_R of G_1 are the level sets of J^η_L.*

Application to a G_η-invariant Hamiltonian system

With the above assumptions and notations, let $H : T^*G \to \mathbb{R}$ be a smooth function, and X_H the associated Hamiltonian vector field on T^*G. Let us assume that H is invariant under the restriction of the action $\widehat{L}$ to the subgroup G_η. Since H is constant on each $\widehat{L}$-orbit of G_η in T^*G, and since these orbits are the level sets of the momentum map $J^\eta_R : T^*G \to \mathscr{G}^*_1 = \mathscr{G}^* \times E^*$, there exists a unique smooth function $\widehat{H} : \mathscr{G}^* \times E^* \to \mathbb{R}$ such that

$$H = \widehat{H} \circ J^\eta_R.$$

Since J^η_R is a Poisson map, the Hamiltonian vector field X_H on T^*G has as its projection by J^η_R on $\mathscr{G}^* \times E^*$ the Hamiltonian vector field $X_{\widehat{H}}$ associated to the Hamiltonian $\widehat{H}$, for the plus Lie–Poisson structure of $\mathscr{G}^* \times E^*$. The corresponding differential equation on $\mathscr{G}^* \times E^*$ is the Euler equation.

5. The rigid body with a fixed point

The assumptions and notations being those of Example 1.3 of section 1, the Hamiltonian of the system is

$$H(\xi) = \frac{1}{2}\Big\langle J_R(\xi),\, I^{-1} \circ J_R(\xi)\Big\rangle - \big\langle {}^t(q_G\xi)F, a\big\rangle.$$

We have denoted by F the gravity force (considered as an element of the dual E^* of E), and by $a \in E$ the vector $\overrightarrow{0G}$, where 0 is the fixed point and G the center of mass of the rigid body in its reference configuration. The linear map $I : \mathscr{G} \to \mathscr{G}^*$ is the inertia operator; it is symmetric definite positive.

The rigid body with a fixed point without gravity effects

The gravity force does not appear in the Hamiltonian either when $F = 0$ (no gravity force), or when $a = \overrightarrow{0G} = 0$ (the fixed point is the center of mass of the body). The Hamiltonian H reduces to

$$H(\xi) = \frac{1}{2}\Big\langle J_R(\xi),\, I^{-1} \circ J_R(\xi)\Big\rangle.$$

It depends only on $J_R(\xi)$; since the $\widehat{L}$-orbits are the level sets of J_R, the Hamiltonian is constant on each $\widehat{L}$-orbit, and can be written as

$$H = \widehat{H} \circ J_R, \quad \text{with } \widehat{H}(M) = \frac{1}{2}\Big\langle M, I^{-1}(M)\Big\rangle, \; (M \in \mathscr{G}^*).$$

The Euler equation is the differential equation on $\mathscr{G}^*$ defined by the Hamiltonian vector field $X_{\widehat{H}}$ associated with $\widehat{H}$, for the plus Lie–Poisson bracket on $\mathscr{G}^*$. It can be written as

$$\frac{dM(t)}{dt} = -\operatorname{ad}^*_{I^{-1}(M(t))} M(t).$$

Following Arnol'd [1], [2], let us define the bilinear map $B : \mathscr{G} \times \mathscr{G} \to \mathscr{G}$ by

$$\langle I \circ B(X, Y), Z \rangle = \langle I(X), [Y, Z] \rangle, \quad X, Y \text{ and } Z \in \mathscr{G}.$$

By the change of variables $\Omega = I^{-1}(M)$, the Euler equation becomes

$$\frac{d\Omega(t)}{dt} = B\left(\Omega(t), \Omega(t)\right).$$

Remark. The variables M and Ω have a natural physical interpretation: M is the angular momentum in the reference frame of the body and Ω the angular velocity, also in the reference frame of the body.

The rigid body with a fixed point and gravity effects

Let us now assume that neither F nor a vanish. The Hamiltonian H is no more invariant by the action $\widehat{L}$ of the full group G, but only by the restriction of that action to the subgroup G_F of G:

$$G_F = \left\{ g \in G \mid {}^t g(F) = F \right\}.$$

We observe that G_F is the stabilizer of F for the action of G on E^* contragredi-ent of the natural action ρ of G on E. According to Theorem 4.1, the semi-direct product $G_1 = G \times_\rho E$ acts on T^*G by a Hamiltonian action on the right, Φ_R. The momentum map J_R^F of that action, which takes its values in $\mathscr{G}^* \times E^*$, can be expressed as

$$J_R^F(\xi) = \left(\widehat{R}_{q(\xi)^{-1}} \xi, {}^t(q_G(\xi)) F \right).$$

The Hamiltonian H can be written as

$$H = \widehat{H} \circ J_R^F,$$

where $\widehat{H} : \mathscr{G}^* \times E^* \to \mathbb{R}$ is given by

$$\widehat{H}(M, P) = \frac{1}{2} \left\langle M, I^{-1}(M) \right\rangle - \langle P, a \rangle.$$

The Euler equation is the Hamilton differential equation associated to the Hamiltonian $\widehat{H}$, the space $\mathscr{G}^* \times E^*$ being equipped with its plus Lie–Poisson bracket. For the 3-dimensional Euclidean vector space E, the scalar product yields a natural identification of E with its dual E^*; once an orientation is chosen on E, the vector product yields another identification of E with the Lie algebra $\mathscr{G}$. By combining these identifications we can consider E, E^*, $\mathscr{G}$ and $\mathscr{G}^*$ as being all the same space. The Euler equation becomes:

$$\frac{dM}{dt} = M \times I^{-1}(M) - P \times a, \qquad \frac{dP}{dt} = P \times I^{-1}(M),$$

where $\times$ denotes the vector product in the 3-dimensional oriented, Euclidean vector space E.

6. The motion of an ideal incompressible fluid

The assumptions are now those of Example 1.4 of section 1. The Lie group G is now the group $\mathrm{Diff}(V, v)$ of volume-preserving diffeomorphisms of V. We assume that no external forces are applied to the fluid. Therefore the Hamiltonian of the system corresponds to the kinetic energy only. It is invariant under the action $\widehat{R}$ of G on T^*G, canonical lift of the action R of G on itself by right translations.

Remark

One may ask why, for the motion of a rigid body with a fixed point, without gravity effects, the Hamiltonian is invariant by the canonical lift of the action of G on itself by left translations, while for the motion of an ideal incompressible fluid, without gravity effects, the Hamiltonian is invariant by the canonical lift of the action of G on itself by right translations. The explanation is the following.

Let us begin by what is common to both systems (the ideal, incompressible fluid and the rigid body). Let $t \mapsto \varphi(t)$ be a smooth curve in the configuration space of the system. We may think about it as of a motion of the system as a function of time. Remember that for each t, $\varphi(t)$ is a map, which sends each material particle x of the system onto its position $\varphi(t)(x)$ in the physical space E when the configuration of the system is $\varphi(t)$. By choosing a particular configuration φ_0 as reference configuration, we may write $\varphi(t) = g(t) \circ \varphi_0$, where $t \mapsto g(t)$ is a smooth curve in the group G. Let h be a fixed element in G. The transforms of $t \mapsto \varphi(t)$ by the left translation L_h and by the right translation R_h are, respectively, $t \mapsto hg(t) \circ \varphi_0$ and $t \mapsto g(t)h \circ \varphi_0$. By applying this to a particular material particle x, and by taking the derivative with respect to t, we obtain the velocity, in the physical space E, of that material particle. Taking half the square of the norm of the velocity and integrating over all the material particles, with respect to the material measure, we obtain the kinetic energy.

Let us now look about what is particular to each system.

For the ideal incompressible fluid which fills a vessel V fixed in space, we see that the velocity field of the fluid, in the physical space, is exactly the same for the motions $t \mapsto g(t) \circ \varphi_0$ and $t \mapsto g(t) \circ h \circ \varphi_0$. The only difference is that the material particle which is at a given point $y \in E$ in the first motion is $x = (g(t) \circ \varphi_0)^{-1}(y)$, while in the second motion it is $(g(t)h \circ \varphi_0)^{-1}(y) = (\varphi_0^{-1} \circ h \circ \varphi_0)^{-1}(x)$. Since the part V of the physical space in which the fluid flows is fixed, and since the density of the fluid is a constant, the kinetic energy is the same in the two motions. As a consequence, the Hamiltonian of the system is invariant under the $\widehat{R}$-action.

In contrast, since the diffeomorphism h is not, in general, a rigid displacement, the velocity fields of the fluid in physical space for the motions $t \mapsto g(t) \circ \varphi_0$ and $t \mapsto hg(t) \circ \varphi_0$ are not the same; therefore the Hamiltonian is not, in general, invariant under the $\widehat{L}$-action.

For the rigid body with a fixed point, the velocity field in the physical space is, as for the ideal fluid, the same in the motions $t \mapsto g(t) \circ \varphi_0$ and $t \mapsto g(t)h \circ \varphi_0$. But the position of the rigid body, in the physical space, is not the same in these two motions: a given point in E may be occupied by a massive particle for one of these motions, and may be empty (or occupied by a lighter particle) in the other motion. Therefore the kinetic energy is not the same in the two motions, and as a consequence, the Hamiltonian of the system is not invariant under the $\widehat{R}$-action.

In contrast, since h is a rigid displacement, the velocity field in the physical space for the motion $t \mapsto hg(t) \circ \varphi_0$ is obtained from the velocity field in the physical space for the motion $t \mapsto g(t) \circ \varphi_0$ by the same rigid displacement, h, which maps the configuration of the rigid body in the motion $t \mapsto g(t) \circ \varphi_0$ onto its configuration in the motion $t \mapsto hg(t) \circ \varphi_0$. In other words, in a frame attached to the rigid body, the velocity fields in these two motions are the same. Therefore the kinetic energy of the system is the same in the two motions. As a consequence, when there are no gravity effects, the Hamiltonian of the system is invariant under the $\widehat{L}$-action.

Let us observe that the Hamiltonian of particular systems may have additional invariance properties. For example, the Hamiltonian of a rigid body with a revolution axis of symmetry, with a fixed point on that axis, is invariant under the restriction of the $\widehat{R}$-action to a one-parameter subgroup of $SO(E, 0)$, made of the rotations around the symmetry axis. The Hamiltonian of an ideal, incompressible fluid which fills a fixed vessel V with a revolution symmetry axis is invariant under the restriction of the $\widehat{L}$-action to a one-parameter subgroup of $\mathrm{Diff}(V, v)$, made of the rotations around that axis; when the vessel V is a spherical cavity, the Hamiltonian of the ideal fluid is invariant under the restriction of the $\widehat{L}$-action to a 3-dimensional subgroup of $\mathrm{Diff}(V, v)$, made of the rigid rotations around the centre of V.

The Lie algebra $\mathscr{G}$ of G can be identified with the space of divergence-free vector fields on V tangent to its boundary ∂V. However, the composition law in this Lie algebra is the *opposite* of the usual bracket of vector fields.

Let us define the pairing $(\alpha, X) \mapsto \langle \alpha, X \rangle$ of a differential 1-form α on V and a divergence-free vector field X on V, by the formula

$$\langle \alpha, X \rangle = \int_V \langle \alpha(x), X(x) \rangle \, dv(x).$$

Using the fact that X is divergence-free (div $X = 0$), one can prove that when α is exact (that means is the differential df of a smooth function f on V), then $\langle \alpha, X \rangle = 0$. When V is a simply connected compact subset of E with a smooth boundary ∂V, the dual space $\mathscr{G}^*$ of $\mathscr{G}$ can be identified with the quotient space $\Omega^1(V)/d\Omega^0(V)$, where $\Omega^1(V)$ is the space of differential 1-forms and $\Omega^0(V)$ the space of smooth functions on V.

We denote by ρ the volumic mass of the fluid. We assume that the fluid is homogeneous; since it is also incompressible, ρ is a constant.

Let X be an element of $\mathscr{G}$ or, in other words, a divergence-free vector field on V tangent to the boundary ∂V. Using the Euclidean structure of E, we can define the differential 1-form $X^\flat$ on V such that, for each point $x \in V$ and each vector $w \in E$,

$$\langle X^\flat(x), w \rangle = (X(x) \mid w),$$

where $(\mid)$ denotes the scalar product in E. We associate with X the differential 1-form $\rho X^\flat$, and we denote by IX the class modulo $d\Omega^0(V)$ of $\rho X^\flat$. The map $I : \mathscr{G} \to \mathscr{G}^*$ so defined is the inertia operator of the system.

The momentum map $J_L : T^*G \to \mathscr{G}^*$ is the map which associates, with each kinematic state $\xi \in T^*G$ of the system, the element IX of $\mathscr{G}^*$, where X is the velocity vector field of the fluid on V for the kinematic state ξ.

Exactly as for the motion of a rigid body around a fixed point, the Euler equation is

$$\frac{dM(t)}{dt} = - \mathrm{ad}^*_{I^{-1}(M(t))} M(t).$$

With the same change of variables as in section 5,

$$X = I^{-1}(M),$$

and the same definition of the bilinear map $B : \mathscr{G} \times \mathscr{G} \to \mathscr{G}$, the Euler equation becomes

$$\frac{dX(t)}{dt} = -B(X(t), X(t)).$$

The explicit expression of B, in terms of the vector product in E and of the gradient and curl operators (denoted by $\overrightarrow{\mathrm{grad}}$ and $\overrightarrow{\mathrm{curl}}$, respectively), is

$$B(X, Y) = \left(\overrightarrow{\mathrm{curl}}\, X \right) \times Y + \overrightarrow{\mathrm{grad}}\, h,$$

where h is the smooth function (unique up to an additive constant) such that the vector field $B(X, Y)$ is divergence-free. That function is solution of the Poisson partial differential equation

$$\Delta h + \mathrm{div} \left(\left(\overrightarrow{\mathrm{curl}}\, X \right) \times Y \right) = 0.$$

The Euler equation is therefore

$$\frac{\partial X(t, x)}{\partial t} + \left(\overrightarrow{\mathrm{curl}}_x X(t, x) \right) \times X(t, x) + \overrightarrow{\mathrm{grad}}_x h = 0.$$

Remark. The smooth function h has a physical interpretation: it is related to the pressure p, the velocity X and the volumic mass ρ of the fluid by the formula

$$h = \frac{p}{\rho} + \frac{\|X\|^2}{2}.$$

7. The Korteweg–de Vries equation as an Euler equation

Let $\mathcal{G}$ be the Lie algebra of smooth vector fields on the circle S^1, with the bracket opposite to the usual bracket of vector fields,

$$\left[f(x)\frac{\partial}{\partial x}, g(x)\frac{\partial}{\partial x} \right] = \left(f'(x)g(x) - g'(x)f(x) \right)\frac{\partial}{\partial x},$$

where x is the angular coordinate on S^1 (defined modulo 2π). The components f and g of the two vector fields on S^1 are considered as 2π-periodic, smooth functions on $\mathbb{R}$.

Let $c : \mathcal{G} \times \mathcal{G} \to \mathbb{R}$ be the bilinear map

$$c\left(f(x)\frac{\partial}{\partial x}, g(x)\frac{\partial}{\partial x} \right) = \int_{S^1} f'(x)g''(x)dx.$$

An integration by parts shows that c is skew-symmetric:

$$c\left(f(x)\frac{\partial}{\partial x}, g(x)\frac{\partial}{\partial x} \right) = -c\left(g(x)\frac{\partial}{\partial x}, f(x)\frac{\partial}{\partial x} \right).$$

The map c satisfies the identity

$$c\left(\left[f(x)\frac{\partial}{\partial x}, g(x)\frac{\partial}{\partial x} \right], h(x)\frac{\partial}{\partial x} \right) + \text{cyclic sum} = 0.$$

The map c is called the *Gelfand–Fuchs cocycle* of the Lie algebra $\mathcal{G}$.

Let $\mathcal{G}_1 = \mathcal{G} \times \mathbb{R}$, equipped with the bracket

$$\left[\left(f(x)\frac{\partial}{\partial x}, a \right), \left(g(x)\frac{\partial}{\partial x}, b \right) \right] = \left(\left[f(x)\frac{\partial}{\partial x}, g(x)\frac{\partial}{\partial x} \right], \right.$$
$$\left. c\left(f(x)\frac{\partial}{\partial x}, g(x)\frac{\partial}{\partial x} \right) \right).$$

With that bracket, $\mathcal{G}_1$ is an infinite-dimensional Lie algebra, called the *Virasoro algebra*.

There exists an infinite-dimensional Lie group G_1 whose Lie algebra is the Virasoro algebra. It is called the *Virasoro–Bott group*, and it is the semi-direct product $\text{Diff}(S^1) \times \mathbb{R}$ of the group of diffeomorphisms of the circle with $\mathbb{R}$, with the composition law

$$(\varphi, a)(\psi, b) = \left(\varphi \circ \psi, a + \int_{S^1} \ln(\varphi \circ \psi)' d(\ln \psi') \right).$$

The dual space $\mathcal{G}_1^*$ of $\mathcal{G}_1$ is the product $\mathcal{G}^* \times \mathbb{R}$, where $\mathcal{G}^*$ is the space of differentiable 1-forms on S^1. The pairing of $\mathcal{G}_1^*$ with $\mathcal{G}_1$ is given by the formula

$$\left\langle (g(x)dx, b), \left(f(x)\frac{\partial}{\partial x}, a \right) \right\rangle = \int_{S^1} g(x)f(x)dx + ab.$$

Let $I : \mathcal{G}_1 \to \mathcal{G}_1^*$ be the linear map

$$I\left(f(x)\frac{\partial}{\partial x}, a \right) = (f(x)dx, a),$$

and let $\widehat{H} : \mathcal{G}^* \to \mathbb{R}$ be the function

$$\begin{aligned}
\widehat{H}(g(x)dx, a) &= \frac{1}{2}\left\langle (g(x)dx, a), I^{-1}(g(x)dx, a) \right\rangle \\
&= \frac{1}{2}\left(\int_{S^1} (g(x))^2\, dx + a^2 \right).
\end{aligned}$$

Let H be the Hamiltonian on the cotangent bundle T^*G_1 of the Virasoro–Bott group G_1, invariant under the $\widehat{R}$-action of G_1, given by

$$H = \widehat{H} \circ J_L,$$

where J_L is the momentum map of the $\widehat{L}$-action of G_1. The corresponding Euler equation on $\mathcal{G}_1^*$, given by the same formulas as in the previous sections, is

$$\frac{\partial}{\partial t}(g(x, t)dx, a(t)) = -\operatorname{ad}^*_{I^{-1}(g(x,t)dx, a(t))}(g(x, t)dx, a(t)).$$

By using the expression of the bracket in $\mathcal{G}_1$, and after several integrations by parts, the Euler equation can be written as

$$\frac{\partial}{\partial t}(g(x, t)dx, a(t)) = \left(b(t)\frac{\partial^3 g(x, t)}{\partial x^3} + 3g(x, t)\frac{\partial g(x, t)}{\partial x}, 0 \right).$$

The second component of that equation is

$$\frac{\partial b(t)}{\partial t} = 0,$$

therefore b is a constant. The first component of the above equation becomes:

$$\frac{\partial g(x, t)}{\partial t} = b\frac{\partial^3 g(x, t)}{\partial x^3} + 3g(x, t)\frac{\partial g(x, t)}{\partial x}.$$

That equation is the famous Korteweg–de Vries equation on the circle.

References

[1]   Arnol'd, V.I., Sur la géométrie différentielle des groupes de Lie de dimension infinie et ses applications à l'hydrodynamique, *Ann. Inst. Fourier, Grenoble* **16** (1966), 319–361.

[2]   Arnol'd, V.I. and Khesin, B.A., *Topological methods in hydrodynamics*, Springer Verlag, New York–Berlin–Heidelberg 1998.

[3]   Euler, L., Mémoires de l'Académie des Sciences de Berlin, 1758 and 1765; quoted in the book by E.T. Whittaker, *A treatise on the analytical dynamics of particles and rigid bodies*, Cambridge University Press, 4th edition, (1988), 144, and in the book [2] by Arnol'd and Khesin, 351.

[4]   Guillemin, V. and Sternberg, S., *Symplectic techniques in physics*, Cambridge University Press, Cambridge 1984.

[5]   Iacob, A. and Sternberg, S., *Coadjoint structures, solitons and integrability*, Lecture notes in Physics, 120, Springer Verlag, Berlin 1980.

[6]   Libermann, P. and Marle, C.-M., *Symplectic geometry and analytical mechanics*, D. Reidel Publishing Company, Dordrecht 1987.

[7]   Marsden, J.E., Ratiu, T., and Weinstein, A., Reduction and Hamiltonian structures on duals of semidirect product Lie algebras, *Contemporary mathematics* **28** (1984), 55–100.

[8]   Marsden, J.E. and Weinstein, A., Reduction of symplectic manifolds with symmetry, *Reports on Mathematical Physics* **5** (1974), 121–130.

[9]   Souriau, J.-M., *Structure des systèmes dynamiques*, Dunod, Paris 1969.

DIRAC FIELDS ON ASYMPTOTICALLY SIMPLE SPACE–TIMES

JEAN–PHILIPPE NICOLAS

Contents

1   Introduction 207
2   The Dirac equation on globally hyperbolic asymptotically flat space–times   207
3   Scattering of Dirac fields on asymptotically simple space–times 212
    3.1   Roger Penrose's k-asymptotically simple space–times 212
    3.2   The scattering operator for massless Dirac fields 213
4   Concluding remarks 215

M. de Gosson (ed.), Jean Leray '99 Conference Proceedings, 205-217.
© *2003 Kluwer Academic Publishers.*

1. Introduction

The notion of global hyperbolicity of a space–time was introduced by Jean Leray in 1953 [11]. Intuitively, a space–time is globally hyperbolic if the Cauchy problem for the wave equation is well posed. This is a fundamental notion in General Relativity since globally hyperbolic space–times are the only class of space–times on which it is meaningful to study the global evolution of fields.

This communication aims at describing some general properties of Dirac fields on globally hyperbolic asymptotically flat and asymptotically simple space–times. It consists of two parts. In section 2, we state results recently obtained in [12] and [15] concerning the Cauchy problem in Sobolev and weighted Sobolev spaces for Dirac fields on globally hyperbolic asymptotically flat space–times. The point of view is similar to that adopted by Madame Choquet–Bruhat in the first talk of this conference; we use the global hyperbolicity to perform a 3 + 1 decomposition of the geometry and write the equation as an evolution system. In section 3, we describe the essential ingredients of a geometric construction, on asymptotically simple space–times, of a scattering operator for the massless Dirac equation and we discuss possible extensions.

2. The Dirac equation on globally hyperbolic asymptotically flat space–times

We work on a 4-dimensional Lorentzian space–time $(\mathcal{M}, g)$ with signature $+ - - -$, oriented, time-oriented, globally hyperbolic and asymptotically flat. Global hyperbolicity implies (see Geroch 1970 [4] and Seifert 1977 [17]) the existence of a time function t globally defined on $\mathcal{M}$ such that its level hypersurfaces Σ_t are a foliation by Cauchy hypersurfaces and all the leaves Σ_t are diffeomorphic to a given 3-manifold Σ (the regularity of the time function and the leaves Σ_t depends on that of the metric; for simplicity, we assume them to be smooth). Therefore, we have $\mathcal{M} \simeq \mathbb{R}_t \times \Sigma$ and we choose to define the product structure along the integral lines of the time-like vector field $\nabla^a t$ which is everywhere orthogonal to the hypersurfaces Σ_t. The metric g can be decomposed into its space-like and time-like parts:

$$g = N^2 t^2 - h,$$

where the lapse function N is defined as

$$Nt = t_a x^a, \quad t^a = \frac{\nabla^a t}{|\nabla t|},$$

and h is the part of g orthogonal to t^a, i.e., $t^a h_{ab} = 0$. The advantage of our choice of product structure is that h is naturally understood as a (time-dependent) metric on Σ. In addition, asymptotic flatness requires that there exists a compact set K on Σ such that $\Sigma \setminus K \simeq \mathbb{R}^3 \setminus \bar{B}(0, 1)$.

In order to specify the regularity of the metric and of solutions to the Dirac equation, we define Sobolev and weighted Sobolev spaces on Σ using a fixed background metric $\tilde{h}$ which is a smooth Riemannian metric on Σ, Euclidian outside a compact set. For $k \in \mathbb{N}$ and $\delta \in \mathbb{R}$, we define the spaces $H^k(\Sigma)$ and $H_\delta^k(\Sigma)$ as the completions of $\mathscr{C}_0^\infty(\Sigma)$ (i.e., the space of smooth functions with compact support on Σ) in the norms

$$\|f\|_{H^k} = \left(\sum_{p=0}^{k} \int_\Sigma \langle \tilde{D}^p f, \tilde{D}^p f \rangle \mathrm{dVol}_{\tilde{h}} \right)^{1/2},$$

$$\|f\|_{H_\delta^k} = \left(\sum_{p=0}^{k} \int_\Sigma \left(1 + r^2\right)^{\delta+p} \langle \tilde{D}^p f, \tilde{D}^p f \rangle \mathrm{dVol}_{\tilde{h}} \right)^{1/2},$$

where $\tilde{D}$, $\mathrm{dVol}_{\tilde{h}}$ and $\langle \cdot, \cdot \rangle$ are the Levi–Civita connection, the volume element and the positive definite inner product on tensors and spinors at a point induced by the metric $\tilde{h}$; for any $x \in \Sigma$, $r(x)$ is the $\tilde{h}$-distance between x and a fixed point $O \in \Sigma$. The spaces $H^0(\Sigma)$ and $H_\delta^0(\Sigma)$ will be denoted, respectively, $L^2(\Sigma)$ and $L_\delta^2(\Sigma)$. We now define families of asymptotically flat space–times on which we shall solve the global Cauchy problem for the Dirac equation:

Definition 2.1. We say that the metric g on $\mathbb{R} \times \Sigma$ is of class (k, δ), $k \in \mathbb{N}^*$, $\delta \in \mathbb{R}$, if g compared to the first-order asymptotic expansion of a Schwarzschild metric satisfies

$$g - \left(\left(1 - \frac{2M}{r}\right) t^2 - \left(1 + \frac{2M}{r}\right) \tilde{h} \right) \in \mathscr{C}^l \left(\mathbb{R}_t; H_\delta^{k-l}(\Sigma) \right),$$

$$\forall l, \ 0 \le l \le k$$

M being a positive constant, and moreover g satisfies the non-degeneracy condition:

(H) There exist two continuous positive functions on $\mathbb{R}$: C_1 and C_2, such that for each $(t, x) \in \mathbb{R} \times \Sigma$, the lapse function $N(t, x)$ and the eigenvalues $\lambda_i(t, x)$, $i = 1, 2, 3$ of $h(t, x)$ as a symmetric form relative to $\tilde{h}$ satisfy $C_1(t) \le N(t, x) \le C_2(t)$ and $C_1(t) \le \lambda_i(t, x) \le C_2(t)$, $i = 1, 2, 3$.

We shall be interested in metrics of class (k, δ), $k \ge 2$, $\delta > -3/2$, which means—using the classic embedding of weighted Sobolev spaces into weighted $\mathscr{C}^k$ spaces (see, for example, [2]).—that the metric is at least continuous and is in effect asymptotically flat, since for all $0 \le l \le k - 2$:

$$\tilde{D}^l \left[g - \left(t^2 - \tilde{h} \right) \right] = 0 \left(r^{-\varepsilon-l} \right) \quad \text{as } r \longrightarrow +\infty,$$

$$\forall \varepsilon; 0 < \varepsilon < \min \left(\delta + 3/2, 1 \right).$$

This is a wide class of asymptotically flat space–times. If one wishes to consider space–times with stronger fall off conditions such as black hole space–times, one can assume $\delta > 0$. In [3] and [10], weaker fall off conditions are considered, corresponding to $\delta > -1/2$.

All such space–times admit a spin structure since, as products of $\mathbb{R}$ with a 3-manifold, they are parallelizable. We choose one and we denote by $\mathbb{S}_{\text{Dirac}}$ the bundle of Dirac spinors on $\mathbb{R} \times \Sigma$. The Dirac equation on $(\mathcal{M}, g)$ is then given by

$$(\mathcal{D} + i\text{m})\,\Psi = 0, \tag{2.1}$$

where $\mathcal{D}$ is the Dirac operator on $(\mathcal{M}, g)$, $m \geq 0$ is the mass of the particle and Ψ is a Dirac spinor. Introducing an orthonormal Lorentz frame $\{e_0 = t^a \partial_a, e_1, e_2, e_3\}$, the vectors e_1, e_2, e_3 are vector fields tangent to Σ_t such that $g(e_i, e_j) = \delta_{ij}$, $1 \leq i, j \leq 3$, the Dirac operator takes the form

$$\mathcal{D} = \sum_{i=0}^{3} e_i \cdot \nabla_{e_i}, \tag{2.2}$$

where "$\cdot$" denotes Clifford multiplication. Modulo a choice of spin frame, the Clifford product by e_0, e_1, e_2, e_3 will be represented as the multiplication by 4×4 matrices $\gamma^0, \gamma^1, \gamma^2, \gamma^3$, called Dirac matrices, such that

$$\gamma^i \gamma^j + \gamma^j \gamma^i = g\left(e_i, e_j\right) \text{Id}_4, \quad 0 \leq i, j \leq 3. \tag{2.3}$$

We choose our spin frame $\{o^A, \iota^A\}$ such that the associated Newman–Penrose tetrad

$$l^a = o^A \bar{o}^{A'}, \qquad n^a = \iota^A \bar{\iota}^{A'}, \qquad m^a = o^A \bar{\iota}^{A'}, \qquad \bar{m}^a = \iota^A \bar{o}^{A'},$$

satisfies

$$t^a = \frac{1}{\sqrt{2}}\left(l^a + n^a\right)$$

i.e., such that

$$\gamma^0 = i \begin{pmatrix} 0 & \text{Id}_2 \\ -\text{Id}_2 & 0 \end{pmatrix}.$$

In such a case the matrices $\gamma^1, \gamma^2, \gamma^3$ are skew-Hermitian and the space-like part of $\mathcal{D}$, called the Dirac–Witten operator

$$\mathcal{D}_W(t) = \sum_{i=1}^{3} \gamma^i \nabla_{e_i}, \quad \text{on } \Sigma_t$$

is formally self-adjoint for the physical positive definite L^2 inner product on Σ_t

$$\langle \Phi, \Psi \rangle_{L^2(\Sigma_t)} = \int_\Sigma \Phi^\dagger \Psi \, d\text{Vol}_{h(t)} .$$

This and property (2.3) imply that the Dirac system (2.1) written in the form of an evolution equation

$$\nabla_{e_0} \Psi = -\gamma^0 \mathcal{D}_W \Psi - i m \gamma^0 \Psi \tag{2.4}$$

is a first order symmetric hyperbolic system on $\mathbb{R} \times \Sigma$. More precisely, with respect to local coordinates x^1, x^2, x^3 on Σ, (2.1) takes the form

$$\frac{\partial \Psi}{\partial t} = \sum_{i=1}^3 A^i(t, x) \frac{\partial \Psi}{\partial x^i} + N(t, x) \left(B(t, x) - i m \gamma^0 \right) \Psi,$$

where A^i, $i = 1, 2, 3$ are 4×4 Hermitian matrices made of coefficients of the metric and B is a 4×4 matrix made of connection coefficients (i.e., first order derivatives of the coefficients of the metric).

Using the theory of hyperbolic partial differential equations introduced by Jean Leray in 1953 [11] and subsequent works by T. Kato [7]–[9], we can solve the Cauchy problem for (2.1) in Sobolev spaces. The theorems for quasi-linear systems in [9] allow us to consider metrics with a weak regularity such as those belonging to our classes (k, δ). Then using powers of the space-like Dirac operator as identifying operators, we extend the result to weighted Sobolev spaces. We have the following theorems whose proofs are given in details in [15]. ($\mathscr{C}_b^k(\Sigma)$ denotes the space of $\mathscr{C}^k$ functions on Σ which are bounded on Σ together with all their derivatives.)

Theorem 2.2. *We assume the metric g to satisfy hypothesis (H) of Definition 2.1 and*

$$g \in \mathscr{C}\left(\mathbb{R}_t; \mathscr{C}_b^1(\Sigma) \right) \cap \mathscr{C}^1\left(\mathbb{R}_t; \mathscr{C}_b^0(\Sigma) \right).$$

Then, for any real number s, for any initial data $\Psi_0 \in L^2(\Sigma, \mathbb{S}_{\text{Dirac}})$, equation (2.1) has a unique solution $\Psi \in \mathscr{C}(\mathbb{R}_t; L^2(\Sigma, \mathbb{S}_{\text{Dirac}}))$, such that

$$\Psi|_{t=s} = \Psi_0 .$$

Moreover, the vector field

$$J = \sum_{i=0}^3 \Psi^\dagger \gamma^0 \gamma^i \Psi e_i \tag{2.5}$$

is a conserved current, i.e., $\nabla_a J^a = 0$. Hence, if we define the charge of the solution at time t as

$$E(t) := \|\Psi(t)\|^2_{L^2(\Sigma_t)} = \int_\Sigma \langle \Psi(t), \Psi(t) \rangle \, d\text{Vol}_{h(t)} = \int_\Sigma |\Psi(t)|^2 \, d\text{Vol}_{h(t)}, \quad (2.6)$$

then $E(t)$ is constant throughout time. If in addition the metric satisfies

$$g \in \mathscr{C}^l\left(\mathbb{R}_t; \mathscr{C}^{k-l}_b(\Sigma)\right), \quad \forall l; \; 0 \le l \le k,$$

for some positive integer k, then for any initial data $\Psi_0 \in H^m(\Sigma; S_{\text{Dirac}})$, $0 \le m \le k-1$, the solution $\Psi \in \mathscr{C}(\mathbb{R}_t; L^2(\Sigma, S_{\text{Dirac}}))$ to (2.1) associated with s and Ψ_0 satisfies

$$\Psi \in \mathscr{C}^l\left(\mathbb{R}_t; H^{m-l}(\Sigma; S_{\text{Dirac}})\right), \quad \forall l; \; 0 \le l \le m.$$

Theorem 2.3. *We assume that g is of class (k, δ), $k \ge 3$, $\delta > -3/2$, and we consider some initial time $s \in \mathbb{R}$.*

1. *For any initial data $\Psi_0 \in L^2_\mu(\Sigma; S_{\text{Dirac}})$, $\mu \in \mathbb{R}$, equation (2.1) has a unique solution $\Psi \in \mathscr{C}(\mathbb{R}_t; L^2_\mu(\Sigma, S_{\text{Dirac}}))$ such that*

$$\Psi|_{t=s} = \Psi_0.$$

(Note that $L^2_\mu \subsetneq L^2$ for $\mu > 0$ but for $\mu < 0$ we have $L^2 \subsetneq L^2_\mu$).

2. *For $\rho \in \mathbb{Z}$, $m \in \mathbb{N}$ such that $0 \le m \le k-1$ and $\rho \ge -1$, if the initial data Ψ_0 belongs to $H^m_\rho(\Sigma; S_{\text{Dirac}})$ then the solution $\Psi \in \mathscr{C}(\mathbb{R}_t; L^2_\rho(\Sigma, S_{\text{Dirac}}))$ associated with Ψ_0 and some initial time s satisfies*

$$\Psi \in \mathscr{C}^l\left(\mathbb{R}_t; H^{m-l}_\rho(\Sigma, S_{\text{Dirac}})\right), \quad \forall l; \; 0 \le l \le m.$$

After solving the Cauchy problem a natural idea for obtaining more precise informations on the propagation of Dirac fields is to develop a scattering theory. We have seen in Theorem 2.2 that the physical L^2 norm of the solutions is conserved throughout time. At each time, the physical norm $\|\cdot\|_{L^2(\Sigma_t)}$ and the standard norm $\|\cdot\|_{L^2(\Sigma)}$ are equivalent. However, the volume element $d\text{Vol}_{h(t)}$ depends on time and therefore so does the $\|\cdot\|_{L^2(\Sigma_t)}$ norm. This implies that the aforementioned norm equivalence is only locally uniform in time. If we choose to work with families of space–times for which we have a sufficiently strong uniform control on the metric in order to guarantee that the norm equivalence is uniform in time, a scattering theory seems accessible. In the next section, we describe a geometrical way of defining a scattering operator for Dirac massless fields on special classes of asymptotically flat globally hyperbolic space–times.

3. Scattering of Dirac fields on asymptotically simple space–times

There are few known examples of time-dependent scattering theories for Dirac fields in curved space–times. To this day, such theories have been developed essentially for space–times describing static spherical black holes. In this type of framework, the existence and asymptotic completeness of classical wave operators at the horizons and (in the massless case) at infinity was obtained in [14]. More recently, the existence of modified wave operators for massive fields at infinity was proved in [6]. The completeness of such modified wave operators remains open.

To my knowledge, no time-dependent scattering theory has been constructed in general non-stationary space–times with a $1/r$ fall off at infinity allowing for the presence of energy. One way of accounting for this is perhaps the difficulty of choosing a free comparison dynamics in a non-stationary situation.

In 1990, J.C. Baez, I.E. Segal and Z.F. Zhou [1] gave a conformal construction on Minkowski space–time of a scattering operator for a non-linear wave equation. An immediately subsequent contribution by L. Hörmander [5] gave a precise analytic meaning to the solution of the characteristic Cauchy problem on the null infinity of compactified Minkowski space–time necessary for their construction. One of the original and interesting features of [1] is that the scattering operator is constructed without any choice of free comparison dynamics; a precise knowledge of the conformal geometry of the space–time and the existence of a conserved current are the only requirements. This seems to indicate that the construction of [1] could be adapted to define a scattering operator for massless Dirac fields on curved space–times for which the conformal geometry can be described by a compactified space–time. Such families of space–times have been studied by R. Penrose under the generic name of asymptotically simple space–times (see, for example, [16] Vol 2, p. 347–358). We first recall briefly their properties and then describe the essential steps of the definition of a scattering operator for massless Dirac fields on such space–times.

3.1. ROGER PENROSE'S K-ASYMPTOTICALLY SIMPLE SPACE–TIMES

A 4-dimensional Lorentzian space–time $(\mathcal{M}, g)$, oriented, time-oriented, globally hyperbolic and asymptotically flat is called k-asymptotically simple, for a positive integer k, if there exists a $\mathscr{C}^{k+1}$ manifold with boundary $\tilde{\mathcal{M}}$, with metric $\tilde{g}$, boundary $\mathfrak{I}$ and a scalar field Ω on $\tilde{\mathcal{M}}$ such that:

(a) $\mathcal{M} = \text{int } \tilde{\mathcal{M}}$;

(b) $\tilde{g} = \Omega^2 g$ in $\mathcal{M}$;

(c) Ω and $\tilde{g}$ are $\mathscr{C}^k$ throughout $\tilde{M}$;

(d) $\Omega > 0$ in $\mathscr{M}$; $\Omega = 0$ on $\mathfrak{J}$, $\tilde{\nabla}\Omega \neq 0$ on $\mathfrak{J}$

(e) every null geodesic in $\mathscr{M}$ acquires a past and future end-point on $\mathfrak{J}$.

Provided that the trace of the energy tensor is zero near $\mathfrak{J}$, $\mathfrak{J}$ is null and is made of two disconnected parts $\mathfrak{J}^+$ and $\mathfrak{J}^-$, called future and past null infinities, which are both diffeomorphic to $\mathbb{R} \times S^2$. There are holes in the boundary of $\tilde{\mathscr{M}}$, corresponding to space-like infinity i_0 and future and past time-like infinities i_+ and i_-. These 'points' are possible singularities of the conformal metric $\tilde{g}$ and are therefore omitted in the definition of $\mathfrak{J}$. We can allow the space–time $(\mathscr{M}, g)$ to contain some energy, i.e., we authorize the metric g to have a Schwarzschild-type $1/r$ fall off at infinity. $\tilde{g}$ is then necessarily singular at i_0, however, this singularity is at space-like infinity and will not propagate inwards, in other words, the singularity at i_0 will not prevent us from propagating smoothly our fields from $\mathfrak{J}^-$ to $\mathfrak{J}^+$. We add one last assumption for the geometry to satisfy: we desire that time-like infinities i_+ and i_- be regular points for $\tilde{g}$, i.e., $\tilde{g}$ is $\mathscr{C}^k$ on $\tilde{\mathscr{M}} \cup i_- \cup i_+$. The idea is that the energy contained in the space–time obtains radiated away to infinity via a source of gravitational radiation. This locally flattens the geometry and guarantees the existence of smooth time-like infinities if this phenomenon is fast enough. The property (e) of asymptotically simple space–times can simply be removed and replaced by the regularity of time-like infinities, this will suffice for our scattering construction to be valid. All the regularity properties at space-like, null or time-like infinities can be specified in terms of the fall off of the Weyl spinor in the corresponding asymptotic directions on $\mathscr{M}$ (*see [16] Vol 2, p. 394–395*).

3.2. THE SCATTERING OPERATOR FOR MASSLESS DIRAC FIELDS

We consider on $(\mathscr{M}, g)$ the Dirac equation (2.1) with $m = 0$ and some Cauchy hypersurface Σ_0 which can be assumed to be one of the leaves of a foliation by the level hypersurfaces $\Sigma_t \simeq t \times \Sigma$ of a global time function t. The massless Dirac equation is conformally invariant, i.e., Ψ (for example, in $\mathscr{C}(\mathbb{R}_t; L^2(\Sigma; \mathbb{S}_{\text{Dirac}}))$) satisfies $\mathscr{D}\Psi = 0$ if and only if the spinor field $\Phi = \Omega^{-1}\Psi$ is a solution of

$$\tilde{\mathscr{D}}\Phi = 0, \tag{3.1}$$

where $\tilde{\mathscr{D}}$ is the Dirac operator associated with the Levi–Civita connection $\tilde{\nabla}$ on $(\mathscr{M}, \tilde{g})$. We consider Σ_0 as embedded in $\tilde{\mathscr{M}}$ and choose some smooth compactly supported (i.e., supported away from i_0) initial data for Φ : $\Phi_0 \in \mathscr{C}_0^\infty(\Sigma_0)$. The compact support of the initial data enables us to stay away from the singularity at i_0 and the regularity of the metric at $\mathfrak{I}$, i_- and i_+ allows us to propagate the smooth solution Φ up to the boundary. This is done by extending the metric smoothly beyond the boundary and defining the solution on a neighbourhood of $\widetilde{\mathcal{M}}$ containing $\mathfrak{I}$, i_- and i_+. We wish to define the traces of Φ on $\mathfrak{I}^-$ and $\mathfrak{I}^+$. To this effect, it is convenient to have a solution Φ which is in $H^1_{\text{loc}}(\widetilde{\mathcal{M}})$; using Theorem 2.2, this is automatically achieved if the metric is $\mathscr{C}^2$ across $\mathfrak{I}$, i_- and i_+, i.e., if $(\mathcal{M}, g)$ is 2-asymptotically simple. We then denote $\Phi^\pm$ the respective traces of Φ on $\mathfrak{I}^\pm$. We now introduce the closed hypersurface

$$\mathscr{S}^+ = \Sigma_0 \cup i_0 \cup \mathfrak{I}^+ \cup i_+$$

and we use the fact that the flux of the current vector must be zero through $\mathscr{S}^+$ (the solution being compactly supported on Σ_0, the finite speed propagation guarantees that Φ, and therefore the vector J, is identically zero near i_0 and the singularity at space-like infinity has no effect here). Choosing a spin frame adapted to the geometry, we obtain

$$\int_{\Sigma_0} |\Phi_0|^2 \, d\text{Vol}_{\Sigma_0, \tilde{g}} = \int_{\mathbb{R} \times S^2} \left(|\Phi_2^+|^2 + |\Phi_3^+|^2 \right) x \omega \tag{3.2}$$

and applying the same procedure in the past,

$$\int_{\Sigma_0} |\Phi_0|^2 \, d\text{Vol}_{\Sigma_0, \tilde{g}} = \int_{\mathbb{R} \times S^2} \left(|\Phi_1^-|^2 + |\Phi_4^-|^2 \right) x \omega; \tag{3.3}$$

besides, we have

$$\int_{\Sigma_0} |\Phi_0|^2 \, d\text{Vol}_{\Sigma_0, \tilde{g}} = \int_{\Sigma_0} |\Psi_0|^2 \, d\text{Vol}_{\Sigma_0, g} = \|\Psi_0\|^2_{L^2(\Sigma_0)}, \tag{3.4}$$

where $|\Psi_0|$ is the Hermitian norm of Ψ_0 induced by the metric g and the unit (for g) future pointing normal vector field T to the space-like hypersurface Σ_0, while $|\Phi_0|$ is the Hermitian norm of Φ_0 induced by $\tilde{g}$ and the unit (for $\tilde{g}$) future pointing normal vector field $\widetilde{T} = \Omega^{-1} T$ to Σ_0. By density of $\mathscr{C}_0^\infty(\Sigma_0)$ in $L^2(\Sigma_0)$, the identities (3.2), (3.3) and (3.4) give us the existence of two unitary operators

$$\widetilde{W}^+ : L^2(\Sigma_0; \mathbb{S}_{\text{Dirac}}) \longrightarrow \mathscr{H}^+$$
$$\Psi_0 \longmapsto \Psi^+ := (0, \Phi_2^+, \Phi_3^+, 0) \tag{3.5}$$

$$\widetilde{W}^- : L^2(\Sigma_0; \mathbb{S}_{\text{Dirac}}) \longrightarrow \mathscr{H}^-$$
$$\Psi_0 \longmapsto \Psi^- := (\Phi_1^-, 0, 0, \Phi_4^-), \tag{3.6}$$

where $\mathcal{H}^-$ and $\mathcal{H}^+$ are the subspaces of $L^2(\mathbb{R} \times S^2 ; x\omega)$ defined by

$$\mathcal{H}^- = \left\{ \Psi \in L^2\left(\mathbb{R} \times S^2 ; \mathbb{C}^4\right) ; \Psi_2 = \Psi_3 = 0 \right\},$$
$$\mathcal{H}^+ = \left\{ \Psi \in L^2\left(\mathbb{R} \times S^2 ; \mathbb{C}^4\right) ; \Psi_1 = \Psi_4 = 0 \right\}.$$

Equalities (3.2) and (3.3) are to be understood as completeness results: all the energy of the solution is contained in its asymptotic profile Ψ^+ on $\mathfrak{I}^+$ and the same is true for the profile Ψ^- on $\mathfrak{I}^-$. The definition of the spaces $\mathcal{H}^+$ and $\mathcal{H}^-$ shows that only two components of Ψ are responsible for the diffusion of energy towards infinity while the two others are responsible for the radiation of energy inwards from infinity.

The operators (3.5) and (3.6) are naturally interpreted as inverse future and past wave operators, provided we can prove the existence of the direct wave operators. This is done by solving the characteristic Cauchy problems for equation (3.1) with data on $\mathfrak{I}^-$ and $\mathfrak{I}^+$, respectively. If we consider on $\mathfrak{I}^-$ some initial data $\Psi^- = (\Phi_1^-, 0, 0, \Phi_4^-) \in H^1(\mathbb{R} \times S^2)$, we have enough regularity to apply the methods of [5] and prove the existence of a unique solution Ψ to Dirac's equation such that

$$\Psi \in \mathscr{C}\left(\mathbb{R}_t ; H^1\left(\Sigma; \mathbb{S}_{\text{Dirac}}\right)\right) \cup \mathscr{C}^1\left(\mathbb{R}_t; L^2(\Sigma; \mathbb{S}_{\text{Dirac}})\right)$$

and the trace of $\Phi = \Omega^{-1}\Psi$ on $\mathfrak{I}^-$ is Ψ^-. The same construction can be made starting from $\mathfrak{I}^+$ and propagating the solution backwards. Using identities (3.2), (3.3) and (3.4), this gives us by density the existence of direct wave operators $W^+ = (\widetilde{W}^+)^{-1}$ and $W^- = (\widetilde{W}^-)^{-1}$. This shows in particular that the traces Φ^+ and Φ^- of Φ on $\mathfrak{I}^+$ and $\mathfrak{I}^-$, respectively, are exactly Ψ^+ and Ψ^-; i.e., only two components of Φ have non-zero traces on $\mathfrak{I}^+$ and only the two others have non-zero traces on $\mathfrak{I}^-$; this is a weak version of the peeling property. We can then define the scattering operator

$$S := \widetilde{W}^+ W^- : \mathcal{H}^- \longrightarrow \mathcal{H}^+$$
$$\Psi^- \longmapsto \Psi^+. \tag{3.7}$$

This scattering operator is the same for all metrics conformally equivalent to g; it describes the complete evolution of the Dirac massless field Φ on $\widetilde{\mathcal{M}}$ from its trace Ψ^- on $\mathfrak{I}^-$ to its trace Ψ^+ on $\mathfrak{I}^+$ and, knowing the conformal factor Ω, allows us to recover the complete evolution of the 'physical' field $\Psi = \Omega\Phi$.

The technical details of the construction outlined here will be found in [13].

4. Concluding remarks

The conformal construction briefly described in the previous section gives us a complete scattering theory for massless Dirac fields in a general family of non-stationary space–times. This theory can be interpreted in terms of the comparison with a free dynamics if one wishes to. In Minkowski space–time, the future free dynamics is easily obtained by translating initial data along outgoing null geodesics $t = r + C$, $C \in \mathbb{R}$, and the past free dynamics corresponds to the translation along the incoming null geodesics $t = -r + C$. We can define a future free dynamics analogously in asymptotically simple space–times by translating along a congruence of outgoing null geodesics near $\mathfrak{I}^+$, and similarly a past free dynamics using a congruence of incoming null geodesics near $\mathfrak{I}^-$. There is a large amount of floppiness in such a definition but it may be used to construct scattering theories for massive equations by modifying these free dynamics.

The conformal construction of S described in the last section is not directly applicable to black hole type space–times such as Schwarzschild or Kerr metrics because these space–times have singular time-like infinities which would prevent us from integrating the current vector over the whole surfaces $\mathscr{S}^+$ and $\mathscr{S}^-$. In such space–times, called weakly asymptotically simple, the conformal construction would only allow us to define an operator S associating future asymptotic profiles to a well chosen family of past asymptotic profiles, well chosen here meaning that these characteristic past initial data should not admit singularities at i_-. In order to show that this family of asymptotic profiles covers all the profiles obtained as traces of finite energy solutions and in order to show that the operator S is unitary, we need to prove that the energy does not remain trapped in compact space-like domains; this is essentially equivalent to developing the complete scattering theory and cannot be read directly on the conformal construction.

The families of vacuum space–times constructed by D. Christodoulou and S. Klainerman [3] do no fit in the framework of 2-asymptotically simple space–times because their regularity on $\mathfrak{I}$ is not quite sufficient. However, it may still be possible to construct a conformal scattering operator in this less regular framework. Ultimately, one would like to see that the scattering operator S is in one to one correspondance with the conformal metric $\tilde{g}$. Work is in progress…

References

[1] Baez, J.C., Segal, I.E., and Zhou, Z.F., The global Goursat problem and scattering for nonlinear wave equations, *J. Funct. Anal.* **93** (1990), 239–269.

[2] Choquet–Bruhat, Y. and Christodoulou, D., Elliptic problems in $H_{s,\delta}$ spaces on manifolds which are euclidian at infinity, *Acta Math.* **146** (1981), 129–150.

[3] Christodoulou, D. and Klainerman, S., *The global nonlinear stability of the Minkowski space*, Princeton Mathematical Series 41, Princeton University Press 1993.

[4] Geroch, R.P., The domain of dependence, *J. Math. Phys.* **11** (1970), 437–449.

[5]   Hörmander, L., A remark on the characteristic Cauchy problem, *J. Funct. Anal.* **93** (1990), 270–277.

[6]   Jin, W.M., Scattering of massive Dirac fields on the Schwarzschild black-hole space–time, *Class. Quantum Grav.* **15** (1998), 3163–3175.

[7]   Kato, T., Linear equations of 'hyperbolic' type, *J. Fac. Sc. Univ. Tokyo* **17** (1970), 241–258.

[8]   _____, Linear equations of 'hyperbolic' type II, *J. Math. Soc. Japan* **25** (1973), 648–666.

[9]   _____, The Cauchy problem for quasi-linear symmetric hyperbolic systems, *Arch. Rational Mech. Anal.* **58** (1975), 181–205.

[10] Klainerman, S. and Nicolò, F., On local and global aspects of the Cauchy problem in general relativity, *Class. Quantum Grav.* **16** (1999), R73–R157.

[11] Leray, J., *Hyperbolic differential equations*, Princeton 1953.

[12] Mason, L.J. and Nicolas, J.-P., Global results for the Rarita-Schwinger equations and Einstein vacuum equations, to appear in *Proc. London Math. Soc.*

[13] _____, Geometric scattering on asymptotically simple space-times, in preparation.

[14] Nicolas, J.-P., Scattering of linear Dirac fields by a spherically symmetric Black-Hole, *Ann. Inst. Henri Poincaré - Physique Théorique* **62**, No. 2 (1995), 145–179.

[15] _____, Dirac fields on asymptotically flat space-times, submitted to *Bulletin de la Société Mathématique de France*.

[16] Penrose, R. and Rindler, W., *Spinors and space–time*, Vol. I and II, Cambridge Monographs on Mathematical Physics, Cambridge University Press 1984-1986.

[17] Seifert, H.J., Smoothing and extending cosmic time functions, *Gen. Rel. and Grav.* **8**, No. 10 (1977), 815–831.

AN EMBEDDING RESULT FOR SOME GENERAL SYMBOL CLASSES IN THE WEYL CALCULUS

JOACHIM TOFT

Contents

1 Introduction 221
2 Preliminaries 223
3 An embedding result in the Weyl calculus 226

M. de Gosson (ed.), Jean Leray '99 Conference Proceedings, 219-233.
© *2003 Kluwer Academic Publishers.*

Abstract. *The paper deals with inclusion relations between s_p and H_s^p. Here s_p is the set of all $a \in \mathcal{S}'$ such that the Weyl operator $a^w(x, D)$ is a Schatten–von Neumann operator on L^2 to the order $p \in [1, \infty]$, and H_s^p is the Sobolev space of distributions with s derivatives in L^p. At the same time we compute the trace norm for $a^w(x, D)$, when a is an arbitrary Gauss function.*

1. Introduction

The theory of pseudo-differential operators has passed many stages since the early papers [9] by Kohn and Nirenberg and [6] by Hörmander. Several important questions in linear PDE theory needed for their solutions various extensions of pseudo-differential calculus, which by now has developed to an indispensable tool also in other fields within analysis, differential geometry and mathematical physics.

A pseudo-differential calculus is a rule which to every element $a = a(x, \xi)$ (which is called a symbol) in an appropriate class of functions on $T^*\mathbf{R}^n = \mathbf{R}^n \oplus \mathbf{R}^n$, where $n < \infty$, associates an operator $\mathrm{Op}(a)$ which acts between some spaces of functions or distributions on $\mathbf{R}^n$. The quantization rule $a \mapsto a^w(x, D)$ in the Weyl calculus of Hörmander (see [7], and Chapter XVIII in [8]) is in a certain sense the most natural one because of its invariance properties with respect to the group of affine canonical transformations on $T^*\mathbf{R}^n$. There is also a close connection between Weyl calculus and the representation theory of the Heisenberg group, and we refer to [11] for more details about this link. Since the Weyl calculus will be in the background of all discussions in this paper we recall its definition. (We refer to Chapter XVIII in [8] for more details.)

Assume that $a \in \mathcal{S}(T^*\mathbf{R}^n)$. (Throughout the paper we will use the notation in [8] for spaces of distributions and test functions.) Then its Weyl quantization (or Weyl operator) $a^w(x, D)$ is the continuous operator on $\mathcal{S}(\mathbf{R}^n)$, defined by

$$a^w(x, D)f(x) \equiv (2\pi)^{-n} \iint a\left(\frac{(x+y)}{2}, \xi\right) f(y)e^{i\langle x-y, \xi\rangle} dy \, d\xi. \quad (1.1)$$

The definition above can be extended to arbitrary $a \in \mathcal{S}'(T^*\mathbf{R}^n)$, in which case one obtains continuous operators from $\mathcal{S}(\mathbf{R}^n)$ to $\mathcal{S}'(\mathbf{R}^n)$. We also remark that every linear operator which is continuous from $\mathcal{S}(\mathbf{R}^n)$ to $\mathcal{S}'(\mathbf{R}^n)$ is the Weyl quantization of some symbol $a \in \mathcal{S}'(T^*\mathbf{R}^n)$.

If $a((x + y)/2, \xi)$ is replaced by $a(x, \xi)$ in the integral in (1.1), then we obtain the standard quantization $a(x, D)$ for a. The connection to the theory of linear partial differential operators becomes particularly obvious if $a(x, \xi) = \sum a_\alpha(x)\xi^\alpha$ is a polynomial in the ξ-variables, since it follows that $a(x, D)f(x) = \sum a_\alpha(x)(D^\alpha f)(x)$, where $D_j = i^{-1}\partial_j$, by Fourier's inversion formula.

That is, for such a one has that $a(x, D)$ is the differential operator with the symbol a in the classical sense.

Many papers in the topic deal with finding necessary and sufficient conditions on the symbols such that the corresponding Weyl operators should be continuous, or even better, compact on L^2. It is then usual that one restricts the considerations to symbols with high regularity assumptions. The aim of this paper is to study L^2-continuity for Weyl operators with minimal regularity conditions on their symbols. When discussing compactness properties we shall make a somewhat detailed study, and finding necessary and sufficient conditions on the symbols, such that the corresponding Weyl operators should belong to $\mathcal{S}_p$, the set of Schatten–von Neumann operators to the order $p \in [1, \infty]$. We recall the definition of the latter space and present some well known facts.

An operator T on $L^2(\mathbf{R}^n)$ is a Schatten–von Neumann operator to the order $p \in [1, \infty]$, if and only if

$$\|T\|_p \equiv \sup \left\| \left((Tf_j, g_j) \right)_{j=1}^\infty \right\|_{l^p} \tag{1.2}$$

is finite. Here the supremum is taken over all orthonormal sequences $(f_j)_{j=1}^\infty$ and $(g_j)_{j=1}^\infty$ in $L^2(\mathbf{R}^n)$, and $(\cdot, \cdot)$ denotes the scalar product on $L^2(\mathbf{R}^n)$. We note that if $T \in \mathcal{S}_p$ for some $p < \infty$, then T is compact on L^2. One has also that $\mathcal{S}_1$, $\mathcal{S}_2$ and $\mathcal{S}_\infty$ are the sets of trace class, Hilbert–Schmidt and continuous operators on L^2, respectively. Roughly speaking, the theory for the $\mathcal{S}_p$-spaces may be considered as an 'L^p-theory' for operator spaces, and we refer to [13] for more facts.

For any $p \in [1, \infty]$, we let $s_p(T^*\mathbf{R}^n)$ be the set of all $a \in \mathcal{S}'(T^*\mathbf{R}^n)$ such that $a^w(x, D)$ belongs to $_p$. It follows from Parseval's formula that $s_2(T^*\mathbf{R}^n) = L^2(T^*\mathbf{R}^n)$. In general there are however no simple characterization of the s_p-spaces in terms of the usual functions and distributions spaces. One is therefore forced to find simple conditions on the symbols in order to guarantee that they should belong to s_p.

Some of these questions are treated in this paper which deals with comparisons between the s_p-spaces and Sobolev spaces. More precisely, we prove that if $f(p) = 2n|1 - 2/p|$, then

$$H^p_{\mu f(p)} (T^*\mathbf{R}^n) \subset s_p (T^*\mathbf{R}^n) \subset H^p_{-\mu f(p')} (T^*\mathbf{R}^n) \tag{1.3}$$

for any $\mu > 1$. Here H^p_s denotes the Sobolev space of distributions with s derivatives in L^p, and p' is the conjugate exponent for p, i.e., p' satisfies $1/p + 1/p' = 1$.

The proof is based on some continuity result for convolutions between s_1 and L^1 where one proves that $s_1 * s_1$ is continuously embedded in L^1, and that $s_1 * L^1$ is continuously embedded in s_1.

It follows from a recent paper of A. Boulkhemair in [3], that the inclusions in (1.3) with f as above is not optimal in the case $p = \infty$. By taking this fact into account we may improve the result above and prove that (1.3) holds for any $p \in [1, \infty]$ and any $\mu > 1$, when f given by $f(p) = 2n|1 - 2/p|$ when $1 \leq p \leq 2$ and $f(p) = n|1 - 2/p|$ when $2 \leq p \leq \infty$.

2. Preliminaries

In this section we recall some basic facts from the operator theory and pseudo-differential calculus which we will need. We omit the proofs since the results may be found in a more detailed version in [14] and [15].

We start by discussing the definition of the symplectic (or twisted) Fourier transform. In some situations it is important to consider $T^*\mathbf{R}^n$ as a symplectic vector space with symplectic form $\sigma(X, Y) = \langle y, \xi \rangle - \langle x, \eta \rangle$, when $X = (x, \xi) \in T^*\mathbf{R}^n$ and $Y = (y, \eta) \in T^*\mathbf{R}^n$. The symplectic Fourier transformation, $\mathcal{F}_\sigma$, on $\mathcal{S}(T^*\mathbf{R}^n)$ is then defined as

$$\hat{a}(X) = \mathcal{F}_\sigma a(X) = \pi^{-n} \int a(Y) e^{2i\sigma(X,Y)} dY. \tag{2.1}$$

Here $dY = dyd\eta$ is the volume form. By a simple application of Fourier's inversion formula, it follows that $\mathcal{F}_\sigma^2$ is the identity on $\mathcal{S}(T^*\mathbf{R}^n)$. Hence $\mathcal{F}_\sigma$ extends in the usual way to a homeomorphism on $\mathcal{S}'(T^*\mathbf{R}^n)$, which restricts to a homeomrphism on $\mathcal{S}(T^*\mathbf{R}^n)$, and to a unitary operator on $L^2(T^*\mathbf{R}^n)$. We note also that

$$\mathcal{F}_\sigma(a * b) = \pi^n \mathcal{F}_\sigma a \mathcal{F}_\sigma b, \qquad \mathcal{F}_\sigma(ab) = \pi^{-n}(\mathcal{F}_\sigma a) * (\mathcal{F}_\sigma b). \tag{2.2}$$

We shall also consider partial symplectic Fourier transforms later on. If $W_0 \subset T^*\mathbf{R}^n$ is a linear symplectic subspace of $T^*\mathbf{R}^n$, then $T^*\mathbf{R}^n$ considered as a symplectic vector space is the direct sum $T^*\mathbf{R}^n = W_0 \oplus W_0^\perp$, where $W_0^\perp$ is the orthogonal complement of W_0 with respect to the symplectic form σ. The partial symplectic Fourier transform $\mathcal{F}_{\sigma, W_0}$ is then obtained by replacing the integral in (2.1) by

$$\mathcal{F}_{\sigma, W_0} a(X) = \pi^{-n_0} \int_{W_0} a(Y_0, X_1) e^{2i\sigma(X_0, Y_0)} dY_0,$$

when $X = (X_0, X_1) \in W_0 \oplus W_0^\perp$. Here $2n_0 = \dim W_0$.

Next we discuss a quantization which in fact is very similar to the Weyl representation. More precisely, assume that $a \in \mathcal{S}'(T^*\mathbf{R}^n)$. Then Aa is the operator on $L^2(\mathbf{R}^n)$ with Schwartz kernel

$$Aa(x, y) = (2\pi)^{-n/2} \int a\left(\frac{(y-x)}{2}, \xi\right) e^{-i\langle x+y, \xi \rangle} d\xi. \tag{2.3}$$

(Here and what follows we identify operators with their corresponding Schwartz kernels.) In the case when a is not an integrable function, Aa is interpreted as the distribution $(\mathcal{K} \circ \mathcal{F}_2^{-1})a$, where $(\mathcal{K}U)(x, y) = U(y-x, -(x+y)/2)$ and $\mathcal{F}_2$ is the partial Fourier transform of

$$f(\xi) = \widehat{f}(\xi) \equiv (2\pi)^{-n/2} \int f(x) e^{-i\langle x, \xi \rangle} dx,$$

with respect to the second variable. It follows in particular that the map $a \mapsto Aa$ is a homeomorphism from $\mathscr{S}'(T^*\mathbf{R}^n)$ to $\mathscr{S}'(\mathbf{R}^{2n})$, which restricts to homeomorphism from $\mathscr{S}(T^*\mathbf{R}^n)$ to $\mathscr{S}(\mathbf{R}^{2n})$, and to a unitary map on L^2, since similar facts hold for $\mathcal{K}$ and $\mathcal{F}_2$.

The main reason for using Aa instead of $a^w(x, D)$, is that it becomes more convenient for the reader who would like to study the subject more deeply in papers like [4], [14] and [15] and some future papers planned by the author. We note also that many positivity problems in the Weyl calculus is in some sense more easy to understand and handle when Aa is used. (Cf. Section 2.1 in [14].)

We observe that continuous questions for the quantizations $a^w(x, D)$ and Aa agree, since it follows from the definitions that the Schwartz kernel for $a^w(x, D)$ is equal to $(2\pi)^{-n/2} Aa(-x, y)$.

Next we discuss some important relations between the twisted Fourier transform and the operator A in (2.3). We start with the following lemma where we list some important properties for the operator A. The proof is omitted, since the result follows easily from the definitions.

Lemma 2.1. *Let A be the operator in (2.3) and let $U = Aa$ where $a \in \mathscr{S}'(T^*\mathbf{R}^n)$. Then the following are true:*

(i) *Let $\check{a}(X) = a(-X)$ for every $X \in T^*\mathbf{R}^n$. Then $\check{U} = A\check{a}$;*

(ii) *$J_{\mathcal{F}}U = A\mathcal{F}_\sigma a = (2\pi)^{n/2} a^w(x, D)$, where $J_{\mathcal{F}}U(x, y) = U(-x, y)$;*

(iii) *For any $X = (x, \xi) \in T^*\mathbf{R}^n$ one has for every $f, g \in \mathscr{S}'(\mathbf{R}^n)$ that $(A^{-1})(f \otimes \overline{g})(\cdot - X) = A^{-1}(f_X \otimes \overline{g_{-X}})$, where $f_X(y) = f(y+x) e^{-i\langle y, \xi \rangle}$. More generally we have*

$$A(a(\cdot - X))(y, z) = (Aa)(y + x, z - x) e^{-i\langle y+z, \xi \rangle};$$

(iv) *The Hilbert space adjoint of Aa is equal to $A\widetilde{a}$, where $\widetilde{a}(X) = \overline{a(-X)}$;*

(v) *The inverse for A is given by*

$$\left(A^{-1}U\right)(x, \xi) = (2\pi)^{-n/2} \int U\left(\frac{y}{2} - x, \frac{y}{2} + x\right) e^{i\langle y, \xi \rangle} dy.$$

We return now to discussing the s_p-spaces from the introduction. Since the Schwartz kernel for $a^w(x, D)$ is equal to $(2\pi)^{-n/2} Aa(-x, y)$, it follows that $a \in s_p$, if and only if $Aa \in \mathscr{I}_p$. We equipp $s_p(T^*\mathbf{R}^n)$ with the norm $\|a\|_{s_p} \equiv \|Aa\|_{\mathscr{I}_p} = (2\pi)^{n/2} \|a^w(x, D)\|_{\mathscr{I}_p}$. (See (1.2).) Since A is homeomorphic on $\mathscr{S}'$, it follows that the mapping $s_p(T^*\mathbf{R}^n) \ni a \mapsto Aa \in \mathscr{I}_p$ is an isometric homeomorphism by the kernel theorem of Schwartz. In particular one has for every $p \in [1, \infty]$ that $s_p(T^*\mathbf{R}^n)$ is a Banach space since this is true for $\mathscr{I}_p$ (see [13]).

In the Propositions 2.2 till 2.5 below we present some more basic properties concerning the s_p-spaces. The proofs are consequences of the discussion above and some well known facts for the p-spaces and may be found in Section 1.4 in [14] or in Section 1.2 in [15].

The first proposition deals with duality properties for the s_p-spaces.

Proposition 2.2. The scalar product $(\cdot, \cdot)$ on $L^2(T^*\mathbf{R}^n)$ extends to a dual form on $s_p(T^*\mathbf{R}^n) \times s_{p'}(T^*\mathbf{R}^n)$ for every $p \in [1, \infty]$. Here and what follows p' (the conjugate exponent for p) satisfies $1/p + 1/p' = 1$. One has

$$|(a, b)| \leq \|a\|_{s_p} \|b\|_{s_{p'}} \quad \text{and} \quad \|a\|_{s_p} = \sup_{0 \neq b \in s_{p'}} |(a, b)| / \|b\|_{s_{p'}}.$$

The same conclusions are true when $(\cdot, \cdot)$ is replaced by $\langle \cdot, \cdot \rangle$, where $\langle a, b \rangle \equiv (a, \bar{b})$.

The next proposition deals with inclusion relations between the s_p-spaces. We let $s_\infty^0(T^*\mathbf{R}^n)$ be the subset of $s_\infty(T^*\mathbf{R}^n)$, consisting of all $a \in \mathscr{S}'(T^*\mathbf{R}^n)$, such that Aa (, or equivalently $a^w(x, D)$) is compact on L^2.

Proposition 2.3. Assume that $1 \leq p_1 \leq p_2 < \infty$. Then the following relation of continuous inclusions is true:

$$\mathscr{S} \subset s_{p_1} \subset s_{p_2} \subset s_\infty^0 \subset s_\infty \subset \mathscr{S}'.$$

One has $\|a\|_{s_\infty} \leq \|a\|_{s_{p_2}} \leq \|a\|_{s_{p_1}}$ and $\|a\|_{L^\infty} \leq (2/\pi)^{n/2} \|a\|_{s_1}$. Moreover, $\mathscr{S}(T^*\mathbf{R}^n)$ is dense in $s_p(T^*\mathbf{R}^n)$ and in $s_\infty^0(T^*\mathbf{R}^n)$ when $p < \infty$, and dense in $s_\infty(T^*\mathbf{R}^n)$ with respect to the weak* topology. One has that $s_2 = L^2$ with equality in norms. The set s_1 is contained in the set of continuous functions on $T^*\mathbf{R}^n$ vanishing at infinity.

We shall next discuss decomposition for elements in the s_p-spaces. We use the notation $\mathscr{B}(T^*\mathbf{R}^n)$ for the set which consists of all sequences (u_j) in $L^2(T^*\mathbf{R}^n)$, such that $Au_j = f_j \otimes g_j$ for every j, where $(f_j), (g_j) \in \mathrm{ON}(\mathbf{R}^n)$. Here $\mathrm{ON}(\mathbf{R}^n)$ is the family of orthonormal sets in $L^2(\mathbf{R}^n)$.

Proposition 2.4. Assume that $a \in s_\infty^0(T^*\mathbf{R}^n)$. Then there exists $\lambda = (\lambda_j) \in l^\infty$ and $(u_j) \in \mathcal{B}(T^*\mathbf{R}^n)$, such that $0 \le \lambda_j \to 0$ as $j \to \infty$ and

$$a = \sum \lambda_j u_j,$$

with convergence in s_∞. One has that (λ_j) and (u_j) are unique except on the order of summation. Moreover, if $p < \infty$, then $a \in s_p$ if and only if $\lambda \in l^p$. It is then true that the sum converges in s_p, and that $\|a\|_{s_p} = \|\lambda\|_{l^p}$.

Proposition 2.5. Any partial symplectic Fourier transform and composition by any affine canonical transformation on $T^*\mathbf{R}^n$ are unitary mappings on $s_p(T^*\mathbf{R}^n)$ and on $s_\infty^0(T^*\mathbf{R}^n)$, for every $1 \le p \le \infty$.

Corollary 2.6. Assume that μ is a Borel measure on $T^*\mathbf{R}^n$ with finite mass $\|\mu\|$. Then the mapping $a \mapsto \mu * a$ on $\mathscr{S}(T^*\mathbf{R}^n)$ extends uniquely to a continuous mapping on $s_p(T^*\mathbf{R}^n)$, for every $p \in [1, \infty]$. One has $\|a * \mu\|_{s_p} \le \|a\|_{s_p}\|\mu\|$.

3. An embedding result in the Weyl calculus

This section is devoted to some studies of convolution products $a * b$ where $a \in s_1(T^*\mathbf{R}^n)$ and b is either in $s_1(T^*\mathbf{R}^n)$ or in $L^1(T^*\mathbf{R}^n)$. In particular we show that $s_1 * s_1 \subset L^1$. As an application of these results we present some conditions on the index s for the Sobolev space $H_s^p(T^*\mathbf{R}^n)$ to contain or to be contained in $s_p(T^*\mathbf{R}^n)$.

We start with the following result.

Proposition 3.1. The convolution on $\mathscr{S}(T^*\mathbf{R}^n)$ extends uniquely to continuous bilinear mappings

$$s_1(T^*\mathbf{R}^n) \times s_1(T^*\mathbf{R}^n) \longrightarrow L^1(T^*\mathbf{R}^n),$$
$$L^1(T^*\mathbf{R}^n) \times s_1(T^*\mathbf{R}^n) \longrightarrow s_1(T^*\mathbf{R}^n)$$

if one requires that $a * b$ is the ordinary convolution when one of the factors is in $\mathscr{S}$. One has the estimates

$$\|a * b\|_{L^1} \le (2\pi)^n \|a\|_{s_1}\|b\|_{s_1}, \quad a, b \in s_1(T^*\mathbf{R}^n) \tag{3.1}$$

and

$$\|a * b\|_{s_1} \le \|a\|_{L^1}\|b\|_{s_1}, \quad a \in L^1(T^*\mathbf{R}^n), b \in s_1(T^*\mathbf{R}^n). \tag{3.2}$$

Proof. The result follows if we prove that (3.1) holds for any $a, b \in \mathscr{S}(T^*\mathbf{R}^n)$ in view of Proposition 2.3, since (3.2) is an immediate consequence of Corollary 2.6. By Proposition 2.4 it follows that it is enough to prove (3.1) when a

and b are rank one elements, and that $\|a\|_{s_1} = \|b\|_{s_1} = 1$. Then $Aa = f_1 \otimes f_2$ and $Ab = g_1 \otimes g_2$ for some unit vectors f_1, f_2, g_1, g_2 in $L^2(\mathbf{R}^n)$. We let w_1 and w_2 be the functions

$$w_1 = A^{-1}(g_1 \otimes f_2) \in L^2(T^*\mathbf{R}^n), \qquad w_2 = A^{-1}(f_1 \otimes g_2) \in L^2(T^*\mathbf{R}^n)$$

and we claim that

$$(a * b)(X) = \left(\frac{\pi}{2}\right)^n w_1\left(\frac{X}{2}\right) w_2\left(\frac{X}{2}\right). \tag{3.3}$$

In fact, writing $X = (x, \xi)$ and recalling Lemma 2.1, we find that

$$\begin{aligned} a * b(X) &= (b, \tilde{a}(\cdot - X)) = (Ab, A(\tilde{a}(\cdot - X))) \\ &= \iint e^{i\langle y+z, \xi\rangle} f_2(y+x) f_1(z-x) g_1(y) g_2(z) dy dz \\ &= (2\pi)^n \widehat{\phi}_x(-\xi) \widehat{\psi}_x(-\xi) \end{aligned}$$

if we set $\phi_x(y) = f_2(y+x) g_1(y)$ and $\psi_x(z) = f_1(z) g_2(z+x)$. Now (2.3), Lemma 2.1 (v) and a simple computation shows that

$$\widehat{\phi}_x(-\xi) = 2^{-n} e^{-i\langle x, \xi\rangle/2} w_1\left(\frac{X}{2}\right), \qquad \widehat{\psi}_x(-\xi) = 2^{-n} e^{i\langle x, \xi\rangle/2} w_2\left(\frac{X}{2}\right).$$

This proves (3.3).

Since A is unitary on L^2 the result follows now from (3.3) and an application of Cauchy–Schwartz inequality. □

Remark 3.2. It follows from the duality properties in Proposition 2.3 together with Propositions 2.3 and 3.1 that the convolution product extends uniquely to continuous mappings $s_1 \times s_\infty \to L^\infty$, $s_1 \times L^\infty \to s_\infty$ and $L^1 \times s_\infty \to s_\infty$, and that

$$\begin{aligned} \|a * b\|_{L^\infty} &\leq \|a\|_{s_1} \|b\|_{s_\infty}, \\ \|a * b\|_{s_\infty} &\leq (2\pi)^n \|a\|_{s_1} \|b\|_{L^\infty}, \\ \|a * b\|_{s_\infty} &\leq \|a\|_{L^1} \|b\|_{s_\infty}. \end{aligned}$$

More generally, if we combine Proposition 3.1 and its proof by some convexity, then one may prove a Young type inequality, such that the convolution product may be uniquely extended to continuous bilinear mappings from $s_p(W) \times s_q(W)$ to $L^r(W)$, and from $s_p(W) \times L^q(W)$ to $s_r(W)$, for any $p, q, r \in [1, \infty]$ such that $1/p + 1/q = 1 + 1/r$. (Cf. Theorem 2.2.3 in [14] or Theorem 1.13 in [15].)

We shall next discuss inclusions between the s_p-spaces and Sobolev spaces. We recall the definition of the latter space. The space $H_s^p(T^*\mathbf{R}^n)$, $s \in \mathbf{R}$, $1 \le p \le \infty$, consists of all $a \in \mathscr{S}'(T^*\mathbf{R}^n)$ such that $(1 + |D|^2)^{s/2}a \in L^p(T^*\mathbf{R}^n)$, and we set $\|a\|_{H_s^p} \equiv \|(1 + |D|^2)^{s/2}a\|_{L^p}$. Here $\varphi(D)$ is multiplication by φ on the symplectic Fourier transform side, when φ is a function on $T^*\mathbf{R}^n$.

Theorem 3.3. *Let* $f(p) = 2n|1 - 2/p|$, *and let* $\mu > 1$. *Then for some constant* $C < \infty$ *one has for any* $a \in \mathscr{S}'(T^*\mathbf{R}^n)$ *that*

$$C^{-1}\|a\|_{H_{-\mu f(p)}^p} \le \|a\|_{s_p} \le C\|a\|_{H_{\mu f(p)}^p}. \tag{3.4}$$

In particular it is true that (1.3) holds.

For the proof we need the following lemma.

Lemma 3.4. *Let* $\lambda = (\lambda_1, \ldots, \lambda_n) \in \mathbf{R}_+^n$, $u_\lambda = e^{-Q}$ *where* $Q(X) = \sum_1^n \lambda_j(x_j^2 + \xi_j^2)$, $X = (x, \xi) \in T^*\mathbf{R}^n$, *and let* $\mathscr{F}_{\sigma,k}$ *be the partial symplectic Fourier transform with respect to the variables* x_k *and* ξ_k. *Then the following are true:*

(1) If $\widetilde{\lambda}_k = (\lambda_1, \ldots, \lambda_{k-1}, \lambda_k^{-1}, \lambda_{k+1}, \ldots, \lambda_n)$, *then* $\mathscr{F}_{\sigma,k}u_\lambda = \lambda_k^{-1}u_{\widetilde{\lambda}_k}$;

(2) One has that

$$\|u_\lambda\|_{s_1} = \left(\frac{\pi}{2}\right)^{n/2} \prod_1^n \max\left(1, \lambda_j^{-1}\right). \tag{3.5}$$

Proof. The first assertion follows immediately by some simple calculations. When proving (2) we assume first that $\lambda = \lambda_0 \equiv (1, \ldots, 1)$. Then

$$u_{\lambda_0} = \left(\frac{\pi}{2}\right)^{n/2} A^{-1}(f \otimes f),$$

where $f(x) = \pi^{-n/4}e^{-|x|^2/2}$. Since $\|f\|_{L^2} = 1$ it follows from the definitions that $\|u_1\|_{s_1} = (\pi/2)^{n/2}$, which proves the assertion when $\lambda = \lambda_0$.

We shall next consider the case $\lambda_j \ge 1$ for every j. Assume first that every $\lambda_j > 1$, and set $1/\lambda = (1/\lambda_1, \ldots, 1/\lambda_n)$ and $\gamma = (\lambda_1 - 1, \ldots, \lambda_n - 1)$. By combining the first part of the lemma by Proposition 2.5 and Proposition 3.1, one obtains

$$\begin{aligned}
\|u_\lambda\|_{s_1} &= \|u_\gamma u_{\lambda_0}\|_{s_1} = \|\mathscr{F}_\sigma(u_\gamma u_{\lambda_0})\|_{s_1} \\
&= \pi^{-n}((\lambda_1 - 1)\cdots(\lambda_n - 1))^{-1}\|u_{1/\gamma} * u_{\lambda_0}\|_{s_1} \\
&\le \pi^{-n}((\lambda_1 - 1)\cdots(\lambda_n - 1))^{-1}\|u_{1/\gamma}\|_{L^1}\|u_{\lambda_0}\|_{s_1} \\
&= \left(\frac{\pi}{2}\right)^{n/2}.
\end{aligned}$$

Hence $\|u_\lambda\|_{s_1} \leq (\pi/2)^{n/2}$. Since $u_\lambda(0) = \|u_\lambda\|_{L^\infty} = 1$, an opposite inequality follows from Proposition 2.3, and we have proved the assertion in the case $\lambda_j > 1$ for every j.

Now we remove the condition that $\lambda_j > 1$ and assume more generally that $\lambda_j \geq 1$ for every j. Let for any $\varepsilon > 0$, $Q_\varepsilon(X) = \sum \lambda_{j,\varepsilon}(x_j^2 + \xi_j^2)$, where $X = (x, \xi)$, and $\lambda_{j,\varepsilon} = \lambda_j + \varepsilon$. Then $e^{-Q_\varepsilon} \to e^{-Q}$ in $\mathscr{S}$ as $\varepsilon \to 0+$. Since $\|e^{-Q_\varepsilon}\|_{s_1} = (\pi/2)^{n/2}$ by the first part of the proof, it follows from Proposition 2.3 that $\|e^{-Q}\|_{s_1} = (\pi/2)^{n/2}$, and the statement follows in this case.

Assume finally that $\lambda \in \mathbf{R}_+^n$ is arbitrary. We may assume that $\lambda_1, \ldots, \lambda_k < 1$ and $\lambda_{k+1}, \ldots, \lambda_n \geq 1$, for some $k \in \{0, \ldots, n+1\}$. Let $\gamma = (\gamma_1, \ldots, \gamma_n)$, where $\gamma_j = 1/\lambda_j$ when $j \leq k$ and $\gamma_j = \lambda_j$ when $j \geq k+1$. Then by repeated use of (1) it follows from Proposition 2.5 that

$$\|u_\lambda\|_{s_1} = (\lambda_1 \cdots \lambda_k)^{-1} \|u_\gamma\|_{s_1} = (\lambda_1 \cdots \lambda_k)^{-1} \left(\frac{\pi}{2}\right)^{n/2}.$$

Here the last equality follows from the first part of the proof, since $\gamma_j \geq 1$ for every j. This completes the proof. $\square$

Remark 3.5. It follows from Section 1.4 in [14] or in Section 1.2 in [15] that if $a \in s_1(T^*\mathbf{R}^n)$, then Aa is positive semi-definite operator on $L^2(\mathbf{R}^n)$, if and only if $\|a\|_{s_1} = (\pi/2)^n a(0)$. Since $u_\lambda(0) = 1$, for every λ, it follows therefore from Lemma 3.4 that Au_λ is a positive semi-definite operator, if and only if $\lambda_j \geq 1$ for every j. Similar facts are valid for $u_\lambda^w(x, D)$, since $u_\lambda^w(x, D) = (2\pi)^{-n/2} Au_{1/\lambda}$, by Lemma 2.1.

Proof of Theorem 3.3. Assume first that $p = 1$. It is enough to prove that (3.4) holds when $a \in \mathscr{S}$. Let $b = (1 + |D|^2)^{n\mu} a$. Then

$$\int_0^\infty e^{-t(1+|D|^2)} t^{n\mu-1} b \, dt = \Gamma(n\mu)a. \tag{3.6}$$

In fact, the Fourier transform on the left hand side is equal to

$$\int_0^\infty e^{-t(1+|X|^2)} t^{n\mu-1} \left(1 + |X|^2\right)^{n\mu} \widehat{a}(X) dt$$
$$= \int_0^\infty e^{-s} s^{n\mu-1} ds \, \widehat{a}(X) = \Gamma(n\mu)\widehat{a}(X). \tag{3.7}$$

Set $u_t(X) = e^{-t|X|^2}$, for any $t > 0$. Then

$$e^{-t|D|^2} b = \pi^{-n} \left(\mathscr{F}_\sigma \left(e^{-t|\cdot|^2}\right)\right) * b = \pi^{-n} t^{-n} u_{1/t} * b.$$

If we insert this into (3.6), then one obtains

$$a = \left(\pi^n \Gamma(n\mu)\right)^{-1} \int_0^\infty e^{-t} t^{n\mu-1-n} \left(u_{1/t} * b\right) dt. \qquad (3.8)$$

Hence, combining the triangle inequality with (3.2), we obtain

$$\|a\|_{s_1} \le \left(\pi^n \Gamma(n\mu)\right)^{-1} \int_0^\infty e^{-t} t^{n\mu-1-n} \left\|u_{1/t} * b\right\|_{s_1} dt$$

$$\le \left(\pi^n \Gamma(n\mu)\right)^{-1} \int_0^\infty e^{-t} t^{n\mu-1-n} \left\|u_{1/t}\right\|_{s_1} \|b\|_{L^1} dt.$$

By Lemma 3.4 it follows that for some constant C_μ, that

$$\|a\|_{s_1} \le C_\mu \int_0^\infty e^{-t} t^{n\mu-1-n} \left(1 + t^n\right) dt \|b\|_{L^1} = C'_\mu \|a\|_{H^1_{2n\mu}},$$

where $C'_\mu = C_\mu \int_0^\infty e^{-t} t^{n\mu-1-n}(1 + t^n) dt$. Since $\mu > 1$ it follows that the last integral is convergent. Hence $C'_\mu < \infty$, and we have proved that $\|a\|_{s_1} \le C\|a\|_{H^1_\mu}$.

If we replace a in (3.6) with $(1 + |D|^2)^{-n\mu} a(X)$, then one obtains

$$\left(1 + |D|^2\right)^{-n\mu} a(X) = \Gamma(n\mu)^{-1} \int_0^\infty e^{-t(1+|D|^2)} t^{n\mu-1} a(X) dt.$$

By applying the L^1-norm and repeating the arguments, where we use (3.1) instead of (3.2) in the integral we obtain $\|a\|_{H^1_{-2n\mu}} \le C\|a\|_{s_1}$ for some constant C. Hence the assertion holds in the case $p = 1$.

If we may repeat the arguments, where we use Remark 3.2 instead of Proposition 3.1, then we obtain for some C that $C^{-1} \|a\|_{H^\infty_{-2n\mu}} \le C\|a\|_{s_1} \le C\|a\|_{H^\infty_{2n\mu}}$, and the assertion follows in the case $p = \infty$. We note also that the assertion is true for $p = 2$, since $s_2 = L^2 = H^2_0$ with equality in the norms.

The proof follows now for general p by interpolation. More precisely, it follows from section IX in [12] that any interpolation result which is valid for the L^p-spaces, holds also for the $\mathscr{I}_p$-spaces. Since the latter spaces are homeomorphic with the s_p-spaces, it follows that the interpolation technique carry over to the s_p-spaces. The result follows now from these facts and standard interpolation results. (See Theorems 4.1.2, 5.1.1 and 6.4.5 in [2].) $\square$

Remark 3.6. A more precise inclusion relation is presented in [14], Theorem 2.2.7, where one discuss inclusions between the s_p-spaces and Besov spaces.

Remark 3.7. Theorem 3.3 may be improved by using Corollary 2.5 in [3]. In fact, from that corollary it follows that $H^\infty_{\mu n}(T^*\mathbf{R}^n)$ is continuous embedded in

$s_\infty(T^*\mathbf{R}^n)$. If we apply Proposition 2.2, then by duality we obtain $s_1(T^*\mathbf{R}^n) \subset H^1_{-\mu n}(T^*\mathbf{R}^n)$. By combining these estimates with Theorem 3.3 and its proof, it follows now that (1.3) and (3.4) holds for any $\mu > 1$, where $f(p) = 2n|1-2/p|$ when $1 \le p \le 2$ and $f(p) = n|1-2/p|$ when $2 \le p \le \infty$.

Remark 3.8. We may also compute $\|u_\lambda\|_{s_2}$ and $\|u_\lambda\|_{s_\infty}$ exactly, where u_λ is the same as in Lemma 3.4. In fact we have that

$$\|u_\lambda\|_{s_2} = \left(\frac{\pi}{2}\right)^{n/2} \prod_1^n \lambda_j^{-1/2}, \qquad \|u_\lambda\|_{s_\infty} = (2\pi)^{n/2} \prod_1^n (1+\lambda_j)^{-1}. \quad (3.9)$$

The first equality follows immediately from the property $\|u_\lambda\|_{s_2} = \|u_\lambda\|_{L^2}$.

For the second inequality we observe that $\|u_\lambda\|_{s_\infty} = \|Au_\lambda\|_{\mathcal{I}_\infty}$ is the operator norm for Au_λ. By some simple calculations, using (2.3), we obtain

$$Au_\lambda(x, y) = \left(\prod_1^n \lambda_j^{-1/2}\right) e^{-Q'(x,y)},$$

where

$$Q'(x, y) = \sum_1^n \left(\lambda_j (x_j - y_j)^2 + \lambda_j^{-1} (x_j + y_j)^2\right) /4.$$

Hence $Au_\lambda(x, y)$ is a Gauss function on $\mathbf{R}^n \oplus \mathbf{R}^n$, i.e., a function of the form e^{-P}, where P is a polynomial of second order on $\mathbf{R}^n \oplus \mathbf{R}^n$. By [10] and the fact that Au_λ is self-adjoint, it follows that

$$\|Au_\lambda\|_{\mathcal{I}_\infty} = \sup |(Au_\lambda f, f)|, \quad (3.10)$$

where it is enough to let the supremum be taken over all Gauss functions f on $\mathbf{R}^n$, such that $\|f\|_{L^2} = 1$. By some straightforward computations one has that the supremum in the right hand side of (3.9) is attained only for $f(x) = \pi^{-n/4}e^{-|x|^2/2}$, and one obtains $\|Au_\lambda\|_{\mathcal{I}_\infty} = (2\pi)^{n/2} \prod_1^n (1 + \lambda_j)^{-1}$, which proves (3.8).

Remark 3.9. Assume that $p \in \{1, 2, \infty\}$. Then Lemma 3.4 and Remark 3.8 actually give us the s_p-norm for any Gauss function e^{-Q}, where Q is a positive definite quadratic form on $\mathbf{R}^n$. In fact, by Lemma 18.6.4 in [8] it follows that for some choice of symplectic coordinates (x, ξ) on $\mathbf{R}^n \oplus \mathbf{R}^n$, one has that $Q(x, \xi) = \sum_1^n \lambda_j(x_j^2 + \xi_j^2)$, where $\lambda_1 \ge \lambda_2 \ge \cdots \ge \lambda_n > 0$ are uniquely determined by Q. The assertion follows now from Proposition 2.5.

Acknowledgements

This paper is based on my Ph.D. thesis (see [14]). I am very grateful to my supervisor Professor Anders Melin for introducing me to this subject, for his support, and for encouragement, as well as constructive criticism. I would also like to thank Professor Lars Hörmander for much valuable advice which led to several improvements of the original manuscript. Finally, I would like to express my gratitude to Professor Maurice de Gosson for letting me participate in the conference.

References

[1] Beals, R. and Fefferman, C., On local solvability of linear partial differential equations, *Ann. of Math.* **97** (1973), 482–498.

[2] Bergh, J. and Löfström, J., *Interpolation spaces*, An Introduction, Springer–Verlag, Berlin–Heidelberg–New York 1976.

[3] Boulkhemair, A., L^2-estimates for pseudodifferential operators, *Ann. Scuola Norm. Sup. Pisa Cl. Sci (4)* **22** (1995), 155–183.

[3] ———, L^2-estimates for Weyl quantization, *J. Funct. Anal.* **165** (1999), 173–204.

[4] Folland, G.B., *Harmonic analysis in phase space*, Princeton U. P., Princeton 1989.

[5] Grossmann, A., Loupias, G., and Stein, E.M., An algebra of pseudo-differential operators and quantum mechanics in phase space, *Ann. Inst. Fourier* **18** (1968), 343–368.

[6] Hörmander, L., Pseudo-differential operators, *Comm. Pure Appl. Math.* **18** (1965), 501–517.

[7] ———, The Weyl calculus of pseudo-differential operators, *Comm. Pure. Appl. Math.* **32** (1979), 359–443.

[8] ———, *The Analysis of linear partial differential operators*, Springer–Verlag, Berlin–Heidelberg–New York–Tokyo 1983-1985.

[9] Kohn, J.J. and Nirenberg, L., On the algebra of pseudo-differential operators, *Comm. Pure Appl. Math.* **18** (1965), 269–305.

[10] Lieb, E.H., Gaussian kernels have only Gaussians maximizers, *Invent. Math.* **102** (1990), 179–208.

[11] Melin, A., Parametrix constructions for some classes of right-invariant differential operators on the Heisenberg group, *Comm. Partial Differential Equations* **6** (1981), 1363–1405.

[12] Reed, M. and Simon, B., *Methods of modern mathematical physics*, Academic Press, London–New York 1979.

[13] Simon, B., *Trace ideals and their applications*, London Math. Soc. Lecture Note Series, Cambridge University Press, Cambridge–London–New York–Melbourne 1979.

[14] Toft, J., *Continuity and positivity problems in pseudo-differential calculus*, Thesis, Department of Mathematics, University of Lund, Lund, 1996.

[15]   _____, Regularisations, decompositions and lower bound problems in the Weyl calculus, *Comm. Partial Differential Equations* **7-8** (2000), 1201–1234.

THE LAGRANGIAN IN SYMPLECTIC MECHANICS

G. M. TUYNMAN

Contents

1   Introduction.. 237
2   Prequantization... 240
3   The postclassical formalism 241
4   Quantum mechanics.. 243
5   Symmetry groups.. 243
6   The physical content of the postclassical formulation 246

M. de Gosson (ed.), Jean Leray '99 Conference Proceedings, 235-247.
© *2003 Kluwer Academic Publishers.*

Abstract. *In this paper we argue that the symplectic description of classical mechanics contains many elements of the Lagrange formulation of classical mechanics, in particular a variational description in terms of an action functional. In order to obtain these results, one has to enlarge the symplectic formulation with part of the prequantization construction: the construction of a principal S^1-bundle with connection over the symplectic manifold, but without the quantization condition. We propose to call this enlargement the 'postclassical formalism.' Apart from mathematical evidence that the post-classical formalism is a (very) useful enlargement of symplectic mechanics, we also argue that this formalism has physical content.*

1. Introduction

One of the first formulae one learns in classical mechanics is Newton's law that force equals mass times acceleration: $F = ma$. In principle the force could depend upon second or higher order time derivatives of position, but usually one restricts attention to forces that depend only upon position and velocity, the first order time derivative of position. One thus obtains a second order differential equation

$$m\frac{d^2q^i}{dt^2} = F_i\left(q, \frac{dq}{dt}\right).$$

If the system describes N particles in our Euclidean space, the position $q = (q^1, \ldots, q^{3N})$ is a vector in $\mathbb{R}^{3N}$. Allowing for planar and linear 'particles' gives us a position vector in $\mathbb{R}^n$ for some $n \in \mathbb{N}$.

The next step in a course on classical mechanics is to restrict attention to conservative systems, i.e., systems for which the force is (minus) the gradient of a potential V depending only upon position: $F_i(q) = -(\partial V/\partial q^i)(q)$. This simplifies the description of such a system because one only has to give a single function of position (the potential V) instead of n functions of position and velocity (the force F). There is one 'disadvantage' to this simplicity: the *form* of the equation

$$m\frac{d^2q^i}{dt^2} = -\frac{\partial V}{\partial q^i}(q) \tag{1}$$

is not invariant under a change of coordinates $q \mapsto \hat{q} = \hat{q}(q)$; only orthogonal changes preserve this form. This non-invariance is often 'explained' by noting that d^2q^i/dt^2 is a contravariant vector, whereas $\partial V/\partial q^i$ is a covariant vector.[1]

---

[1] This is more or less implicit in the height of the indices (part of the Einstein summation convention which we will use): in d^2q^i/dt^2 the index is a superscript, whereas in $\partial V/\partial q^i$ it is a subscript.

The Lagrange formalism is one way to overcome this problem of non-invariance. First one introduces n new variables $v = (v^1, \ldots, v^n)$ and a function $L : \mathbb{R}^n \times \mathbb{R}^n \to \mathbb{R}$ by $L(q, v) = \sum_i (1/2) m (v^i)^2 - V(q)$. Next one shows that the n equations (1) are equivalent to the $2n$ Euler–Lagrange equations

$$\frac{dq^i}{dt} = v^i \quad \text{and} \quad \frac{\partial L}{\partial q^i} = \frac{d}{dt} \frac{\partial L}{\partial v^i}.$$

The form of these equations *is* invariant under general coordinate changes $q \mapsto \hat{q}(q)$, provided one changes the new coordinates v according to $v^i \mapsto \hat{v}^i(q, v) = (\partial \hat{q}^i / \partial q^j) v^j$. The transformation $(q, v) \mapsto (\hat{q}, \hat{v})$ is exactly the coordinate transformation in a tangent bundle in which the v are the fibre coordinates. One thus can consider the more general situation in which one has an n-dimensional manifold Q as configuration space and an arbitrary function $L : TQ \to \mathbb{R}$, in this context called a Lagrangian. The form-invariance of the Euler–Lagrange equations implies that the local solutions of these equations glue together to a global solution curve in TQ.

This general Lagrange formalism has advantages and disadvantages, of which we will mention some.

(1) Not every Lagrangian L will give rise to solutions of the Euler–Lagrange equations. For instance, the Lagrangian $L(q, v) = q$ gives rise to the equation $1 = 0$.

(2) Several Lagrangians can give rise to the same solutions. For instance, adding $(\partial u / \partial q^i) v^i$ to any Lagrangian (with u an arbitrary function of q) does not change the solutions. Common wisdom says that adding a total time derivative will not change the solutions (note that $du/dt = (\partial u / \partial q^i) v^i$). However, I do not know how to interpret this wisdom in its generality because adding a total time derivative of an arbitrary function of q *and* v can change the solutions, but not necessarily.

(3) The Euler–Lagrange equations can be described by a variational principle. For two fixed end points $q_0, q_1 \in Q$ and two times $t_0, t_1 \in \mathbb{R}$ one denotes by $\mathscr{C}$ the set of all paths $\gamma : [t_0, t_1] \to Q$ such that $\gamma(t_0) = q_0$ and $\gamma(t_1) = q_1$. Then the lifted curve $(\gamma, d\gamma/dt)$ in TQ is a solution of the Euler–Lagrange equations if and only if the path γ is a stationary point for the action functional $A : \mathscr{C} \to \mathbb{R}$ defined by

$$A(\gamma) = \int_{t_0}^{t_1} L\left(\gamma, \frac{d\gamma}{dt}\right) dt. \tag{2}$$

(4) The phase factor $e^{iA(\gamma)}$ (an element of the unit circle $\mathbf{S}^1 \subset \mathbf{C}$) plays an important role in Feynman path integral quantization.

Another way of dealing with the problem of the non-invariance of the form of the equations (1) is the Hamilton formalism. One introduces n new variables $p = (p_1, \ldots, p_n)$ and a function $H : \mathbb{R}^n \times \mathbb{R}^n \to \mathbb{R}$ by $H(q, p) = \sum_i (1/2m)(p_i)^2 + V(q)$. And then one shows that the n equations (1) are equivalent to the $2n$ Hamilton equations

$$\frac{dq^i}{dt} = \frac{\partial H}{\partial p_i} \quad \text{and} \quad \frac{dp_i}{dt} = -\frac{\partial H}{\partial q^i}.$$

The form of these (first order differential) equations *is* invariant under general coordinate changes $q \mapsto \hat{q}(q)$, provided one changes the new coordinates p according to $p_i \mapsto \hat{p}_i(q, p) = (\partial q^j / \partial \hat{q}^i) p_j$. The transformation $(q, p) \mapsto (\hat{q}, \hat{p})$ is exactly the coordinate transformation in a cotangent bundle in which the p are the fibre coordinates. One thus can consider the more general situation in which one has an n-dimensional manifold Q as configuration space and an arbitrary function $H : T^*Q \to \mathbb{R}$, in this context called a Hamiltonian. The form-invariance of the Hamilton equations implies that the local solutions of these equations glue together to a global solution curve in T^*Q. Noting that a cotangent bundle has a natural symplectic structure ω and that the Hamilton equations can be described as the flow of the (unique) vector field ξ_H given by the equation

$$\iota(\xi_H)\omega + dH = 0, \tag{3}$$

one can consider an even more general situation in which one has an arbitrary symplectic manifold (M, ω) and a function $H : M \to \mathbb{R}$ (a Hamiltonian). In this way one arrives at the symplectic formulation of classical mechanics.

As for the Lagrange formulation, the symplectic formulation has advantages and disadvantages, of which we mention some.

(1) Any Hamilton function $H : M \to \mathbb{R}$ gives rise to solutions of the Hamilton equations, simply because any vector field admits a flow.

(2) Two Hamilton functions give rise to the same solutions if and only if they differ by a constant. [2]

(3) It seems to be impossible to describe the Hamilton equations by means of a variational principle: if we fix both endpoints in the symplectic manifold, we have twice too many boundary conditions for the equations, and if we fix only the initial point, we obtain boundary terms in the integral. And in the general symplectic formulation we do not have a canonical splitting in position q and momentum p coordinates.

---

[2] We tacitly assume that all our manifolds are connected.

To end this section we come back to the Lagrange formalism with a configuration space Q and a Lagrangian $L : TQ \to \mathbb{R}$. With these data one can construct a map $\mathcal{L}e : TQ \to T^*Q$ called the Legendre transformation; in local coordinates it is given by $\mathcal{L}e(q, v) = (q, p_i = \partial L/\partial v^i)$. In general this map is neither injective nor surjective. If however this map is a diffeomorphism, one can define a function $H : T^*Q \to \mathbb{R}$ by $H = (v^i \partial L/\partial v^i - L) \circ \mathcal{L}e^{-1} = (p_i v^i - L) \circ \mathcal{L}e^{-1}$. It turns out that the solutions of the Euler–Lagrange equations for L on TQ correspond (under the Legendre transformation) to the solutions of the Hamilton equations for H on T^*Q. Moreover, the inverse Legendre transformation is given in local coordinates by $\mathcal{L}e^{-1}(q, p) = (q, v^i = \partial H/\partial p_i)$ and the Lagrangian L can be recovered as $L(q, v) = (p_i \partial H/\partial p_i - H) \circ \mathcal{L}e = (p_i v^i - H) \circ \mathcal{L}e$.

2. Prequantization

When one looks at points 1 and 2 in the list of (dis)advantages, the symplectic formalism comes out ahead of the Lagrange formalism. On the other hand, the Lagrange formalism has the advantage of the variational description by means of the action functional and its related importance in Feynman path integral quantization. We will show that by extending the symplectic formulation, one can obtain similar results there. The key ingredient is the procedure called Prequantization. If ω is a closed 2-form on a manifold M, one can form the group Per(ω) of periods of ω, obtained by integrating ω over all 2-cycles (2-dimensional surfaces without boundary) in M. As a subgroup of the additive group of reals, there are three distinct possibilities:

(i)   Per(ω) = {0}, which is equivalent to ω being exact (by de Rham duality),

(ii) Per(ω) = $d\mathbb{Z}$ for some generator $d \in \mathbb{R}^{>0}$, or

(iii) Per(ω) is dense in $\mathbb{R}$ (with respect to the usual topology).

Case (i) can be included in case (ii) by allowing d to be zero; these two cases are summarized by saying that the group of periods is discrete. Depending upon convention, if d belongs to $\mathbb{Z}$ or $2\pi \mathbb{Z}$ the form ω is said to represent an integral cohomology class. If Per(ω) is discrete, one can construct a principal fibre bundle $\pi : Y \to M$ over the manifold M with structure group $\mathbb{R}$ mod D for any discrete subgroup D of $\mathbb{R}$ containing Per(ω). Such a bundle can be equipped with a connection α whose curvature equals ω (note that for a 1-dimensional abelian structure group, the curvature of a connection α is given by $d\alpha$). Using the natural notion of equivalence of principal bundles with connection, one can show that inequivalent bundles with connection are parametrized by $H^1(M, \mathbb{R} \bmod D) = \text{Hom}(\pi_1(M) \to \mathbb{R} \bmod D)$. Since the group $\mathbb{R}$ mod D is isomorphic to the circle $\mathbf{S}^1$ (we ignore the possibility $D = \{0\}$), we can summarize this as follows.

Proposition. If the group of periods of a closed 2-form ω on a manifold M is discrete, there exists a principal S^1-bundle $\pi : Y \to M$ with connection α whose curvature equals ω. The various inequivalent possibilities are parametrized by $H^1(M, S^1) = \text{Hom}(\pi_1(M) \to S^1)$.

According to Souriau [2], prequantization of a symplectic manifold (M, ω) is the construction of a principal S^1-bundle $\pi : Y \to M$ with connection α whose curvature is equal to ω and such that integration of α over a fibre equals $2\pi\hbar$. This requires that the generator of D is equal to $2\pi\hbar$, which requires in turn that the generator d of the group of periods belongs to $2\pi\hbar\mathbb{Z}$. According to Kostant [1], prequantization is the construction of a complex line bundle $\pi : L \to M$ over M with connection ∇ whose curvature equals $\omega/\hbar$. This requires that $\omega/\hbar$ represents an integral cohomology class (in the $2\pi\mathbb{Z}$ convention). These two points of view of prequantization are completely equivalent, the link being given by taking the associated complex line bundle according to the canonical representation of S^1 on $\mathbb{C}$.

3. The postclassical formalism

Our generalization of the symplectic formulation is the first part of Souriau's prequantization procedure: the construction of a principal $\mathbb{R}$ mod D-bundle $\pi : Y \to M$ with connection α whose curvature equals the symplectic form ω. This requires that the group $\text{Per}(\omega)$ be discrete, but not that its generator belongs to $2\pi\hbar\mathbb{Z}$. We propose the name 'postclassical formalism' for this generalization.[3] We now describe some of the features of this generalization; proofs can be found in [3], [4] and [5].

(1) For each Hamiltonian $H : M \to \mathbb{R}$ there exists a (unique) vector field η_H on Y with the following properties:

   (a) $\mathcal{L}(\eta_H)\alpha = 0$, where $\mathcal{L}(\eta_H)$ denotes the Lie derivative in the direction of η_H,

   (b) $\pi_*\eta_H = \xi_H$, where ξ_H is the Hamilton vector field (3),

   (c) $\alpha(\eta_H) = \pi^*H$.

(2) The map $H \mapsto \eta_H$ is an irreducible injective representation of the Poisson algebra $C^\infty(M)$ of smooth function on M to vector fields on Y; it is surjective when we restrict attention to connection preserving vector fields on Y.

---

[3] This name has been coined by J. Koiller when I grumbled that the name prequantization was terribly ill adapted because no quantization condition is involved.

(3) If (q, p) are local Darboux coordinates on M and ϕ a local coordinate on $\mathbb{R} \bmod D$, the vector field η_H is given as $\eta_H = \xi_H + (H - p_i \partial H / \partial p_i) \partial / \partial \phi$. One should recognize the function $L = p_i \partial H / \partial p_i - H$ as the Lagrangian associated to the Hamiltonian H. Of course this is only a formal Lagrangian because we are not (necessarily) in a cotangent bundle, and even in that case we need a condition on H to assure that the (inverse) Legendre transformation is a diffeomorphism. It follows that the (local) flow of η_H is given by the flow of the Hamilton vector field ξ_H for the coordinates (q, p) and by

$$\phi(t_1) = \phi(t_0) - \int_{t_0}^{t_1} L\big(q(t), p(t)\big) dt$$

for the circle coordinate. The action functional A given in (2) thus appears in the expression for the flow of the circle coordinate ϕ.

(4) Under a gauge transformation $\phi \mapsto \widehat{\phi} = \phi + u(q, p)$ (another choice for the local coordinates on Y), the local expression of η_H changes to $\eta_H = \xi_H + (H - p_i \partial H / \partial p_i + \xi_H u) \partial / \partial \phi$. In other words, the Lagrangian L changes with the amount $-\xi_H u$. Since ξ_H is the vector field generating the time evolution of the system, this means that the (local) Lagrangian changes with the total time derivative $-du/dt$. It follows that, in this framework, the freedom to change the (local) Lagrangian by a total time derivative reflects a (local) gauge transformation. Note that here we can take the total time derivative of an arbitrary function on the symplectic manifold, contrary to the Lagrange formalism, where not all total time derivatives gave the same equations.

(5) Using the bundle Y, we can obtain the Hamilton equations by means of a variational principle. We fix two points m_0 and m_1 in the symplectic manifold, and we denote by $\mathscr{C}$ the space of all paths in M starting at m_0 and finishing in m_1. Finally we fix once and for all a reference path $\gamma_o \in \mathscr{C}$. We now can define a kind of action functional A on with values in the circle $\mathbf{S}^1 \cong \mathbb{R} \bmod D$ as follows. For any path $\gamma \in \mathscr{C}$ one computes the holonomy $\mathrm{hol}(\gamma \cdot \gamma_o^{-1}) \in \mathbb{R} \bmod D$ of the loop $\gamma \cdot \gamma_o^{-1}$, the concatenation of γ with the opposite of the reference path γ_o. One also computes the integral $\int_{\gamma \cdot \gamma_o^{-1}} H \, dt$. Then the action functional A is defined by

$$A(\gamma) = \mathrm{hol}\left(\gamma \cdot \gamma_o^{-1}\right) - \left(\int_{\gamma \cdot \gamma_o^{-1}} H \, dt \bmod D\right) \in \mathbb{R} \bmod D.$$

If all 'ingredients' lie within a Darboux coordinate system, the holonomy is given by $\int_{\gamma \cdot \gamma_o^{-1}} p_i \, dq^i \bmod D$. Hence $A(\gamma)$ is given by $\int_{\gamma \cdot \gamma_o^{-1}} p_i \, dq^i - H \, dt \bmod D = \int_{\gamma \cdot \gamma_o^{-1}} L \, dt \bmod D$. It thus should come as no surprise that one can show that γ is a solution of the equations of motion if and only if it is a critical point for the action functional A. It follows immediately that A has *no* critical points if the end point m_1 does not lie on the integral curve of ξ_H passing through m_0.

4. Quantum mechanics

In the previous section we have shown that many features of the Lagrange formalism are also present in the postclassical formalism. In this section we will show that there also is a *formal* correspondence between the postclassical formalism and nonrelativistic quantum dynamics.

Let $\mathcal{H}$ be a Hilbert space and $\mathbf{H}$ a self-adjoint operator on $\mathcal{H}$. If we see $\mathcal{H}$ as the Hilbert space of a (nonrelativistic) quantum mechanical system and $\mathbf{H}$ as the quantum Hamiltonian, then the evolution is governed by the Schrödinger equation

$$-i\frac{\partial \psi}{\partial t} = \mathbf{H}\psi.$$

It is well known that a physical state is represented, not by a point of $\mathcal{H}$, but by a complex line in $\mathcal{H}$, i.e., the physical states are represented by the projective Hilbert space $P\mathcal{H}$. Now any separable projective Hilbert space (even infinite-dimensional) carries a natural symplectic structure, derived from the Fubini–Study metric or, if one wishes, induced by the scalar product on $\mathcal{H}$. Moreover, the quantum Hamiltonian $\mathbf{H}$ defines a real valued function $H : P\mathcal{H} \to \mathbb{R}$ by the formula $H(\mathbf{C}\psi) = \langle \psi, \mathbf{H}\psi \rangle / \langle \psi, \psi \rangle$, i.e., the expectation value of $\mathbf{H}$ in the state ψ. Given this situation, we can try to perform the extension to the postclassical formalism. It turns out that the unit sphere $S\mathcal{H} \subset \mathcal{H}$, which projects naturally onto $P\mathcal{H}$, is a principal $\mathbf{S}^1 = \mathbb{R}$ mod 2π bundle admitting a connection (essentially the scalar product on $\mathcal{H}$) whose curvature is the Fubini–Study symplectic form. Moreover, the vector field η_H on $S\mathcal{H}$ is the restriction to $S\mathcal{H}$ of the linear vector field $\psi \mapsto i\mathbf{H}\psi$ on $\mathcal{H}$. In other words, the flow of η_H is exactly the flow of the Schrödinger equation. The inescapable conclusion is that nonrelativistic quantum dynamics is a special case of our postclassical formalism.

5. Symmetry groups

The postclassical formalism gains even more respectability when we consider symmetry groups. Given a symplectic manifold (M, ω), a symmetry group is a Lie group G acting smoothly on M and preserving the symplectic form ω. If we fix a point $m_o \in M$, we can consider the evaluation map $f : G \to M$ defined as $f(g) = g(m_o)$. Using this map, we can pull back the symplectic form ω to a closed 2-form $\mu = f^*\omega$ on G. Since the action of G on M preserves ω, it follows that μ is a left-invariant closed 2-form on G (invariant under the left action of G on itself). But the de Rham cohomology of left-invariant forms on G is isomorphic to the Lie algebra cohomology $H^*(\mathfrak{g}, \mathbb{R})$ of its Lie algebra $\mathfrak{g}$ (with values in the trivial $\mathfrak{g}$-module $\mathbb{R}$). We thus obtain a cohomology class $[\mu] \in H^2(\mathfrak{g}, \mathbb{R})$. Moreover, one can show that this cohomology class does not depend upon the choice of the point $m_o \in M$, provided M is connected. We conclude that the fact that G is a symmetry group of (M, ω) implies that we have a well defined cohomology class in $H^2(\mathfrak{g}, \mathbb{R})$. But, this cohomology space classifies the central extensions of $\mathfrak{g}$ by $\mathbb{R}$, i.e., the Lie algebras $\widehat{\mathfrak{g}}$ with $\mathbb{R}$ in their center in such a way that $\widehat{\mathfrak{g}}/\mathbb{R}$ is isomorphic to $\mathfrak{g}$. In other words, a symmetry group G of (M, ω) provides us with a well defined central extension $\widehat{\mathfrak{g}}$ of $\mathfrak{g}$ by $\mathbb{R}$.

We now can ask two questions: does there exist a unique corresponding central extension of the Lie group G, and what is the role of this central extension for the symplectic manifold M? The answer to the first question is in general negative. If we consider the symplectic manifold $(\mathbf{T}^2, d\theta_1 \wedge d\theta_2)$ (i.e., the 2-torus with the 'canonical' symplectic form) with the action of the abelian group $\mathbf{T}^2$ on the manifold $\mathbf{T}^2$, then the corresponding central extension of $\mathfrak{g} = \mathbb{R}^2$ is the Heisenberg algebra. And there is no central extension of $\mathbf{T}^2$ which corresponds to this algebra extension.[4] On the other hand, let us consider the action of the group $SO(3)$ on the 2-sphere $\mathbf{S}^2$ equipped with its natural volume form (which is symplectic). Here the associated central extension of $\mathfrak{g}$ (the Lie algebra $su(2)$) is the trivial one: $\widehat{\mathfrak{g}} = \mathfrak{g} \times \mathbb{R}$. But there are two different central extensions of $SO(3)$ by $\mathbf{S}^1$ corresponding to the trivial extension of its Lie algebra: the trivial extension $SO(3) \times \mathbf{S}^1 = SU(2) \times \mathbf{S}^1/\{(\mathrm{id}, 1), (-\mathrm{id}, 1)\}$ and $U(2) = SU(2) \times \mathbf{S}^1/\{(\mathrm{id}, 1), (-\mathrm{id}, -1)\}$.[5] We conclude that in general neither existence nor uniqueness of a corresponding central extension of G is guaranteed.

The second question is rather philosophical in nature, so one can debate its answers. However, there is a very natural interpretation of this central extension in the postclassical formalism. Before we can give this interpretation, we need to introduce the concept of a momentum map. If G is a symmetry group of (M, ω), we can associate to each element X of its Lie algebra $\mathfrak{g}$ the fundamental vector field X^M on M. Since G preserves ω, we have $\mathcal{L}(X^M)\omega = 0$. Since ω is closed, this implies that all 1-forms $\iota(X^M)\omega$ are closed. One says that there exists a momentum map if all these 1-forms are exact, i.e., if all the fundamental vector fields are Hamilton vector fields. More precisely, the momentum map J is a map $J : M \to \mathfrak{g}^*$ such that for all $X \in \mathfrak{g}$ we have

$$\iota(X^M)\omega + d\langle J, X \rangle = 0,$$

---

[4] Of course, if we go over to the simply connected cover $\mathbb{R}^2$ of $\mathbf{T}^2$, there does exist such a group extension: the Heisenberg group.

[5] If we consider central extensions by $\mathbb{R}$ instead of by $\mathbf{S}^1$, then there is only the trivial extension.

i.e., X^M is the Hamilton vector field for the function $\langle J, X \rangle : M \to \mathbb{R}$. Let us now suppose that we are in the postclassical formalism, i.e., that we have a principal $\mathbf{S}^1$-bundle $Y \to M$ with connection α such that $d\alpha = \omega$. One can show the following property: if there exists a momentum map, then there exists a unique Lie group central extension $\widehat{G}$ of G by $\mathbf{S}^1$ which acts smoothly on Y in such a way that $\mathbf{S}^1$ acts as the structure group of Y, which preserves the connection form α and which is compatible with the G-action on M. Vice versa, if such a central extension exists, there exists a momentum map. Modulo the existence of a momentum map, we see that the central extension of $\mathfrak{g}$ is the natural counter part of the principal bundle $Y \to M$.

The analogy with non-relativistic quantum mechanics can be extended to symmetry groups. In quantum mechanics, a symmetry group is a group acting on the space of physical states, i.e., on the projective Hilbert space $P\mathcal{H}$ and preserving transition probabilities (the transition probability is more or less the absolute value squared of the scalar product on $\mathcal{H}$). According to a theorem of E. Wigner, to each quantum symmetry corresponds a unitary or anti-unitary transformation of $\mathcal{H}$ compatible with the symmetry action on $P\mathcal{H}$, and this (anti-)unitary transformation is unique up to a phase factor (an element of $\mathbf{S}^1$). If G is a connected group of quantum symmetries, the corresponding transformations of $\mathcal{H}$ must be unitary. The collection of all these transformations forms a central extension $\widehat{G}$ of G by $\mathbf{S}^1$ which acts on the unit sphere $S\mathcal{H}$ in such a way that $\mathbf{S}^1$ acts as multiplication by scalars, which preserves the scalar product and which is compatible with the G-action on $P\mathcal{H}$. We have already argued that $S\mathcal{H}$ corresponds to the bundle Y in the postclassical formalism; now we can argue that the transition probability is the analogue of the symplectic form and that the scalar product is the analogue of the connection. And then the situation is the same with respect to symmetry groups.

This analogy can be made even more precise. If a diffeomorphism of $P\mathcal{H}$ preserves transition probabilities, it is induced by a unitary transformation. And thus it preserves both the symplectic form and the complex structure of $P\mathcal{H}$. Since $P\mathcal{H}$ is simply connected, the momentum map always exists. Thus, according to the postclassical formalism, there exists a unique central extension acting on $S\mathcal{H}$. Uniqueness of this extension then tells us that it must be the extension given by Wigner's theorem.[6] In finite dimensions one can prove the converse: a diffeomorphism of $P\mathcal{H}$ which preserves the symplectic form and which preserves the complex structure (i.e., which is biholomorphic) is necessarily induced by a unitary transformation. Hence it preserves the transition probabilities. And again Wigner and the postclassical formalism give the same central extension.

---

[6] Modulo the technical problem whether Wigner's extension is a Lie group which acts smoothly.

6. The physical content of the postclassical formulation

In the preceding sections we have given circumstantial evidence of a mathematical nature that it might be natural to extend the symplectic formulation of classical mechanics to the postclassical formulation. For those who are not yet convinced of the importance of the postclassical formulation of classical mechanics, we now claim that it has a physical content as well. In most cases of physical interest, the symplectic manifold M is simply connected. Hence the bundle Y with its connection α is unique. It follows that in those cases the symplectic manifold contains all the necessary information for the postclassical formalism. However, there are (at least) two physically interesting situations in which the symplectic manifold is not simply connected and in which there exist several inequivalent choices for the bundle with connection (Y, α).

The first example is the phase space of a system of N identical particles in $\mathbb{R}^3$. Here the symplectic manifold M is the cotangent bundle $M = T^*Q$ of a configuration space $Q = (\mathbb{R}^{3N} \setminus \Delta)/\mathfrak{S}_N$, i.e., Q is the space $\mathbb{R}^{3N}$ minus the 'diagonal' points in which two or more particles are at the same place in $\mathbb{R}^3$, quotiented by the group $\mathfrak{S}_N$ of permutations of N elements. It turns out that there exist two inequivalent bundles with connection (Y, α), characterized by the two characters of $\mathfrak{S}_N$: the identity character and the signature character. It now so happens that in nature there also exist two different kind of identical particles: bosons and fermions. In quantum mechanics they are characterized by symmetric and antisymmetric wave functions, i.e., by wave functions which transform under the action of $\mathfrak{S}_N$ according to one of the two characters of this group.

The second example we have in mind is the Aharonov–Bohm experiment. Here the phase space $M = T^*Q$ is the cotangent bundle of the configuration space Q which is $\mathbb{R}^3$ minus the z-axis. The idea is that the z-axis represents an infinitely long, infinitely thin solenoid. When a current is applied to the solenoid, a constant magnetic field proportional to the current appears inside the solenoid, while outside the solenoid there is still a zero magnetic field. By excluding the z-axis from the configuration space we have a zero magnetic field in Q, independent of the current through the solenoid. In other words, we have a phase space which is independent of the current. The classification tells us that there is a circle of inequivalent bundles with connection (Y, α) for this symplectic manifold. And again, experimental results, the actual Aharonov–Bohm experiments, show that there is a circle of inequivalent physical systems. More precisely, there exists a value I_o for the current through the solenoid such that the experimental results are different if and only if the current I *modulo* I_o is different. Plugging physical constants into the construction and classification of the bundle with connection (Y, α) shows that the inequivalent bundles are classified by $\mathbb{R}$ mod I_o.

In both examples we see that the description by the symplectic manifold is not able to distinguish between the different physical systems, whereas the postclassical formalism *is* able to distinguish between them. We claim (without proof) that this is universal: each time a physical system is described by a phase space which allows for non-equivalent bundles with connection (Y, α), then there are other physical systems described by the same phase space but distinguished by the choice for (Y, α). [7] And vice versa, if different physical systems are described by the same symplectic manifold, then they can be distinguished by inequivalent bundles with connection.

References

[1] Kostant, B., *Quantization and unitary representations*, Lectures in modern analysis and applications III, Taam, C.T. (eds.), Springer Verlag, Berlin–New York 1970, LNM 170.

[2] Souriau, J.-M., *Structure des Systèmes dynamiques*, Dunod, (1970), Paris. *Structure of Dynamical Systems, A Symplectic View of Physics*, Birkhäuser–Boston–Basel 1997, PM 149.

[3] Tuynman, G.M. and Wiegerinck, W.W.A.J., Central extensions and physics, *J. Geom. Phys.* **4** (1987), 207–258.

[4] Tuynman, G.M., Prequantization is irreducible, *Indagationes Mathematicae* **9** (1998), 607–618.

[5] ———, Un principe variationnel pour les variétés symplectiques, *Comptes Rendus de l'Académie des Sciences de Paris* **58** (1998), 747–750.

---

[7] Note that we say 'distinguish' which does not imply that the postclassical formalism can describe (quantitively) the experimental results!

GEOMETRY OF SOLUTION SPACES OF SPACES OF YANG–MILLS EQUATIONS

JĘDRZEJ ŚNIATYCKI

Contents

1   Introduction 251
2   Statement of results 253
3   Regularity of the constraint set 256
4   Gauge symmetries 261
5   Reduction 264

M. de Gosson (ed.), Jean Leray '99 Conference Proceedings, 249-265.
© *2003 Kluwer Academic Publishers.*

Abstract. *We show that the set of solutions of Yang–Mills equations in the Minkowski space–time, with the Cauchy data $A \in H^{k+1}(\mathbb{R}^3)$, $E \in H^k(\mathbb{R}^3)$, $k \geq 3$, is a smooth principal fibre bundle over the reduced phase space with structure group consisting of the gauge symmetries approaching the identity at infinity.*

Résumé. *Nous démontrons que l'ensemble des solutions de les équations de Yang–Mills dans l'espace-temps de Minkowski, pour les données de Cauchy $A \in H^{k+1}(\mathbb{R}^3)$, $E \in H^k(\mathbb{R}^3)$, $k \geq 3$, est un espace fibré ayant pour base l'espace de phase reduit et pour groupe structural le groupe de symétries de jauge approchant l'identité à l'infini.*

Key words: *Banach manifolds, Hilbert–Lie groups, non-linear partial differential equations, Yang–Mills fields.*

1. Introduction

We consider here the Yang–Mills–Dirac theory for the internal symmetry group given by a compact classical group G. The Lie algebra $\mathfrak{g}$ admits an Ad-invariant positive definite metric. The Yang–Mills potential A_μ describes a connection in the principal fibre bundle $\mathbb{R}^4 \times G$ with respect to the trivialization given by the product structure. It is a $\mathfrak{g}$-valued 1-form on $\mathbb{R}^4$. The curvature form of the connection given by A_μ is described by the Yang–Mills field $F_{\mu\nu} = \partial_\mu A_\nu - \partial_\nu A_\mu + [A_\mu, A_\nu]$, where $[\cdot, \cdot]$ denotes the Lie bracket in $\mathfrak{g}$. The $3 + 1$ splitting of the space–time leads to a decomposition of the Yang–Mills potential A_μ into its time component A_0 and the spatial component $A = (A_1, A_2, A_3)$. Similarly, the Yang–Mills field $F_{\mu\nu}$ can be decomposed into its electric components $E = (F_{01}, F_{02}, F_{03})$ and the magnetic components $B = (F_{23}, F_{31}, F_{12})$. From the point of view of the evolution equations it is convenient to consider A, E and B as $\mathfrak{g}$-valued time-dependent vector fields on $\mathbb{R}^3$, and A_0 as a $\mathfrak{g}$-valued time dependent function on $\mathbb{R}^3$. The vector potential A describes the induced connection in $\mathbb{R}^3 \times G$. The matter is described by a Dirac spinor field Ψ, that is a time dependent map from $\mathbb{R}^3$ to $\mathbb{C}^4 \otimes V$, where V is the space of a representation of G. The field equations split into the evolution equations

$$\dot{A} = E + \operatorname{grad} A_0 - [A_0, A]$$
$$\dot{E} = -\operatorname{curl} B - [A\times, B] - [A_0, E] + \Psi^\dagger \gamma^0 \gamma T_a \Psi T^a$$
$$\dot{\Psi} = -\gamma^0 \left(\gamma^j \partial_j + \mathrm{im} + \gamma^0 A_0 + \gamma^j A_j \right) \Psi$$

and the constraint equation

$$\operatorname{div} E + [A; E] - \Psi^\dagger T_a \Psi T^a = 0,$$

where $[\cdot\,;\cdot]$ is the Lie bracket in $\mathfrak{g}$ combined with the Euclidean scalar product of vector fields in $\mathbb{R}^3$, $\{T^a\}$ is an orthonormal basis in $\mathfrak{g}$, and the Latin indices are lowered in terms of the Ad-invariant metric in $\mathfrak{g}$.

In the preceding papers, [11] and [14], we have studied the structure of the set of solutions of the constraint equations for minimally coupled Yang–Mills and Dirac equations with the Cauchy data (A, E) for the Yang–Mills field in the space

$$P_{YM} = H^2\left(\mathbb{R}^3, \mathfrak{g} \otimes \mathbb{R}^3\right) \times H^1\left(\mathbb{R}^3, \mathfrak{g} \otimes \mathbb{R}^3\right),$$

and the Cauchy data Ψ for the Dirac field in

$$P_D = H^2\left(\mathbb{R}^3, \mathbb{C}^4 \otimes V\right).$$

The aim of this paper is to extend the results obtained there to the Cauchy data in higher Sobolev spaces:

$$(A, E) \in P_{YM}^k = H^{k+1}\left(\mathbb{R}^3, \mathfrak{g} \otimes \mathbb{R}^3\right) \times H^k\left(\mathbb{R}^3, \mathfrak{g} \otimes \mathbb{R}^3\right),$$

and

$$\Psi \in P_D^k = H^{k+1}\left(\mathbb{R}^3, \mathbb{C}^4 \otimes V\right),$$

where k is an integer > 1. Hence, the extended phase space of the Yang–Mills–Dirac system under consideration is

$$P^k = P_{YM}^k \times P_D^k.$$

All the results obtained here for minimally interacting Yang–Mills and Dirac fields are also valid for pure Yang–Mills fields. In this case one can consider P^k to be just P_{YM}^k.

The time derivatives of A_0 do not appear in the evolution equations. Hence, the evolution of A_0 is not determined by the initial data and can be arbitrarily fixed by a gauge condition. In absence of matter fields, the Yang–Mills equations, supplemented by the temporal gauge condition

$$A_0 = 0,$$

are

$$\dot{A} = E, \qquad \dot{E} = -\operatorname{curl} B - [A \times, B], \qquad \operatorname{div} E = -[A; E].$$

For these equations, the global existence, uniqueness and smooth dependence of solutions on the Cauchy data in P_{YM}^k, $k \geq 3$, was proved by Eardley and

Moncrief, [4]. This implies that, for $k \geq 3$, the space of global solutions of the Yang–Mills equations in temporal gauge, with the Cauchy data in P_{YM}^k, has the same geometric structure as the constraint set

$$C_{YM}^k = \left\{ (A, E) \in P_{YM}^k \mid \operatorname{div} E + [A; E] = 0 \right\}.$$

Combining this with the results obtained in the present paper we obtain our

Main Theorem. *For $k \geq 3$, the space of global solutions of the Yang–Mills equations in the Minkowski space–time, supplemented by the temporal gauge condition, with the Cauchy data in $H^{k+1}(\mathbb{R}^3, \mathfrak{g} \otimes \mathbb{R}^3) \times H^k(\mathbb{R}^3, \mathfrak{g} \otimes \mathbb{R}^3)$ is a smooth principal fibre bundle with a Hilbert–Lie structure group and a weakly symplectic base space.*

It would be of interest to extend this result for pure Yang–Mills fields to $k = 1, 2$. In these cases, we have no global existence results for Yang–Mills equations. Klainerman and Machedon, [8], proved the global existence and uniqueness of solutions of Yang–Mills equations in the local Sobolev space

$$\left\{ (A, E) \in H_{\mathrm{loc}}^2 \left(\mathbb{R}^3, \mathfrak{g} \otimes \mathbb{R}^3 \right) \times H_{\mathrm{loc}}^1 \left(\mathbb{R}^3, \mathfrak{g} \otimes \mathbb{R}^3 \right) \right\}.$$

However, in this space we have no results on the structure of the constraint set.

For minimally interacting Yang–Mills and Dirac fields, one cannot rely on the energy estimates because the energy of the Dirac field is not positive definite, and the global existence and uniqueness theorems have not been proved. Therefore, we cannot identify the space of global solutions of the field equations with the constraint set. This problem will be studied elsewhere. Nevertheless, in this paper we investigate the structure of the constraint set for minimally interacting Yang–Mills and Dirac fields because it does not pose any additional difficulties. Moreover, as soon as the global existence theorems for minimally interacting Yang–Mills and Dirac fields are obtained, our results will imply the analogous results for the appropriate spaces of the global solutions of the field equations.

2. Statement of results

Theorem 2.1. *For minimally interacting Yang–Mills and Dirac fields, the constraint set*

$$C^k = \left\{ (A, E, \Psi) \in P^k \mid \operatorname{div} E + [A; E] - \Psi^\dagger T_a \Psi T^a = 0 \right\}$$

is a C^∞-submanifold of the Banach space

$$P^{k,0} = \left\{ (A, E, \Psi) \in P^k \mid \operatorname{div} E \in L^{6/5} \left(\mathbb{R}^3, \mathfrak{g} \right) \cap H^k \left(\mathbb{R}^3, \mathfrak{g} \right) \right\}$$

with the norm

$$\|(A, E, \Psi)\|_{P^{k,0}} = \|A\|_{H^{k+1}(\mathbb{R}^3, \mathfrak{g})} + \|E\|_{H^k(\mathbb{R}^3, \mathfrak{g})} + \|\Psi\|_{H^{k+1}(\mathbb{R}^3, \mathbb{C}^4 \otimes V)}$$
$$+ \|\operatorname{div} E\|_{L^{6/5}(\mathbb{R}^3, \mathfrak{g})} + \|\operatorname{div} E\|_{H^k(\mathbb{R}^3, \mathfrak{g})}.$$

The manifold topology of C^k coincides with the topology induced by its embedding into P^k. Similarly, in the absence of Dirac fields, the constraint set

$$C_{YM}^k = \left\{ (A, E) \in P_{YM}^k \mid \operatorname{div} E + [A; E] = 0 \right\}$$

is a C^∞-submanifold of the Banach space

$$P_{YM}^{k,0} = \left\{ (A, E) \in P_{YM}^k \mid \operatorname{div} E \in L^{6/5}\left(\mathbb{R}^3, \mathfrak{g}\right) \cap H^k\left(\mathbb{R}^3, \mathfrak{g}\right) \right\}$$

with the norm

$$\|(A, E)\|_{P_{YM}^{k,0}} = \|A\|_{H^{k+1}(\mathbb{R}^3, \mathfrak{g})} + \|E\|_{H^k(\mathbb{R}^3, \mathfrak{g})}$$
$$+ \|\operatorname{div} E\|_{L^{6/5}(\mathbb{R}^3, \mathfrak{g})} + \|\operatorname{div} E\|_{H^k(\mathbb{R}^3, \mathfrak{g})},$$

and its manifold topology coincides with the topology inherited from P_{YM}^k.

As in [14], the proof of this theorem is based on the following result.

Theorem 2.2. *For each $A \in H^{k+1}(\mathbb{R}^3, \mathfrak{g} \otimes \mathbb{R}^3)$, the covariant divergence operator $\operatorname{Div}_A = \operatorname{div} + [A, \cdot]$ maps*

$$D_{k+1} = \left\{ e \in H^{k+1}\left(\mathbb{R}^3, \mathfrak{g} \otimes \mathbb{R}^3\right) \mid \operatorname{div} e \in L^{6/5}\left(\mathbb{R}^3, \mathfrak{g}\right) \right\}$$

onto

$$B_k = H^k\left(\mathbb{R}^3, \mathfrak{g}\right) \cap L^{6/5}\left(\mathbb{R}^3, \mathfrak{g}\right).$$

Time independent gauge transformations are given by maps $\varphi : \mathbb{R}^3 \to G$ acting on the configurations (A, E, Ψ) via the transformation law

$$A \longmapsto \varphi A \varphi^{-1} + \varphi \operatorname{grad} \varphi^{-1}, \qquad E \longmapsto \varphi E \varphi^{-1}, \qquad \Psi \longmapsto \varphi \Psi.$$

The gauge symmetry group $GS(P^k)$ of P^k is the connected group of time independent gauge transformations which preserve the space P^k. By assumption, $G \subset gl(n, \mathbb{R})$ for some positive integer n and $GS(P^k)$ can be considered as a space of maps $\varphi : \mathbb{R}^3 \to gl(n, \mathbb{R})$.

Theorem 2.3. *A mapping* $\varphi : \mathbb{R}^3 \to G$ *is in* $GS(P^k)$ *if and only if it has finite norm*

$$\|\varphi\|^2_{\mathcal{B}^{k+2}(\mathbb{R}^3, \mathrm{gl}(n,\mathbb{R}))} = \int_{B_1(0)} \mathrm{tr}(\varphi^T \varphi) d_3 x + \|\mathrm{grad}\, \varphi\|^2_{H^{k+1}(\mathbb{R}^3 \otimes \mathbb{R}^3, \mathfrak{g})},$$

where $B_1(0)$ *is the unit ball in* $\mathbb{R}^3$ *centered at the origin, and* φ^T *is the transpose of* φ *in* $\mathrm{gl}(n, \mathbb{R})$. *With the topology given by this norm,* $GS(P^k)$ *is a Hilbert–Lie group. The action of* $GS(P^k)$ *in* P^k *is continuous and proper.*

The closure, with respect to the norm $\|\varphi\|^2_{\mathcal{B}^{k+2}(\mathbb{R}^3, \mathrm{gl}(n,\mathbb{R}))}$, of the group of smooth maps $\varphi : \mathbb{R}^3 \to G$ which differ from the identity only in compact sets is a normal subgroup $GS(P^k)_0$ of the gauge symmetry group $GS(P^k)$.

Theorem 2.4. *The action of* $GS(P^k)_0$ *in* P^k *is free and proper.*

The extended phase space $P^k = P^k_{YM} \times P^k_D$ is weakly symplectic, with the weak symplectic form $\omega = d\theta$, where

$$\langle \theta(A, E, \Psi) \mid (a, e, \psi) \rangle = \int_{\mathbb{R}^3} \left(E \cdot a + \Psi^\dagger \psi \right) d_3 x.$$

The infinitesimal action in P^k of an element ξ of the Lie algebra $gs(P^k)$ of $GS(P^k)$ is given by the vector field

$$\xi_P(A, E, \Psi) = \left(-D_A \xi, -[E, \xi], \Psi^\dagger \xi \right),$$

where

$$D_A \xi = d\xi + [A, \xi]$$

is the covariant differential of ξ with respect to the connection A. The action of $GS(P^k)$ in P^k preserves the 1-form θ. Since $\omega = d\theta$, this action is Hamiltonian with an equivariant momentum map $\mathcal{J}$ such that, for every $\xi \in gs(P^k)$,

$$\langle \mathcal{J}(A, E, \Psi) \mid \xi \rangle = \langle \theta \mid \xi_P(A, E, \Psi) \rangle = \int_{\mathbb{R}^3} \left(-E \cdot D_A \xi + \Psi^\dagger \xi \Psi \right) d_3 x.$$

Theorem 2.5. *The constraint set* C^k *coincides with the zero level of the momentum map* $\mathcal{J}$ *restricted to the Lie algebra* $gs(P^k)_0$ *of* $GS(P^k)_0$. *That is,*

$$(A, E, \Psi) \in C^k \iff \langle \mathcal{J}(A, E, \Psi) \mid \xi \rangle = 0 \quad \forall \xi \in gs\left(P^k \right)_0.$$

The pull-back ω_{C^k} of ω to C^k has involutive kernel ker ω_{C^k}. The reduced phase space $\check{P}^k$ is defined as the set of equivalence classes of points in C^k under the equivalence relation $p \simeq p'$ if and only if there is a piece-wise smooth curve in C^k with the tangent vector contained in ker ω_{C^k}. If ker ω_{C^k} is a distribution, it is clearly involutive and the equivalence classes coincide with integral manifolds of ker ω_{C^k}. We denote by $\rho^k : C^k \to \check{P}^k$ the canonical projection associating to each $p \in C^k$ its equivalence class containing p.

Theorem 2.6. *The reduced phase space $\check{P}^k$ coincides with the set $C^k/GS(P^k)_0$ of the orbits of the $GS(P^k)_0$-action in C^k,*

$$\check{P}^k = C^k/GS\left(P^k\right)_0.$$

It is a quotient manifold of C^k endowed with a weak Riemannian metric induced by the L^2 scalar product in P^k, and with a 1-form $\check{\theta}$ such that

$$\rho^*\check{\theta} = \theta_{C^k},$$

where θ_{C^k} is the pull-back of θ to C^k. Moreover, $\check{\omega} = d\check{\theta}$ is weakly symplectic and

$$\rho^*\check{\omega} = \omega_{C^k}.$$

The constraint manifold C^k is a principal fibre bundle over $\check{P}^k$ with structure group $GS(P^k)_0$.

3. Regularity of the constraint set

In this section we outline the proofs of Theorems 2.1 and 2.2. Recall that

$$P^k = \left\{(A, E, \Psi) \in H^{k+1}\left(\mathbb{R}^3, \mathfrak{g} \otimes \mathbb{R}^3\right)\right.$$

$$\left. \times H^k\left(\mathbb{R}^3, \mathfrak{g} \otimes \mathbb{R}^3\right) \times H^{k+1}\left(\mathbb{R}^3, \mathbb{C}^4 \otimes V\right)\right\},$$

is a Hilbert space with the norm

$$\|(A, E, \Psi)\|_{P^k} = \|A\|_{H^{k+1}(\mathbb{R}^3, \mathfrak{g}\otimes\mathbb{R}^3)} + \|E\|_{H^k(\mathbb{R}^3, \mathfrak{g}\otimes\mathbb{R}^3)} + \|\Psi\|_{H^{k+1}(\mathbb{R}^3, \mathbb{C}^4\otimes V)},$$

and

$$P^{k,0} = \left\{(A, E, \Psi) \in P^k \mid \operatorname{div} E \in L^{6/5}\left(\mathbb{R}^3, \mathfrak{g}\right) \cap H^k\left(\mathbb{R}^3, \mathfrak{g}\right)\right\}$$

is a dense subspace of P^k. It is a Banach space with the norm

$$\|(A, E, \Psi)\|_{P^{k,0}} = \|A\|_{H^{k+1}(\mathbb{R}^3,\mathfrak{g})} + \|E\|_{H^k(\mathbb{R}^3,\mathfrak{g})} + \|\Psi\|_{H^{k+1}(\mathbb{R}^3,\mathbb{C}^4\otimes V)}$$
$$+ \|\operatorname{div} E\|_{L^{6/5}(\mathbb{R}^3,\mathfrak{g})} + \|\operatorname{div} E\|_{H^k(\mathbb{R}^3,\mathfrak{g})} .$$

The constraint set

$$C^k = \left\{ (A, E, \Psi) \in P^k \mid \operatorname{div} E = -[A; E] + \Psi^\dagger T_a \Psi T^a \right\}$$

is a closed subset of P^k.

Lemma 3.1. *The constraint set C^k is contained in $P^{k,0}$.*

Proof. Proof that $C^1 \subset P^{1,0}$ was given in [14] on the basis of the estimates obtained in [5]. Since $C^k = C^1 \cap P^k$ and $P^{k,0} = P^{1,0} \cap P^k$, it follows that $C^k \subset P^{k,0}$. ☐

Lemma 3.1 implies that C^k can be identified with the zero level of the smooth map

$$f_k : P^{k,0} \longrightarrow B_k : (A, E, \Psi) \longmapsto \operatorname{div} E + [A; E] - \Psi^\dagger T_a \Psi T^a,$$

where

$$B_k = H^k\left(\mathbb{R}^3, \mathfrak{g}\right) \cap L^{6/5}\left(\mathbb{R}^3, \mathfrak{g}\right).$$

It follows from the Implicit Function Theorem that

$$C^k = f_k^{-1}(0)$$

is a submanifold of $P^{k,0}$ if, for each $(A, E, \Psi) \in C^k$, $D_{(A,E,\Psi)} f_k$ maps $P^{k,0}$ onto B_k and its kernel splits, [9]. For each $(A, E, \Psi) \in C^k$ and $(a, e, \psi) \in P^{k,0}$,

$$D_{(A,E,\Psi)} f_k(a, e, \psi) = \operatorname{div} e + [A; e] + [E; a] - \psi^\dagger \left(I \otimes T^a\right) \Psi T_a$$
$$- \Psi^\dagger \left(I \otimes T^a\right) \psi T_a.$$

Hence, $D_{(A,E,\Psi)} f_k$ is onto B_k if, for every $A \in H^{k+1}(\mathbb{R}^3, \mathfrak{g} \otimes \mathbb{R}^3)$, the covariant divergence operator $\operatorname{Div}_A = \operatorname{div} + [A, \cdot]$ maps

$$E_k = \left\{ e \in H^k\left(\mathbb{R}^3, \mathfrak{g} \otimes \mathbb{R}^3\right) \mid \operatorname{div} e \in L^{6/5}\left(\mathbb{R}^3, \mathfrak{g}\right) \cap H^k\left(\mathbb{R}^3, \mathfrak{g}\right) \right\}$$

onto B_k. However, according to Theorem 2.2, for each $A \in H^{k+1}(\mathbb{R}^3, \mathfrak{g} \otimes \mathbb{R}^3)$, the covariant divergence operator $\operatorname{Div}_A = \operatorname{div} + [A, \cdot]$ maps

$$D_{k+1} = \left\{ e \in H^{k+1}\left(\mathbb{R}^3, \mathfrak{g} \otimes \mathbb{R}^3\right) \mid \operatorname{div} e \in L^{6/5}\left(\mathbb{R}^3, \mathfrak{g}\right) \right\}$$

onto B_k. Since $D_{k+1} \subset E_k$, it implies that $D_{(A,E,\Psi)}f_k$ maps E_k onto B_k. Moreover, if $(a, e, \psi) \in P^k$ and

$$D_{(A,E,\Psi)}f_k(a, e, \psi) = 0,$$

then div $e \in L^{6/5}(\mathbb{R}^3, \mathfrak{g}) \cap H^k(\mathbb{R}^3, \mathfrak{g})$ and $(a, e, \psi) \in P^{k,0}$. Hence, ker $Df_{k(A,E,\Psi)}$ is a closed subspace of $P^{k,0}$. Since the P^k-scalar product is continuous and non-degenerate in $P^{k,0}$, it follows that the P^k-orthogonal complement of ker $Df_{k(A,E,\Psi)}$ is closed in $P^{k,0}$ and its intersection with ker $Df_{k(A,E,\Psi)}$ is zero. Thus, the assumptions of the Implicit Function Theorem are satisfied and $C^k = f_k^{-1}(0)$ is a smooth submanifold of $P^{k,0}$. Clearly, the same result holds in absence of the Dirac fields.

It remains to show that the manifold topology and the subspace topology in C^k are the same.

The manifold topology of C^k is the same as the topology induced by the embedding $C^k \hookrightarrow P^{k,0}$. Since the topology of $P^{k,0}$ is finer than the topology induced in $P^{k,0}$ by its embedding into P^k, it follows that the manifold topology of C^k is finer than the topology induced by the embedding $C^k \hookrightarrow P^k$.

Let $p_n = (A_n, E_n, \Psi_n)$ be a sequence in C^k convergent in P^k to $p = (A, E, \Psi)$. Since C^k is closed in P^k, as the zero level of a continuous function, then $p \in C^k$. It was shown in [14] that the convergence of the sequence p_n in P^1 implies the convergence of div E_n to div E in $L^{6/5}(\mathbb{R}^3, \mathfrak{g})$. For $k > 1$, the convergence of the sequence p_n in P^k implies convergence in P^1. Hence, div E_n converges to div E in $L^{6/5}(\mathbb{R}^3, \mathfrak{g})$, which implies that p_n converges to p in $P^{k,0}$. Therefore, the topology of C^k induced by its embedding into P^k is the same as the topology of C^k induced by its embedding into $P^{k,0}$. It follows that the manifold topology of C^k and its topology induced by the embedding into P^k coincide. Thus, to complete the proof of Theorem 2.1 we need to prove Theorem 2.2.

It follows from the results of Ref. [5] that the Laplace operator restricted to the space

$$F_{k+2} = \left\{ v \in H^{k+2}\left(\mathbb{R}^3, \mathfrak{g}\right) \mid \Delta v \in L^{6/5}\left(\mathbb{R}^3, \mathfrak{g}\right) \right\}$$

is onto

$$B_k = H^k\left(\mathbb{R}^3, \mathfrak{g}\right) \cap L^{6/5}\left(\mathbb{R}^3, \mathfrak{g}\right).$$

Hence, taking $e = \operatorname{grad} v$, we see that the divergence operator maps

$$D_{k+1} = \left\{ e \in H^{k+1}\left(\mathbb{R}^3, \mathfrak{g} \otimes \mathbb{R}^3\right) \mid \operatorname{div} e \in L^{6/5}\left(\mathbb{R}^3, \mathfrak{g}\right) \right\}$$

onto B_k. The covariant divergence operator differs from the divergence by a multiplication operator $e \mapsto [A; e]$. We show that, for $A \in H^{k+1}(\mathbb{R}^3, \mathfrak{g} \otimes \mathbb{R}^3)$, the operator $D_{k+1} \rightarrow B_k : e \mapsto [A; e]$ is compact. This will imply that $\mathrm{Div}_A = \mathrm{div} + [A, \cdot]$ from D_{k+1} to B_k is semi-Fredholm. Hence it has closed range, [7]. The proof will be completed by showing that the annihilator in B'_k of the range of Div_A vanishes.

Lemma 3.2. *For $k > 1$ and $A \in H^{k+1}(\mathbb{R}^3, \mathfrak{g} \otimes \mathbb{R}^3)$, the map*

$$H^{k+1}\left(\mathbb{R}^3, \mathfrak{g} \otimes \mathbb{R}^3\right) \longrightarrow L^{6/5}\left(\mathbb{R}^3, \mathfrak{g}\right) : e \mapsto [A; e]$$

is compact.

Proof. For $k > 1$, we have $H^{k+1}(\mathbb{R}^3, \mathfrak{g} \otimes \mathbb{R}^3) \subset H^2(\mathbb{R}^3, \mathfrak{g} \otimes \mathbb{R}^3)$ and the compactness of the map

$$H^k\left(\mathbb{R}^3, \mathfrak{g} \otimes \mathbb{R}^3\right) \longrightarrow L^{6/5}\left(\mathbb{R}^3, \mathfrak{g}\right) : e \mapsto [A; e]$$

follows from the compactness of

$$H^2\left(\mathbb{R}^3, \mathfrak{g} \otimes \mathbb{R}^3\right) \longrightarrow L^{6/5}\left(\mathbb{R}^3, \mathfrak{g}\right) : e \mapsto [A; e]$$

for $A \in H^2(\mathbb{R}^3, \mathfrak{g} \otimes \mathbb{R}^3)$. This result was proved in [14]. $\square$

Lemma 3.3. *For $k > 1$, and $A \in H^{k+1}(\mathbb{R}^3, \mathfrak{g} \otimes \mathbb{R}^3)$, the map*

$$H^{k+1}\left(\mathbb{R}^3, \mathfrak{g}\right) \longrightarrow H^k\left(\mathbb{R}^3, \mathfrak{g}\right) : e \mapsto [A; e]$$

is compact.

Proof of this result was given in [14]. $\square$

It remains to show that the annihilator of the range of Div_A in the dual B'_k of B_k vanishes.

Since B_k contains the Schwarz space S of rapidly decaying smooth functions on $\mathbb{R}^3$, it follows that elements of the dual B'_k of B_k are tempered distributions. For each $\mu \in B'_k$, and $e \in D_{k+1}$,

$$\langle \mu \mid \mathrm{Div}_A e \rangle = \langle \mu \mid \mathrm{div}\, u + [A; e] \rangle = -\langle \mathrm{grad}\, \mu + [A, \mu] \mid e \rangle,$$

where the last equality follows from the ad-invariance of the metric in $\mathfrak{g}$. Hence μ annihilates the range of Div_A if and only if

$$\mathrm{grad}\, \mu + [A, \mu] = 0$$

in the sense of distributions.

Lemma 3.4. *For $k > 1$ and $A \in H^{k+1}(\mathbb{R}^3, \mathfrak{g} \otimes \mathbb{R}^3)$, let $\mu \in B'_k$ satisfy the distribution equation $\mathrm{grad}\, \mu + [A, \mu] = 0$. Then μ extends to a continuous linear functional on B_{k-1}.*

Proof. It follows from the results of Eardley and Moncrief, [5], that the Laplace operator Δ from

$$F_{k+2} = \left\{ v \in H^{k+2}\left(\mathbb{R}^3, \mathfrak{g}\right) \mid \Delta v \in L^{6/5}\left(\mathbb{R}^3, \mathfrak{g}\right) \right\}$$

is onto

$$B_k = H^k\left(\mathbb{R}^3, \mathfrak{g}\right) \cap L^{6/5}\left(\mathbb{R}^3, \mathfrak{g}\right),$$

and every $h \in B_k$ can be expressed in the form $h = \Delta v$, where $v \in F_{k+2}$ is the convolution of h with the Green's function

$$K(x) = \frac{1}{4\pi \, |x|}$$

for the Laplace operator in $\mathbb{R}^3$. That is

$$v = K * h.$$

Moreover, for $k > 1$,

$$\|\operatorname{grad}(K * h)\|_{H^k(\mathbb{R}^3, \mathfrak{g} \otimes \mathbb{R}^3)} \leq C \left\{ \|h\|_{H^{k-1}(\mathbb{R}^3, \mathfrak{g})} + \|h\|_{L^{6/5}(\mathbb{R}^3, \mathfrak{g})} \right\} \leq C \, \|h\|_{B_{k-1}}.$$

Evaluating $\mu \in B_k'$ on $h = K * v$, we obtain

$$\begin{aligned}
\langle \mu \mid h \rangle &= \langle \mu \mid \Delta v \rangle = - \langle \operatorname{grad} \mu \mid \operatorname{grad} v \rangle = \langle [A, \mu] \mid \operatorname{grad} v \rangle \\
&= \langle [A, \mu], \operatorname{grad}(K * h) \rangle = - \langle \mu, [A; \operatorname{grad}(K * h)] \rangle,
\end{aligned}$$

where the last equality follows from the ad-invariance of the metric in $\mathfrak{g}$. Hence,

$$\begin{aligned}
|\langle \mu \mid h \rangle| &\leq \|\mu\|_{B_k'} \|[A; \operatorname{grad}(K * h)]\|_{B_k} \\
&= \|\mu\|_{B_k'} \left\{ \|[A; \operatorname{grad}(K * h)]\|_{H^k(\mathbb{R}^3, \mathfrak{g} \otimes \mathbb{R}^3)} \right. \\
&\qquad \left. + \|[A; \operatorname{grad}(K * h)]\|_{L^{6/5}(\mathbb{R}^3, \mathfrak{g})} \right\} \\
&\leq C \|\mu\|_{B_k'} \|A\|_{H^{k+1}(\mathbb{R}^3, \mathfrak{g} \otimes \mathbb{R}^3)} \|\operatorname{grad}(K * h)\|_{H^k(\mathbb{R}^3, \mathfrak{g} \otimes \mathbb{R}^3)}.
\end{aligned}$$

Taking into account the above estimate for $\|\operatorname{grad}(K * h)\|_{H^k(\mathbb{R}^3, \mathfrak{g} \otimes \mathbb{R}^3)}$ we obtain

$$|\langle \mu \mid f \rangle| \leq C \|A\|_{H^{k+1}(\mathbb{R}^3, \mathfrak{g} \otimes \mathbb{R}^3)} \|\mu\|_{B_k'} \|h\|_{B_{k-1}}.$$

This implies that μ extends to a continuous linear form on B_k. $\qquad \square$

Lemma 3.5. *For $A \in H^2(\mathbb{R}^3, \mathfrak{g} \otimes \mathbb{R}^3)$, let $\mu \in B_1'$ satisfy the distribution equation* $\operatorname{grad} \mu + [A, \mu] = 0$. *Then $\mu = 0$.*

Proof of this result was given in Ref. [14, Lemmas 4 and 5].   □
This completes the proof of Theorem 2.2. Hence, Theorem 2.1 is also proved.

4. Gauge symmetries

The aim of this section is to outline basic steps in the proofs of Theorems 2.3 and 2.4. The infinitesimal action in P^k of an element ξ of the Lie algebra $gs(P^k)$ of $GS(P^k)$ is given by

$$A \longmapsto -\operatorname{grad} \xi - [A, \xi], \qquad E \longmapsto -[E, \xi], \qquad \Psi \longmapsto \Psi^\dagger \xi.$$

Since the Yang–Mills potential A in P^k is of Sobolev class H^{k+1} it follows that $\xi \in gs(P^k)$ only if $\operatorname{grad} \xi \in H^{k+1}(\mathbb{R}^3, \mathfrak{g} \otimes \mathbb{R}^3)$. Conversely, it can be shown that, if $\operatorname{grad} \xi \in H^{k+1}(\mathbb{R}^3, \mathfrak{g} \otimes \mathbb{R}^3)$, then the infinitesimal action of ξ preserves P^k.

The space

$$B^{k+2}\left(\mathbb{R}^3, \mathfrak{g}\right) = \left\{\xi : \mathbb{R}^3 \longrightarrow \mathfrak{g} \mid \operatorname{grad} \xi \in H^{k+1}\left(\mathbb{R}^3, \mathfrak{g} \otimes \mathbb{R}^3\right)\right\}$$

is the intersection of $k + 2$ Beppo Levi spaces, see Refs. [1] and [3]. It can be topologized with the norm

$$\|\xi\|_{\mathcal{B}^{k+2}(\mathbb{R}^3, \mathfrak{g})} = \int_{B_1(0)} \xi^\dagger \xi \, d_3 x + \|\operatorname{grad} \xi\|_{H^{k+1}(\mathbb{R}^3, \mathfrak{g} \otimes \mathbb{R}^3)}, \qquad (4.1)$$

where $B_1(0)$ denotes the unit ball in $\mathbb{R}^3$ centered at the origin. Its properties were discussed in the appendix of Ref. [11]. In particular, we proved the following

Lemma 4.1. *Let U, V and W be normed vector spaces, $f : \mathbb{R}^3 \to U$, $g : \mathbb{R}^3 \to V$ and $f \cdot g$ a pointwise multiplication with values in W. For $p = 1, 2$, and $q \geq 2$, the following estimates hold*

$$\|f \cdot g\|_{H^p(\mathbb{R}^3, W)} \leq c_p \|f\|_{\mathcal{B}^q(\mathbb{R}^3, U)} \|g\|_{H^p(\mathbb{R}^3, V)},$$
$$\|f \cdot g\|_{\mathcal{B}^q(\mathbb{R}^3, W)} \leq c_q' \|f\|_{\mathcal{B}^q(\mathbb{R}^3, U)} \|g\|_{\mathcal{B}^q(\mathbb{R}^3, V)}.$$

The range of the parameter p in the first estimate above can be increased to all positive integers.

Corollary. *Let U, V and W be normed vector spaces, $f : \mathbb{R}^3 \to U$, $g : \mathbb{R}^3 \to V$ and $f \cdot g$ a pointwise multiplication with values in W. For every integer $p > 2$ and $q \geq p$,*

$$\|f \cdot g\|_{H^p(\mathbb{R}^3, W)} \leq c_p \|f\|_{\mathcal{B}^q(\mathbb{R}^3, U)} \|g\|_{H^p(\mathbb{R}^3, V)}.$$

Proof. Since

$$\|g\|_{\mathcal{B}^p(\mathbb{R}^3, V)} \leq \|g\|_{H^p(\mathbb{R}^3, V)}$$

it follows that

$$
\begin{aligned}
\|f \cdot g\|_{H^p(\mathbb{R}^3, W)} &\leq \|f \cdot g\|_{H^2(\mathbb{R}^3, W)} + \|f \cdot g\|_{\mathcal{B}^p(\mathbb{R}^3, W)} \\
&\leq c_2 \|f\|_{\mathcal{B}^q(\mathbb{R}^3, U)} \|g\|_{H^2(\mathbb{R}^3, V)} + c'_p \|f\|_{\mathcal{B}^p(\mathbb{R}^3, U)} \|g\|_{\mathcal{B}^p(\mathbb{R}^3, V)} \\
&\leq c_2 \|f\|_{\mathcal{B}^q(\mathbb{R}^3, U)} \|g\|_{H^p(\mathbb{R}^3, V)} + c'_p \|f\|_{\mathcal{B}^q(\mathbb{R}^3, U)} \|g\|_{H^p(\mathbb{R}^3, V)} \\
&\leq c_p \|f\|_{\mathcal{B}^q(\mathbb{R}^3, U)} \|g\|_{H^p(\mathbb{R}^3, V)},
\end{aligned}
$$

where $c_p = (c_2 + c'_p)$.   □

It follows that, for each $\xi \in B^{k+2}(\mathbb{R}^3, \mathfrak{g})$ and $(A, E, \Psi) \in P^k$, the image of the infinitesimal action of ξ is contained in P^k. Therefore,

$$gs\left(P^k\right) = B^{k+2}\left(\mathbb{R}^3, \mathfrak{g}\right).$$

Using estimates in Lemma 3.1 we can prove the propositions given below. Proofs are almost identical to the proofs given in [11].

Proposition. The Lie algebra $gs(P^k)$ is a Hilbert–Lie algebra with the norm (4.1). The action of $gs(P^k)$ in P^k is continuous.

The gauge symmetry group $GS(P^k)$ of P^k is the connected Hilbert–Lie group with the Lie algebra $gs(P^k)$. For $k = 1$, it was constructed in [11] following the approach of Segal, [12], for the Lie algebra $H^3(\mathbb{R}^3, \mathfrak{g})$. The construction for $k > 1$ is outlined below. A much bigger gauge group was considered by Eichhorn, [6].

Let $C_c^\infty(\mathbb{R}^3, \mathfrak{g})$ denote the space of all smooth maps from $\mathbb{R}^3$ to $\mathfrak{g}$ which are constant outside compact sets, and let $C_0^\infty(\mathbb{R}^3, \mathfrak{g})$ be the subspace of compactly supported maps. From the decomposition results of [1] it follows that

$$gs(P^k) = gs\left(P^k\right)_0 \oplus \mathfrak{g}, \tag{4.2}$$

where $gs(P^k)_0$ is the completion of $C_0^\infty(\mathbb{R}^3, \mathfrak{g})$ in the norm (4.1), and $\mathfrak{g}$ represents constant maps from $\mathbb{R}^3$ to $\mathfrak{g}$. It follows from Theorem A.2 of [11] that elements of $gs(P^k)$ are C^k-maps from $\mathbb{R}^3$ to $\mathfrak{g}$, and $C_c^\infty(\mathbb{R}^3, \mathfrak{g})$ is dense in $gs(P^k)$.

Let $C_c^\infty(\mathbb{R}^3, G)$ denote the space of all smooth maps from $\varphi : \mathbb{R}^3 \to G$ which are constant outside a compact set. It forms a group under pointwise multiplication. One-parameter subgroups of $C_c^\infty(\mathbb{R}^3, G)$ are of the form $\exp(t\xi)$, where ξ is in the dense subalgebra $C_c^\infty(\mathbb{R}^3, \mathfrak{g})$ of $gs(P^k)$. The topology of

$gs(P^k)$ induces a uniform structure in $C_c^\infty(\mathbb{R}^3, G)$, with a neighbourhood basis at the identity consisting of the sets

$$N_\epsilon = \left\{ \exp(t\xi) \mid \xi \in C_c^\infty\left(\mathbb{R}^3, G\right), \|\xi\|_{B^{k+2}(\mathbb{R}^3,\mathfrak{g})} < \epsilon \right\},$$

with $\epsilon > 0$.

Proposition. The mapping $\exp\xi \mapsto (\exp\xi)^{-1}$ is uniformly continuous relative to N_1. That is, for every $\epsilon > 0$, there exists $\delta > 0$ such that, for every $\exp\xi \in N_1$,

$$(\exp\xi)^{-1} N_\delta \exp\xi \subseteq N_1.$$

By a result of Ref. [16], this proposition implies that the completion of $C_c^\infty(\mathbb{R}^3, G)$ in this uniform structure is a toplolgical group. It is a Hilbert–Lie group, whose Lie algebra is canonically isomorphic to the Hilbert–Lie algebra $gs(P^k)$. In view of this we adopt the following

Definition. The group $GS(P^k)$ of gauge symmetries of P^k is the completion of the group $C_c^\infty(\mathbb{R}^3, G)$ in the uniform structure defined by the topology of its Lie algebra $gs(P^k)$.

Proposition. The gauge symmetry group $GS(P^k)$ is connected.

By assumption, $G \subset gl(n, \mathbb{R})$ so that $GS(P^k)$ can be considered as a space of maps $\varphi : \mathbb{R}^3 \to gl(n, \mathbb{R})$. An alternative description of the topology of $GS(P^k)$ is given by the following

Proposition. A sequence $\varphi_m \in GS(P^k)$ is convergent to $\varphi \in GS(P^k)$ if and only if it converges to φ in the topology defined by the norm

$$\|\varphi\|_{\mathcal{B}^{k+2}(\mathbb{R}^3, gl(n,\mathbb{R}))} = \int_{B_1(0)} \varphi^T \varphi \, d_3x + \|\text{grad}\,\varphi\|_{H^{k+1}(\mathbb{R}^3, gl(n,\mathbb{R})\otimes\mathbb{R}^3)}. \quad (4.3)$$

Proposition. The action of $GS(P^k)$ in P^k is continuous and proper.

The propositions above are basic steps in proving Theorem 2.3, where the norm given by (4.3) was denoted by $\|\varphi\|_{\mathcal{B}^{k+2}}^2$. Their proofs, based on estimates in Lemma 3.1, are nearly identical to the corresponding proofs given in [11] for the case $k = 1$.

Let $C_0^\infty(\mathbb{R}^3, G)$ be the normal subgroup of $C_c^\infty(\mathbb{R}^3, G)$ consisting of maps $\varphi : \mathbb{R}^3 \to G$ which differ from the identity only in compact sets. Its closure in $GS(P^k)$ is a connected, closed, normal subgroup $GS(P^k)_0$ of $GS(P^k)$. The connectedness of $GS(P^k)_0$ can be proved in the same way as the connectedness of $GS(P^k)$. The algebra $gs(P^k)_0$ appearing in the decomposition (4.2) is the Lie algebra of $GS(P^k)_0$. The action of $GS(P^k)_0$ in P^k is proper because it is a closed subgroup of $GS(P^k)$ and the action of $GS(P^k)$ is proper.

In the case when $k = 1$, a proof that the action of $GS(P^k)$ in P^k is free was given in Ref. [14]. It is based on the existence, for $A \in H^2(\mathbb{R}^3, g \otimes \mathbb{R}^3)$, of a gauge transformation Γ such that the transformed gauge potential

$$\widetilde{A} = \Gamma A \Gamma^{-1} + \Gamma \operatorname{grad} \Gamma^{-1}$$

satisfies the Cronström gauge condition

$$\widetilde{A}(x) \cdot x = 0,$$

see [4] and [15]. If $A \in H^{k+1}(\mathbb{R}^3, g \otimes \mathbb{R}^3)$, then $\widetilde{A}$ is also in $H^{k+1}(\mathbb{R}^3, g \otimes \mathbb{R}^3)$, and the proof for $k = 1$ extends without any changes to $k > 1$. Hence, Theorem 2.4 is valid for $k \geq 1$.

5. Reduction

The extended phase space P^k is weakly symplectic, with the weak symplectic form $\omega = d\theta$, where

$$\langle \theta(A, E, \Psi) \mid (a, e, \psi) \rangle = \int_{\mathbb{R}^3} \left(E \cdot a + \Psi^\dagger \psi \right) d_3 x.$$

The action of $GS(P^k)$ in P^k preserves the 1-form θ. Since $\omega = d\theta$, this action is Hamiltonian with an equivariant momentum map $\mathcal{J}$ such that, for every $\xi \in gs(P^k)$

$$\langle \mathcal{J}(A, E, \Psi) \mid \xi \rangle = \int_{\mathbb{R}^3} \left(-E \cdot D_A \xi + \Psi^\dagger \xi \Psi \right) d_3 x.$$

Using Stokes' theorem we obtain

$$\langle \mathcal{J}(A, E, \Psi) \mid \xi \rangle = \int_{\mathbb{R}^3} \left\{ (\operatorname{div} E + [A; E]) \xi + \Psi^\dagger \xi \Psi \right\} d_3 x$$

$$+ \lim_{r \to \infty} \int_{S_r(0)} nE \, d_2 S,$$

where $S_r(0)$ is the sphere of radius r centered at the origin, nE is the normal component of E and $d_2 S$ is the surface area form. For $\xi \in gs(P^k)_0$, the second term on the right hand side vanishes. Hence,

$$\langle \mathcal{J}(A, E, \Psi) \mid \xi \rangle = \int_{\mathbb{R}^3} \left\{ (\operatorname{div} E + [A; E]) \xi + \Psi^\dagger \xi \Psi \right\} d_3 x \quad \forall \xi \in gs\left(P^k\right)_0.$$

The coefficient of ξ in the integral is the left hand side of the constraint equation. Taking into account the Fundamental Theorem in the Calculus of Variations we obtain

$$(A, E, \Psi) \in C^k \iff \langle \mathcal{J}(A, E, \Psi) \mid \xi \rangle = 0 \quad \forall \xi \in gs\left(P^k\right)_0.$$

This proves Theorem 2.5.

Theorem 2.6 was proved in [14] in the case $k = 1$. The proof is based on the existence of slices for proper actions of Lie groups, see Refs. [10] and [13], and the results established in Ref. [2] for momentum maps of equivariant actions of Hilbert–Lie groups. The proof extends to $k > 1$ without changes. This completes an outline of the proof of Theorem 2.6.

References

[1]   Aikawa, H., On weighted Beppo Levi functions. Integral representations and behaviour at infinity, *Analysis* **9** (1989), 323–346.

[2]   Arms, J., Marsden, J.E., and Moncrief, V., Symmetry and bifurcation of momentum maps, *Comm. Math. Phys.* **90** (1981), 361–372.

[3]   Deny, J. and Lions, J.L., Les éspaces du type Beppo Levi, *Ann. Inst. Fourier* **5** (1955), 305–370.

[4]   Eardley, D.M. and Moncrief, V., The global existence of Yang-Mills-Higgs fields in 4-dimensional Minkowski space. II. Completion of the proof. *Comm. Math. Phys.* **83** (1982), 193–212.

[5]   ——, The global existence of Yang-Mills-Higgs fields in 4-dimensional Minkowski space. I. Local existence and smoothness properties, *Comm. Math. Phys.* **83** (1982), 171–191.

[6]   Eichhorn, J., Gauge theory of open manifolds of bounded geometry, *Ann. Global Anal. Geom.* **11** (1993), 253–300.

[7]   Kato, T., *Perturbation theory for linear operators*, Springer Verlag, Berlin–Heidelberg–New York 1984.

[8]   Klainerman, S. and Machedon, M., Finite energy solutions of the Yang-Mills equations in R^{3+1}, *Ann. Math.* **142** (1995), 39–119.

[9]   Lang, S., *Differential and Riemannian manifolds*, Springer, New York 1995.

[10] Palais, R., On the existence of slices for actions of non-compact Lie groups, *Ann. Math.* **73** (1961), 295–323.

[11] Schwarz, G. and Śniatycki, J., Gauge symmetries of an extended phase space for Yang-Mills and Dirac fields, *Ann. Inst. Henri Poincaré* **66** (1996), 109–136.

[12] Segal, I., The Cauchy problem for the Yang-Mills equations, *J. Funct. Anal.* **33** (1979), 175–194.

[14] Śniatycki, J., Regularity of constraints and reduction in the Minkowski space Yang-Mills-Dirac theory, *Ann. Inst. Henri Poincaré*, **70** (1999), 277–293.

[13] Śniatycki, J., Schwarz, G., and Bates, L., Yang-Mills and Dirac fields in a bag, constraints and reduction, *Comm. Math. Phys.* **176** (1996), 95–115.

[15] Uhlenbeck, K., Removable singularities in Yang-Mills fields, *Comm. Math. Phys.* **83** (1982), 11–29.

[16] Weil, A., *Sur les éspaces a structures uniformes*, Act. Sci. Ind., 551, Hermann, Paris 1938.

PART III

SHEAVES AND SPECTRAL SEQUENCES

LA THEÓRIE DES RÉSIDUS SUR UN ESPACE ANALYTIQUE COMPLEXE

VINCENZO ANCONA AND BERNARD GAVEAU

À la mémoire de Monsieur Jean Leray dont la théorie des résidus et la théorie des faisceaux sont à la base de ce travail

M. de Gosson (ed.), Jean Leray '99 Conference Proceedings, 269-279.

Résumé. *Nous rappelons brièvement la théorie des résidus de Leray et nous montrons comment il est possible de l'étendre au cas d'un espace analytique complexe quelconque.*

Une formule fondamentale de l'Analyse mathématique est la formule de Cauchy. Si $f(z)$ est une fonction holomorphe dans un domaine D, si γ est une chaîne de D bordant un domaine Δ dans D et z est un point de Δ

$$f(z) = \frac{1}{2\pi} \int_\gamma \frac{f(z')}{z' - z} \, dz.$$

Cette formule repose sur les deux faits suivants:

(i) $f(z')(z' - z)dz$ est une différentielle fermée sur D, ayant un pôle simple en $z' = z$

(ii) γ borde une Δ chaîne dont l'intersection avec $\{z\}$ est z.

La généralisation à deux variables a été entreprise par Poincaré [11] et par Picart–Simart [12], il y a plus d'un siècle. C'est Leray qui, dans son article fondamental [7] de 1959, donne la formulation générale dans le cadre d'une variété analytique complexe sous la forme que voici. Soit X une variété analytique complexe, D une hypersurface lisse. Une forme différentielle ω au voisinage d'un point de D où une équation locale de D est donnée par $z = 0$ a des pôles d'ordre ≤ 1 (ou encore pôles logarithmique) si ω s'écrit en voisinage de ce point comme

$$\omega = \frac{dz}{z} \wedge \Psi + \varphi, \tag{1}$$

où Ψ et φ sont des formes C^∞ au voisinage de ce point. La forme $\Psi|_D$ est appelée alors forme résidu et Leray montre que si ω est fermée, $\Psi|_D$ est fermée. Il montre que la cohomologie $H^k(X \backslash D)$ peut se calculer à l'aide des formes de X ayant des pôles d'ordre ≤ 1 sur D, (comme espace quotient de formes fermées de X à pôle d'ordre ≤ 1 sur D, modulo les différentielles de telles formes) et que l'application

$$H^k(X \backslash D) \longrightarrow H^{k-1}(D),$$
$$[\omega] \longrightarrow [\Psi|_D] \tag{2}$$

est bien définie. Il montre aussi que la transposée de l'application (2) en homologie

$$H_{k-1}(D) \longrightarrow H_k(X \backslash D) \tag{3}$$

se construit de la façon suivante: pour toute chaîne Γ de dimension $k-1$ de D, on associe une chaîne Δ située dans un petit voisinage de D de dimension $k+2$ dont l'intersection avec D est Γ

$$\Delta \cdot D = \Gamma. \tag{4}$$

L'application transposée (3) consiste à associer à la class d'homologie de Γ sur D, la classe d'homologie de ∂D dans $(X \backslash D)$.

La définition des pôles d'ordre ≤ 1 selon la formule (1) et la propriété (4) d'intersection généralisent exactement les deux propriétés permettant d'établir la formule des résidus de la façon suivante

$$\frac{1}{2i\pi} \int_{\Gamma} \Psi|_D = \int_{\partial \Delta} \omega. \tag{5}$$

C'est cette formule qui permet à Leray de démontrer la formule de Cauchy–Fantappié–Leray, à revoir la représentation intégrale d'une fonction holomorphe dans un convexe D de $\mathbb{C}^n$ par une intégrale sur le bord de D de puissances inverses d'ordre n de l'équation du plan tangent au bord. Leray applique sa théorie des résidus au cas de la quadrique d'incidence de $\mathbb{C}^n$, i.e., de la quadrique formée par les couples (x, ξ) avec $\zeta \in \mathbb{P}^n$ et $\zeta_0 - \sum_{i=1}^{n} \zeta_i x_i = 0$.

Supposons toujours que X est une variété analytique complexe et que D est un diviseur à croisements normaux. Cela signifie qu'au voisinage U d'un point $m \in D$, il existe un système de coordonnées locales $\{z_1, \ldots, z_n\}$ nulles en m sur X tel que dans ce voisinage, $D \cap U$ s'écrit

$$D \cap U = \bigcup_{i=1}^{p} \{z_1 = 0\}. \tag{6}$$

Donc $D \cup U$ est réunion de p hypersurfaces analytiques en position générales. Une forme ω dans un tel voisinage U est dite avoir des pôles logarithmiques si elle s'écrit comme combinaison de formes du type

$$\frac{dz_{i_1}}{z_{i_1}} \wedge \cdots \wedge \frac{dz_{i_k}}{z_{i_k}} \wedge \Psi + \varphi \tag{7}$$

avec $\{i_1, \ldots, i_k\} \subset \{1, \ldots, p\}$ et Ψ et φ des formes différentielles lisses sur U.

Par exemple, prenons $X = \mathbb{C}^n$, D la réunion des hypersurfaces $\{z_1 = 0\}$ $\cup \cdots \cup \{z_n = 0\}$ et la forme différentielle

$$\omega \frac{dz_1}{z_1} \wedge \cdots \wedge \frac{dz_n}{z_n} \wedge \Psi, \tag{8}$$

où Ψ est une forme $\mathbb{C}^\infty$ près de 0.

Dans [5] et [6], il est démontré que la cohomologie $H^k(X \backslash D)$ peut se calculer à l'aide des formes de X de type (7), c'est à dire, que la cohomologie $H^k(X \backslash D)$ est l'espace-quotient de l'espace des formes de type (7) qui sont formées modulo les différentielles de formes de type (7). A priori, le théorème de De Rham nous disait que la cohomologie de $X \setminus D$ est l'espace quotient des formes $\mathbb{C}^\infty$ sur $X \setminus D$ d-fermés, par les différentielles de formes $\mathbb{C}^\infty$ sur $X \setminus D$. Nous noterons désormais par $\mathscr{E}_X^k$ le faisceau des formes $\mathbb{C}^\infty$ de degré k sur X, $\mathscr{E}_X^k \langle \log D \rangle$ le faisceau des formes de degré k sur X à pôles logarithmiques sur D. Les faisceaux définissent des complexes $(\mathscr{E}_X^{\cdot}, d)$ et $(\mathscr{E}_X^{\cdot} \langle \log D \rangle, d)$ où d est la différentielle ordinaire.

Le complexe $(\mathscr{E}_X^{\cdot}, d)$ est une résolution fine du faisceau $\mathbb{C}_X$ (c'est le lemme de Poincaré)

$$0 \longrightarrow \mathbb{C}_X \longrightarrow \mathscr{E}_X^0 \overset{d}{\longrightarrow} \mathscr{E}_X^1 \overset{d}{\longrightarrow} \cdots \longrightarrow \mathscr{E}_X^p \overset{d}{\longrightarrow} \cdots .$$

Soit

$$j : X^0 \equiv X \setminus D \hookrightarrow X.$$

Evidemment $\mathscr{E}_X^{\cdot} \langle \log D \rangle$ est un sous-complexe du complexe de De Rham $\mathscr{E}_{X^0}^{\cdot}$ sur X^0, plus précisement, on a un morphisme de complexes

$$\mathscr{E}_X^p \langle \log D \rangle \longrightarrow j_* \mathscr{E}_{X^0}^p. \tag{9}$$

En utilisant cette terminologie un peu plus systématique, nous pouvons dire que le morphisme (9) induit un isomorphisme des faisceaux de cohomologie de ces complexes, c'est à dire, pour tout ouvert $U \subset X$

$$H^p \left(U, \mathscr{E}_X^{\cdot} \langle \log D \rangle \right) \simeq H^p \left(U^0, j_* \mathscr{E}_{X^0}^{\cdot} \right) \simeq H^p \left(U^0, \mathbb{C} \right). \tag{10}$$

Dans (9), la seconde égalite est le théorème de De Rham standard, et la première égalité est le resultat de [6]. Notons que (9) introduit la définition des faisceaux de cohomologie des complexes considérés, notion définie par Leray dans son article [8] de facon très générale afin d'étudier de nouveaux invariants des applications continues entre espaces topologiques. La démonstration de [6] repose sur un argument de suite spectrale que nous n'exposerons pas ici.

La notion de résidu d'une forme telle que (7) devient beaucoup plus problématique. Une notion naturelle serait de définir le résidu de ω au point $p \in D$ au voisinage duquel ω s'écrit selon (7) par

$$\Psi|_{\{z_{i_1} = 0, \dots, z_{i_k} = 0\}}. \tag{11}$$

C'est cette notion qui est à l'origine de la formule de Cauchy à plusieurs variables pour des formes telles que (8) avec $\Psi = f$, f holomorphe

$$f(0) = \frac{1}{(2i\pi)^n} \int_\delta f(z) \bigwedge_{i=1}^n \frac{dz_i}{z_i}$$

où δ est le cycle $\{|z_1| = r_1, \dots, |z_n| = r_n\}$.

La définition correcte du résidu consiste à introduire la filtration par le poids $\mathscr{E}_X^{\cdot} \langle \log D \rangle$ essentiellement par l'ordre du pôle (le nombre k figurant dans (7)). C'est cette définition qui permet à la fois de définir le résidu généralisée, de démontrer l'isomorphisme (9) et d'introduire la structure de Hodge mixte sur la cohomologie de $X \backslash D$ lorsque X est Kählérienne compacte.

De nombreuses tentatives ont été faites pour généraliser la théorie des résidus au cas où D présenterait des singularités. Leray lui-même, suivant Poincaré et Picard–Simart [11], [12] a traité le cas où D a des singularités quadratiques isolées. Mais sa méthode ne s'étend pas plus loin. Il nous faut donc pour étendre les résultats de Leray, changer complètement notre point de vue. Dans ce qui suit, nous allons expliquer comment il est possible d'étendre la théorie des résidus au cas où X est un espace analytique complexe quelconque, et D est un sous-ensemble analytique complexe avec $X \backslash D$ ouvert dense de X. Nous n'indiquerons pas les démonstrations, pour lesquelles nous renvoyons à [1], [2], [3] et [4]. De plus, nous nous contenterons ici de définir des complexes de formes à pôles logarithmiques et de d'énoncer l'analogue du résultat (9), à savoir que la cohomologie de $X \backslash D$ peut se calculer grâce aux formes à pôles logarithmiques que nous allons introduire. La théorie des résidus à proprement parler paraîtra dans [1], [2], [3] et [4].

La question préliminaire qui se pose à nous est la définition de la notion des formes différentielles sur un espace analytique complexe quelconque. Nous voulons dons définir des complexes $(\Lambda_X^{\cdot}, d)$ qui sont fins et qui sont des résolutions de $\mathbb{C}_X$

$$0 \longrightarrow \mathbb{C}_X \longrightarrow \Lambda_X^0 \overset{d}{\longrightarrow} \Lambda_X^1 \longrightarrow \cdots \longrightarrow \Lambda_X^p \longrightarrow \cdots \qquad (12)$$

tels que en dehors d'un sous-ensemble analytique propre fermé de X, contenant les singularités de X, la résolution fine (12) se réduise à la résolution de De Rham standard.

Commençons par un exemple simple. Supposons que X soit un espace analytique complexe, et supposons qu'il existe un sous-ensemble analytique complexe $E \subset X$ avec $X \backslash E$ ouvert dense de X, E contenant $\mathrm{Sing}\, X$ et E est une variété analytique complexe. Soit

$$p : \widetilde{X} \longrightarrow X$$

une modification lisse de X le long de E obtenue par le procédé d'Hironaka. Donc $\widetilde{X}$ est lisse, $p^{-1}(E)$ est un diviseur à croisements normaux de $\widetilde{X}$ et $p : \widetilde{X} \backslash \widetilde{E} \to X \backslash E$ est un isomorphisme. Nous aurons aussi un diagramme

$$
\begin{array}{ccc}
\widetilde{E} & \overset{\tilde{i}}{\longrightarrow} & \widetilde{X} \\
{\scriptstyle p}\downarrow & & \downarrow{\scriptstyle p} \\
E & \overset{i}{\longrightarrow} & X.
\end{array}
\qquad (13)
$$

Supposons encore que $\widetilde{E}$ soit lisse et définissons

$$\Lambda_X^k = p^* \mathscr{E}_{\widetilde{X}}^k \oplus i_* \mathscr{E}_E^k \oplus (i \circ p)_* \mathscr{E}_{\widetilde{E}}^{k-1} \tag{14}$$

où $\mathscr{E}_Y^k$ désigne le faisceau des formes C^∞ de degré k sur Y, Y étant une variété analytique, donc lisse. Définissons la différentielle $d : \Lambda_X^k \to \Lambda_X^{k+1}$ par

$$d(\omega, \sigma, \gamma) = \left(d\omega, d\sigma, d\gamma + (-1)^k \left(\widetilde{\imath}^* - p^* \sigma \right) \right) \tag{15}$$

où $\widetilde{\imath} : \widetilde{E} \to \widetilde{X}$ est l'injection.

D'autre part, définissons l'injection

$$\mathbb{C}_X \longrightarrow \Lambda_X^0, \qquad c \longrightarrow (c, c, 0).$$

Il est facile de vérifier que (Λ_X^k, d) est un complexe de faisceaux fins. Le résultat fondamental est que c'est aussi une résolution du faisceau constant $\mathbb{C}_X$. Cela résulte du fait que si U est un ouvert assez petit de X, alors $p^{-1}(U)$ se rétracte sur $p^{-1}(U) \cap \widetilde{E}$, ce qui est une conséquence du théorème de Lojasiewicz sur la triangulation des espaces analytiques [10].

Nous allons maintenant étendre cette construction au cas général d'un espace analytique complexe de la façon suivante. A tout espace analytique complexe X nous associons un ensemble $\mathscr{R}(X)$ de complexes $(\Lambda_X^\cdot, d)$ avec les propriétés suivantes:

(i) chaque complexe $(\Lambda_X^\cdot, d)$ est une résolution du faisceau constant $\mathbb{C}_X$ sur X par des faisceaux fins;

(ii) $\Lambda_X^k = 0$ si $k < 0$ ou si $k > 2 \dim_{\mathbb{C}} X$;

(iii) il existe un ensemble analytique propre $E \subset X$ avec $X \backslash E$ ouvert dense lisse de X, tel que $(\Lambda_X^\cdot |_{X \backslash E}, d)$ soit la résolution de De Rham standard sur la variété lisse $X \backslash E$;

(iv) si X est une variété complexe, la résolution de De Rham standard $(\mathscr{E}_X^\cdot, d)$ de $\mathbb{C}_X$ est un élément de l'ensemble $\mathscr{R}(X)$.

De plus, si $f : X \to Y$ est un morphisme d'espaces analytiques complexes, nous lui associerons un ensemble $\mathscr{R}(f)$ de morphismes de complexes entre certains $\Lambda_Y^\cdot$ et certains $\Lambda_X^\cdot$. Nous nommerons *pull-back* associés à f, les éléments de $\mathscr{R}(f)$, dont les propriétés sont les suivantes:

(i') pour tout $(\Lambda_Y^\cdot) \in \mathscr{R}(Y)$, il existe $(\Lambda_X^\cdot) \in \mathscr{R}(X)$ et un morphisme $\varphi : \Lambda_Y^\cdot \to \Lambda_X^\cdot$ élément de $\mathscr{R}(f)$, donc un pull-back associé à f;

(ii') si $\alpha, \beta : \Lambda_Y^\cdot \to \Lambda_X^\cdot$ sont deux pull-backs associés à f, alors $\alpha = \beta$;

(iii') si $f : X \to Y, g : X \to Z$ sont deux morphismes et $\varphi : \Lambda_Y^{\cdot} \to \Lambda_X^{\cdot}, \Psi : \Lambda_Z^{\cdot} \to \Lambda_Y^{\cdot}$ des pull-backs associés à f et g respectivement, alors $\varphi \circ \Psi : \Lambda_Z^{\cdot} \to \Lambda_X^{\cdot}$ est un pull-back associé à $g \circ f$. L'identité de $\Lambda_X^{\cdot}$ est un pull-back associé à Id_X;

(iv') si $\Lambda_{X,1}^{\cdot}$ et $\Lambda_{X,2}^{\cdot}$ sont deux éléments de $\mathcal{R}(X)$ il existe un $\Lambda_{X,3}^{\cdot}$ de $\mathcal{R}(X)$ et deux pull-backs $\Lambda_{X,1}^{\cdot} \to \Lambda_{X,3}^{\cdot}$ et $\Lambda_{X,2}^{\cdot} \to \Lambda_{X,3}^{\cdot}$ associés à l'identité de X;

(v') si $f : X \to Y$ est un morphisme de variétés complexes, le pull-back de De Rham de formes $f^* : \mathcal{E}_Y^{\cdot} \to \mathcal{E}_X^{\cdot}$ est dans la famille $\mathcal{R}(f)$.

Le résultat fondamental de [1], [2], [3] et [4] s'énonce ainsi. Il est possible de construire, par récurrence sur les dimensions, les ensembles $\mathcal{R}(X)$ et les pull-backs de la façon suivante: tout $(\Lambda_X^{\cdot}) \in \mathcal{R}(X)$, est spécifié par:

(α) la donnée d'un sous-ensemble analytique E, propre contenant les singularités de X, tel que $X \backslash E$ soit ouvert dense de X;

(β) la donnée d'une modification $p : \widetilde{X} \to X$ avec $\widetilde{X}$ lisse, obtenue par éclatements le long de E avec $\widetilde{E} = p^{-1}(E)$ diviseur à croisements normaux de $\widetilde{X}$ et $p : \widetilde{X} \backslash \widetilde{E} \to X \backslash E$ isomorphisme. On a alors un diagramme du type (13);

(γ) la donnée de deux éléments $(\Lambda_E^{\cdot}) \in \mathcal{R}(E)$ et $(\Lambda_{\widetilde{E}}^{\cdot}) \in \mathcal{R}(\widetilde{E})$ et de deux pull-backs $\varphi : \Lambda_X^{\cdot} \to \Lambda_{\widetilde{E}}^{\cdot}$ et $\Psi : \mathcal{E}_{\widetilde{X}}^{\cdot} \to \Lambda_{\widetilde{E}}^{\cdot}$ associés à $p \mid \widetilde{E}$ et à $\tilde{\imath} : \widetilde{E} \hookrightarrow \widetilde{X}$, respectivement;

(δ) $(\Lambda_X^{\cdot}, d)$ est alors donné par la formule

$$\Lambda_X^k = p_* \mathcal{E}_{\widetilde{X}}^k \oplus i_* \Lambda_E^k \oplus (i \circ p)_* \Lambda_{\widetilde{E}}^{k-1} \tag{16}$$

avec la différentielle

$$d(\omega, \sigma, \gamma) = \left(d\omega, d\sigma, d\gamma + (-1)^k (\Psi(\omega) - \varphi(\sigma)) \right). \tag{17}$$

On peut montrer que pour tout E comme ci-dessus tout $(\Lambda_X^{\cdot}) \in \mathcal{R}(E)$, il existe une modification $p : (\widetilde{X}, \widetilde{E}) \to (X, E)$ et un élément $(\Lambda_X^{\cdot}) \in \mathcal{R}(X)$ dont la seconde composante est $i_* \Lambda_E^{\cdot}$.

Nous pouvons maintenant commencer la construction des faisceaux de formes à pôles logarithmiques.

Supposons d'abord X lisse et Q un ensemble analytique de X.

Nous définissons

$$\mathcal{E}_X^{\cdot} \langle \log Q \rangle = 0 \quad \text{si } Q = X. \tag{18}$$

Si $Q \neq X$, nous considérons une modification $\pi : (\widetilde{X}, \widetilde{Q}) \to (X, Q)$ avec $\widetilde{X}$ lisse, $\widetilde{Q}$ diviseur à croisement normal et nous poserons

$$\mathscr{E}_X^{\cdot}\langle \log Q \rangle = \pi_* \mathscr{E}_{\widetilde{X}}^{\cdot}\langle \log \widetilde{Q} \rangle. \tag{19}$$

Comme $\pi|_{\widetilde{X}\backslash\widetilde{Q}}$ est un isomorphisme de $\widetilde{X}\backslash\widetilde{Q}$ sur $X\backslash Q$, il est clair que la cohomologie du complexe de sections de $\mathscr{E}_X^{\cdot}\langle \log Q \rangle$ est la cohomologie de $X\backslash Q$.

Dans le cas général, soit (X, Q) un couple formé d'un espace analytique X et d'un sous-espace Q. Un morphisme $f : (X, Q) \to (Y, R)$ de couples est un morphisme $f : X \to Y$ d'espaces analytiques avec $f(Q) \subset R$ et $f(X\backslash Q) \subset Y\backslash R$.

Notons $X^0 = X\backslash Q$. Lorsque $Q = X$, nous définissons $\Lambda_X^{\cdot}\langle \log Q \rangle = 0$ comme dans (18). Nous allons définir pour tout couple (X, Q) un ensemble $\mathscr{R}(X\langle \log Q \rangle)$ de complexes $(\Lambda_X^{\cdot}\langle \log Q \rangle, d)$ et pour tout morphisme de couples $f : (X, Q) \to (Y, R)$ un ensemble de morphismes de complexes $\mathscr{R}(f)$ entre certains $(\Lambda_X^{\cdot}\langle \log R \rangle)$ et certains $(\Lambda_X^{\cdot}\langle \log Q \rangle)$ appelés pull-backs avec les propriétés suivantes

(i)p $\Lambda_X^k\langle \log Q \rangle = 0$ si $k < 0$ ou $k > 2\dim_{\mathbb{C}} X$;

(ii)p pour tout $\Lambda_X^{\cdot}\langle \log Q \rangle$, il existe un ensemble E tel que $X\backslash E$ soit un ouvert dense lisse de X et tel que $\Lambda_X^{\cdot}\langle \log Q \rangle|_{X\backslash E}$ soit $\mathscr{E}_{X\backslash E}^{\cdot}\langle \log Q\backslash E \rangle$ défini comme ci-dessus;

(iii)p les $\Lambda_X^k\langle \log Q \rangle$ sont fins;

(iv)p si X est lisse, les $\Lambda_X^{\cdot}\langle \log Q \rangle$ définis en (19) sont dans $\mathscr{R}(X\langle \log Q \rangle)$;

(v)p les pull-backs ont des propriétés analogues à celles détaillées en (i)'–(v)' ci-dessus.

Le second résultat de [1], [2], [3] et [4] s'énonce ainsi.

Il est possible de construire par récurrence sur les dimensions, les ensembles $\mathscr{R}(X\langle \log Q \rangle)$ de la façon suivante: un élément $(\Lambda_X^{\cdot}\langle \log Q \rangle) \in \mathscr{R}(X\langle \log Q \rangle)$ est spécifié par les données suivantes;

$(\alpha)_p$ la donnée d'un sous-ensemble analytique E propre contenant les singularités de X avec $X\backslash E$ ouvert dense de X;

$(\beta)_p$ la donnée d'une modification $p : \widetilde{X} \to X$, avec $\widetilde{X}$ lisse obtenue par éclatements le long de E, et $\widetilde{E} = p^{-1}(E)$ étant un diviseur à croisements normaux avec un diagramme (13);

$(\gamma)_p$ la donnée de trois éléments

$$\mathscr{E}_X^{\cdot} \langle \log \widetilde{Q} \rangle, \quad \widetilde{Q} = p^{-1}(Q),$$
$$\Lambda_E^{\cdot} \langle \log(E \cap Q) \rangle \in \mathscr{R}\left(E \langle \log E \cap Q \rangle\right),$$
$$\Lambda_{\widetilde{E}}^{\cdot} \langle \log\left(\widetilde{E} \cap \widetilde{Q}\right) \rangle \in \mathscr{R}\left(\widetilde{E} \langle \log \widetilde{E} \cap \widetilde{Q} \rangle\right)$$

et des pull-backs

$$\Psi : \mathscr{E}_{\widetilde{X}}^{\cdot} \langle \log \widetilde{Q} \rangle \longrightarrow \Lambda_{\widetilde{E}}^{\cdot} \langle \log \widetilde{E} \cap \widetilde{Q} \rangle,$$
$$\varphi : \Lambda_E^{\cdot} \langle \log(E \cap Q) \rangle \longrightarrow \Lambda_{\widetilde{E}}^{\cdot} \langle \log \widetilde{E} \cap \widetilde{Q} \rangle$$

associés aux morphismes de paires respectifs

$$\widetilde{\imath} : \left(\widetilde{E}, \widetilde{E} \cap \widetilde{Q}\right) \longrightarrow \left(\widetilde{X}, \widetilde{Q}\right), \qquad |p_E : \left(\widetilde{E}, \widetilde{E} \cap \widetilde{Q}\right) \longrightarrow (E, E \cap Q).$$

$(\delta)_p$ Le complexe $\Lambda_E^{\cdot} \langle \log Q \rangle$ est donné par la formule

$$\begin{aligned}
\Lambda_X^k \langle \log Q \rangle = {}& p_* \mathscr{E}_{\widetilde{X}}^k \langle \log \widetilde{Q} \rangle \oplus i_* \Lambda_E^k \langle \log(Q \cap E) \rangle \\
& \oplus (p \circ \widetilde{\imath})_* \Lambda_{\widetilde{E}}^{k-1} \langle \log\left(\widetilde{Q} \cap \widetilde{E}\right) \rangle
\end{aligned} \tag{20}$$

la différentielle est donnée par l'analogue de la formule (17), grâce aux pull-backs Ψ et φ.

On peut montrer que pour tout E comme indiqué dans $(\alpha)_p$, toute modification $(\widetilde{X}, \widetilde{E}) \to (X, E)$ comme dans $(\beta)_p$, et tous complexes $\mathscr{E}_{\widetilde{X}}^k \langle \log \widetilde{Q} \rangle$ et $\Lambda_E^{\cdot} \langle \log(Q \cap E) \rangle$, il existe un complexe $\Lambda_{\widetilde{E}}^{\cdot} \langle \log(\widetilde{Q} \cap \widetilde{E}) \rangle$ et des pull-backs Ψ et φ, donnant lieu a un complexe $\Lambda_X^{\cdot} \langle \log Q \rangle$ comme dans l'équation (19).

Nous pouvons maintenant énoncer la généralisation du résultat (9). Considérons le morphisme de complexes de faisceaux

$$\Lambda_X^{\cdot} \langle \log Q \rangle \, j_* \Lambda_{X^0}^{\cdot} \tag{21}$$

où $X^0 = X \backslash Q$ et $j : X^0 \to X$ est l'injection. Ici $\Lambda_{X^0}^{\cdot}$ est un complexe de faisceaux de l'ensemble $\mathscr{R}(X)$ construit grâce à $E \backslash Q$. En fait

$$\Lambda_{X^0}^{\cdot} = \Lambda_X^{\cdot}|_{X^0}.$$

Le théorème principal est pour tout complexe $\Lambda_X^{\cdot} \langle \log Q \rangle$, le morphisme de complexes (21) induit un isomorphisme en cohomologie, donc pour tout ouvert U de X (en particulier X lui-même), on a

$$H^k\left(U, \Lambda_X^{\cdot} \langle \log Q \rangle\right) \simeq H^k\left(U, j_* \Lambda_{X^0}^{\cdot}\right) \tag{22}$$

$$\simeq H^k\left(U^0, \mathbb{C}\right), \tag{23}$$

où le second isomorphisme (23) est dû à la construction dans le cas général d'un espace analytique complexe donnée plus haut qui permet de calculer la cohomologie de $\mathbb{C}$ sur U^0, comme cohomologie du complexe des sections sur U^0 du complexe de faisceaux $(\Lambda_{X^0}^{\cdot})$ qui est une résolution fine de $\mathbb{C}_{U^0}$.

Le dernier théorème permet de calculer la cohomologie de toute variété algébrique projective ouverte, Z (lisse ou non), parce qu'une telle variété est du type $Z = X^0$ où X est une variété projective.

D'autre part, il est possible de définir dans notre contexte la notion de résidu. Plus précisément, il faut d'abord définir une filtration des complexes $\Lambda_X^{\cdot}\langle\log Q\rangle$ qui induit une filtration de la cohomologie. Le résidu apparaît alors comme un morphisme défini sur les gradués de la cohomologie à valeur des cohomologies de certains sous-espaces $\widetilde{Q}_a$ à croisements normaux. Tout ceci sera expliqué en detail dans [1], [2], [3] et [4].

Références

[1] Ancona, V. and Gaveau, B., The De Rham complex of a reduced analytic space, in: *Contribution to complex analysis and analytic geometry*, Skoda, H. and Trépeau, J.M. (eds.), Vieweg 1994.

[2] _____, Théorèmes de De Rham sur un espace analytique, *Revue Roumaine de Mathématiques pures et appliquées* **38** (1993), 579–594.

[3] _____, *Families of differential forms on complex spaces* (to appear).

[4] _____, *Differential forms on singular complex spaces and mixed Hodge structure* (to appear).

[5] Deligne, P., Theories de Hodge II, III, *Publi. Math. IHES* **40** (1971), 5–58 et **44** (1974), 5–77 .

[6] Griffiths, P.A. and Schmidt, W., Recent developments in Hodge theory: a discussion of techniques and results, *Proceedings of the International Colloquium on Discrete Subgroups of Lie Groups (Bombay 1973)*, Oxford University Press 1975.

[7] Leray, J., Problème de Cauchy III, *Bull. Soc. Math. france* **87** (1985), 81–180.

[8] _____, *Cours de topologie algébrique*, 1940–1945.

[9] _____, L'anneau spectral et l'anneau filtré d'homologie d'un espace localement compact et d'une application continue, *J. Maths Pures et appl.* **39** (1950), 1–139.

[10] Lojasiewicz, S., Triangulation of semi analytic sets, *Ann. Scuola. Norm. Sup. pisa (3)* **18** (1964), 449–474.

[11] Poincaré, H., Sur les résidus des intégrales doubles, *Acta Math.* **9** (1887), 321–380.

[12] Picart, E. and Simart, G., *Théorie des fonctions algébriques de deux variables indépendantes*, Republiées par Chelsea Publishing Compagny (1971) à partir de l'édition originale publiée en 1887 et 1906 par Gauthier-Villars (Paris).

DERIVATION OF EXACT TRIPLES AND LERAY–KOSZUL SPECTRAL SEQUENCES

B. BENDIFFALAH

Contents

1   Introduction 283
2   Derivation of exact triples 285
3   Proofs 287
4   Massey exact couples 291
5   Browder–Eckmann–Hilton theorem 292
6   Exact triangles 295
7   The limit triangle 296
8   Convergence 299

M. de Gosson (ed.), Jean Leray '99 Conference Proceedings, 281-302.
© *2003 Kluwer Academic Publishers.*

1. Introduction

We introduce the new concept of 'exact triple' and show how is attached, a Leray–Koszul spectral sequence generalizing those of Massey exact couples. We generalize also a result of Browder–Eckmann–Hilton, computing the limit of Bockstein spectral sequences. The letter A means a unitary associative algebra over a commutative ring K. Modules, without any supplementary precisions, are left A-modules and morphisms are modules morphisms. Differential modules and differential morphisms deal with the algebra $A[d]/(d^2)$. For us, a Leray–Koszul spectral sequence is a sequence of differential modules $(E^n, d^n)_n$ such that $E^{n+1} = H(E^n, d^n)$ (homology), $n \geq 0$. A limit module, usually denoted by E^∞, is defined; if the spectral sequence is stationary ($\exists n, \forall N \geq n, d^N = 0$), then: $E^n = E^{n+1} = \cdots = E^\infty$.

Spectral sequences have appeared in Algebraic Topology, since the forties, with Jean Leray's works [6], [7] and [8] about the homology of fibre bundles and will be modelled by Jean–Louis Koszul with a vengeance [5]. The first important result being with any doubt the Leray–Serre theorem [11]. Since, we find spectral sequences in all mathematical theories using homological methods [2] and [9]. In most cases, the spectral sequence comes from a judicious filtered differential module $M = M^\circ \supset M' \supset M'' \supset \cdots$. We have then $E^\circ = H(\mathrm{gr}\, M)$ (homology of the 'graded module' $\mathrm{gr}(M) = M^\circ/M' \oplus M'/M'' \oplus \cdots$) and it exists a filtration over $H(M)$ such that $E^\infty = \mathrm{gr}\, H(M)$: i.e., *the terms of the spectral sequence can be considered as obstructions modules for the existence of an isomorphism* $H(\mathrm{gr}\, M) = \mathrm{gr}\, H(M)$. Considerations that were generalized by A. Grothendieck to additive functors between abelian categories [4, VIII Th. 9.3]. An exact couple over a module $\mathbf{E}$ is a special exact triangle of modules $\mathbf{C}$:

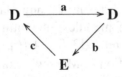

Figure 1.

The exact couples were introduced by Massey [10] in order to study the homotopy type of CW-complexes. Comparatively with Cartan–Eilenberg's exposition [2], this is a more synthetic and conceptual theory which allowed several generalizations [3], [12] and new applications, e.g., [1]. Let us remark that $\mathbf{d} = \mathbf{bc}$ is a differential map for $\mathbf{E}$. The interest of the exact couple $\mathbf{C}$ is its derivability into another exact couple $\mathbf{C}^1$ over $\mathbf{E}^1 = H(\mathbf{E}, \mathbf{d})$:

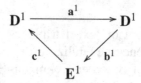

Figure 2.

where $\mathbf{D}^1 = \mathrm{Im}(\mathbf{a})$ and, the morphisms $\mathbf{a}^1$ and $\mathbf{c}^1$ are induced by $\mathbf{a}$ and $\mathbf{c}$; concerning the morphism $\mathbf{b}^1$, it is an application induced by the linear relation $\mathbf{ba}^{-1}$. By iterating the process, one finds a sequence of exact couples $(\mathbf{C}^n)_{n \geq 0}$ but, especially, the *Massey spectral sequence*: $(\mathbf{E}, \mathbf{d}), (\mathbf{E}^1, \mathbf{d}^1), (\mathbf{E}^2, \mathbf{d}^2), \ldots$. A construction of Massey shows that any spectral sequence associated to a filtration, derives from a certain exact couple.

In [1], W. Browder used Massey's exact couples to obtain a functorial modelization for higher Bockstein operators. His main lemma is the explicit computation, for any prime p, of the limit module of the Bockstein's spectral sequence $(E^n)_n$ associated to a differential group C (so $E^\circ = H(C)$). He found [1, Pr. 1.1]:

$$E^\infty = \frac{H(C)}{TH(C)} \otimes \frac{\mathbf{Z}}{p\mathbf{Z}}, \tag{1.1}$$

where $TH(C)$ denotes the torsion sub-group of $H(C)$. With Massey's theory it is very easy to guess the behavior of a spectral sequence: in fact, the spectral sequence of the above exact couple $\mathbf{C}$ is controlled by its endomorphism $\mathbf{a}$. This is the reason why B. Eckmann and P.J. Hilton extended the definition of exact couples to very general abelian categories [3, Th. 4.16]. Moreover, they generalized the above Browder's result: for any Massey spectral sequence, they gave a little exact sequence of modules integrating the limit module E^∞ and two 'asymptotic modules' U and I ([3, Th. 4.16], [4, Th. 5.3 (5.12), p. 278]).

In this article, a generalization of Massey's exact couples is given by 'exact triples', i.e., exact triangles with the following form:

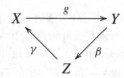

Figure 3.

where t is a fixed central element of A. We are going to expose the following results:

— In sections 2 and 3, we explain how to generate a Leray–Koszul spectral sequence from any exact triple $\mathcal{T}$, by 'iterated derivation': $\mathcal{T} \Rightarrow \mathcal{T}' \Rightarrow \mathcal{T}'' \Rightarrow \cdots$.

— Exact couples are special instances of exact triple theory. Even if both possible derivations give different sequences of triangles, the spectral sequence of the exact triple coincides with Massey's one (section 4).

— We introduce a new module called the 'triangular invariant' t which satisfies: *if $t = 0$, then the spectral sequence of T derives from an exact couple* (section 4).

— In section 5, we generalize the Browder–Eckmann–Hilton theorem, giving a little exact sequence involving the limit module $Z^{(\infty)}$ and some avatars of Eckmann–Hilton modules.

— We point out special properties concerning the 'convergence of exact triples', connecting the triangular invariant t and the limit $Z^{(\infty)}$ (sections 6, 7 and 8).

2. Derivation of exact triples

Any $A[t]$-module is defined by a certain couple (X, t_X), where: X is a A-module (the 'underlying A-module' of the $A[t]$-module) and t_X is an endomorphisme of X (corresponding to the action of t). For simplicity, every morphism of $A[t]$-modules $g : (X, t_X) \to (Y, t_Y)$ will be denoted by $g : X \to Y$ and we will denote tg, the morphism $gt_X = t_Y g : X \to Y$.

Definition 2.1. An exact triple $\mathcal{T}$ is a triangle of $A[t]$-morphisms

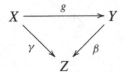

Figure 4.

such that the triangle in diagram 4 is exact.

Definition 2.2. Obviously, the endomorphism $d^\circ = \beta g \gamma$ is a differential map over Z: we will note Z' its homology module. The triangular invariant associated to the triple exact $\mathcal{T}$ is the module

$$\mathbf{t}(\mathcal{T}) = \frac{\mathrm{Ker}(g\gamma)}{\mathrm{Im}(\beta g)}. \tag{2.1}$$

Example. Let **C** be an exact couple. With the notations of section 1 (fig. 1) and these of the Definition 2.1 (fig. 4), we get an exact triple $\mathcal{T}$ with the following construction: first, we convert the A-modules **D** and **E** into $A[t]$-modules X and Z by endowing, respectively, **D** with the endomorphism $t_X = \mathbf{a}$ and **E** with the endomorphism null; then, we define $Y = X$; **b** and **c** are obviously $A[t]$-morphisms β and γ. Finally, by putting $g = 1_X$, we actually obtain $\mathbf{a} = tg$. Besides, since the differential map **d** coincides with d° over $Z = \mathbf{E}$, we obtain: $Z' = \mathbf{E}^1$. One notices that the triangular invariant of $\mathcal{T}$ is zero.

We arrive at the key property of exact triples:

Lemma 2.3. *(Key Lemma) In a natural way, we have a derived exact triple $\mathcal{T}'$:*

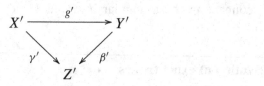

Figure 5.

By iterating, we obtain the 'derived sequence' of exact triples $\mathcal{T}^{(n)}$ $(n \geq 0)$:

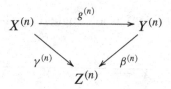

Figure 6.

Theorem 2.4 (Main Theorem). *For any exact triple $\mathcal{T}$, it exists a canonical Leray–Koszul spectral sequence $(Z^{(n)}, d^{(n)})_n$, with $d^{(n)} = \beta^{(n)} g^{(n)} \gamma^{(n)}$ $(n \geq 0)$.*

Corollary 2.5. We have the isomorphisms:

$$Z^{(n)} = \frac{\gamma^{-1}\left(t^n X + \mathrm{Ker}(g)\right)}{\beta\left(\mathrm{Ker}\left(t_Y^n\right) \cap \mathrm{Im}(g)\right)}, \quad (n \geq 0),$$

where the morphisms $\beta^{(n)}$ and $\gamma^{(n)}$ are induced by β and γ.

Corollary 2.6.

$$Z^{(\infty)} = \frac{\bigcap_m \gamma^{-1}(t^m X + \mathrm{Ker}(g))}{\bigcup_n \beta(\mathrm{Ker}(t_Y^n) \cap \mathrm{Im}(g))}.$$

Corollary 2.7. The triangular invariant satisfies: $\forall n \geq 0$, $\mathbf{t}(\mathcal{T}^{(n)}) = \mathbf{t}(\mathcal{T})$.

The following paragraph is dedicated to the proofs of the key lemma, the main theorem and its corollaries.

3. Proofs

Definition 3.1. We define the $A[t]$-morphism $g' : X' \to Y'$, 'derived from g', by composing three morphisms:

$$X' = tX + \mathrm{Ker}(g) \twoheadrightarrow \mathrm{Im}(tg) \xrightarrow{\cong} \frac{X}{\mathrm{Ker}(tg)} \hookrightarrow Y' = \frac{Y}{\mathrm{Ker}(t_Y) \cap \mathrm{Im}(g)}, \quad (3.1)$$

where both ends are induced by g and the middle isomorphism by $(tg)^{-1}$.

Thus the key lemma states: *with the morphism $g' : X' \to Y'$ and these induced by β and γ, β' and γ', the triangle $\mathcal{T}'$ is an exact triple.* Its proof needs the following two lemmas:

Lemma 3.2. *If the exact triple $\mathcal{T}$ verifies $tg = 0$, then we have an exact triangle:*

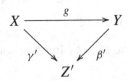

Figure 7.

Proof of Lemma 3.2. Let us precise the homology module $Z' = H(Z, d^\circ)$. Since $tg = 0$, the morphism γ is surjective and the morphism β is injective. We then obtain:

$$Z' = \frac{\mathrm{Ker}(\beta g \gamma)}{\mathrm{Im}(\beta g \gamma)} = \frac{\mathrm{Ker}(g\gamma)}{\mathrm{Im}(\beta g)} = \frac{(\gamma)^{-1}\mathrm{Ker}(g)}{(\beta)\,\mathrm{Im}(g)}.$$

Since $\gamma((\gamma)^{-1}\mathrm{Ker}(g)) = \mathrm{Ker}(g)$ and $\gamma(\beta\,\mathrm{Im}(g)) = 0$, the morphism γ induces a surjective morphism $\gamma_1 : Z' \to \mathrm{Ker}(g)$, whose kernel remains to be precised:

$$\mathrm{Ker}(\gamma_1) = \frac{\mathrm{Ker}(\gamma) \cap (\gamma)^{-1}\,\mathrm{Ker}(g) + (\beta)\,\mathrm{Im}(g)}{(\beta)\,\mathrm{Im}(g)}$$

$$= \frac{\mathrm{Ker}(\gamma)}{(\beta)\,\mathrm{Im}(g)} = \frac{\mathrm{Im}(\beta)}{(\beta)\,\mathrm{Im}(g)} \cong \frac{Y}{\mathrm{Im}(g)};$$

the latest isomorphism is induced by β^{-1}. We find the morphisms γ' and β'. $\square$

The following lemma works without any assumptions on the exact triple $\mathcal{T}$.

Lemma 3.3. *The morphism $\bar{g}$, obtained by the compositions of the three morphisms $\mathrm{Ker}(tg) \hookrightarrow X \xrightarrow{g} Y \twoheadrightarrow \mathrm{Coker}(tg)$, defines an exact sequence:*

$$0 \longrightarrow \mathrm{Ker}\left(tg'\right) \longrightarrow \mathrm{Ker}(tg) \xrightarrow{\theta} \mathrm{Coker}(tg) \longrightarrow \mathrm{Coker}\left(tg'\right) \longrightarrow 0. \quad (3.2)$$

Proof of the Lemma 3.3. We have:

$$\mathrm{Ker}(\bar{g}) = g^{-1}\,\mathrm{Im}(tg) \cap \mathrm{Ker}(tg) = (tX + \mathrm{Ker}(g)) \cap \mathrm{Ker}(tg),$$

$$\mathrm{Coker}(\bar{g}) = \frac{Y}{g\,\mathrm{Ker}(tg) + \mathrm{Im}(tg)} = \frac{Y}{\mathrm{Ker}(t_Y) \cap \mathrm{Im}(g) + \mathrm{Im}(tg)}.$$

On one hand,

$$\mathrm{Ker}\left(tg'\right) = X' \cap \mathrm{Ker}(tg) = \mathrm{Ker}(\bar{g}).$$

On the other hand, the computation

$$\mathrm{Im}\left(tg'\right) = \frac{\mathrm{Im}(tg) + \mathrm{Ker}(t_Y) \cap \mathrm{Im}(g)}{\mathrm{Ker}(t_Y) \cap \mathrm{Im}(g)}$$

implies the identity $\mathrm{Coker}(tg') = \mathrm{Coker}(\bar{g})$. $\square$

Let us prove at present time the key lemma: for that, we have a short exact sequence

$$0 \longrightarrow \mathrm{Coker}(tg) \xrightarrow{\beta_0} Z \xrightarrow{\gamma_0} \mathrm{Ker}(tg) \longrightarrow 0$$

(simply read on the exact triple $\mathcal{T}$) and a morphism $\bar{g} : \mathrm{Ker}(tg) \to \mathrm{Coker}(tg)$ by Lemma 3.3; that is to say: we have an exact triple $\bar{\mathcal{T}}$ with morphism $\bar{g}$ as top edge and module Z as opposite vertex, such that $t\bar{g} = 0$. By noticing the equality of the differential maps over Z, $\beta_0 \bar{g} \gamma_0 = \beta g \gamma$, we find, with the help of the Lemma 3.2, a short exact sequence $0 \to \mathrm{Coker}(\bar{g}) \to Z' \to \mathrm{Ker}(\bar{g}) \to 0$. Therefore, the use of the exact sequence (3.2) provides us a short exact sequence $0 \to \mathrm{Coker}(tg') \to Z' \to \mathrm{Ker}(tg') \to 0$. $\square$

We deduce the existence of a Leray–Koszul spectral sequence: (Z, d), (Z', d'), $(Z'', d''), \ldots$ (Main theorem) whose terms are going to be precised.

Lemma 3.4. *We have:* $X^{(n)} = t^n X + \mathrm{Ker}(g)$, $Y^{(n)} = Y/\mathrm{Ker}(t_Y^n) \cap \mathrm{Im}(g)$
and the morphism $g^{(n)} : X^{(n)} \to Y^{(n)}$ *is defined by the composition of the morphism:*

$$X^{(n)} \xrightarrow{g} \mathrm{Im}\left(t^n g\right) \overset{(t^n g)^{-1}}{\cong} \frac{X}{\mathrm{Ker}\left(t^n g\right)} \overset{g}{\hookrightarrow} Y^{(n)}. \tag{3.3}$$

In particular, for any $n \in \mathbb{N}$, *the map* $g^{(n)}$ *is induced by the relation* $g t^{-n}$.
Moreover, we have some isomorphisms:

$$\mathrm{Ker}\left(g^{(n)}\right) = \mathrm{Ker}(g),$$

$$\mathrm{Im}\left(g^{(n)}\right) \cong \mathrm{Im}\left(t^n g\right) \quad and \tag{3.4}$$

$$\mathrm{Coker}\left(g^{(n)}\right) = \mathrm{Coker}(g).$$

Proof of Lemma 3.4. This is an easy induction: we assume that the lemma is true for some $n \geq 1$. We have:

$$X^{(n+1)} = t X^{(n)} + \mathrm{Ker}(g) = t\left(t^n X + \mathrm{Ker}(g)\right) + \mathrm{Ker}(g)$$
$$= t^{n+1} X + \mathrm{Ker}(g).$$

In the same way:

$$Y^{(n+1)} = \left(Y^{(n)}\right)' = Y^{(n)}/\mathrm{Ker}\left(t_{Y^{(n)}}\right) \cap \mathrm{Im}\left(g^{(n)}\right),$$

where

$$\mathrm{Ker}\left(t_{Y^{(n)}}\right) \cap \mathrm{Im}(g)$$
$$= \frac{\left\{y \in Y \mid t y \in \mathrm{Ker}\left(t_Y^n\right) \cap \mathrm{Im}(g)\right\} \cap \mathrm{Im}(g) + \mathrm{Ker}\left(t_Y^n\right) \cap \mathrm{Im}(g)}{\mathrm{Ker}\left(t_Y\right) \cap \mathrm{Im}\left(g^{(n)}\right)}$$
$$= \frac{\mathrm{Ker}\left(t_Y^{n+1}\right) \cap \mathrm{Im}(g)}{\mathrm{Ker}\left(t_Y\right) \cap \mathrm{Im}\left(g^{(n)}\right)}.$$

The map $g^{(n+1)} = (g^{(n)})' : X^{(n+1)} \to Y^{(n+1)}$ is obtained by the composition of the following:

$$X^{(n+1)} \xrightarrow{g^n} \mathrm{Im}\left(t g^{(n)}\right) \overset{(t g^{(n)})^{-1}}{\cong} \frac{X^{(n)}}{\mathrm{Ker}\left(t g^{(n)}\right)} \xrightarrow{g^{(n)}} Y^{(n+1)}. \tag{3.5}$$

By induction, $\mathrm{Ker}(tg^{(n)}) = \mathrm{Ker}(tg) \cap X^{(n)}$, thus the middle isomorphism of (3.5) is isomorphic to the isomorphism coming from the compositions:

$$\mathrm{Im}\left(t^{n+1}g\right) \cong \mathrm{Im}\left(tg^{(n)}\right) \overset{(tg^{(n)})^{-1}}{\cong} \frac{X^{(n)}}{\mathrm{Ker}\left(tg^{(n)}\right)}$$

$$\cong \frac{\mathrm{Ker}(tg) + t^n X}{\mathrm{Ker}(tg)} \overset{t^{-n}}{\cong} \frac{X}{\mathrm{Ker}\left(t^{n+1}g\right)}.$$

Moreover, with this isomorphism both morphisms of (3.5) induced by $g^{(n)}$, that is to say by the relation $t^{-n}g$ (induction), become morphisms induced by g. $\square$

Lemma 3.5. *We have canonical isomorphisms:*

$$Z^{(n)} = \frac{\gamma^{-1}\left(t^n X + \mathrm{Ker}(g)\right)}{\beta\left(\mathrm{Ker}\left(t_Y^n\right) \cap \mathrm{Im}(g)\right)}, \quad (n \in \mathbb{N})$$

and the morphisms $\beta^{(n)}$ and $\gamma^{(n)}$ are induced by β and γ.

Proof of Lemma 3.5. There, we proceed by induction too: we assume that the lemma is true for a fixed integer $n \geq 0$. We have over $Z^{(n)}$ the differential map $d^{(n)} = \beta^{(n)} g^{(n)} \gamma^{(n)}$:

$$\forall z \in \gamma^{-1} X^{(n)}, {}^{(n)}\left(z \bmod \beta\left(\mathrm{Ker}\left(t_Y^n\right) \cap \mathrm{Im}(g)\right)\right)$$
$$= \beta g^{(n)} \gamma(z) \bmod \beta\left(\mathrm{Ker}\left(t_Y^n\right) \cap \mathrm{Im}(g)\right).$$

One deduces its image:

$$\mathrm{Im}\left(d^{(n)}\right) = \beta g^{(n)}\left(X^{(n)} \cap \mathrm{Ker}(gt)\right) \bmod \beta\left(\mathrm{Ker}\left(t_Y^n\right) \cap \mathrm{Im}(g)\right)$$

$$= \frac{\beta\left(\mathrm{Ker}\left(t_Y^{n+1}\right) \cap \mathrm{Im}(g)\right)}{\beta\left(\mathrm{Ker}\left(t_Y^n\right) \cap \mathrm{Im}(g)\right)}$$

$$\mathrm{Im}\left(d^{(n)}\right) = \beta g^{(n)}\left(X^{(n)} \cap \mathrm{Ker}(gt)\right) \bmod \beta\left(\mathrm{Ker}\left(t_Y^n\right) \cap \mathrm{Im}(g)\right)$$

$$= \beta \mathrm{Ker}\left(t_Y^{n+1}\right) \cap \mathrm{Im}(g)/\beta\left(\mathrm{Ker}\left(t_Y^n\right) \cap \Im(g)\right)$$

as its kernel:

$$\mathrm{Ker}\left(d^{(n)}\right) = \left\{ \begin{matrix} z \in \gamma^{-1}\left(X^{(n)}\right) \bmod \beta\left(\mathrm{Ker}\left(t_Y^n\right) \cap \mathrm{Im}(g)\right) \\ \beta g^{(n)} \gamma(z) \in \beta\left(\mathrm{Ker}\left(t_Y^n\right) \cap \mathrm{Im}(g)\right) \end{matrix} \right\}$$

we have actually the equivalences:

$$\beta g^{(n)} \gamma(z) \in \beta\left(\mathrm{Ker}\left(t_Y^n\right) \cap \mathrm{Im}(g)\right) \iff \gamma(z) \in t^{n+1} X + \mathrm{Ker}(g)$$

so we find:

$$\mathrm{Ker}\left(d^{(n)}\right) = \gamma^{-1}\left(t^{n+1}X + \mathrm{Ker}(g)\right)/\beta\left(\mathrm{Ker}\left(t_Y^n\right)\cap\mathrm{Im}(g)\right).$$

Concerning the morphisms $\gamma^{(n+1)}$ and $\beta^{(n+1)}$ respectively induced by $\gamma^{(n)}$ and $\beta^{(n)}$ (by construction), we see that they are respectively induced by γ et β (by induction). $\qquad\square$

We immediately deduce the limit of the spectral sequence of $\mathcal{T}$ (Corollary 2.6). Let us finally show the Corollary 2.7. By induction, the following computation is sufficient:

$$\mathbf{t}\left(\mathcal{T}'\right) = \frac{\mathrm{Ker}\left(g'\gamma'\right)}{\mathrm{Im}\left(\beta'g'\right)} = \frac{\gamma^{-1}\,\mathrm{Ker}(g)\cap\mathrm{Ker}(\beta g\gamma)}{\beta\,\mathrm{Im}(g)+\mathrm{Im}(\beta g\gamma)} = \frac{\gamma^{-1}\,\mathrm{Ker}(g)}{\beta\,\mathrm{Im}(g)} = \mathbf{t}(\mathcal{T}). \qquad\square$$

4. Massey exact couples

Let us give an exact couple $\mathbf{C}$ with the notations of section 1 (fig. 1) following the example of the same section, we get an exact triple $\mathcal{T}$.

Proposition 4.1. The spectral sequence of the exact triple $\mathcal{T}$ coincides with the Massey spectral sequence of the exact couple $\mathbf{C}$. In particular, both spectral sequences have same limit: $Z^{(\infty)} = E^{\infty}$.

Proof. The Massey spectral sequence $(E^n, d^n)_n$ of the exact couple $\mathbf{C}$ is fully described in the article of B. Eckmann and P.J. Hilton [3] (in the more conceptual frame of abelian categories; see also [4, Th. 1.2, p. 259]):

$$\forall n \geq 0, \quad E^n = \frac{\mathbf{c}^{-1}\,\mathrm{Im}\,(\mathbf{a}^n)}{\mathbf{b}\,\mathrm{Ker}\,(\mathbf{a}^n)}, \qquad d^n = \mathbf{ba}^{-n}\mathbf{c}. \tag{4.1}$$

By replacing $\mathbf{a}$, $\mathbf{b}$ and $\mathbf{c}$ by tg, β and γ, respectively, we have:

$$\forall n \geq 0, \quad E^n = \frac{\gamma^{-1}\left(t^n X\right)}{\beta\left(\mathrm{Ker}\left(t_X^n\right)\right)} = \frac{\gamma^{-1}\left(t^n X + \mathrm{Ker}(g)\right)}{\beta\left(\mathrm{Ker}\left(t_X^n\right)\cap\mathrm{Im}(g)\right)} \quad (g = 1_X); \tag{4.2}$$

we conclude with the Lemma 3.4: $\forall n \geq 0$, $E^n = Z^{(n)}$, $d^n = d^{(n)}$. $\qquad\square$

Conversely, we have [12]:

Proposition 4.2. The spectral sequence of any exact triple derives from a certain Whitehead semi-exact couple.

Construction: With the usual notations, let us consider an exact triple $\mathcal{T}$. We successively define:

$\mathbf{E} = Z$; the product by t induces an endomorphism a of $\mathbf{D} = \mathrm{Im}(g)$; the composition $g\gamma$ defines a morphism $\mathbf{c} : \mathbf{E} \to \mathbf{D}$; the morphism β restricts itself in a morphism $\mathbf{b} : \mathbf{D} \to \mathbf{E}$. Since we have that

$$\mathrm{Im}(\mathbf{c}) = g\,\mathrm{Im}(\gamma) = g(\mathrm{Ker}(tg)) = \mathrm{Im}(g) \cap \mathrm{Ker}(t_Y) = \mathrm{Ker}(\mathbf{a})$$

and

$$\mathrm{Ker}(\mathbf{b}) = \mathrm{Ker}(\beta) \cap \mathrm{Im}(g) = \mathrm{Im}(tg) \cap \mathrm{Im}(g) = \mathrm{Im}(tg)\,\mathrm{Im}(\mathbf{a}),$$

we deduce an exact sequence $\mathbf{E} \overset{\mathbf{c}}{\to} \mathbf{D} \overset{\mathbf{a}}{\to} \mathbf{D} \overset{\mathbf{b}}{\to} \mathbf{E}$, which withdraws into itself in a Whitehead semi-exact couple $\mathbf{C}$. To show that the spectral sequence of $\mathcal{T}$ is the Massey–Whitehead spectral sequence of $\mathbf{C}$, it suffices, by induction, to insure that the semi-exact couple $\mathbf{C}'$ associated to the exact triple $\mathcal{T}'$ by the same construction is isomorphic to the derived semi-exact couple $\mathbf{C}^1$. We already have, by construction, $\mathbf{E}' = \mathbf{E}^1$. Following (3.5), the product by t induces an isomorphism $f : \mathrm{Im}(g') \to \mathrm{Im}(tg)$: from there, one obtains an isomorphism $\mathbf{f} : \mathbf{C}' \to \mathbf{C}^1$, if one shows the commutativity of the three square diagrams. Checking this is easy. As an example, we have the following diagram for the top square (all the arrows are induced by t):

$$
\begin{array}{ccc}
\mathbf{D}' & \overset{f}{\longrightarrow} & \mathbf{D}^1 \\
{\scriptstyle a'}\downarrow & & \downarrow{\scriptstyle a^1} \\
\mathbf{D} & \overset{f}{\longrightarrow} & \mathbf{D}
\end{array}
$$

Proposition 4.3. If the triangular invariant of an exact triple $\mathcal{T}$ is zero, then it exists a Massey exact couple whose spectral sequence is that of $\mathcal{T}$. Conversely, if it exists an exact couple $\mathbf{C}$ like in fig. 1 with the same spectral sequence as $\mathcal{T}$ (and such that $D = a^{+\infty}D \oplus a_{-\infty}D$), then $\mathbf{t}(\mathcal{T}) = 0$.

Proof. Effectively, since the triangular invariant being zero, by going back to the above construction, we have $\mathrm{Ker}(\mathbf{c}) = \mathrm{Ker}(g\gamma) = \mathrm{Im}(\beta g) = \mathrm{Im}(\mathbf{b})$. Let us remark that $\mathbf{C}'$ is again an exact couple thanks to the Corollary 3 of §2 which asserts that $\mathbf{t}(\mathcal{T}') = 0$. $\qquad\square$

5. Browder–Eckmann–Hilton theorem

We denote by $A[[t]]$, the K-algebra of series in the variable t with coefficients in A. Let us recall that, for any $A[t]$-module U, we have a 'complete tensor product' and a 'continuous' Hom functor:

$$A[[t]] \underset{A[t]}{\hat{\otimes}} U = \lim_{\overleftarrow{m}} \frac{A[[t]]}{(t^m)} A[[t]] \underset{A[t]}{\otimes} U,$$

$$\mathrm{Hom}_{A[t]}(A[[t]], U) = \lim_{\overrightarrow{m}} \mathrm{Hom}_{A[t]} \left(\frac{A[[t]]}{(t^m)}, U \right).$$

Definition 5.1. The morphism g induces a morphism $g^{(\infty)} : X^{(\infty)} \to Y^{(\infty)}$ where

$$X^{(\infty)} = \mathrm{Ker} \begin{cases} X \longrightarrow A[[t]] \underset{A[t]}{\hat{\otimes}} \mathrm{Im}(g) \\ x \longmapsto (1 \otimes g(x))_{m \geq 0} \end{cases},$$

$$Y^{(\infty)} = \mathrm{Coker} \begin{cases} \mathrm{Hom}_{A[t]}(A[[t]], \mathrm{Im}(g)) \longrightarrow Y \\ \varphi \longmapsto \varphi(1) \end{cases}.$$

The following theorem generalizes the result on exact couples, of Browder–Eckmann–Hilton [1, 3]. We use the following notations for any $A[t]$-module U:

$$t^{+\infty}U = \bigcap_{n \geq 0} t^n U, \qquad t_{-n}U = \mathrm{Ker}\,(t_U^n) \qquad \text{and} \qquad t_{-\infty}U = \bigcup_{n \geq 0} t_{-n}U.$$

Theorem 5.2. *We have a triangular complex (called 'limit triangle') $\mathcal{T}^{(\infty)}$:*

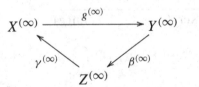

Figure 8.

where $\beta^{(\infty)}$ and $\gamma^{(\infty)}$ are induced by β et γ. It is exact at $Z^{(\infty)}$ and the morphism tg induces some isomorphisms:

$$\mathrm{Ker}\left(\beta^{(\infty)}\right) = X/g^{-1} = (t_{-\infty} \mathrm{Im}(g))$$

and

$$\mathrm{Coker}\left(\gamma^{(\infty)}\right) = t\left(t^{+\infty} \mathrm{Im}(g)\right).$$

Remark 5.3. The example of Eckmann–Hilton [3, 6.11] shows that, in general, the identity $tt^{+\infty} \mathrm{Im}(g) = t^{+\infty} \mathrm{Im}(g)$ is false.

Proof Theorem 5.2. We have:

$$\lim_{\leftarrow} \frac{A[[t]]}{(t^m)} \underset{A[t]}{\otimes} \mathrm{Im}(g) \subset \prod_{m=0}^{+\infty} \frac{A[[t]]}{(t^m)} \underset{A[t]}{\otimes} \mathrm{Im}(g)$$

$$= \prod_{m=0}^{+\infty} \frac{A[t]}{(t^m)} \underset{A[t]}{\otimes} \mathrm{Im}(g) = \prod_{m=0}^{+\infty} \frac{\mathrm{Im}(g)}{\mathrm{Im}\,(t^m g)};$$

from that there follows the expression:

$$X^{(\infty)} = \bigcap_{m=0}^{+\infty} (t^m X + \mathrm{Ker}(g)) = \bigcap_{m=0}^{+\infty} X^{(m)} = \lim_{\overleftarrow{m}} \left(X^{(m)} \longrightarrow X \right).$$

In the same way we have:

$$\lim_{\rightarrow} \mathrm{Hom}_{A[t]} \left(\frac{A[[t]]}{(t^m)}, \mathrm{Im}(g) \right) = \sum_{m=0}^{+\infty} \mathrm{Hom}_{A[t]} \left(\frac{A[t]}{(t^m)}, \mathrm{Im}(g) \right) \subset \mathrm{Im}(g).$$

One deduces

$$Y^{(\infty)} = \frac{Y}{\mathrm{Im}(g) \cap \bigcup_{n=0}^{+\infty} \mathrm{Ker}\,(t_Y^n)} = \lim_{\overrightarrow{m}} \left(Y \longrightarrow Y^{(m)} \right).$$

Let us consider, at present time, the spectral sequence of $\mathcal{T}$: its limit $Z^{(\infty)}$ is given by the main theorem (Theorem 2.4). We obtain a short exact sequence (with morphisms deduced from β and γ):

$$0 \longrightarrow \frac{Y}{\mathrm{Ker}(\beta) + t_{-\infty} Y \cap \mathrm{Im}(g)}$$

$$\longrightarrow Z^{\infty} \longrightarrow \mathrm{Im}(\gamma) \cap \bigcap_{m=0}^{+\infty} (t^m X + \mathrm{Ker}(g)) \longrightarrow 0.$$

Since $\mathrm{Ker}(\beta) = \mathrm{Im}(tg)$ and $\mathrm{Im}(\gamma) = \mathrm{Ker}(tg)$, we deduce an exact sequence

$$Y^{(\infty)} \xrightarrow{\beta^{(\infty)}} Z^{(\infty)} \xrightarrow{\gamma^{(\infty)}} X^{(\infty)},$$

with, on the one hand, an isomorphism

$$\mathrm{Ker}\left(\beta^{(\infty)} \right) \xleftarrow{\cong} \frac{X}{g^{-1} t_{-\infty}\,\mathrm{Im}(g)}$$

induced from the map tg and, on the other hand,

$$tg\left(X^{(\infty)} \right) = tg \bigcap_{m=0}^{+\infty} \left(g^{-1} g t^n X \right) = tg g^{-1} \bigcap_{m=0}^{+\infty} \left(t^n\,\mathrm{Im}(g) \right) = t\left(t^{+\infty}\,\mathrm{Im}(g) \right),$$

so we get an isomorphism induced by tg : $\mathrm{Coker}(\gamma^{(\infty)}) \cong t(t^{+\infty}\,\mathrm{Im}(g))$. □

Remark 5.4. For any $n \in \mathbb{N}$, it is easy to verify that the spectral sequence of $\mathcal{T}$ is stationary from the rank n if and only if $X^{(n)} \cap \mathrm{Ker}(tg) = X^{(n+1)} \cap \mathrm{Ker}(tg) = X^{(n+2)} \cap \mathrm{Ker}(tg) = \cdots$ or, in an equivalent way, if and only if $\mathrm{Im}(t^n g) \cap \mathrm{Ker}(t_Y) = \mathrm{Im}(t^{n+1} g) \cap \mathrm{Ker}(t_Y) = \mathrm{Im}(t^{n+2} g) \cap \mathrm{Ker}(t_Y) = \cdots$. For example, if the filtration on Y by the sequence of modules $(Im(t^m g))_m$ stops from the rank n, then it is the same for the spectral sequence.

Corollary 5.5. If for some $n \geq 0$ we have $t^n g = 0$, then the spectral sequence of $\mathcal{T}$ is stationary and the limit is the triangular invariant: $Z^{(\infty)} = \mathbf{t}(\mathcal{T})$.

Proof. It is obvious:

$$Z^{(n)} = \frac{\gamma^{-1}(\mathrm{Ker}(g))}{\beta(\mathrm{Im}(g))} = \frac{\mathrm{Ker}(g\gamma)}{\mathrm{Im}(\beta g)} = \mathbf{t}(\mathcal{T}).$$

$\square$

6. Exact triangles

In this section, we study the special case of an exact triple that is an exact triangle; this study is motivated by the Theorem 7.1. We keep the previous notations.

Definition 6.1. The morphism g is said to be 'exact' if t induces an automorphism of $\mathrm{Im}(g)$.

That means: $\mathrm{Ker}(tg) = \mathrm{Ker}(g)$ and $\mathrm{Im}(tg) = \mathrm{Im}(g)$, i.e., the exact triple $\mathcal{T}$ is an exact triangle. In particular, the differential map $d^\circ = \beta g \gamma$ is null and $Z' = Z$. Since the morphism g' is induced by the relation $t^{-1} g$, we have the commutative diagram:

$$
\begin{array}{ccc}
tX + \mathrm{Ker}(g) & \xrightarrow{\;tg'\;} & \dfrac{Y}{\mathrm{Im}(g) \cap \mathrm{Ker}(t_Y)} \\[2mm]
\downarrow & & \uparrow \\[2mm]
tX + \mathrm{Ker}(tg) & \xrightarrow{\;(tg)'\;} & \dfrac{Y}{\mathrm{Im}(tg) \cap \mathrm{Ker}(t_Y)} \\[2mm]
\downarrow & & \uparrow \\[2mm]
X & \xrightarrow{\;\;g\;\;} & Y
\end{array}
\qquad (6.1)
$$

where the vertical morphisms are inclusions and projections. We deduce that the morphism g is exact if and only if $g = tg'$; in such a case, we have: $(tg)' = g = tg'$.

Lemma 6.2. *If g is exact then $t^n g^{(m)}$ are exact for all $n \geq 0, m \geq 0$.*

In particular, the spectral sequence of $\mathcal{T}$ is stationary and we have the identities:

$$\forall m, \forall n, \quad \left(t^m g\right)^{(n)} = t^m g^{(n)} = \begin{cases} t^{m-n} & \text{if } m \geq n \\ g^{(n-m)} & \text{otherwise.} \end{cases} \tag{6.2}$$

Proof. We assume that g is exact. By induction, it suffices to prove that both tg and g' are exact. The exactness of tg is easy to see:

$$\mathrm{Ker}\left(t^2 g\right) = t^{-1}\,\mathrm{Ker}(tg) = t^{-1}\,\mathrm{Ker}(g) = \mathrm{Ker}(tg),$$

$$\mathrm{Im}\left(t^2 g\right) = t\,\mathrm{Im}(tg) = t\,\mathrm{Im}(g) = \mathrm{Im}(tg).$$

Otherwise we have $\mathrm{Ker}(g') = \mathrm{Ker}(g)$ and $\mathrm{Ker}(tg') = \mathrm{Ker}(g)$. Hence $\mathrm{Ker}(g') = \mathrm{Ker}(tg')$. We also have $\mathrm{Im}(tg') = \mathrm{Im}(g)$. Now, $\mathrm{Im}(g') = \mathrm{Im}(g)$ $(\mathrm{Ker}(t_Y) \cap \mathrm{Im}(g) = 0$ by exactness of g). Therefore $\mathrm{Im}(tg') = \mathrm{Im}(g')$ and g' is exact. $\square$

Theorem 6.3. *If the morphism $g^{(n)}$ is exact, for some integer n, then the spectral sequence of $\mathcal{T}$ is stationary up the rank n.*

We will show in section 8, that the limit $Z^{(\infty)}$ of the spectral sequence of $\mathcal{T}$ is the triangular invariant $\mathbf{t}(\mathcal{T})$.

Remark 6.4. It is useful to remark that the exactness is connected to Fitting's property:

$$\forall n \in \mathbb{N}, \quad g^{(n)} \text{ est exact} \iff \mathrm{Im}(g) = t^n\,\mathrm{Im}(g) \oplus t_{-n}\,\mathrm{Im}(g). \tag{6.3}$$

Corollary 6.5. *If the module $\mathrm{Im}(g)$ is noetherian, then the spectral sequence of $\mathcal{T}$ is stationary (and the limit module is the triangular invariant).*

We rely on the following result for non noetherian cases.

Theorem 6.6. *Let $P(t) = p_\omega t^\omega + p_{\omega+1} t^{\omega+1} + \cdots + p_d t^d \in A[t]$ be a non zero polynomial such that its valuative term p_ω is invertible. If $P(t)g = 0$ then $g^{(\omega)}$ is exact and the spectral sequence of $\mathcal{T}$ is stationary up the rank $\omega = \mathrm{val}(P)$.*

7. The limit triangle

We use the notations of the previous sections. Here, we wish to study the exactness property of the limit triangle $\mathcal{T}^{(\infty)}$.

Lemma 7.1. *The triangle* $\mathcal{T}^{(\infty)}$ *is an exact triple if and only if*

$$t^{+\infty} \operatorname{Im}(g) \cap t_{-\infty} \operatorname{Im}(g) \subset \operatorname{Ker}(t_Y)$$

and

$$\operatorname{Im}(g) + \operatorname{Ker}(t_Y) = t^{+\infty} \operatorname{Im}(g) + t_{-\infty} \operatorname{Im}(tg) + \operatorname{Ker}(t_Y).$$

The lemma is written in the more compact form:

$$\operatorname{Im}(tg) = tt^{+\infty} \operatorname{Im}(g) \oplus t_{-\infty} \operatorname{Im}(tg). \tag{7.1}$$

Proof. Since $\operatorname{Ker}(tg^{(\infty)}) = \{x \in X^{(\infty)} \mid x \in g^{-1}t_{-\infty} \operatorname{Im}(g)\} = X^{(\infty)} \cap g^{-1}t_{-\infty} \operatorname{Im}(g)$, we have $\operatorname{Ker}(tg^{(\infty)}) = g^{-1}(t^{+\infty} \operatorname{Im}(g) \cap t_{-\infty} \operatorname{Im}(g))$ and, applying g, we find that the condition $\operatorname{Ker}(tg^{(\infty)}) = \operatorname{Im}(\gamma^{(\infty)})$ is equivalent to the first of the lemma. In the same way, since

$$\operatorname{Im}\left(tg^{(\infty)}\right) = t\left(t^{+\infty} \operatorname{Im}(g)\right) + t_{-\infty} \operatorname{Im}(g)/t_{-\infty} \operatorname{Im}(g)$$

we obtain:

$$\operatorname{Ker}\left(\beta^{(\infty)}\right) \ = \ \operatorname{Im}\left(tg^{(\infty)}\right)$$
$$\Longleftrightarrow \operatorname{Im}(tg) + t_{-\infty} \operatorname{Im}(g) = t\left(t^{+\infty} \operatorname{Im}(g)\right) + t_{-\infty} \operatorname{Im}(g)$$
$$\Longleftrightarrow \operatorname{Im}(tg) = \operatorname{Im}(tg) \cap \left(tt^{+\infty} \operatorname{Im}(g) + t_{-\infty} \operatorname{Im}(g)\right)$$
$$\Longleftrightarrow \operatorname{Im}(tg) = t\left(t^{+\infty} \operatorname{Im}(g)\right) + \operatorname{Im}(tg) \cap t_{-\infty} \operatorname{Im}(g)$$
$$\Longleftrightarrow t \operatorname{Im}(g) = t\left(t^{+\infty} \operatorname{Im}(g)\right) + t\left(t_{-\infty} \operatorname{Im}(g)\right)$$

hence an equivalence with the second condition of the Lemma 7.1. ☐

Lemma 7.2. *If* $\mathcal{T}^{(\infty)}$ *is an exact triple then:*

$$t^n t^{+\infty} \operatorname{Im}(g) = tt^{+\infty} \operatorname{Im}(g) \quad \text{for any } n \geq 1.$$

Proof. By the previous lemma, it is sufficient to verify that the module $t^n t^{+\infty} \operatorname{Im}(g) \oplus t_{-\infty} \operatorname{Im}(tg)$ contains the submodule $\operatorname{Im}(tg)$. As a matter of fact:

$$t^n t^{+\infty} \operatorname{Im}(g) \oplus t_{-\infty} \operatorname{Im}(tg) = t^{n-1}(\operatorname{Im}(tg)) + t_{-\infty} \operatorname{Im}(tg) \supset \operatorname{Im}(tg).$$

☐

Lemma 7.3. *If* $\mathcal{T}^{(\infty)}$ *is an exact triple then:* $t^{+\infty} \operatorname{Im}(g) = tt^{+\infty} \operatorname{Im}(g) \oplus t^{+\infty}t_{-\infty} \operatorname{Im}(g).$

Proof. From the previous lemmas, we have:

$$\forall n \geq 1, \quad t^n \operatorname{Im}(g) = tt^{+\infty} \operatorname{Im}(g) \oplus t^n t_{-\infty} \operatorname{Im}(g).$$

Then, we take the intersection over the set of indices $\{n \geq 1\}$. ☐

Theorem 7.4. *If $\mathcal{T}^{(\infty)}$ is an exact triple then it is an exact triangle.*

In particular (cf. §3), the spectral sequence of $\mathcal{T}^{(\infty)}$ is immediately stationary:

$$Z^{(\infty)} = \left(Z^{(\infty)}\right)' = \left(Z^{(\infty)}\right)'' = \left(Z^{(\infty)}\right)''' = \cdots = \left(Z^{(\infty)}\right)^{(\infty)}. \quad (7.2)$$

Moreover we have the identities: $g^{(\infty)} = t(g^{(\infty)})' = t^2(g^{(\infty)})'' = t^3(g^{(\infty)})''' = \cdots$.

Proof of Theorem 7.4. On the one hand:

$$\forall n \geq 0, \quad \mathrm{Ker}\left(t^n g^{(\infty)}\right) = \left\{x \in X^{(\infty)} \mid x \in g^{-1}t_{-\infty}\,\mathrm{Im}(g)\right\} = \mathrm{Ker}\left(g^{(\infty)}\right).$$

Let us show, one the other hand, that $\mathrm{Im}(tg^{(\infty)}) = \mathrm{Im}(g^{(\infty)})$, e.g., $tt^{+\infty}\,\mathrm{Im}(g) + t_{-\infty}\,\mathrm{Im}(g) = t^{+\infty}\,\mathrm{Im}(g) + t_{-\infty}\,\mathrm{Im}(g)$. Indeed, from the lemma above, we have:

$$\begin{aligned}
tt^{+\infty}\,\mathrm{Im}(g) + t_{-\infty}\,\mathrm{Im}(g) &= tt^{+\infty}\,\mathrm{Im}(g) + t^{+\infty}t_{-\infty}\,\mathrm{Im}(g) + t_{-\infty}\,\mathrm{Im}(g) \\
&= t^{+\infty}\,\mathrm{Im}(g) + t_{-\infty}\,\mathrm{Im}(g).
\end{aligned}$$

$\square$

Theorem 7.5. *If $\mathcal{T}^{(\infty)}$ is an exact triple then, for any $n \geq 0$, $(\mathcal{T}^{(n)})^{(\infty)}$ is an exact triple too and $(\mathcal{T}^{(n)})^{(\infty)} = (\mathcal{T}^{(\infty)})^{(n)}$.*

With the Theorem 7.4, one deduces the

Corollary 7.6. *If $\mathcal{T}^{(\infty)}$ is an exact triple then, for all $n \geq 0$, the exact triple $(\mathcal{T}^{(n)})^{(\infty)}$ is an exact triangle, $(g^{(n)})^{(\infty)} = t^n g^{(\infty)}$ and $(Z^{(n)}_\bullet)^{(\infty)} = Z^{(\infty)}$.*

Proof of Theorem 7.5. In order to show that each triangle $(\mathcal{T}^{(n)})^{(\infty)}$ $(n \geq 1)$ is an exact triple, we have to verify the following formulas:

$$\forall n \geq 1, \quad \mathrm{Im}\left(tg^{(n)}\right) = tt^{+\infty}\,\mathrm{Im}\left(g^{(n)}\right) \oplus t_{-\infty}\,\mathrm{Im}\left(tg^{(n)}\right). \quad (7.3)$$

These are nothing but the identity (7.2), simply transposed for the morphisms $g^{(n)}$. Starting with (7.1), we have:

$$\forall n \geq 1, \quad \mathrm{Im}\left(t^{n+1}g\right) = t^{n+1}t^{+\infty}\,\mathrm{Im}(g) \oplus t^n t_{-\infty}\,\mathrm{Im}(tg)$$

that is to say, by Lemma 7.2:

$$\forall n \geq 1, \quad \mathrm{Im}\left(t^{n+1}g\right) = tt^{+\infty}\,\mathrm{Im}(g) \oplus t^n t_{-\infty}\,\mathrm{Im}(tg).$$

Recall that for any $A[t]$-module U we have the two following properties $(n \geq 0)$: $t^{+\infty}(t^n U) = t^{+\infty}U$ and $t^n t_{-\infty}U = t_{-\infty}(t^n U)$. So we obtain:

$$\mathrm{Im}\left(t^{n+1}g\right) = tt^{+\infty}\,\mathrm{Im}\left(t^n g\right) \oplus t_{-\infty}\,\mathrm{Im}\left(t^{n+1}g\right), \quad (n \geq 1)$$

which is exactly (7.3), since we have isomorphisms $\text{Im}(g^{(n)}) \cong \text{Im}(t^n g)$ for any $n \geq 0$.

Let us show that $(\mathcal{T}^{(n)})^{(\infty)} = (\mathcal{T}^{(n)})^{(\infty)}$ for any $n \geq 0$: since $(g^{(n)})^{(\infty)}$ is induced by $g^{(n)}$, the point is to verify that $(X^{(n)})^{(\infty)} = (X^{(\infty)})^{(n)}$ and that $(Y^{(n)})^{(\infty)} = (Y^{(\infty)})^{(n)}$. That is to say, by Theorem 7.4, that $(X^{(n)})^{(\infty)} = X^{(\infty)}$ and that $(Y^{(n)})^{(\infty)} = Y^{(\infty)}$. Since $\text{Ker}(g^{(n)}) = \text{Ker}(g)$ for all $n \geq 0$, we finally get:

$$\left(X^{(n)}\right)^{(\infty)} = \bigcap_m \left(t^{m+n} X + t^m \text{Ker}(g) + \text{Ker}(g)\right) = X^{(\infty)}.$$

In the same way, we could show the identities $(Y^{(n)})^{(\infty)} = Y^{(\infty)}$ $(n \geq 0)$, by using the canonical isomorphism $\text{Coker}(g^{(n)}) = \text{Coker}(g)$. □

Remark 7.7. If $\mathcal{T}^{(\infty)}$ is an exact triple then $\text{Ker}(g) \subset \text{Ker}(g^{(\infty)}) \subset \text{Ker}(tg)$ and $\text{Coker}(g^{(\infty)}) = \text{Coker}(g)$.

8. Convergence

This paragraph is devoted to the interpretation of the limit $Z^{(\infty)}$ of the spectral sequence of $\mathcal{T}$: under a 'convergence assumption' (Definition 8.4), we find an isomorphism $Z^{(\infty)} = \mathbf{t}(\mathcal{T})$ (Theorem 8.5, Corollary 8.6). We conclude with convergence criterions (Corollaries 8.7 and 8.8).

Definition 8.1. The modules $t^{+\infty} \text{Im}(g)$ and $\text{Im}(g)/t_{-\infty} \text{Im}(g)$ are called 'Eckmann–Hilton modules' of $\mathcal{T}$. These are nearly the modules I et U of [3]: in the particular case of an exact couple, $I = t^{+\infty} \text{Im}(g)$ but $U = t_{-\infty} \text{Im}(g)$.

We call 'asymptotic morphism' the morphism induced by the identity $\mathbf{a}(\mathcal{T})$: $t^{+\infty} \text{Im}(g) \to \text{Im}(g)/t_{-\infty} \text{Im}(g)$; its image is $\text{Im}\,\mathbf{a}(\mathcal{T}) = \text{Im}(g^{(\infty)})$.

Proposition 8.2. We have a canonical isomorphism $Z^{(\infty)} = \mathbf{t}(\mathcal{T})$ if and only if the scalar t induces some automorphisms in the two Eckmann–Hilton modules.

Proof. From $Z^{(\infty)} = \gamma^{-1}(g^{-1}t^{+\infty} \text{Im}(g))/\beta(t_{-\infty} \text{Im}(g))$ (section 2) we get the equivalence:

$$Z^{(\infty)} = \mathbf{t}(\mathcal{T}) \Longleftrightarrow \text{Ker}(g\gamma) = \gamma^{-1}\left(g^{-1}t^{+\infty} \text{Im}(g)\right)$$

and $\text{Im}(\beta g) = \beta(t_{-\infty} \text{Im}(g))$. On the one hand, we have:

$$\text{Ker}(g\gamma) = \gamma^{-1}\left(g^{-1}t^{+\infty} \text{Im}(g)\right) \Longleftrightarrow 0 = t^{+\infty} \text{Im}(g) \cap \text{Ker}(t\gamma)$$

i.e., t induces a monomorphism in $t^{+\infty} \operatorname{Im}(g)$. But then, t commutes with intersections and also induces a surjection:

$$tt^{+\infty} \operatorname{Im}(g) = t^{+\infty} t \operatorname{Im}(g) = t^{+\infty} \operatorname{Im}(g).$$

On the other hand, the sequence of equivalences

$$\operatorname{Im}(\beta g) = \beta t_{-\infty} \operatorname{Im}(g) \iff \operatorname{Im}(g) = t_{-\infty} \operatorname{Im}(g) + \operatorname{Im}(tg)$$

$$\iff t \frac{\operatorname{Im}(g)}{t_{-\infty} \operatorname{Im}(g)} = \frac{\operatorname{Im}(g)}{t_{-\infty} \operatorname{Im}(g)}$$

shows that $\operatorname{Im}(\beta g) = \beta t_{-\infty} \operatorname{Im}(g)$ if and only if the scalar t induces an epimorphism in $\operatorname{Im}(g)/t_{-\infty} \operatorname{Im}(g)$, i.e., an automorphism since it is already injective.   □

Proposition 8.3. If $Z^{(\infty)} = \mathbf{t}(\mathcal{T})$, then the morphism $g^{(\infty)}$ is exact and $\operatorname{Ker}(g^{(\infty)}) = \operatorname{Ker}(g)$. Moreover, we have an isomorphism: $\operatorname{Im}(g^{(\infty)}) \cong t^{+\infty} \operatorname{Im}(g)$.

Proof. We always have

$$\operatorname{Ker}\left(tg^{(\infty)}\right) = \operatorname{Ker}\left(g^{(\infty)}\right) = g^{-1}\left(t^{+\infty} \operatorname{Im}(g) \cap t_{-\infty} \operatorname{Im}(g)\right).$$

If $Z^{(\infty)} = \mathbf{t}(\mathcal{T})$, then $t^{+\infty} \operatorname{Im}(g) \cap t_{-\infty} \operatorname{Im}(g) = 0$. From there,

$$\operatorname{Ker}\left(g^{(\infty)}\right) = \operatorname{Ker}(g)$$

and

$$\operatorname{Im}\left(g^{(\infty)}\right) = t^{+\infty} \operatorname{Im}(g) \oplus t_{-\infty} \operatorname{Im}(g)/t_{-\infty} \operatorname{Im}(g) \cong t^{+\infty} \operatorname{Im}(g).$$

Moreover, the identity $tt^{+\infty} \operatorname{Im}(g) + t_{-\infty} \operatorname{Im}(g) = t^{+\infty} \operatorname{Im}(g) + t_{-\infty} \operatorname{Im}(g)$ implies that $\operatorname{Im}(tg^{(\infty)}) = \operatorname{Im}(g^{(\infty)})$.   □

In short, we have two implications:
— $\mathcal{T}^\infty$ is an exact triple $\implies$ the asymptotic morphism is surjective (Remark 7.7.);
— $Z^{(\infty)} = \mathbf{t}(\mathcal{T}) \implies$ the asymptotic morphism is injective (Proposition 8.3.);
In both cases, the morphism $g^{(\infty)}$ is exact and t induces an automorphism in $\operatorname{Im} g^{(\infty)}$.

Definition 8.4. The exact triple $\mathcal{T}$ is said 'convergent to $\mathcal{T}^{(\infty)}$' if the Eckmann–Hilton modules are isomorphic via $\mathbf{a}(\mathcal{T})$.

If $\mathcal{T}$ is convergent to $\mathcal{T}^{(\infty)}$, we have the isomorphisms of Propositions 8.2 and 8.3. Let us remark, besides, the equivalence of the following assertions:

(i)   the exact triple $\mathcal{T}$ converges to $\mathcal{T}^{(\infty)}$;

(ii) $\mathrm{Ker}(g^{(\infty)}) = \mathrm{Ker}(g)$ and $\mathrm{Coker}(g^{(\infty)}) = \mathrm{Coker}(g)$;

(iii) $\mathrm{Im}(g) = t^{+\infty}\,\mathrm{Im}(g) \oplus t_{-\infty}\,\mathrm{Im}(g)$.

By using the isomorphism $\mathrm{Im}(g^{(n)}) \cong t^n\,\mathrm{Im}(g)$ $(n \geq 0)$, we see that, if the exact triple $\mathcal{T}$ converge to $\mathcal{T}^{(\infty)}$, then the exact triple $\mathcal{T}^{(n)}$ converges to $(\mathcal{T}^{(\infty)})^{(n)}$.

Theorem 8.5. *The exact triple $\mathcal{T}$ is convergent to $\mathcal{T}^{(\infty)}$ if and only if $\mathcal{T}^{(\infty)}$ is an exact triple and* $Z^{(\infty)} = \mathbf{t}(\mathcal{T})$.

Proof. If the exact triple $\mathcal{T}$ is convergent to $\mathcal{T}^{(\infty)}$, then the $(i) \Longrightarrow (iii)$ above immediately implies the formula (7.1) and $\mathcal{T}^{(\infty)}$ is an exact triple. Moreover, since $t^{+\infty}\,\mathrm{Im}(g) \cap t_{-\infty}\,\mathrm{Im}(g) = 0$, we have $t^{+\infty}\,\mathrm{Im}(g) \cap \mathrm{Ker}(t_Y) = 0$ and t induces an automorphism in $t^{+\infty}\,\mathrm{Im}(g)$ (Proof of Proposition 8.2): $tt^{+\infty}\,\mathrm{Im}(g) = t^{+\infty}\,\mathrm{Im}(g)$; the scalar t also induces an automorphism of the other Eckmann–Hilton module:

$$t\,\mathrm{Im}(g) + t_{-\infty}\,\mathrm{Im}(g) = tt^{+\infty}\,\mathrm{Im}(g) \oplus t_{-\infty}\,\mathrm{Im}(g)$$
$$= t^{+\infty}\,\mathrm{Im}(g) \oplus t_{-\infty}\,\mathrm{Im}(g) = \mathrm{Im}(g),$$

i.e.,

$$t\,(\mathrm{Im}(g)/t_{-\infty}\,\mathrm{Im}(g)) = \mathrm{Im}(g)/t_{-\infty}\,\mathrm{Im}(g).$$

We conclude with $Z^{(\infty)} = \mathbf{t}(\mathcal{T})$.
The converse is more immediate: if $\mathcal{T}^{(\infty)}$ is an exact triple, then we have

$$\mathrm{Coker}\left(g^{(\infty)}\right) = \mathrm{Coker}(g)$$

by the Remark 7.7. And, if $Z^{(\infty)} = \mathbf{t}(\mathcal{T})$, then $\mathrm{Ker}(g^{(\infty)}) = \mathrm{Ker}(g)$ from Proposition 8.3. We end the proof with $(ii) \Longrightarrow (i)$ above: the exact triple $\mathcal{T}$ converges to $\mathcal{T}^{(\infty)}$. $\qquad\square$

Corollary 8.6. *If the exact triple $\mathcal{T}$ converges to $\mathcal{T}^{(\infty)}$, then we have a short exact sequence:* $0 \to \mathrm{Coker}(g) \to Z^{(\infty)} \to \mathrm{Ker}(g) \to 0$.

In particular, in case of convergence, we have: *if g is an isomorphism (resp. a monomorphism, an epimorphism), then:* $Z^{(\infty)} = 0$ *(resp. $Z^{(\infty)} \cong \mathrm{Coker}\,g$, $Z^{(\infty)} \cong \mathrm{Ker}\,g$)*. This purpose may be applied to Massey exact couples: with the Proposition 4.1, the Corollary 8.6 is reduced to a vanishing criterion for the limit $\mathbf{E}^\infty$ of the Massey spectral sequence $(\mathbf{E}^n)_n$:

$$\mathbf{D} = t^{+\infty}\mathbf{D} \oplus t_{-\infty}\mathbf{D} \Longrightarrow \mathbf{E}^\infty = 0. \qquad (8.1)$$

Second particular case: if $t^{+\infty} \operatorname{Im}(g) = 0$ and $t_{-\infty} \operatorname{Im}(g) = \operatorname{Im}(g)$, then $\mathcal{T}$ is convergent to $\mathcal{T}^{(\infty)}$, $Z^{(\infty)} = \mathbf{t}(\mathcal{T})$ and $g^{(\infty)}$ is the zero morphism $\operatorname{Ker}(g) \to \operatorname{Coker}(g)$.

Finally, if the module $\operatorname{Im}(g)$ is Noetherian, we know by Fitting's lemma that: $\operatorname{Im}(g) = t^n \operatorname{Im}(g) \oplus t_{-n} \operatorname{Im}(g)$ for some $n \in \mathbb{N}$ big enough; this decomposition implies the convergence of $\mathcal{T}$ and we have:

Corollary 8.7. If $\operatorname{Im}(g)$ is noetherian, then the spectral sequence of $\mathcal{T}$ is stationary and its limit $Z^{(\infty)}$ is the triangular invariant $\mathbf{t}(\mathcal{T})$.

We also have the following criterion (cf. Theorem 6.6):

Corollary 8.8. Let $P(t) = p_\omega t^\omega + p_{\omega+1} t^{\omega+1} + \cdots + p_d t^d \in A[t]$ be a non zero polynomial such that its valuative term p_ω is invertible and $P(t)g = 0$. Then, $\mathcal{T}$ is convergent to $\mathcal{T}^{(\infty)}$ and, in particular, $Z^{(\infty)} = \mathbf{t}(\mathcal{T})$.

References

[1] Browder, W., Torsion in H-spaces, *Annals of Math.* **74** n° 1 (1961), 24–51.

[2] Cartan, H. and Eilenberg, S., *Homological algebra*, Princeton Mathematical Series, Princeton university press 1956.

[3] Eckmann, B. and Hilton, P.J., Exact couples in an abelian category, *Journal of Algebra* **3** (1966), 38–87.

[4] Hilton, P.J. and Stammbach, U., *A course in homological algebra*, Graduate Texts in Math. 2nd edition, Springer–Verlag New-York 1997.

[5] Koszul, J.L., Sur les opérateurs de dérivation dans un anneau, *C.R.A.S. Paris* 28 Juillet (1947), 217–219.

[6] Leray, J., L'anneau d'homologie d'une représentation, *C.R.A.S. Paris* 12 Juin (1946), 1366–1368.

[7] _____, Structure de l'anneau d'homologie d'une représentation, *C.R.A.S. Paris* 17 Juin (1946), 1419–1422.

[8] _____, L'homologie d'un espace fibré dont la fibre est connexe, *Jour. Math. Pures Appl.* **29** (1950), 1–139.

[9] Lyndon, R.C., The cohomology theory of groups extensions, *Duke Math. J.* **15** (1948), 271–292.

[10] Massey, W.S., Exacts couples in algebraic topology, *Annals of Math.* **56** n° 2 September (1952), 363–396.

[11] Serre, J.P., Homologie singulière des espaces fibrés. Applications, *Annals of Math.* **54** (1951), 425–505.

[12] Whitehead, J.H.C., The G-dual of a semi-exact couple, *Proc. London Math. Soc.* (3) **3** (1953), 385–416.

PART IV

ELLIPTIC OPERATORS; INDEX THEORY

LE NOYAU DE LA CHALEUR DES OPÉRATEURS SOUS-ELLIPTIQUES DES GROUPES D'HEISENBERG

RICHARD BEALS, BERNARD GAVEAU, AND PETER GREINER

À la memoire de Jean Leray

Contents

1   L'opérateur sous-elliptique du groupe d'Heisenberg 307
2   Motivations 307
3   La difficulté fondamentale: Absence d'uniformité et de régularité
    des asymptotiques 309
4   Les groupes d'Heisenberg généraux 311
5   Dérivation du noyau de la chaleur 312
6   Intégration de l'équation d'Hamilton–Jacobi (5.3) 313
7   Solution fondamentale $p_u(x, t)$ 313
8   Bicaractéristiques classiques et problème de contrôle associé 314
    8.1 1er cas: $x'' \neq 0$ 315
    8.2 2ème cas: $x'' = 0$ 315
9   Asymptotique, pour les temps petits, du noyau de la chaleur 316
10 Estimation uniforme de $p_u(x, t)$ 317
11 Extension au cas des variétés de Heisenberg 318
12 Extension au cas des variétés de Heisenberg 318

M. de Gosson (ed.), Jean Leray '99 Conference Proceedings, 305-320.
© *2003 Kluwer Academic Publishers.*

Résumé. *Nous décrivons le noyau de la chaleur, son estimation asymptotique pour temps petit et son estimation uniforme dans le cas des groupes d'Heisenberg quelconques.*

1. L'opérateur sous-elliptique du groupe d'Heisenberg

Étant donnés trois variables réelles (x_1, x_2, t), considérons les champs de vecteurs

$$X_1 = \frac{\partial}{\partial x_1} + 2x_2\frac{\partial}{\partial t}, \qquad X_2 = \frac{\partial}{\partial x_2} - 2x_1\frac{\partial}{\partial t}$$

et l'opérateur du second ordre sous-elliptique

$$\Delta = \frac{1}{2}\left(X_1^2 + X_2^2\right). \tag{1.1}$$

Comme le crochet $[X_1, X_2] = -4\frac{\partial}{\partial t}$, les champs X_1, X_2, $[X_1, X_2]$ engendrent l'espace vectoriel $\mathbb{R}^3$ en tout point et il résulte de [15] que l'opérateur Δ est hypoelliptique, c'est à dire, que si Δf est indéfiniment dérivable, il en est de même de f.

Soit u une variable positive. Nous allons considérer l'équation de la chaleur associée à Δ,

$$\frac{\partial}{\partial u} = \Delta, \tag{1.2}$$

ainsi que ses généralisations à des groupes d'Heisenberg plus généraux. L'étude de l'opérateur Δ et de ses généralisations a fait l'objet de nombreux travaux ([1], [5], [6], [10], [11], [12], [14], [16], [20]).

La solution fondamentale de l'équation de la chaleur été obtenue par une méthode probabiliste dans [12], [13], par la théorie des représentations dans [17] et par le calcul de Laguerre dans [8].

2. Motivations

Avant d'entreprendre l'exposé des résultats les plus récents sur le sujet, expliqués en detail dans [5], il peut être utile de rappeler quelques raisons pour étudier les opérateurs tels que ceux figurant dans les définitions (1.1) et (1.2).

(i) Ces opérateurs sont les opérateurs sous-elliptiques, non elliptiques les plus simples.

La fonction de Green de (1.1) a été trouvée par Folland [10] et elle est égale à

$$g(x, t) = C\left(\left(x_1^2 + x_2^2\right)^2 + t^2\right)^{-1/2}.$$

C'est d'ailleurs la fonction la plus simple homogène de degré -2 pour les dilatations

$$(x, t) \longrightarrow \left(\lambda x, \lambda^2 t\right).$$

Le noyau de la chaleur, [8], [12], [13], [17], est donnée par une intégrale non explicitement calculable

$$p_u(x, t) = \frac{1}{(2\pi u)^2} \int_{-\infty}^{+\infty} \exp\left(\frac{i\tau t - \tau |x|^2 \coth 2\tau}{u}\right) \frac{2\tau}{sh 2\tau} d\tau. \quad (2.1)$$

Le noyau de la chaleur donne le calcul symbolique complet de Δ.

(ii) Les opérateurs Δ et leurs généralisations sont des opérateurs invariants sur les groupes non Abéliens les plus simples, ici le groupe d'Heisenberg. X_1, X_2 sont les champs invariants à gauche.

(iii) Ce sont aussi les opérateurs sous-elliptiques qui sont associés aux équations de Cauchy–Riemann tangentielles des frontières strictement pseudo-convexes de $\mathbb{C}^n$ [1], [11].

Ici $\mathbb{R}^3$ est plongé dans $\mathbb{C}^2$ par

$$\text{Im } w = |z|^2 \quad z = x_1 + ix_2, \text{ Re } w = t$$

et l'opérateur $\overline{Z} = \frac{1}{2}(X_1 - i X_r)$ est l'opérateur de Hans Lewy non résoluble.

En général, un bord strictement pseudo-convexe est ponctuellement un groupe d'Heisenberg de la même façon qu'une variété Riemannienne est ponctuellement Euclidienne.

(iv) La théorie des opérateurs pseudo-différentiels associés à de tels opérateurs sous-elliptiques est assez délicate et peu explicite [8].

(v) Ces opérateurs interviennent dans l'étude des diffusions dégénérées en théorie des processus stochastiques. Un autre exemple fameux est l'opérateur sous-elliptique associé à une diffusion de Langevin [12], [13], [18].

(vi) L'saymptotique pour temps petits de $p_u(x, t)$ n'est pas uniforme dans aucun voisinage de la source, ce qui conduit à des difficultés importantes [12], [13].

(vii) Ces opérateurs sont associés à une géométrie dite sous-Riemannienne. Les principaux concepts de cette géométrie ont été introduites dans [12], [13], pour l'étude de l'opérateur Δ du groupe d'Heisenberg et généralisés dans [14] à une classe d'opérateurs sous-elliptiques sur des fibrés.

Ces opérateurs sont aussi liés à la théorie du contrôle optimal. Par exemple, les bicaractéristiques de l'opérateur (1.1) ont leur projection en (x_1, x_2) qui minimisent la longeur d'arc pour une aire balayée par le rayon vecteur fixée. Le type de phénomène est général [14], par exemple, pour les opérateurs horizontaux associés à une connexion, [9], [11].

3. La difficulté fondamentale: Absence d'uniformité et de régularité des asymptotiques

Reprenons l'équation (1.2). L'idée générale serait d'étudier $p_u(x, t)$, solution fondamentale de (1.2), par un développement asymptotique pour u petit du type

$$V(x, t, u) \exp\left(-S_u(x, t)\right).$$

Si l'on introduit un tel développement dans l'équation, on découvre comme d'habitude que S doit satisfaire une équation d'Hamilton Jacobi

$$\frac{\partial S}{\partial u} + H(\nabla S, x) = 0, \tag{3.1}$$

où $H(\xi_1, \theta; x, t)$ est le symbole de Δ

$$H = \frac{1}{2}\left((\xi_1, 2x_2\theta)^2 + (\xi_2, 2x_1\theta)^2\right) \tag{3.2}$$

ξ, θ, ètant les moments conjuguées de x, t. Cependant, un calcul direct [12], [13], montre que $S(x, t, u)$ (qui est l'action le long des bicaractéristiques issues de 0 et arrivant en (x, t) à l'instant u) est une fonction continue, mais non différentiable dans aucun voisinage de 0 (elle n'est pas C^1 sur l'axe des t). De plus le préfacteur V, que l'on peut calculer directement par l'asymptotique de la formule exacte (2.1) a le comportement suivant pour u petit

$$V(x, t, u) \sim \begin{cases} u^{-3/2} & \text{si } |x| \neq 0, \\ u^{-2} & \text{si } |x| = 0 \end{cases} \tag{3.3}$$

(voir [12], [13]).

Nous en déduisons donc que

1^0) l'asymptotique de $p_u(x, t)$ pour $\to 0^+$ n'est pas uniforme dans aucun voisinage de 0,

2^0) lorsque nous calculons le noyau de Green $g(x, t)$ par la formule, nous ne pouvons par échanger $p_u(x, t)$ et son asymptotique $\exp(-S(x, t/u))$ et donc, le noyau de Green n'est pas $1/S|_{u=1}$, puisque d'une part nous savons qu'il est $(|x|^4 + t^2)^{-1/2}$, ce qui n'a rien voir avec S et d'autre part le fait que S ne soit pas C^1 sur l'axe t et l'étude précise de son comportement montre que

$$\Delta S \backsim \delta(x_1)\delta(x_2)$$

ce qui ne peut en aucun cas reconstituer la fonction de Dirac en $x_1 = x_2 = t = 0$, qui est $\delta(x_1)\delta(x_2)\delta(t)$.

La comparaison avec le cas elliptique est elle-aussi très instructive: si Δ est un Laplacien Riemannien, le noyau de la chaleur de

$$\frac{\partial}{\partial u} = \Delta$$

a son comportement asymptotique en

$$p_u(x) \backsim \frac{1}{u^{n/2}} e^{-d^2/2u}$$

où d est la distance Riemannienne de x à l'origine, source de chaleur $p_u(x)$ et n est la dimension et ce comportement asymptotique est uniforme pour $|x|$ voisin de 0. La non uniformité ne s'introduira qu'aux points conjugués de l'origine le long des géodésiques issues de l'origine, mais une boule assez petite de centre l'origine ne contient aucun point conjugué.

Cela implique aussi que d^2 est fonction C^∞ près de l'origine. Enfin la fonction de Green

$$g(x) = \int_0^{+\infty} p_u(x)du \backsim \frac{1}{d^{n-2}}$$

de sorte que le comportement de g près de l'origine est obtenu en échangeant $p_u(x)$ et son asymptotique pour u petit, ce qui est naturel vu l'uniformité de ce comportement asymptotique pour $|x|$ petit. C'est pour toutes ces raisons que, dans le cas elliptique, les développements de Hadamard et, ce qui n'est pas équivalent, les développements asymptotiques pseudo-différentiels, donnent des solutions fondamentales exactes en analytique ou des paramétrix modulo des termes infiniment régularisants, dans le contexte C^∞.

Toutes ces constructions échouent, précisèment pour ces raisons dans le cas sous-elliptique, parce que les concepts sous-Riemanniens pris au pied de la lettre, ne sont pas adaptés pour trouver les solutions fondamentales où les paramétrix des cas sous-elliptiques. La géométerie sous-Riemannienne est une extension trop naïve du cas Riemannien. Le paradoxe est qu'elle fournit des estimations asymptotiques correctes, mais non uniformes.

Notre problème qui fut posé dès l'article [12] (2), puis dans [8], fut d'étendre le formalisme au cas des variétés afin de produire leur noyau de la chaleur. La solution de ce problème ne pouvait pas reposer sur des techniques probabilistes ou de représentations de groupes, tandis que les méthodes pseudo-différentielles restaient trop lourdes pour être effectives. Dans le cas le plus simple du noyau de Green, ce problème fut résolu dans [2]. C'est la solution de ce problème, pour le cas du noyau de la chaleur, que nous esquisserons ici.

4. Les groupes d'Heisenberg généraux

Nous définissons un groupe d'Heisenberg H_n comme l'ensemble des points $(x, t) \in \mathbb{R}^{2n} \times \mathbb{R}$ avec la loi de groupe

$$(x, t) * (x', t') = \left(x + x', t + t' + \sum_{j=1}^{n} 2a_j \left(x_{2j} x'_{2j-1} - x'_{2j} x_{2j-1} \right) \right)$$

où les a_j sont des nombres positifs que nous ordonnerons selon $0 < a_1 \leq a_2 \leq \cdots \leq a_n$. Les constantes a_j sont importantes pour le cas des variétés qui sont strictement pseudo-convexes où elles varient de point en point. Ce sont les valeurs de la forme de Lévi. L'espace H_n est un groupe de Lie nilpotent d'ordre 2. Les champs invariants à gauche sont

$$X_{2j-1} = \frac{\partial}{\partial x_{2j-1}} + 2a_j x_{2j} \frac{\partial}{\partial t}$$

$$X_{2j-1} = \frac{\partial}{\partial x_{2j}} - 2a_j x_{2j-1} \frac{\partial}{\partial t}$$

et l'opérateur

$$\Delta = \frac{1}{2} \sum_{j=1}^{n} \left(X_{2j-1}^2 + X_{2j}^2 \right) \tag{4.1}$$

est sous-elliptique, hypoelliptique de symbole

$$H(\xi, \theta; x, t) = \frac{1}{2} \sum_{j=1}^{n} \left(\zeta_{2j-1}^2 + \zeta_{2j}^2 \right) \tag{4.2}$$

où ζ_{2j-1}, ζ_{2j} sont les symboles de X_{2j-1}, X_{2j}

$$\zeta_{2j-1} = \xi_{2j-1} + 2a_j x_{2j} \theta, \qquad \zeta_{2j} = \xi_{2j} - 2a_j x_{2j-1} \theta$$

Δ étant invariant à gauche, il suffit d'étudier la solution fondamentale de l'équation de la chaleur $\partial/\partial u = \Delta$ de source en 0. Notons $p_u(x, t)$ cette solution fondamentale.

5. Dérivation du noyau de la chaleur

Nous essayons une formule

$$p_u(x, t) = \frac{1}{u^{n+1}} \int \exp\left(-\frac{1}{u} f(x, t; \tau)\right) V(x, t, \tau) d\tau \qquad (5.1)$$

où τ est un paramètre supplémentaire, la puissance de u en facteur étant destinée à rendre compte de l'homogénéité. Un calcul formel nous donne le résultat suivant

$$\left(\Delta - \frac{\partial}{\partial u}\right)\left(\frac{e^{-f/u} V}{u^{n+1}}\right) = \frac{e^{-f/u}}{u^{n+3}}[H(\nabla f, x, t) - f]$$

$$+ \frac{e^{-f/u}}{u^{n+2}}[-(Xf, XV) - (\Delta f)V + (n+1)V] \quad (5.2)$$

$$+ \frac{e^{-f/u}V}{u^{n+1}} \Delta V.$$

Imposons que f satisfasse l'équation d'Hamilton–Jacobi non standard qui suit:

$$H(\nabla f, x, t) = f - \tau \frac{\partial f}{\partial \tau}. \qquad (5.3)$$

Cela réduit l'équation (5.2) à

$$\left(\Delta - \frac{\partial}{\partial u}\right)\left(\frac{e^{-f/u} V}{u^{n+1}}\right) = \frac{e^{-f/u}}{u^{n+2}}\left[\tau \frac{\partial V}{\partial \tau} + (Xf, XV) + (\Delta f)V - nV\right]$$

$$+ \frac{e^{-f/u}\Delta V}{u^{n+1}} + \frac{\partial}{\partial \tau}\left(\frac{\tau V e^{-f/u}}{u^{n+2}}\right). \qquad (5.4)$$

Imposons que V satisfasse l'équation de transport du premier ordre

$$\tau \frac{\partial V}{\partial \tau} + (XF, XV) + (\Delta f) - nV = 0. \qquad (5.5)$$

Nous obtenons alors

$$\left(\Delta - \frac{\partial}{\partial u}\right) p_u(x, V) = \int e^{-f/u} \frac{\Delta V}{u^{n+1}} d\tau \qquad (5.6)$$

pourvu qu'il ait été vérifié que

$$\int_{-\infty}^{+\infty} \frac{\partial}{\partial \tau}\left(\frac{\tau V e^{-f/u}}{u^{n+2}}\right) d\tau = 0. \qquad (5.7)$$

6. Intégration de l'équation d'Hamilton–Jacobi (5.3)

La redéfinition $f = \tau g$ réduit cette équation à l'équation standard

$$H(\nabla g, x, t) + \frac{\partial g}{\partial \tau} = 0$$

que nous résoudrons par la méthode standard des bicaractéristiques $(x(s), t(s), \xi(s), t(s))$ mais avec les conditions au bord non standard

$$\begin{cases} x(0) = 0, & \theta(0) = -i\tau, \\ x(1) = x, & t(1) = t \end{cases} \tag{6.1}$$

(τ réel). C'est pourquoi nous nommerons *bicaractéristiques non standard* celles qui satisfont les conditions au bord non standard (6.1). *Donc nous n'imposons rien pour $t(0)$, en particulier, nous ne voulons pas $t(0) = 0$, mais nous imposons une valeur imaginaire pure du moment conjugué θ de t qui reste d'ailleurs constant dans le mouvement bicaractéristique puisque l'Hamiltonien H, symbole principal de Δ, ne dépend pas explicitement de t.*

Ce système Hamiltonien des bicaractéristiques peut être explicitement résolu et son action $g(x, t, \tau)$ calculée par

$$g(x, t, \tau) = \theta(0) t(0) + \int_0^1 \left[(\xi, \dot{x}) + \theta \dot{t} - H \right] ds$$

et fournit donc f

$$\begin{cases} f(x, t, \tau) = i\tau t + \sum_{j=1}^n a_j \tau \coth\left(2 a_j \tau\right) r_j^2, \\ r_j^2 = x_{2j-1}^2 + x_{2j}^2. \end{cases} \tag{6.2}$$

7. Solution fondamentale $p_u(x, t)$

Clairement Δf, f étant donnée par l'equation (6.2), est indépendante de (x, t). L'équation de transport (5.5) peut être alors résolue par une fonction V indépendante de (x, t), pourvu que $V = V(\tau)$ satisfasse

$$\tau \frac{\partial V}{\partial t} + (\Delta f - n) V = 0$$

dont la solution immédiate est

$$V(\tau) = C \prod_{j=1}^n \left(\frac{2 a_j \tau}{sh 2 a_j \tau} \right),$$

où C est une constante de normalisation. Mais alors $\Delta V \equiv 0$ et de plus (5.7) est vrai de sorte que la solution fondamentale $p_u(x, t)$ est donnée par

$$p_u(x, t) = \frac{1}{(2\pi u)^{n+1}} \int_{-\infty}^{+\infty} \exp\left(-\frac{f(x, t, \tau)}{u}\right) V(\tau) d\tau \qquad (7.1)$$

la constante $(2\pi u)^{-(n+1)}$ étant telle que $p_u(x, t) \to \delta$ si $u \to 0^+$. (Voir [1] pour des généralisations tout groupe nilpotent d'ordre 2.)

8. Bicaractéristiques classiques et problème de contrôle associé

L'estimation d'une intégrale telle que (7.1) pour u petit fait intervenir les points critiques de la phase, donc la solution en τ de l'équation $\partial f / \partial \tau = 0$ (pour x, t fixés).

Il est facile de voir que ces points critiques, notés $\tau_c(x, t)$ sont tels que $\tau(0) = 0$, et sont donc pour ces valeurs $\tau = \tau_c(x, t)$ que la bicaractéristique correspondante joint le point $(0, 0)$ au point (x, t) en le temps unité donc que les bicaractéristiques satisfont les conditions au bord

$$\begin{cases} x(0) = 0, & t(0) = 0, \\ x(1) = x, & t(1) = t. \end{cases} \qquad (8.1)$$

Nous nommerons standard de telles bicaractéristiques.

La fonction suivante, qui est essentiellement $\partial f / \partial \tau$

$$\mu(\varphi) = \frac{2\varphi - \sin 2\varphi}{2 \sin^2 \varphi} \qquad (8.2)$$

jouera le rôle fondamental, parce qu'elle permet de classifier les bicaractéristiques joignant $(0, 0)$ à (x, t) en un temps unité, et parce qu'elle permet la calculer leur moment initial $\theta(0)$ par l'équation

$$t = \sum_{j=1}^{n} a_j \mu\left(2a_j \theta(0)\right) r_j^2. \qquad (8.3)$$

La fonction $\mu(\varphi)$ a ses pôles en les points $m\pi$, m entier $\neq 0$ et dans chaque intervalle $]m, \pi, (m+1)\pi[$, elle a un unique minimum, les valeurs minima allant en croissant lorsque m croît.

La fonction $n = 1$ ou lorsque les a_j sont tous égaux, nous voyons d'après l'équation (8.3) que lorsque $t / \sum r_j^2$ croît, le nombre de bicaractéristiques standard joignant 0 a (x, t) croît et que si $x = 0, t \neq 0$, il y a un nombre infini de bicaractéristiques.

Plus précisément chaque fois que le rapport $t / \sum r_j^2$ franchit certains seuils critiques, deux bicaractéristiques standard nouvelles apparaissent.

Dans le cas général, supposons que

$$0 < a_1 \leq \cdots \leq a_p < a_{p+1} = \cdots = a_n \qquad (8.4)$$

et écrivons aussi

$$x = (x', x''), \qquad x' = (x_1, \ldots, x_{2p}), \qquad x'' = (x_{2p+1}, \ldots, x_{2n}). \qquad (8.5)$$

Remarquons que la fonction f (6.2) a ses pôles sur l'axe imaginaire, le pôle le plus voisin de l'origine de partie imaginaire positive étant $\pi/2a_n$, d'où la nécessité de distinguer les a_k qui sont égaux à a_n dans la formule (8.4). D'où aussi la discussion que voici:

8.1. 1er CAS: $x'' \neq 0$

(i) f a un point critiqué noté $i\theta_c(x, t)$ unique dans l'intervalle $[0, i\pi/2a_n[$. Ce point critique correspond à la géodésique d'action minimale joignant 0 au point (x, t) en un temps unité

(ii) lorsque $|x''|$ tend vers 0, d'autres points critiques apparaissent par couples et la fonction f à un nombre fini de points critiques sur l'axe imaginaire qui correspondent à des géodésiques non minimisantes pour l'action. De plus, θ_c tend vers $\pi/2a_n$.

8.2. 2ème CAS: $x'' = 0$

Nommons E_p la fonction que voici:

$$E_p(x, t) = t - \sum_{j=1}^{p} a_j \mu \left(\frac{2\pi a_j}{2a_n} \right) r_j^2. \qquad (8.6)$$

Cette fonction est obtenue en considérant l'équation (8.3), en y remplaçant $\theta(0)$ par $\theta_c(x, t)$ dans le second membre, $\pi/2a_n$ étant la valeur maximale de $\theta_c(x, t)$ vers laquelle tend θ_c lorsque $|x''| \to 0$.

(i) Supposons $x'' = 0$, $E_p(x, t) < 0$.

Il existe alors un unique $i\theta_c(x, t)$ avec $0 < \theta_c < \pi/2a_n$ solution de

$$t = \sum_{j=1}^{p} a_j \mu \left(2a_j \theta_c \right) r_j^2. \qquad (8.7)$$

Ce θ_c correspond à la géodésique d'action minimale.

(ii) Supposons $E_p(x, t) \geq 0$.

Choisissons $\xi_{2p+1}(0), \ldots, \xi_{2n}(0)$ de sorte que

$$\frac{1}{2} \left| \xi''(0) \right|^2 = \frac{\pi}{2a_n} E_p(x, t). \tag{8.8}$$

On peut alors construire une géodésique d'action minimale et dont le moment conjugué initial $\xi''(0)$ satisfait (8.8). Il y a évidemment une infinité de telles géodésiques et les points (x, t) tels que $x'' = 0$, $E_p(x, t) > 0$ sont caustiques de 0. Evidemment, il est possible aussi de trouver une géodésique allant de 0 à (x, t) avec $\xi''(0) = 0$. Une telle géodésique satisfait $x''(s) = 0$ pour $0 \leq s \leq 1$ et elle reste donc dans le sous-groupe d'Heisenberg

$$H_p = \left\{ (x, t) \in H_n / x'' = 0 \right\}. \tag{8.9}$$

A cause de cette contrainte supplémentaire, une telle géodésique ne minimise plus l'action.

(iii) *Dans tous les cas, l'action minimale est donnée par la formule*

$$S(x, t) = 2\theta_c(x, t) \left[t + \sum_{j=1}^{n} a_j \cot g \left(2a_j \theta_c(x, t) \right) r_j^2 \right] \tag{8.10}$$

(dans le cas où $x'' = 0$ et $E_p > 0$, on prendra $\theta_c = \pi/2a_n$ dans la formule précédente).

La fonction S est donc continue partout.

9. Asymptotique, pour les temps petits, du noyau de la chaleur

Les notations sont définies au secion 8.

Désignons par $C_p = \{ (x, t) \in H_p / E_p(x, t) > 0 \}$ où H_p désigne le sous-groupe d'Heisenberg défini en (8.9).

(i) $x'' \neq 0$:

Soit $\theta_c(x, t)$ l'unique solution dans $[0, \pi/2a_n[$ de l'équation

$$|t| = \sum_1^n a_j \mu \left(2a_j \theta_c \right) r_j^2.$$

Alors

$$p_u(x, t) \sim \frac{1}{(2\pi u)^{n+1/2}} \Theta(x, t) \exp \left(-\frac{S(x, t)}{u} \right) \tag{9.1}$$

où

$$\Theta(x, t) = \frac{1}{(f''(x, t, i\theta_c(x, t)))^{1/2}} \prod_{j=1}^{n} \left(\frac{2a_j\theta_c}{\sin 2a_j\theta_c} \right) \qquad (9.2)$$

(ii) $x'' = 0 \ (x, t) \notin C_p$:

L'estimation asymptotique (9.1) a encore lieu

(iii) $x'' = 0 \ (x, t) \in C_p$:

Alors

$$p_u(x, t) \sim \frac{1}{(2\pi)^n} \frac{1}{(n - p - 1)!} \frac{\exp(-S(x, t)/u)}{u^{2n-p}} \Theta_p(x, t) \qquad (9.3)$$

où $\Theta_p(x, t)$ est donné par

$$\Theta_p(x, t) = \left(\frac{\pi}{2a_n} \right)^{n-p} |E_p|^{n-p-1} \prod_{j=1}^{p} \left(\frac{2a_j\pi/2a_n}{\sin 2a_j\pi/2a_n} \right). \qquad (9.4)$$

10. Estimation uniforme de $p_u(x, t)$

L'extension, au cas des variétés de Heisenberg, du comportement asymptotique du noyau de la chaleur nécessite une estimation du noyau de la chaleur des groupes d'Heisenberg, *uniforme par rapport aux variables (x, t, u) et par rapport aux constantes a_j, $j = 1, \ldots, n$* en particulier lorsque ces constantes peuvent devenir égales. La raison en est que les constantes a_j varieront de point en point sur une variété d'Heisenberg. Un calcul assez compliqué (voir [5]) donne l'estimation uniforme suivante

$$p_u(x, t) \leq C \frac{e^{-S(x,t)/u}}{u^{n+1}a_n} \prod_{j=1}^{n-1} \text{Min} \left\{ \frac{a_n}{a_n - a_j}, 1 + \frac{S(x, t)}{u} \right\} \qquad (10.1)$$

où C est une constante absolue.

Des estimations uniformes plus précises peuvent être obtenues dans le cas isotrope $a_1 = \cdots = a_n$

$$p_u(x, t) \leq C \frac{e^{-S(x,t)/u}}{u^{n+1}} \times \left[\text{Min} \left(1 + \frac{S(x, t)}{u}, \frac{\sqrt{S(x, t)}}{|x|} \right) \right]^{n-1}$$
$$\times \text{Min} \left[1 \left(\frac{u}{|x| \sqrt{S(x, t)}} \right)^{1/2} \right]. \qquad (10.2)$$

Une comparaison de (10.2) avec les estimations déduites du comportement asymptotique montre que l'estimation (10.2) est très précise. La difficulté des

estimations (10.1) et (10.2) provient du fait que l'intégrale qui donne la valeur
de $p_u(x, t)$

$$\frac{1}{u^{n+1}} \int \exp\left(-\frac{1}{u} f(x, t, \tau)\right) \prod_{j=1}^{n} \left(\frac{2a_j \tau}{sh2a_j \tau}\right) d\tau$$

doit être estimée en déplaçant le contour d'intégration de façon à le faire passer
par le point critique $i\theta_c$ de f, mais l'uniformité nécessite une estimation qui
est indépendante de la distance $|\theta_c - \pi/2a_n|$ qui peut être aussi petite que
l'on veut, alors qu'en même temps $\pi/2a_n$ est un point singulier essentiel de
$\exp(-f/u)$, puisque c'est un pôle de f, tout en étant aussi un pôle du préfacteur $1/sh2a_j \tau$ pouvant éventuellement devenir multiple lorsque plusieurs a_j
deviennent égaux à a_n.

11. Extension au cas des variétés de Heisenberg

Signalons rapidement comment étendre ces résultats aux variétés de Heisenberg.
Dans ce cas, ni le théorie des représentations, ni la théorie probabiliste des
diffusions ne permettent d'aboutir. La méthode suivie consiste à obtenir un
développement de type Hadamard

$$p_u(x, t \mid x_0, t_0) \backsim \sum_{k \geq 0} \frac{1}{u^{n+1-k}} \int_{-\infty}^{+\infty} e^{-f/u} V_k(x, t, \tau) d\tau,$$

où les f, V_k sont les fonctions de x, t, x_0, t_0, τ_0. f est une action satisfaisant

$$H(\nabla f, x, t) = f - \tau \frac{\partial f}{\partial \tau},$$

et les V_k sont obtenues en résolvant par récurrence des équations de transport
successives le long des bicaractéristiques de f, dépendant de $V_0, \ldots, V_{k-1}$. Les
conditions des bicaractéristiques sont non standard.

Une conséquence de ce type de formule est que $p_u(x, t \mid x_0, t_0)$ est analytique réelle lorsque la variété d'Heisenberg est analytique.

De tels résultats ont été obtenus pour les fonctions de Green dans [2] et pour
le noyau de la chaleur dans [6].

12. Extension au cas des variétés de Heisenberg

Des résultats ont été obtenus pour les cas faiblement pseudo-convexes

$$\text{Im } z_{n+1} = \left(|z_1|^2 + \cdots + |z_n|^2\right)^k$$

δ pour tout entier k dans [4] pour le cas des fonctions de Green et de leur uniformisation. Le cas du noyau de la chaleur semble cependant beaucoup plus difficile que celui de l'inverse Δ^{-1} de l'operateur sous-elliptique. Notons ici que la méthode employée dans le cas dégénéré est complètement différente de celle expliquée ici, ou de celle de [1], [2], [5]. En fait, il est montré, dans l'Appendice de [5] que la méthode précédente, adaptée au noyau de Green, fournit non pas une paramétrix, mais seulement une pseudo-paramétrix, c'est à dire, un noyau P tel que

$$\Delta P = \delta - S,$$

où S est une intégale singulière. En d'autres termes, les méthodes semi-classiques sont ineffectives dans le cas plus dégénéré que l'ordre 2, et cela pour des raisons profondes.

D'autres résultats concernant les opérateurs elliptiques dégénérés sont expliqués dans [7].

Références

[1] Beals, R., Gaveau, B., and Greiner, P., The Green function of model two step hypoelliptic operators and the analysis of certain tangential Cauchy-Riemann complexes, *Adv. Math.* **121** (1996), 288–345.

[2] _____, Complex Hamiltonian mechanics and parametrices for subelliptic Laplacians, I, II, III, *Bull. Sci. Math.* **121** (1997), 1–36, 97–149, 195–259.

[3] _____, Subelliptic geometry, *AMS Contemporary Mathematics* **212** (1998), 25–39.

[4] _____, On a geometric formula for the fundamental solution of subelliptic laplaciens, *Maths. Nach.* **181** (1996), 81–163.

[5] _____, *Hamilton Jacobi theory and the heat kernel on Heisenberg groups* (to appear).

[6] Beals, R., Gaveau, B., Greiner, P., and Kannai, Y., *The heat kernel on Heisenberg manifolds* (in preparation).

[7] _____, Exact fundamental solutions for a class of degenerate elliptic operators, *Comm. PDE* **24** (1999), 719–742.

[8] Beals, R. and Greiner, P., *Calculus on Heisenberg manifolds*, Ann. Math. studies 119, Princeton Uni. Press 1988.

[9] Bellaiche, A. and Risler, J. (eds), *Subriemannian geometry*, Birhkaüser, Basel 1996.

[10] Folland, G.-B., A fundamental solution for a subelliptic operator, *Bull. Am. Math. Soc.* **78** (1973), 373–376.

[11] Folland, G.-B. and Stein, E.-M., Estimates for the $\bar{\partial}_b$-complex and analysis on the Heisenberg group, *Comm. Pure. Appl. Math.* **27** (1974), 429–522.

[12] Gaveau, B., Holonomie stochastique et représentations du groupe d'Heisenberg, *CR Acad. Sci. Paris* **280** (1975), 571–572;

Principe de moindre action et propagation de la chaleur sur le groupe d'Heisenberg, *CR Acad. Sci. Paris* **281** (1975), 237–329;
Principe de moindre action, propagation de la chaleur et estimées sous-elliptiques pour les groupes nilpotents de rang 2, *CRAS. Sci. Paris* **282** (1976), 563–566;
Propagation de la chaleur, quantification, bicaractéristiques et caustiques sur les groupes nilpotents de rang 2, *CRAS Paris* **282** (1976), 865–868.

[13] _____, Principe de moindre action et propagation de la chaleur et estimées sous-elliptiques sur certains groupes nilpotents, *Acta Math.* **139** (1977), 95–153.

[14] _____, Systèmes dynamiques associés à certains opérateurs hypoelliptiques, *Bull. Sci. Math.* **102** (1978), 203–229.

[15] Hörmander, L., Hypoelliptic second order differential equations, *Acta Math.* **119** (1967), 147–171.

[16] Hueber, H. and Müller, D., Asymptotics for some Green kernels on the Heisenberg group and the Martin boundary, *Math Ann.* **283** (1989), 97–119.

[17] Hulanicki, A., The distribution of energy in the Brownian motion in the Gaussian field and analytic-hypoellipticity of certain subelliptic operators on the Heisenberg group, *Studia Math.* (1976), 165–173.

[18] Kac, M. and Van Moerbeke, P., *Aspects probabilistes de la théorie du potentiel*, Presses de l'Université de Montréal 1968.

[19] Strichartz, R., Subriemannian geometry, *J. Diff. Geom.* **24** (1986), 221–263- correction, ibid. **30** (1989), 595–596.

[20] Varopoulos, N., Analysis on nilpotent groups, *J. Funct. Anal.* **66** (1986), 406–431.

THE GEOMETRY OF CAUCHY DATA SPACES

B. BOOSS–BAVNBEK, K. FURUTANI AND K.P. WOJCIECHOWSKI

This report is dedicated to the memory of Jean Leray

Contents

1 Introduction 323
    1.1   Topological, geometric, and physics motivation 323
    1.2   Various concepts of Cauchy data spaces 323
    1.3   The function of symplectic geometry 324
2 Symplectic functional analysis 325
3 The analysis of operators of Dirac type 329
    3.1   The general setting 329
    3.2   Analysis tools: Green's formula 330
    3.3   Analysis tools: the Unique Continuation Property 330
4 Cauchy data spaces and the Calderón projection 332
    4.1   Invertible extension 333
    4.2   The Poisson operator and the Calderón projection 334
    4.3   Calderón and Atiyah–Patodi–Singer projection 335
    4.4   Twisted orthogonality of Cauchy data spaces 338
5 Cauchy data spaces and maximal and minimal domains 339
    5.1   The natural Cauchy data space 339
    5.2   Criss-cross reduction 342
6 Non-Lagrangian half-spaces and index theory 344
    6.1   Index theory 344
    6.2   The Bojarski conjecture 345
    6.3   Generalizations for global boundary conditions 345
    6.4   Pasting formulas 346
7 Family versions: The spectral flow and the Maslov index 347
    7.1   Spectral flow and the Maslov index 347
    7.2   Correction formula for the spectral flow 348
8 The boundary reduction and the gluing of determinants 348
    8.1   Three determinant concepts 349
    8.2   The Scott–Wojciechowski formula 350
    8.3   An adiabatic pasting formula 352

M. de Gosson (ed.), Jean Leray '99 Conference Proceedings, 321-354.
© *2003 Kluwer Academic Publishers.*

Abstract. *First we summarize two different concepts of Cauchy data ('Hardy') spaces of elliptic differential operators of first order on smooth compact manifolds with boundary: the L^2-definition by the range of the pseudo-differential Calderón–Szegö projection and the 'natural' definition by projecting the kernel into the (distributional) quotient of the maximal and the minimal domain. We explain the interrelation between the two definitions. Second we give various applications for the study of topological, differential, and spectral invariants of Dirac operators and families of Dirac operators on partitioned manifolds.*

1. Introduction

1.1. TOPOLOGICAL, GEOMETRIC, AND PHYSICS MOTIVATION

After Jean Leray's pioneering work of the 40s, topologists have been studying the cutting and pasting of manifolds using Betti numbers, long exact (co)homology sequences, and spectral sequences. Roughly speaking, their goal is to control the replacement of simple parts by complex ones and vice versa when constructing or decomposing a manifold. The corresponding hypersurfaces appear as a kind of 1-codimensional fixed points. The shift between manifold and hypersurface is a tenet of this branch.

Similar ideas have been around in differential geometry for long; e.g., the celebrated Gauss–Bonnet theorem of the curvature integrals of manifolds with boundary and the Morse theory for the decomposition of manifolds. The shift to codimension 1 is also widely exploited in modern differential geometry, e.g., in Donaldson's work on 4-manifolds as boundaries of 5-manifolds and in the Seiberg–Witten theory (for a recent survey see [8, Part IV] and [19]). The same goes for modern quantum field theory of gauge invariant fields which are pure gauge at the boundary of, e.g., a 4-ball, or when one sums over all compact (Euclidean) four-geometries and matter field configurations on both sides of a fixed three-surface Σ (see, e.g., [11] and in particular [23]).

Here we shall restrict ourselves to the study of the index, the determinant, and the spectral flow of Dirac operators and families of Dirac operators on partitioned manifolds. These invariants can be coded by the intersection geometry of the *Cauchy data spaces* ('Hardy classes' in complex and Clifford analysis) along the partitioning hypersurface. Various kinds of gluing formulas can be obtained for them.

1.2. VARIOUS CONCEPTS OF CAUCHY DATA SPACES

The literature treats the Cauchy data spaces in slightly different ways. One is to establish the Cauchy data spaces as L^2-closures of smooth sections over the partitioning hypersurface Σ, coming from the restriction to the boundary of all smooth solutions ('monogenic functions' in Clifford analysis) over one of the parts M_j of a partitioned manifold $M = M_0 \cup_\Sigma M_1$. As for the Dirac operator, this Cauchy data space can be represented as the range of the L^2-extension of the pseudo-differential Calderón projection ('Szegö projection' in Clifford analysis) and established as a Lagrangian subspace of the symplectic Hilbert space $L^2(-\Sigma) \dotplus L^2(\Sigma)$.

The pseudo-differential approach involves a machinery which makes it too heavy and inflexible for an adequate study of continuous families of operators and other topological considerations. It imposes unnecessary limitations and assumptions in spite of the fact that L^2-spaces are more amenable than distributional spaces. However, this approach also has its merits: it leads to operational parametrices and boundary integrals in the theory of elliptic boundary value problems with important applications. In section 4 we summarize the details of the pseudo-differential (L^2-)approach.

A more elementary approach to the Cauchy data spaces is to establish them as subspaces of the symplectic Hilbert space $\beta := D_{\max}/D_{\min}$ of natural boundary values, i.e., as the boundary values of sections belonging to the maximal domain $D_{\max}$ of the operator. One can embed β as a non closed subspace into the distribution space $H^{-1/2}(-\Sigma) \dotplus H^{-1/2}(\Sigma)$ and get a surprisingly simple proof of the closedness of the corresponding (distributional) Cauchy data space — without resorting to pseudo-differential operator calculus. In this way it is also easy to obtain the continuity of the Cauchy data spaces for continuous families of operators. In section 5 we explain how the results of the technically more simple distributional approach can be transferred to the more customary L^2-approach.

1.3. THE FUNCTION OF SYMPLECTIC GEOMETRY

The 'complementarity' of the *skew-symmetry* of Clifford multiplication and spin geometry on the one hand and the *symmetry* of the induced differential operators on the other were a puzzle until Leray in his famous monograph [26] *explained* why aspects of symmetric differential operators can be best understood in the framework of symplectic geometry.

In Part II we shall elaborate the conjured *Leray's view* and use the Lagrangian property of the Cauchy data spaces to discuss the gluing formulas for spectral invariants in this setting. Section 6 shows how index theory may be considered as expressing the index of a non symmetric elliptic differential operator over a closed partitioned manifold (which is a 'quantum' variable, defined by the multiplicity of the 0-eigenvalues) by the index of the Fredholm pair of Cauchy data spaces from both sides of the separating hypersurface, measuring their 'non-Lagrangianess' (which is a 'classical' variable, defined by the geometry of the solution spaces). This is the Bojarski conjecture, proved in [14] for chiral Dirac type operators on even-dimensional partitioned manifolds. Various generalizations for global (elliptic) boundary conditions will be discussed.

In section 7 we give a family version of the Bojarski conjecture: the Yoshida–Nicolaescu theorem. It relates the spectral flow of a continuous one-parameter family of (total) Dirac type operators to the Maslov index of the corresponding family of Cauchy data spaces. This is another case of quasi-classical approximation, which is further underlined by the fact that the 'quantum' variable *spectral flow* has a topological meaning only in infinite dimensions, whereas the 'classical' variable *Maslov index* has a non-trivial topological meaning already in finite dimensions.

In section 8 we explain the function of the Cauchy data spaces in the boundary reduction and in the adiabatic gluing formula for the determinant regularized by the ζ-function. Even more than the index and the spectral flow, the determinant has remained a puzzle and it is still a challenge to understand the different character of the invariants *over the whole manifold* and in terms of the *Cauchy data spaces*.

An example: the determinant of a Dirac operator over a closed manifold or a manifold with boundary is clearly not defined as a true product of the eigenvalues (which go to $\pm\infty$) but requires a kind of regularization. A recent result is owed to Scott and Wojciechowski [34]. Formally, it is almost identical with a variant of the Bojarski conjecture for the index and the Yoshida–Nicolaescu theorem for the spectral formula: the Scott–Wojciechowski formula (our Theorem 8.1) says that the ζ-regularized determinant of the Dirac operator over a compact smooth manifold with boundary, subject to a Lagrangian global pseudo-differential boundary condition 'of Atiyah–Patodi–Singer type', is equal to a Fredholm determinant defined over the Cauchy data space along the boundary, i.e., a true infinite product of eigenvalues which go rapidly to 1.

Part I: The Functional Analysis of Cauchy Data Spaces

2. Symplectic functional analysis

We fix the following notation.

Let $(\mathcal{H}, \langle \cdot, \cdot \rangle)$ be a separable real Hilbert space with a fixed symplectic form ω, i.e., a skew–symmetric bounded bilinear form on $\mathcal{H} \times \mathcal{H}$ which is non-degenerate. Let $J : \mathcal{H} \to \mathcal{H}$ denote the corresponding almost complex structure defined by

$$\omega(x, y) = \langle Jx, y \rangle \qquad (2.1)$$

with $J^2 = -\mathrm{Id}$, ${}^t\!J = -J$, and $\langle Jx, Jy \rangle = \langle x, y \rangle$. Here ${}^t\!J$ denotes the transpose of J with regard to the (real) inner product $\langle \cdot, \cdot \rangle$. Let $\mathcal{L} = \mathcal{L}(\mathcal{H})$ denote the set of

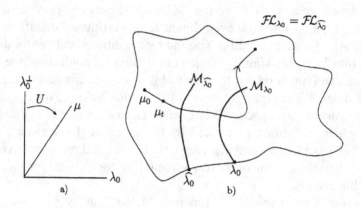

Figure 1. a) The generation of $\mathcal{L} = \{\mu = U(\lambda_0^{\perp}) \mid U \in \mathcal{U}(\mathcal{H})\}$, b) One curve and two Maslov cycles in $\mathcal{FL}_{\lambda_0} = \mathcal{FL}_{\widehat{\lambda_0}}$.

all Lagrangian subspaces of $\mathcal{H}$ (i.e., $\lambda = (J\lambda)^{\perp}$, or, equivalently, let λ coincide with its annihilator λ^0 with respect to ω). The topology of $\mathcal{L}$ is defined by the operator norm of the orthogonal projections onto the Lagrangian subspaces.

Let $\lambda_0 \in \mathcal{L}$ be fixed. Then any $\mu \in \mathcal{L}$ can be obtained as the image of $\lambda_0^{\perp}$ under a suitable unitary transformation

$$\mu = U\left(\lambda_0^{\perp}\right)$$

(see also Figure 1a). Here we consider the real symplectic Hilbert space $\mathcal{H}$ as a complex Hilbert space by J. The group $\mathcal{U}(\mathcal{H})$ of unitary operators of $\mathcal{H}$ acts transitively on $\mathcal{L}$, i.e., the mapping

$$\rho : \mathcal{U}(\mathcal{H}) \longrightarrow \mathcal{L}$$
$$U \longmapsto U\left(\lambda_0^{\perp}\right) \tag{2.2}$$

is surjective and defines a principal fibre bundle with the group of orthogonal operators $\mathcal{O}(\lambda_0)$ as structure group.

Example 2.1. (a) In finite dimensions one considers the space $\mathcal{H} := \mathbb{R}^n \oplus \mathbb{R}^n$ with the symplectic form

$$\omega\left((x, \xi), (y, \eta)\right) := -\langle x, \eta \rangle + \langle \xi, y \rangle \quad \text{for } (x, \xi), (y, \eta) \in \mathcal{H}.$$

To emphasize the finiteness of the dimension we write $\mathrm{Lag}(\mathbb{R}^{2n}) := \mathcal{L}(\mathcal{H})$. For linear subspaces of $\mathrm{Lag}(\mathbb{R}^{2n})$ one has

$$l \in \mathcal{L} \iff \dim l = n \quad \text{and } l \subset l^0,$$

i.e., Lagrangian subspaces are true half-spaces which are maximally isotropic ('isotropic' means $l \subset l^0$).

One finds $\text{Lag}(\mathbb{R}^{2n}) \cong U(n)/O(n)$ with the fundamental group

$$\pi_1 \left(\text{Lag}\left(\mathbb{R}^{2n} \right), \lambda_0 \right) \cong \mathbf{Z}.$$

The mapping is given by the 'Maslov index' of loops of Lagrangian subspaces which can be described as an intersection index with the 'Maslov cycle'. There is a rich literature on the subject, see, e.g., the seminal paper [1], the systematic review [17], or the cohomological presentation [21].

(b) Let $\{\varphi_k\}_{k \in \mathbb{Z} \setminus \{0\}}$ be a complete orthonormal system for $\mathcal{H}$. We define an almost complex structure, and so by (2.1) a symplectic form, by

$$J\varphi_k := \text{sign}(k)\varphi_{-k}.$$

Then the spaces $\mathcal{H}_- := \text{span}\{\varphi_k\}_{k<0}$ and $\mathcal{H}_+ := \text{span}\{\varphi_k\}_{k>0}$ are complementary Lagrangian subspaces of $\mathcal{H}$.

In infinite dimensions, the space $\mathcal{L}$ is contractible owing to Kuiper's theorem (see [8, Part I]) and therefore topologically not interesting. Also, we need some restrictions to avoid infinite-dimensional intersection spaces when counting intersection indices. Therefore we replace $\mathcal{L}$ by a smaller space. This problem can be solved as first suggested in Swanson [38]: by relating symplectic functional analysis with the space $\text{Fred}(\mathcal{H})$ of Fredholm operators, we obtain finite dimensions for suitable intersection spaces and at the same time topologically highly non-trivial objects.

Definition 2.2. (a) The space of *Fredholm pairs* of closed infinite–dimensional subspaces of $\mathcal{H}$ is defined by

$$\text{Fred}^2(\mathcal{H}) := \{(\lambda, \mu) \mid \dim \lambda \cap \mu < +\infty \text{ and } \lambda + \mu \subset \mathcal{H} \text{ closed}$$
$$\text{and } \dim \mathcal{H}/(\lambda + \mu) < +\infty\}.$$

(b) The *Fredholm–Lagrangian Grassmannian* of a real symplectic Hilbert space $\mathcal{H}$ at a fixed Lagrangian subspace λ_0 is defined as

$$\mathcal{FL}_{\lambda_0} := \left\{ \mu \in \mathcal{L} \mid (\mu, \lambda_0) \in \text{Fred}^2(\mathcal{H}) \right\}.$$

(c) The *Maslov cycle* of λ_0 in $\mathcal{H}$ is defined as

$$\mathcal{M}_{\lambda_0} := \mathcal{FL}_{\lambda_0} \setminus \mathcal{FL}_{\lambda_0}^{(0)},$$

where $\mathcal{FL}_{\lambda_0}^{(0)}$ denotes the subset of Lagrangians intersecting λ_0 transversally, i.e., $\mu \cap \lambda_0 = \{0\}$.

Recall the following algebraic and topological characterization of Lagrangian Fredholm pairs (see [9], inspired by [13, Part 2, Lemma 2.6]).

Proposition 2.3. (a) Let λ, $\mu \in \mathcal{L}$ and let π_λ, π_μ denote the orthogonal projections of $\mathcal{H}$ onto λ respectively μ. Then

$$(\lambda, \mu) \in \text{Fred}^2(\mathcal{H}) \iff \pi_\lambda + \pi_\mu \in \text{Fred}(\mathcal{H}) \iff \pi_\lambda - \pi_\mu \in \text{Fred}(\mathcal{H}).$$

(b) The fundamental group of $\mathcal{FL}_{\lambda_0}$ is $\mathbf{Z}$, and the mapping of the loops in $\mathcal{FL}_{\lambda_0}(\mathcal{H})$ onto $\mathbf{Z}$ is given by the *Maslov index*

$$\mathbf{mas} : \pi_1\left(\mathcal{FL}_{\lambda_0}(\mathcal{H})\right) \xrightarrow{\cong} \mathbf{Z}.$$

To define the Maslov index, one needs a systematic way of counting, adding and subtracting the dimensions of the intersections $\mu_s \cap \lambda_0$ of the curve $\{\mu_s\}$ with the Maslov cycle $\mathcal{M}_{\lambda_0}$. We refer to [9], inspired by [30] for a functional analytical definition for continuous curves without additional assumptions.

Remark 2.4. (a) By identifying $\mathcal{H} \cong \lambda_0 \otimes \mathbf{C} \cong \lambda_0 \oplus \sqrt{-1}\,\lambda_0$, we split in [10] any $U \in \mathcal{U}(\mathcal{H})$ into a real and imaginary part

$$U = X + \sqrt{-1}Y$$

with $X, Y : \lambda_0 \to \lambda_0$. Let $\mathcal{U}(\mathcal{H})^{\text{Fred}}$ denote the subspace of unitary operators which have a Fredholm operator as real part. This is the total space of a principal fibre bundle over the Fredholm Lagrangian Grassmannian $\mathcal{FL}_{\lambda_0}$ as base space and with the orthogonal group $\mathcal{O}(\lambda_0)$ as structure group. The projection is given by the restriction of the trivial bundle $\rho : \mathcal{U}(\mathcal{H}) \to \mathcal{L}$ of (2.2). This bundle

$$\mathcal{U}(\mathcal{H})^{\text{Fred}} \xrightarrow{\rho} \mathcal{FL}_{\lambda_0}$$

may be considered as the infinite-dimensional generalization of the well studied bundle $U(n) \to \text{Lag}(\mathbb{R}^{2n})$ for finite n and provides an alternative proof of the homotopy type of $\mathcal{FL}_{\lambda_0}$.

(b) The Maslov index for curves depends on the specified Maslov cycle $\mathcal{M}_{\lambda_0}$. It is worth emphasizing that two equivalent Lagrangian subspaces λ_0 and $\widehat{\lambda}_0$ (i.e., $\dim \lambda_0/(\lambda_0 \cap \widehat{\lambda}_0) < +\infty$) always define the same Fredholm Lagrangian Grassmannian $\mathcal{FL}_{\lambda_0} = \mathcal{FL}_{\widehat{\lambda}_0}$ but may define different Maslov cycles $\mathcal{M}_{\lambda_0} \neq \mathcal{M}_{\widehat{\lambda}_0}$. The induced Maslov indices may also become different

$$\mathbf{mas}\left(\{\mu_s\}_{s\in[0,1]}, \lambda_0\right) - \mathbf{mas}\left(\{\mu_s\}_{s\in[0,1]}, \widehat{\lambda}_0\right) \overset{\text{in general}}{\neq} 0 \qquad (2.3)$$

(see [9, Proposition 3.1 and section 5]). However, if the curve is a loop, then the Maslov index does not depend on the choice of the Maslov cycle. From this property it follows that the difference in (2.3), beyond the dependence on λ_0 and $\widehat{\lambda}_0$, depends only on the initial and end points of the path $\{\mu_s\}$ and may be considered as the infinite-dimensional generalization $\sigma_{\text{Hör}}(\mu_0, \mu_1; \lambda_0, \widehat{\lambda}_0)$ of the Hörmander index. It plays a part as the transition function of the universal covering of the Fredholm Lagrangian Grassmannian (see also Figure 1b).

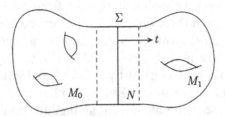

Figure 2. The partition of $M = M_0 \cup_\Sigma M_1$.

3. The analysis of operators of Dirac type

We fix the notation and recall basic properties of operators of Dirac type.

3.1. THE GENERAL SETTING

Let M be a compact smooth Riemannian *partitioned* manifold

$$M = M_0 \cup M_1, \quad \text{where } M_0 \cap M_1 = \partial M_0 = \partial M_1 = \Sigma$$

and Σ a hypersurface (see Figure 2). We assume that $M \setminus \Sigma$ does not have a closed connected component (i.e., Σ intersects any connected component of M_0 and M_1). Let

$$A : C^\infty(M; S) \longrightarrow C^\infty(M; S)$$

be an *operator of Dirac type* acting on sections of a Hermitian bundle S of Clifford modules over M, i.e., $A = \mathbf{c} \circ \nabla$ where $\mathbf{c}$ denotes the Clifford multiplication and ∇ is a connection for S which is compatible with $\mathbf{c}$ ($\nabla \mathbf{c} = 0$). From the compatibility assumption it follows that A is symmetric and essentially self-adjoint over M.

For even $n = \dim M$ the splitting $C\ell(M) = C\ell^+(M) \oplus C\ell^-(M)$ of the Clifford bundles induces a corresponding splitting of $S = S^+ \oplus S^-$ and a *chiral decomposition*

$$A = \begin{pmatrix} 0 & A^- = (A^+)^* \\ A^+ & 0 \end{pmatrix}$$

of the *total* Dirac operator. The *chiral Dirac operators* $A^\pm$ are elliptic but not symmetric, and for that reason they may have non-trivial indices which provide us with important topological and geometric invariants.

Here we assume that all metric structures of M and S are product in a collar neighborhood $N = [-1, 1] \times \Sigma$ of Σ. Then

$$A|_N = \sigma \left(\frac{\partial}{\partial t} + \mathcal{B} \right), \tag{3.1}$$

where t denotes the normal coordinate (running from M_0 to M_1) and $\mathscr{B}$ denotes the canonically associated Dirac operator over Σ, called the *tangential operator*. We have a similar product formula for the chiral Dirac operator. Here the point of the product structure is that then σ and $\mathscr{B}$ do not depend on the normal variable. We note that σ is defined by Clifford multiplication by dt. It is a unitary mapping $L^2(\Sigma; S|_\Sigma) \to L^2(\Sigma; S|_\Sigma)$ with $\sigma^2 = -\text{Id}$ and $\sigma\mathscr{B} = -\mathscr{B}\sigma$. In the non product case, there are certain ambiguities in defining a 'tangential operator' which we shall not discuss here (but see also Formula (3.3)).

3.2. ANALYSIS TOOLS: GREEN'S FORMULA

For notational economy, we set $X := M_1$. For greater generality, we consider the chiral Dirac operator

$$A^+ : C^\infty\left(X; S^+\right) \longrightarrow C^\infty\left(X; S^-\right)$$

and write A^- for its formally adjoint operator. The corresponding results follow at once for the total Dirac operator.

Lemma 3.1. *Let $\langle \cdot, \cdot \rangle_\pm$ denote the scalar product in $L^2(X; S^\pm)$. Then we have*

$$\left\langle A^+ f_+, f_- \right\rangle_- - \left\langle f_+, A^- f_- \right\rangle_+ = -\int_\Sigma (\sigma \gamma_\infty f_+, \gamma_\infty f_-)\, d\,\text{vol}_\Sigma$$

for any $f_\pm \in C^\infty(X; S^\pm)$.

Here

$$\gamma_\infty : C^\infty\left(X; S^\pm\right) \longrightarrow C^\infty\left(\Sigma; S^\pm|_\Sigma\right) \tag{3.2}$$

denotes the restriction of a section to the boundary. This is not problematic within the smooth category.

3.3. ANALYSIS TOOLS: THE UNIQUE CONTINUATION PROPERTY

One of the basic properties of an operator of Dirac type A is the weak Unique Continuation Property (UCP). For $M = M_0 \cup_\Sigma M_1$, it guarantees that there are no *ghost* solutions of $Au = 0$, i.e., there are no solutions which vanish on M_0 and have non-trivial support in the interior of M_1. This property is also called UCP *from open subsets* or *across any hypersurface*. For Euclidean (classical) Dirac operators the property follows from Holmgren's uniqueness theorem for scalar elliptic operators with real analytic coefficients (see, e.g., Taylor [39, Proposition 4.3]).

In [14, Chapter 8], the reader will find a very simple proof of the weak UCP for operators of Dirac type. We refer to [5] for a further slight simplification and a broader perspective (see also [3]). The proof does not use advanced arguments of the Aronszajn/Cordes type regarding the diagonal and real form of the principal symbol of the Dirac Laplacian, but only the following well known generalization of the product property of Dirac type operators.

Lemma 3.2. *Let Σ be a closed hypersurface of M with orientable normal bundle. Let t denote a normal variable with fixed orientation such that a bicollar neighborhood N of Σ is parametrized by $[-\varepsilon, +\varepsilon] \times \Sigma$. Then any operator of Dirac type can be rewritten in the form*

$$A|_N = \mathbf{c}(dt) \left(\frac{\partial}{\partial t} + B_t + C_t \right), \tag{3.3}$$

where B_t is a self-adjoint elliptic operator on the parallel hypersurface Σ_t, and $C_t : S|_{\Sigma_t} \to S|_{\Sigma_t}$ a skew-symmetric operator of 0th order, actually a skew-symmetric bundle homomorphism.

To prove the weak UCP we basically follow the standard lines of the UCP literature. Let $u \in C^\infty(M; S)$ be a solution of $Au = 0$ that vanishes on an open subset ω of M. Then it vanishes on the whole connected component of the manifold. This is to be shown. First we localize and convexify the situation and we introduce spherical coordinates (see Figure 3). Without loss of generality we may assume that ω is maximal, namely the union of all open subsets where u vanishes. If the solution u does not vanish on the whole connected component containing ω, we consider a point $x_0 \in \operatorname{supp} u \cap \partial\omega$. We choose a point p inside of ω such that the ball around p with radius $r := \operatorname{dist}(x_0, p)$ is contained in $\overline{\omega}$. We call the coordinate running from p to x_0 the *normal* coordinate and denote it by t. The boundary of the ball around p of radius r is a hypersphere and will be denoted by $\mathcal{S}_{p,0}$. It goes through x_0 which has a normal coordinate $t = 0$. Correspondingly, we have larger hyperspheres $\mathcal{S}_{p,t} < M$ for $0 \le t \le T$ with $T > 0$ sufficiently small. In such a way we have parametrized an annular region $N_T := \{\mathcal{S}_{p,t}\}_{t\in[0,T]}$ around p of width T and inner radius r, ranging from the hypersphere $\mathcal{S}_{p,0}$ which is contained in $\overline{\omega}$, to the hypersphere $\mathcal{S}_{p,T}$ which cuts deeply into $\operatorname{supp} u$, if $\operatorname{supp} u$ is not empty.

Next, we replace the solution u by a section

$$v(t, y) := \varphi(t)u(t, y) \tag{3.4}$$

with a smooth bump function φ with $\varphi(t) = 1$ for $t \le 0.8\,T$ and $\varphi(t) = 0$ for $t \ge 0.9T$. Then $\operatorname{supp} v$ is contained in N_T. More precisely, it is contained in the annular region $N_{0.9T}$. Moreover, $\operatorname{supp}(Av)$ is contained in the annular region $0.8T \le t \le 0.9T$.

The weak UCP follows immediately from the following lemma.

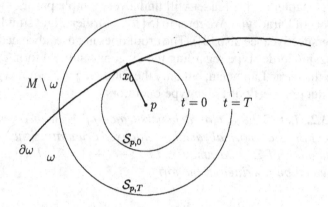

Figure 3. Local specification for the Carleman estimate.

Lemma 3.3. *Let* $A : C^\infty(M; E) \to C^\infty(M; E)$ *be a linear elliptic differential operator of order 1 which can be written on* N_T *in the product form (3.3). Let* v *denote a section made from a solution* u *as in (3.4).*

   (a) *Then for* T *sufficiently small there exists a constant* C *such that the Carleman inequality*

$$R \int_{t=0}^{T} \int_{\mathscr{S}_{p,t}} e^{R(T-t)^2} \|v(t, y)\|^2 \, dy \, dt$$

$$\leq C \int_{t=0}^{T} \int_{\mathscr{S}_{p,t}} e^{R(T-t)^2} \|Av(t, y)\|^2 \, dy \, dt \qquad (3.5)$$

holds for any real R *sufficiently large.*

   (b) *If (3.5) holds for any sufficiently large* $R > 0$, *then* u *is equal* 0 *on* $N_{T/2}$.

4.   Cauchy data spaces and the Calderón projection

To explain the L^2-Cauchy data spaces we recall three additional, somewhat delicate and not widely known properties of operators of Dirac type on compact manifolds with boundary from [14]:

1. the invertible extension to the double;

2. the Poisson type operator and the Calderón projection; and

3. the twisted orthogonality of the Cauchy data spaces for chiral and total Dirac operators which gives the Lagrangian property in the symmetric case (i.e., for the total Dirac operator).

The idea and the properties of the Calderón projection were announced in Calderón [16] and proved in Seeley [35] in great generality. In the following, we restrict ourselves to constructing the Calderòn projection for operators of Dirac type (or, more generally, elliptic differential operators of first order) which simplifies the presentation substantially.

4.1. INVERTIBLE EXTENSION

First we construct the *invertible double*. Clifford multiplication by the inward normal vector gives a natural clutching of S^+ over one copy of X with S^- over a second copy of X to a smooth bundle $\widetilde{S^+}$ over the closed double $\widetilde{X}$. The product forms of A^+ and $A^- = (A^+)^*$ fit together over the boundary and provide a new operator of Dirac type

$$\widetilde{A^+} := A^+ \cup A^- : C^\infty \left(\widetilde{X}; \widetilde{S^+}\right) \longrightarrow C^\infty \left(\widetilde{X}; \widetilde{S^-}\right).$$

Clearly $(A^+ \cup A^-)^* = A^- \cup A^+$; hence index $\widetilde{A^+} = 0$. It turns out that $\widetilde{A^+}$ is invertible with a pseudo-differential elliptic inverse $(\widetilde{A^+})^{-1}$. Of course A^+ is not invertible and $r^+(\widetilde{A^+})^{-1}e^+A^+ \neq \mathrm{Id}$, where $e^+ : L^2(X; S^+) \to L^2(\widetilde{X}; \widetilde{S^+})$ denotes the extension-by-zero operator and $r^+ : H^s(\widetilde{X}; \widetilde{S^+}) \to H^s(X; S^+)$ the natural restriction operator for Sobolev spaces, s real.

Example 4.1. In the simplest possible two-dimensional case we consider the Cauchy–Riemann operator $\bar{\partial} : C^\infty(D^2) \to C^\infty(D^2)$ over the disc D^2, where $\bar{\partial} = \frac{1}{2}(\partial_x + i\partial_y)$. In polar coordinates out of the origin, this operator has the form $\frac{1}{2}e^{i\varphi}(\partial_r + (i/r)\partial_f)$. Therefore, after some small smooth perturbations (and modulo the factor $1/2$), we assume that $\bar{\partial}$ has the following form in a certain collar neighborhoods of the boundary:

$$\bar{\partial} = e^{i\varphi}\left(\partial_r + i\partial_f\right).$$

Now we construct the invertible double of $\bar{\partial}$. By E^k, $k \in \mathbb{Z}$, we denote the bundle, which is obtained from two copies of $D^2 \times \mathbb{C}$ by the identification $(z, w) = (z, z^k w)$ near the equator. We obtain the bundle E^1 by gluing two halves of $D^2 \times \mathbb{C}$ by $\sigma(\varphi) = e^{i\varphi}$ and E^{-1} by gluing with the adjoint symbol. In such a way we obtain the operator

$$\widetilde{\bar{\partial}} := \bar{\partial} \cup (\bar{\partial})^* : C^\infty \left(S^2; H^1\right) \longrightarrow C^\infty \left(S^2; H^{-1}\right)$$

over the whole 2-sphere.

Let us analyze the situation more carefully. We fix $N := (-\varepsilon, +\varepsilon) \times S^1$, a bicollar neighborhoods of the equator. The formally adjoint operator to $\bar{\partial}$ has the form:

$$(\bar{\partial})^* = e^{-i\varphi}\left(-\partial_t + i\partial_f + 1\right)$$

$(t = r - 1)$ in this cylinder. A section of H^1 is a couple (s_1, s_2) such that in N:

$$s_2(t, \varphi) = e^{i\varphi} s_1(t, \varphi).$$

The couple $(\bar{\partial} s_1, (\bar{\partial})^* s_2)$ is a smooth section of H^{-1}. To show that, we check the equality $(\bar{\partial})^* s_2 = e^{-i\varphi} \bar{\partial} s_1$. We have in the neighborhoods N:

$$\left(\bar{\partial}\right)^* s_2 = \left(\bar{\partial}\right)^* \left(e^{i\varphi} s_1\right) = e^{-i\varphi} \left(\partial_t + i\partial\varphi + 1\right) \left(e^{i\varphi} s_1\right)$$

$$= \partial_t s_1 + i e^{-i\varphi} \partial_f \left(e^{i\varphi} s_1\right) + s_1 = \left(\partial_t + i\partial_f\right) s_1$$

$$= e^{-i\varphi} e^{i\varphi} \left(\partial_t + i\partial_f\right) s_1 = e^{-i\varphi} \left(\bar{\partial} s_1\right).$$

Then the operator $\bar{\partial} \cup \bar{\partial}^*$ becomes injective and index $(\bar{\partial} \cup \bar{\partial}^*) = 0$.

4.2. THE POISSON OPERATOR AND THE CALDERÓN PROJECTION

Next we investigate the solution spaces and their traces at the boundary. For a total or chiral operator of Dirac type over a smooth compact manifold with boundary and for any real s we define the *null space*

$$\ker\left(A^+, s\right) := \left\{f \in H^s\left(X; S^+\right) \mid A^+ f = 0 \text{ in } X \setminus \Sigma\right\}.$$

The null spaces consist of sections which are distributional for negative s; by elliptic regularity they are smooth in the interior; in particular they possess a smooth restriction on the hypersurface $\Sigma_\varepsilon = \{\varepsilon\} \times \Sigma$ parallel to the boundary Σ of X at a distance $\varepsilon > 0$. By a Riesz operator argument they can be shown to also possess a trace over the boundary. Of course, that trace is no longer smooth but belongs to $H^{s-\frac{1}{2}}(\Sigma; S^+|_\Sigma)$. More precisely, we have the following well known *General Restriction theorem* (for a proof see, e.g., [14, Chapters 11 and 13]):

Theorem 4.2. (a) *Let $s > 1/2$. Then the restriction map γ_∞ of (3.2) extends to a bounded map*

$$\gamma_s : H^s\left(X; S^+\right) \longrightarrow H^{s-\frac{1}{2}}\left(\Sigma; S^+|_\Sigma\right). \tag{4.1}$$

(b) *For $s \leq 1/2$, the preceding reduction is no longer defined for arbitrary sections but only for solutions of the operator A^+: let $f \in \ker(A^+, s)$ and let $\gamma_{(\varepsilon)} f$ denote the well defined trace of f in $C^\infty(\Sigma_\varepsilon; S^+|_\Sigma)$. Then, as $\varepsilon \to 0_+$, the sections $\gamma_{(\varepsilon)} f$ converge to an element $\gamma_s f \in H^{s-\frac{1}{2}}(\Sigma; S^+|_\Sigma)$.*
(c) *For any $s \in \mathbb{R}$ the mapping*

$$\mathcal{K} := r_+ \left(\widetilde{A^+}\right)^{-1} \gamma_\infty^* \sigma : C^\infty\left(\Sigma; S^+|_\Sigma\right) \longrightarrow C^\infty\left(X; S^+\right)$$

extends to a continuous map $\mathcal{K}^{(s)} : H^{s-1/2}(\Sigma; S^+|_\Sigma) \to H^s(X; S^+)$ *with*

$$\text{range } \mathcal{K}^{(s)} = \ker\left(A^+, s\right).$$

In the preceding theorem, $\widetilde{A^+}$ denotes the invertible double of A^+, r_+ denotes the restriction operator $r_+ : H^s(\widetilde{X}; \widetilde{S}^+) \to H^s(X; S^+)$ and γ_∞^* the dual of γ_∞ in the distributional sense. The composition

$$\mathcal{P}\left(A^+\right) := \gamma_\infty \circ \mathcal{K} : C^\infty\left(X; S^+\right) \longrightarrow C^\infty\left(\Sigma; S^+|_\Sigma\right) \qquad (4.2)$$

is called the *(Szegö–)Calderón projection*. It is a pseudo-differential projection (idempotent, but in general not orthogonal). We denote by $\mathcal{P}(A^+)^{(s)}$ its extension to the sth Sobolev space over Σ.

We now have three options of defining the corresponding *Cauchy data* (or *Hardy*) *spaces*:

Definition 4.3. For all real s we define

$$\Lambda\left(A^+, s\right) := \gamma_s\left(\ker\left(A^+, s\right)\right),$$

$$\Lambda^{\text{clos}}\left(A^+, s\right) := \overline{\gamma_\infty\left\{f \in C^\infty\left(X; S^+\right) \mid A^+ f = 0 \text{ in } X \setminus \Sigma\right\}}^{H^{s-\frac{1}{2}}(\Sigma; S^+|_\Sigma)},$$

$$\text{and } \Lambda^{\text{Cald}}\left(A^+, s\right) := \text{range } \mathcal{P}\left(A^+\right)^{(s-\frac{1}{2})}.$$

The range of a projection is closed; the inclusions of the Sobolev spaces are dense; and range $\mathcal{P}(A^+) = \gamma_\infty\{f \in C^\infty(X; S^+) \mid A^+ f = 0 \text{ in } X \setminus \Sigma\}$, as shown in [14]. So, the second and the third definition of the Cauchy data space coincide. Moreover, for $s > 1/2$ one has $\Lambda(A^+, s) = \Lambda^{\text{Cald}}(A^+, s)$. This equality can be extended to the L^2-case ($s = 1/2$, see also Theorem 5.3 below), and remains valid for any real s, as proved in Seeley [35, Theorem 6]. For $s \leq 1/2$, the result is somewhat counter intuitive (see also Example 4.5(b) in the following Subsection). We have:

Proposition 4.4. *For all* $s \in \mathbb{R}$

$$\Lambda\left(A^+, s\right) = \Lambda^{\text{clos}}\left(A^+, s\right) = \Lambda^{\text{Cald}}\left(A^+, s\right).$$

4.3. CALDERÓN AND ATIYAH–PATODI–SINGER PROJECTION

The Calderón projection is closely related to another projection determined by the 'tangential' part of A: Let $\mathcal{B}$ denote the *tangential* symmetric elliptic differential operator over Σ in the product form

$$(A, \text{ respectively}) \ A^+ = \sigma\left(\partial_t + \mathcal{B}\right) \text{ in a collar neighbourhood of } \Sigma \text{ in } X.$$

It has discrete real eigenvalues and a complete system of L^2-orthonormal eigensections. Let $\Pi_{\geq}(\mathcal{B})$ denote the spectral (Atiyah–Patodi–Singer) projection onto the subspace $L_+(\mathcal{B})$ of $L^2(\Sigma; S^+|_\Sigma)$ spanned by the eigensections corresponding to the non negative eigenvalues of $\mathcal{B}$. It is a pseudo-differential operator and its principal symbol p_+ is the projection onto the eigenspaces of the principal symbol $b(y, \zeta)$ of $\mathcal{B}$ corresponding to non negative eigenvalues. It turns out that it coincides with the principal symbol of the Calderón projection.

We call the space of pseudo-differential projections with the same principal symbol p_+ the *Grassmannian* $\mathcal{G}r_{p_+}$. It has enumerable many connected components; two projections P_1, P_2 belong to the same component, if and only if the *virtual codimension*

$$\mathbf{i}(P_2, P_1) := \text{index}\{P_2 P_1 : \text{range } P_1 \longrightarrow \text{range } P_2\} \tag{4.3}$$

of P_2 in P_1 vanishes; the higher homotopy groups of each connected component are given by Bott periodicity.

Example 4.5. (a) For the Cauchy–Riemann operator on the disc $D^2 = \{|z| \leq 1\}$, the Cauchy data space is spanned by the eigenfunctions $e^{ik\theta}$ of the tangential operator ∂_θ over $S^1 = [0, 2\pi]/\{0, 2\pi\}$ for non negative k. So, the Calderón projection and the Atiyah–Patodi–Singer projection coincide in this case.

(b) Next we consider the cylinder $X^R = [0, R] \times \Sigma_0$ with $A_R = \sigma(\partial_t + B)$. Here B denotes a symmetric elliptic differential operator of first order acting on sections of a bundle E over Σ_0, and σ a unitary bundle endomorphism with $\sigma^2 = -\text{Id}$ and $\sigma B = -B\sigma$. Let B be invertible (for the ease of presentation). Let $\{\varphi_k, \lambda_k\}$ denote B's spectral resolution of $L^2(\Sigma_0, E)$ with

$$\cdots \lambda_{-k} \leq \cdots \lambda_{-1} < 0 < \lambda_1 \leq \cdots \lambda_k \leq \cdots$$

Then

$$\begin{cases} B\varphi_k = \lambda_k \varphi_k & \text{for all } k \in \mathbb{Z} \setminus \{0\}, \\ \lambda_{-k} = -\lambda_k, \sigma(\varphi_k) = \varphi_{-k}, \text{ and } \sigma(\varphi_{-k}) = -\varphi_k & \text{for } k > 0. \end{cases} \tag{4.4}$$

We consider

$$f \in \ker(A_R, 0) = \text{span}\left\{e^{-\lambda_k t}\varphi_k\right\}_{k \in \mathbb{Z} \setminus \{0\}} \text{ in } L^2\left(X^R\right)$$

$$= \ker A_{R\max} \text{ (kernel of maximal extension)}.$$

It can be written in the form

$$f(t, y) = f_>(t, y) + f_<(t, y), \tag{4.5}$$

where

$$f_<(t, y) = \sum_{k<0} a_k e^{-\lambda_k t} \varphi_k(y) \quad \text{and} \quad f_>(t, y) = \sum_{k>0} a_k e^{-\lambda_k t} \varphi_k(y).$$

Because of

$$\langle f, f \rangle_{L^2(X^R)} < +\infty \iff \langle f_<, f_< \rangle < +\infty \quad \text{and} \quad \langle f_>, f_> \rangle < +\infty,$$

the coefficients a_k satisfy the conditions

$$\sum_{k<0} |a_k|^2 \frac{e^{-2\lambda_k R} - 1}{2|\lambda_k|} < +\infty \quad \text{or, equivalently,} \quad \sum_{k<0} |a_k|^2 \frac{e^{2|\lambda_k|R}}{|\lambda_k|} < +\infty \quad (4.6)$$

and

$$\sum_{k>0} |a_k|^2 \frac{1 - e^{-2\lambda_k R}}{2\lambda_k} < +\infty \quad \text{or, equivalently,} \quad \sum_{k>0} \frac{|a_k|^2}{\lambda_k} < +\infty. \quad (4.7)$$

We consider the Cauchy data space $\Lambda(A_R, 0)$ consisting of all $\gamma(f)$ with $f \in \ker(A_R, 0)$. Here $\gamma(f)$ denotes the trace of f at the boundary

$$\Sigma = \partial X^R = -\Sigma_0 \sqcup \Sigma_R,$$

where Σ_R denotes a second copy of Σ_0. According to the spectral splitting $f = f_> + f_<$, we have

$$\gamma(f) = \left(s_<^0, s_<^R \right) + \left(s_>^0, s_>^R \right),$$

where

$$s_>^0 = f_>(0), \qquad s_<^0 = f_<(0), \qquad s_>^R = f_>(R), \qquad s_<^R = f_<(R).$$

Because of (4.6), we have

$$\left(s_<^0, s_>^R \right) \in C^\infty (\Sigma_0 \cup \Sigma_R).$$

Because of (4.7), we have

$$\left(s_>^0, s_<^R \right) \in H^{-1/2}(\Sigma_0 \cup \Sigma_R).$$

Recall that

$$\sum a_k \varphi_k \in H^s(\Sigma_0) \iff \sum |a_k|^2 |k|^{2s/(m-1)} < +\infty$$

and $|\lambda_k| \sim |k|^{\frac{1}{m-1}}$ for $k \to \pm\infty$, where $m-1$ denotes the dimension of Σ_0.

One notices that the estimate (4.6) for the coefficients of $s_<^0$ is stronger than the assertion that $\sum_{k<0} |a_k|^2 |\lambda_k|^N < +\infty$ for all natural N. Thus our estimates confirm that not each smooth section can appear as initial value over Σ_0 of a solution of $A_R f = 0$ over the cylinder. To sum up the example, the space $\Lambda(A_R, 0)$ can be written as the graph of an unbounded, densely defined, closed operator $T : \text{dom } T \to H^{-1/2}(\Sigma_R)$, mapping $s_<^0 + s_>^0 =: s^0 \mapsto s^R := s_<^R + s_>^R$ with dom $T < H^{-1/2}(\Sigma_0)$. To obtain a closed subspace of $L^2(\Sigma)$ one takes the range $\Lambda(A_R, 1/2)$ of the L^2-extension $\mathscr{P}(A_R)^{(0)}$ of the Calderón projection. It coincides with $\Lambda(A_R, 0) \cap L^2(\Sigma)$ by Proposition 4.4. In Theorem 5.3 we show without use of the pseudo-differential calculus why the intersection $\Lambda(A_R, 0) \cap L^2(\Sigma)$ must become closed in $L^2(\Sigma)$. See also Theorem 8.1 for another description of the Cauchy data space $\Lambda(A_R, 1/2)$, namely as the graph of a unitary elliptic pseudo-differential operator of order 0.

Since $\Sigma = -\Sigma_0 \sqcup \Sigma_R$, the tangential operator takes the form $\mathscr{B} = B \oplus (-B)$ and we obtain from (4.4)

$$\text{range } \Pi_>(\mathscr{B})^{(0)} = L_+(\mathscr{B}) = \text{span}_{L^2(\Sigma)}\{(\varphi_k, \sigma\,(\varphi_k))\}_{k>0}.$$

For comparison, we have in this example

$$\text{range } \mathscr{P}(A_R)^{(0)} = \Lambda\left(A_R, \frac{1}{2}\right) = \text{span}_{L^2(\Sigma)}\left\{\left(\varphi_k, e^{-\lambda_k R}\varphi_k\right)\right\}_{k \in \mathbf{Z}\setminus\{0\}},$$

hence $L_+(\mathscr{B})$ and $\Lambda(A_R, 1/2)$ are transversal subspaces of $L^2(\Sigma)$. On the half-infinite cylinder $[0, \infty) \times \Sigma$, however, we have only one boundary component Σ_0. Hence

$$\text{range } \Pi_>(B)^{(0)} = \text{span}_{L^2(\Sigma_0)}\{\varphi_k\}_{k>0} = \lim_{R \to \infty} \text{range } \mathscr{P}(A_R)^{(0)}.$$

One can generalize the preceding example: For any smooth compact manifold X with boundary Σ and any real $R \geq 0$, let X^R denote the stretched manifold

$$X^R := ([-R, 0] \times \Sigma) \cup_\Sigma X.$$

Assuming product structures with $A = \sigma(\partial_t + \mathscr{B})$ near Σ gives a well defined extension A_R of A. Nicolaescu [28] proved that the Calderón projection and the Atiyah–Patodi–Singer projection coincide up to a finite-dimensional component in the *adiabatic limit* ($R \to +\infty$ in a suitable setting). Even for finite R and, in particular for $R = 0$, one has the following interesting result. It was first proved in Scott [32] (see also Grubb [22] and Wojciechowski [20], Appendix who both offered different proofs).

Lemma 4.6. *For all $R \geq 0$, the difference $\mathscr{P}(A_R) - \Pi_\geq(\mathscr{B})$ is an operator with a smooth kernel.*

4.4. TWISTED ORTHOGONALITY OF CAUCHY DATA SPACES

Green's formula (in particular the Clifford multiplication σ in the case of Dirac type operators) provides a symplectic structure for $L^2(\Sigma; S|_\Sigma)$ for linear symmetric elliptic differential operators of first order on a compact smooth manifold X with boundary Σ. For elliptic systems of second order differential equations, various interesting results have been obtained in the 70s by exploiting the symplectic structure of corresponding spaces (see, e.g., [25]). Restricting oneself to first order systems, the geometry becomes very clear and it turns out that the Cauchy data space $\Lambda(A, 1/2)$ is a Lagrangian subspace of $L^2(\Sigma; S|_\Sigma)$. More generally, in [14] we described the orthogonal complement of the Cauchy data space of the chiral Dirac operator A^+ by

$$\sigma^{-1}\left(\Lambda\left(A^-, \frac{1}{2}\right)\right) = \left(\Lambda\left(A^+, \frac{1}{2}\right)\right)^\perp. \tag{4.8}$$

We obtained a short exact sequence

$$0 \longrightarrow \sigma^{-1}\left(\Lambda\left(A^-, s\right)\right) \hookrightarrow H^{s-\frac{1}{2}}\left(\Sigma; S^+|_\Sigma\right) \xrightarrow{\mathcal{K}^{(s)}} \ker\left(A^+, s\right) \longrightarrow 0.$$

For the total (symmetric) Dirac operator this means:

Proposition 4.7. The Cauchy data space $\Lambda(A, 1/2)$ of the *total* Dirac operator is a Lagrangian subspace of the Hilbert space $L^2(\Sigma; S|_\Sigma)$ equipped with the symplectic form $\omega(\varphi, \psi) := (\sigma\varphi, \psi)$.

The preceding result has immediate applications in index theory, see section 6 below.

5. Cauchy data spaces and maximal and minimal domains

We give a systematic presentation of the boundary reduction of the solution spaces, inspired by M. Krein's construction of the maximal space of boundary values for closed symmetric operators (see [8] and [9]). In this section we stay in the real category and do not assume product structure near $\Sigma = \partial X$ unless otherwise stated. The operator A needs not be of Dirac type. We only assume that it is a linear elliptic symmetric differential operator of first order.

5.1. THE NATURAL CAUCHY DATA SPACE

Let A_0 denote the restriction of A to the space $C_0^\infty(X; S)$ of smooth sections with support in the interior of X. As mentioned above, there is no natural choice of the order of the Sobolev spaces for the boundary reduction. Therefore, a systematic treatment of the boundary reduction may begin with the minimal closed extension $A_{\min} := \overline{A_0}$ and the adjoint $A_{\max} := (A_0)^*$ of A_0. Clearly, $A_{\max}$ is the maximal closed extension. This gives

$$D_{\min} := \operatorname{dom}(A_{\min}) = \overline{C_0^\infty(X; S)}^{\mathcal{G}} = \overline{C_0^\infty(X; S)}^{H^1(X;S)}$$

and

$$D_{max} := \text{dom}\,(A_{max})$$
$$= \left\{ u \in L^2(X; S) \mid A\,u \in L^2(X; S) \text{ in the sense of distributions} \right\}.$$

Here, the superscript $\mathcal{G}$ means the closure in the graph norm which coincides with the 1st Sobolev norm on $C_0^\infty(X; S)$. We form the space β of *natural boundary values* with the *natural trace map* γ in the following way:

$$D_{max} \xrightarrow{\gamma} D_{max}/D_{min} =: \beta$$
$$x \longmapsto \gamma(x) = [x] := x + D_{min}.$$

The space β becomes a symplectic Hilbert space with the scalar product induced by the graph norm

$$(x, y)_{\mathcal{G}} := (x, y) + (A\,x, A\,y) \tag{5.1}$$

and the symplectic form given by Green's form

$$\omega([x], [y]) := (A\,x, y) - (x, A\,y) \quad \text{for } [x], [y] \in \beta. \tag{5.2}$$

It is easy to show that ω is non degenerate. We define the *natural Cauchy data space* $\Lambda(A) := \gamma(\ker A_{max})$. Let us assume that A has a self-adjoint L^2-extension with a compact resolvent. Then it is a Fredholm operator. Let us choose such an extension and denote its domain by D. Such an extension always exists. Take for instance $A_{\mathcal{P}(A)}$, the operator A with domain

$$\text{dom}\,A_{\mathcal{P}(A)} := \left\{ f \in H^1(X; S) \mid \mathcal{P}(A)^{(1/2)}\,(f|_\Sigma) = 0 \right\},$$

where $\mathcal{P}(A)$ denotes the Calderón projection defined in (4.2).

Proposition 5.1. (a) The Cauchy data space $\Lambda(A)$ is a closed Lagrangian subspace of β and belongs to the Fredholm–Lagrangian Grassmannian $\mathcal{FL}_{\lambda_0}$ at $\lambda_0 := \gamma(D)$.

(b) For arbitrary domains D with $D_{min} < D < D_{max}$ and $\gamma(D)$ Lagrangian, the extension $A_D := A_{max}|_D$ is self-adjoint. It becomes a Fredholm operator, if and only if the pair $(\gamma(D), \Lambda(A))$ of Lagrangian subspaces of β becomes a Fredholm pair.

(c) Let $\{C_t\}_{t\in I}$ be a continuous family (with respect to the operator norm) of bounded self-adjoint operators. Here the parameter t runs within the interval $I = [0, 1]$. We assume the non-existence of inner solutions ('weak UCP') for all operators $A^* + C_t$, i.e.,

$$D_{min} \cap \ker\,(A^* + C_t, 0) = \{0\} \quad \text{for all } t \in [0, 1]. \tag{5.3}$$

Then the spaces $\gamma(\ker(A^* + C_t, 0))$ of Cauchy data vary continuously in β.

Here, for $t \in I$ the spaces β_t are all naturally identified with β.

Remark 5.2. (a) It is an astonishing aspect of symplectic functional analysis that the proof of the preceding proposition is completely elementary (see [8, Proposition 3.5 and Theorem 3.8]).

(b) Clearly, $D_{\max}$ and $D_{\min}$ are $C^\infty(X)$-modules, and so the space β is a $C^\infty(\Sigma)$-module. This shows that β is *local* in the following sense: If Σ decomposes into r connected components $\Sigma = \Sigma_1 \sqcup \cdots \sqcup \Sigma_r$, then β decomposes into

$$\beta = \bigoplus_{j=1}^{r} \beta_j,$$

where

$$\beta_j := \gamma\left(\{f \in D_{\max} \mid \text{supp } f < N_j\}\right)$$

with suitable collar neighborhoods N_j of Σ_j. Note that each β_j is a closed symplectic subspace of β and therefore a symplectic Hilbert space.

(c) For elliptic differential operators of first order, the weak unique continuation property discussed in Subsection 3.3 implies unique continuation from hypersurfaces and so the property (5.3): if a section f belongs to $D_{\min}$, it vanishes at the whole boundary. So it can be extended to the closed double by 0. By elliptic regularity this extension is smooth, so $f|_X = 0$ by weak UCP.

Now consider a section f which vanishes on one connected component of the boundary. The arguments of Subsection 3.3 extend to this case and we obtain once again $f|_X = 0$ by weak UCP. For operators of Dirac type, this follows also from the early results by Aronszajn and Cordes (for a recent review and generalization see [3]). We combine this result with the preceding Remark b. Let us assume that the operator A satisfies the weak unique continuation property (from each connected component of the boundary as explained). For simplicity we assume that the boundary consists of three connected components. Then the natural Cauchy data space $\Lambda(A)$ intersects transversally each of the 'faces'

$$\Lambda(A) \cap (\{0\} \times \beta_2 \times \beta_3) = \Lambda(A) \cap (\beta_1 \times \{0\} \times \beta_3)$$
$$= \Lambda(A) \cap (\beta_1 \times \beta_2 \times \{0\}) = \{0\}. \quad (5.4)$$

In finite dimensions, this would contradict the Lagrangian property of $\Lambda(A)$ in the full symplectic Hilbert space $\beta = \beta_1 \dotplus \beta_2 \dotplus \beta_3$ for dimension considerations: because of the transversality condition, any such space $l := \Lambda(A)$ can be written as the graph of an injective linear mapping $C : \beta_1 \to \beta_2 \times \beta_3$, so $l = \text{graph } C$ and $\dim l = \dim \beta_1$. Without loss of generality we assume that $\dim \beta_1 \leq \dim \beta_j$ for $j = 2, 3$. That implies $2 \dim l \neq \dim \beta_1 \times \beta_2 \times \beta_3$. So l is not Lagrangian.

There is nothing disquieting in this remark because β becomes only finite-dimensional when X is an interval where the number of connected components of the boundary is limited to two.

By Theorem 4.2(a) and, alternatively and in greater generality, by Hörmander [24, Theorem 2.2.1 and the Estimate (2.2.8), p. 194] the space β is naturally embedded in the distribution space $H^{-1/2}(\Sigma; S|_\Sigma)$. Under this embedding we have $\Lambda(A) = \Lambda(A, 0)$, where the last space was defined in Definition 4.3.

If the metrics are product close to Σ, we can give a more precise description of the embedding of β, namely as a *graded* space of distributions. Let $\{\varphi_k, \lambda_k\}$ be a spectral resolution of $L^2(\Sigma)$ by eigensections of $\mathfrak{B}$. (Here and in the following we do not mention the bundle S). Once again, for simplicity, we assume ker $\mathfrak{B} = \{0\}$ and have $\mathfrak{B}\varphi_k = \lambda_k\varphi_k$ for all $k \in \mathbb{Z} \setminus \{0\}$, and $\lambda_{-k} = -\lambda_k$, $\sigma(\varphi_k) = \varphi_{-k}$, and $\sigma(\varphi_{-k}) = -\varphi_k$ for $k > 0$. In [9, Proposition 7.15] (see also [15] for a more general setting) it was shown that

$$\beta = \beta_- \oplus \beta_+$$

with

$$\beta_- := \overline{[\{\varphi_k\}_{k<0}]}^{H^{1/2}(\Sigma)} \quad \text{and} \quad \beta_+ := \overline{[\{\varphi_k\}_{k>0}]}^{H^{-1/2}(\Sigma)}. \tag{5.5}$$

Then β_- and β_+ are Lagrangian and transversal subspaces of β.

5.2. CRISS-CROSS REDUCTION

Let us define two Lagrangian and transversal subspaces $L_\pm$ of $L^2(\Sigma)$ in a similar way, namely by the closure in $L^2(\Sigma)$ of the linear span of the eigensections with negative, respectively. with positive eigenvalue. Then L_+ is dense in β_+, and β_- is dense in L_-. This anti-symmetric relation may explain some of the well observed delicacies of dealing with spectral invariants of continuous families of Dirac operators.

Moreover, $\gamma(D_{\mathrm{aps}}) = \beta_-$, where

$$D_{\mathrm{aps}} := \left\{ f \in H^1(X) \mid \Pi_> (f|_\Sigma) = 0 \right\} \tag{5.6}$$

denotes the domain corresponding to the Atiyah–Patodi–Singer boundary condition. Note that a series $\sum_{k<0} c_k \varphi_k$ may converge to an element $\varphi \in L^2(\Sigma)$ without converging in $H^{1/2}(\Sigma)$. Therefore such $\varphi \in L_-$ can not appear as trace at the boundary of any $f \in D_{\mathrm{max}}$.

Recall Proposition 5.1 and note that $(\beta_-, \Lambda(A))$ is a Fredholm pair.

This can all be achieved without the symbolic calculus of pseudo-differential operators. Therefore one may ask how the preceding approach to Cauchy data

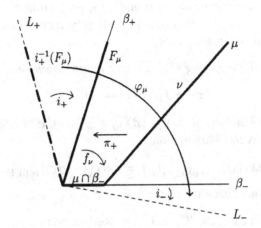

Figure 4. The mapping $\tau : \mathscr{FL}_{\beta_-}(\beta) \to \mathscr{FL}_{L_-}(L)$.

spaces and boundary value problems via the maximal domain and our symplectic space β is related to the approach via the Calderón projection, which we reviewed in the preceding section. How can results from the distributional theory be translated into L^2-results?

To relate the two approaches we recall a fairly general symplectic 'Criss-Cross' Reduction theorem from [10, Theorem 1.2]. Let β and L be symplectic Hilbert spaces with symplectic forms ω_β and ω_L, respectively. Let

$$\beta = \beta_- \dot{+} \beta_+ \qquad \text{and} \qquad L = L_- \dot{+} L_+$$

be direct sum decompositions by transversal (not necessarily orthogonal) pairs of Lagrangian subspaces. We assume that there exist continuous, injective mappings

$$i_- : \beta_- \longrightarrow L_- \qquad \text{and} \qquad i_+ : L_+ \longrightarrow \beta_+$$

with dense images and which are compatible with the symplectic structures, i.e.,

$$\omega_L\,(i_-(x), a) = \omega_\beta\,(x, i_+(a)) \qquad \text{for all } a \in L_+ \text{ and } x \in \beta_-.$$

Let $\mu \in \mathscr{FL}_{\beta_-}(\beta)$, e.g., $\mu = (\mu \cap \beta_-) \dot{+} v$ with a suitable closed v. Let us define (see also Figure 4)

$$\tau(\mu) := i_- \left(\mu \cap \beta_-\right) \dot{+} \operatorname{graph}\left(\varphi_\mu\right), \tag{5.7}$$

where

$$\varphi_\mu : i_+^{-1}\left(F_\mu\right) \longrightarrow L_-$$
$$x \longmapsto i_- \circ f_v \circ i_+(x).$$

Here F_μ denotes the image of μ under the projection π_+ from μ to β_+ along β_- and $f_\nu : F_\mu \to \beta_-$ denotes the uniquely determined bounded operator which yields ν as its graph. Then:

Theorem 5.3. *The mapping* (5.7) *defines a continuous mapping*

$$\tau : \mathcal{FL}_{\beta_-}(\beta) \longrightarrow \mathcal{FL}_{L_-}(L)$$

which maps the Maslov cycle $\mathcal{M}_{\beta_-}(\beta)$ *of* β_- *into the Maslov cycle* $\mathcal{M}_{L_-}(L)$ *of* L_- *and preserves the Maslov index*

$$\mathbf{mas}\left(\{\mu_s\}_{s\in[0,1]}, \beta_-\right) = \mathbf{mas}\left(\{\tau(\mu_s)\}_{s\in[0,1]}, L_-\right)$$

for any continuous curve $[0, 1] \ni s \mapsto \mu_s \in \mathcal{FL}_{\beta_-}(\beta)$.

In the product case, the 'Criss-Cross' Reduction theorem implies for our two types of Cauchy data that all results proved in the theory of natural boundary values (β-theory) remain valid in the L^2-theory. In particular we have:

Corollary. The $L^2(\Sigma)$ part $\Lambda(A) \cap L^2(\Sigma)$ of the natural Cauchy data space $\Lambda(A)$ is closed in $L^2(\Sigma)$. Actually, it is a Lagrangian subspace of $L^2(\Sigma)$ and it forms a Fredholm pair with the component L_-, defined at the beginning of this subsection.

Part II: Cauchy Data Spaces and Spectral Invariants

6. Non-Lagrangian half-spaces and index theory

6.1. INDEX THEORY

Recall that the index of a Fredholm operator measures its non-symmetry: it is defined by the difference between the dimension of the kernel (the null space) of the operator and the dimension of the kernel of the adjoint operator (= the codimension of the range). So, the index vanishes for self-adjoint Fredholm operators. For an elliptic differential operator A on a closed manifold M the index is finite and depends only on the homotopy class of the principal symbol σ of the operator over the cotangent sphere bundle S^*M. Therefore, the index always vanishes on odd-dimensional manifolds. On even-dimensional manifolds one has the Atiyah–Singer Index theorem which expresses the index in explicit topological terms, involving the Todd class defined by the Riemannian structure of M and the Chern class defined by gluing two copies of a bundle over S^*M by σ. It has turned out that various topological invariants of manifolds can be expressed by the index of naturally defined operators of Dirac type. In the index theory of closed manifolds one mostly studies the chiral and not the total Dirac operator (which is symmetric for compatible connections).

6.2. THE BOJARSKI CONJECTURE

The *Bojarski conjecture* gives quite a different description of the index of an elliptic operator over a closed partitioned manifold $M = M_0 \cup_\Sigma M_1$. It relates the 'quantum' quantity index with a 'classical' quantity, the Fredholm intersection index of the Cauchy data spaces from both sides along the hypersurface Σ. It was suggested in [4] and proved in [14] for operators of Dirac type.

Proposition 6.1. Let M be a partitioned manifold as before and let $\Lambda(A_j, 1/2)$ denote the L^2-Cauchy data spaces, $j = 0, 1$. Then

$$\text{index } A = \text{index } \left(\Lambda \left(A_0, \frac{1}{2} \right), \Lambda \left(A_1, \frac{1}{2} \right) \right).$$

Recall that

$$\text{index} \left(\Lambda \left(A_0, \frac{1}{2} \right), \Lambda \left(A_1, \frac{1}{2} \right) \right)$$

$$:= \dim \left(\Lambda \left(A_0, \frac{1}{2} \right) \cap \Lambda \left(A_1, \frac{1}{2} \right) \right)$$

$$- \dim \left(L^2 \left(\Sigma; S|_\Sigma \right) / \left(\Lambda \left(A_0, \frac{1}{2} \right) + \Lambda \left(A_1, \frac{1}{2} \right) \right) \right).$$

It is equal to $\mathbf{i}(\text{Id} - \mathcal{P}(A_1), \mathcal{P}(A_0))$ where $\mathcal{P}(A_j)$ denotes the corresponding Calderón projections. The proof of the proposition depends on the unique continuation property for Dirac operators and the Lagrangian property of the Cauchy data spaces, more precisely the chiral twisting property (4.8).

6.3. GENERALIZATIONS FOR GLOBAL BOUNDARY CONDITIONS

On a smooth compact manifold X with boundary Σ, the solution spaces $\ker(A, s)$ depend on the order s of differentiability and they are infinite-dimensional. To obtain a finite index one must apply suitable boundary conditions (see [14] for local and global boundary conditions for operators of Dirac type). In this report, we restrict ourselves to boundary conditions of Atiyah–Patodi–Singer type, i.e., $P \in \mathcal{G}r(A) := \mathcal{G}r_{p_+}$, and consider the extension

$$A_P : \text{dom}(A_P) \longrightarrow L^2(X; S) \tag{6.1}$$

of A defined by the domain

$$\text{dom}(A_P) := \left\{ f \in H^1(X; S) \mid P^{(0)}(f|_\Sigma) = 0 \right\}. \tag{6.2}$$

It is a closed operator in $L^2(X; S)$ with finite-dimensional kernel and cokernel. We have an explicit formula for the adjoint operator

$$(A_P)^* = A_{\sigma(\mathrm{Id}-P)\sigma^*}.\tag{6.3}$$

In agreement with Proposition 5.1(b), the preceding equation shows that an extension A_P is self-adjoint, if and only if $\ker P^{(0)}$ is a Lagrangian subspace of the symplectic Hilbert space $L^2(\Sigma; S|_\Sigma)$.

Let us recall the *Boundary Reduction formula for the Index* of (global) elliptic boundary value problems discussed in [13] (inspired by [37], see [14] for a detailed proof for Dirac operators). Like the Bojarski conjecture, the point of the formula is that it gives an expression for the index in terms of the geometry of the Cauchy data in the symplectic space of all (here L^2) boundary data.

Proposition 6.2.

$$\mathrm{index}\, A_P = \mathrm{index}\left\{ P\mathscr{P}(A) : \Lambda\left(A, \frac{1}{2}\right) \longrightarrow \mathrm{range}\left(P^{(0)}\right)\right\}.$$

6.4. PASTING FORMULAS

We shall close this section by mentioning a slight modification of the Bojarski conjecture/theorem, namely a non-additivity formula for the splitting of the index over partitioned manifolds.

Theorem 6.3. *Let P_j be projections belonging to $\mathscr{G}\mathrm{r}(A_j)$, $j = 0, 1$. Then*

$$\mathrm{index}\, A = \mathrm{index}\,(A_0)_{P_0} + \mathrm{index}\,(A_1)_{P_1} - \mathbf{i}\,(P_1, \mathrm{Id} - P_0).$$

It turns out that the boundary correction term $\mathbf{i}(P_1, \mathrm{Id} - P_0)$ equals the index of the operator $\sigma(\partial_t + \mathscr{B})$ on the cylinder $[0, 1] \times \Sigma$ with the boundary conditions P_0 at $t = 0$ and P_1 at $t = 1$.

Remark 6.4. (a) In this section we have not always distinguished between the *total* and the *chiral* Dirac operator because all the discussed index formulas are valid in both cases.

(b) Important index formulas for (global) elliptic boundary value problems for operators of Dirac type can also be obtained without analyzing the concept and the geometry of the Cauchy data spaces (see, e.g., the celebre Atiyah–Patodi–Singer Index theorem, [2] and [14, Chapter 22], or [31] for a recent survey of index formulas where there is no trace of the Calderón projection). The basic reason is that the index is an invariant represented by a local density inside the manifold plus a correction term which lives on the boundary and may be local or non-local as well. However, these formulas do not explain the simple origin of the index or the spectral flow, namely that *all* index information is naturally coded by the geometry of the Cauchy data spaces. To us it seems necessary to use the Calderón projection in order to understand (not calculate) the index of an elliptic boundary problem and the reason for the non locality, respectively, locality.

7. Family versions: The spectral flow and the Maslov index

7.1. SPECTRAL FLOW AND THE MASLOV INDEX

Let $\{A_t\}_{t \in [0,1]}$ be a continuous family of (from now on always *total*) Dirac operators with the same principal symbol and the same domain D. To begin with, we do not distinguish between the case of a closed manifold (when D is just the first Sobolev space and all operators are essentially self-adjoint) and the case of a manifold with boundary (when D is specified by the choice of a suitable boundary value condition).

We consider the *spectral flow* $\mathbf{sf}\{A_{t,D}\}$. Roughly speaking, it is the difference between the number of eigenvalues which change the sign from $-$ to $+$ as t goes from 0 to 1, and the number of eigenvalues which change the sign from $+$ to $-$. It can be defined in a satisfactory, purely functional analytical way, following a suggestion by J. Phillips (see [8] and [30]). We want a pasting formula for the spectral flow. To achieve that, one replaces the spectral flow of a continuous 1-parameter family of self-adjoint Fredholm operators, which is a 'quantum' quantity, by the *Maslov index* of a corresponding path of Lagrangian Fredholm pairs, which is a 'quasi-classical' quantity. The idea is owed to Floer and was worked out subsequently by Yoshida in dimension 3, by Nicolaescu in all odd dimensions, and pushed further by Cappell, Lee and Miller, Daniel and Kirk, and many other authors. For a survey, see [9] and [20].

Let us fix the space β for the family. By Proposition 5.1(c) the corresponding family $\{\Lambda(A_t)\}$ of natural Cauchy data spaces is continuous. In [8] we obtained the *General Boundary Reduction Formula* for the spectral flow which gives a family version of the Bojarski conjecture (our Proposition 6.1):

Theorem 7.1. *The spectral flow of the family* $\{A_{t,D}\}$ *and the Maslov index*

$$\mathbf{mas}\left(\{\Lambda(A_t)\}, \gamma(D)\right)$$

are well defined and we have

$$\mathbf{sf}\{A_{t,D}\} = \mathbf{mas}\left(\{\Lambda(A_t)\}, \gamma(D)\right). \tag{7.1}$$

We have various corollaries for the spectral flow on closed manifolds with fixed hypersurface (see [9]). For product structures near Σ, one can apply Theorem 5.3 and obtain an L^2-version of the preceding theorem which gives a new proof and a slight generalization of the Yoshida–Nicolaescu formula (for details see [10, section 3]):

Theorem 7.2.

$$\mathbf{sf}\{A_0 + C_t\} = \mathbf{mas}\left(\left\{\Lambda_t^0 \cap L^2(-\Sigma) + \Lambda_t^1 \cap L^2(\Sigma)\right\}, L_-\right)$$

$$=: \mathbf{mas}\left(\left\{\Lambda_t^0 \cap L^2(-\Sigma)\right\}, \left\{\Lambda_t^1 \cap L^2(\Sigma)\right\}\right),$$

where the last expression is given by the formula of the Maslov index of Fredholm pairs of two curves.

Remark 7.3. Nor here is it compelling to use the symplectic geometry of the Cauchy data spaces (see Remark 6.4(b)). Actually, deep gluing formulas can and have been obtained for the spectral flow by coding relevant information not in the full infinite-dimensional Cauchy data spaces but in families of Lagrangian subspaces of suitable finite-dimensional symplectic spaces, like the kernel of the tangential operator (see [18] and [19]).

7.2. CORRECTION FORMULA FOR THE SPECTRAL FLOW

Let D, D' with $D_{\min} < D, D' < D_{\max}$ be two domains such that both $\{A_{t,D}\}$ and $\{A_{t,D'}\}$ become families of self-adjoint Fredholm operators. We assume that D and D' differ only by a finite dimension, more precisely that

$$\dim \gamma(D)/\gamma(D) \cap \gamma(D') = \dim \gamma(D')/\gamma(D) \cap \gamma(D') < +\infty. \quad (7.2)$$

Then we find from Theorem 7.1 (for details see [9, Theorem 6.5]):

$$\mathbf{sf}\left(\{A_{t,D}\}\right) - \mathbf{sf}\left(\{A_{t,D'}\}\right)$$

$$= \mathbf{mas}\left(\{\Lambda(A_t)\}, \gamma(D')\right) - \mathbf{mas}\left(\{\Lambda(A_t)\}, \gamma(D)\right)$$

$$= \sigma_{\mathrm{Hör}}\left(\Lambda(A_0), \Lambda(A_1); \gamma(D'), \gamma(D)\right) \quad (7.3)$$

(cf. Remark 2.4(b)). The assumption (7.2) is rather restrictive. The pair of domains, for instance, defined by the Atiyah–Patodi–Singer projection and the Calderón projection, may not always satisfy this condition. For the present proof, however, it seems indispensable.

8. The boundary reduction and the gluing of determinants

There are competing concepts of the Fredholm determinant and the ζ-function regularized determinant. Following the condensed presentation of [6], this section presents the recent Scott–Wojciechowski formula under the perspective of Cauchy data spaces (the 'Bojarski approach').

8.1. THREE DETERMINANT CONCEPTS

Let us begin with the most simple integral of statistical mechanics, the *partition function* which is the model for all quadratic functionals:

$$Z(\beta) := \int_{\Gamma} e^{-\beta(Tx,x)} \, dx. \tag{8.1}$$

To begin with, let dim $\Gamma = d < \infty$ and β real with $\beta > 0$, and assume that T is a strictly positive, symmetric endomorphism. In suitable coordinates we evaluate the Gaussian integrals and find

$$Z(\beta) = \pi^{d/2} \cdot \beta^{-d/2} \cdot (\det T)^{-1/2}.$$

Two fundamental problems arise when we try to take a Dirac operator for T and all sections in a bundle S over a compact manifold M for Γ according to the Feynman recipes. What if T is not > 0? And what if $d = +\infty$ (i.e., if M is not a finite set of points)? To get around the first problem, one proceeds as follows: We decompose $\Gamma = \Gamma_+ \times \Gamma_-$ and $T = T_+ \oplus T_-$ with $T_+, -T_-$ strictly positive in $\Gamma_\pm$ and dim $\Gamma_\pm = d_\pm$. Formally, we obtain by a suitable path in the complex plane approaching $\beta = 1$:

$$Z(1) = \pi^{\zeta/2} \, e^{\pm i \frac{\pi}{4}(\zeta - \eta)} (\det |T|)^{-1/2}, \tag{8.2}$$

with $\zeta := d_+ + d_-$ and $\eta := d_+ - d_-$.

We shall not discuss the various stochastic ways of evaluating the integral when $d = +\infty$, but present two other concepts of the determinant.

From the point of view of functional analysis, the only natural concept is the *Fredholm determinant* of bounded operators acting on a separable Hilbert space of the form e^{α} or, more generally, Id $+ \alpha$ where α is of trace class. We recall the formulas

$$\det{}_{Fr} e^{\alpha} = e^{\mathrm{Tr}\,\alpha} \quad \text{and} \quad \det{}_{Fr}(\mathrm{Id} + \alpha) = \sum_{k=0}^{\infty} \mathrm{Tr} \wedge^k \alpha. \tag{8.3}$$

The Fredholm determinant is notable for obeying the product rule in difference to other generalizations of the determinant to infinite dimensions, where the error term of the product rule leads to new invariants.

Clearly, the parametrix (or Green's function) of a Dirac operator leads to operators for which the Fredholm determinant can be defined, but the relevant information about the spectrum of the Dirac operator does not seem sufficiently maintained. Note also that Quillen and Segal's construction of the *determinant line bundle* is based on the concept of the Fredholm determinant, but does not lead to *numbers* when the bundle is non trivial.

A third concept is the ζ-*function regularized determinant*, based on Ray and Singer's observation that formally

$$\det T = \prod \lambda_j = \exp\left\{\sum \ln \lambda_j e^{-z \ln \lambda_j}\big|_{z=0}\right\} = e^{-\frac{d}{dz}\zeta_T(z)|_{z=0}},$$

where

$$\zeta_T(z) := \sum_{j=1}^{\infty} \lambda_j^{-z} = \frac{1}{\Gamma(z)} \int_0^{\infty} t^{z-1} \operatorname{Tr} e^{-tT} \, dt.$$

For a *positive definite self-adjoint elliptic* operator T of second order, acting on sections of a Hermitian vector bundle over a closed manifold M of dimension m, Seeley [36] has shown that the function $\zeta_T(z)$ is holomorphic for $\Re(z)$ sufficiently large and can be extended meromorphically to the whole complex plane with $z = 0$ no pole.

The preceding definition does not apply immediately to the Dirac operator A which has infinitely many positive λ_j and negative eigenvalues $-\mu_j$. As an example, consider the operator

$$A_a := -i\frac{d}{dx} + a : C^{\infty}\left(S^1\right) \longrightarrow C^{\infty}\left(S^1\right)$$

with $A_a \varphi_k = k\varphi_k + a\varphi_k$ where $\varphi_k(x) = e^{ikx}$. It follows that $\operatorname{spec} A_a = \{k + a\}_{k \in \mathbb{Z}}$.

Choosing the branch $(-1)^{-z} = e^{-i\pi z}$, we find

$$\zeta_A(z) = \sum \lambda_j^{-z} + \sum (-1)^{-z} \mu_j^{-z}$$
$$= \frac{1}{2}\left\{\zeta_{A^2}\left(\frac{z}{2}\right) + \eta_A(z)\right\} + \frac{1}{2}e^{-i\pi z}\left\{\zeta_{A^2}\left(\frac{z}{2}\right) - \eta_A(z)\right\},$$

where $\eta_A(z) := \sum_{j=1}^{\infty} \lambda_j^{-z} + \mu_j^{-z}$. Thus:

$$\zeta_A'(0) = \frac{1}{2}\zeta_{A^2}'(0) - \frac{i\pi}{2}\left\{\zeta_{A^2}(0) - \eta_A(0)\right\}$$

and

$$\det_z A = e^{-\zeta_A'(0)} = e^{i\pi/2\{\zeta_{A^2}(0) - \eta_A(0)\}} \cdot e^{-\frac{1}{2}\zeta_{A^2}'(0)}. \tag{8.4}$$

8.2. THE SCOTT–WOJCIECHOWSKI FORMULA

In the one-dimensional case various authors obtained formulas for the ζ-regularized determinant of a system of ordinary differential equations subject to linear boundary conditions in terms of the usual determinant of the (finite) matrices defining these boundary conditions, see, e.g., [12], [17] and [27]. Here, we shall review recent progress in higher dimensions. In 1995 it was shown by Wojciechowski that the ζ-regularized determinant can also be defined for certain self-adjoint Fredholm extensions of the Dirac operator on a compact manifold with boundary, namely when the domain is defined by a projection belonging to the *smooth, self-adjoint* Grassmannian

$$\mathcal{Gr}^*_\infty(A) = \left\{ P \in \mathcal{Gr}(A) \mid P \text{ is self-adjoint, } P - \mathcal{P}(A) \text{ is smoothing} \right.$$
$$\left. \text{and range} \left(P^{(0)} \right) \text{ is Lagrangian in } L^2 \left(\Sigma; S|_\Sigma \right) \right\}. \tag{8.5}$$

We refer to [40] for the details of the delicate estimates needed for establishing the three involved invariants in that case.

Since then, Scott and Wojciechowski have established a formula which relates the ζ-determinant and the Fredholm determinant (see [33] and [34]).

Theorem 8.1. *Let* A *be a Dirac operator over an odd-dimensional compact manifold* M *with boundary* Σ *and let* $P \in \mathcal{Gr}^*_\infty(A)$. *Then the range of the Calderón projection* $\mathcal{P}(A)$ *(the Cauchy data space* $\Lambda(A, 1/2)$*) and the range of* P *can be written as the graphs of unitary, elliptic operators of order* 0, K, *respectively.* T *which differ from the operator* $(B^+ B^-)^{-1/2} B^+ : C^\infty(\Sigma; S^+|_\Sigma) \to C^\infty(\Sigma; S^-|_\Sigma)$ *by a smoothing operator. Moreover,*

$$\det{}_\zeta A_P = \det{}_\zeta A_{\mathcal{P}(A)} \cdot \det{}_{Fr} \frac{1}{2} \left(\mathrm{Id} + KT^{-1} \right). \tag{8.6}$$

In geometric terms, the key to understanding the preceding theorem is that the determinant line bundle, parametrized by the projections belonging to the smooth self-adjoint Grassmannian, is *trivial* so that one can attribute complex numbers (up to a multiple) to the canonical determinant section. This may explain why earlier attempts at relating the concept of the ζ-determinant with the Fredholm determinant had to be content with discussing the metric of the determinant bundle in terms of the ζ-determinant, and why the break through in understanding the mutual relation required a concept of boundary reduction.

Remark 8.2. (a) The variation of the modulus of the ζ-determinant contains a truly global term and can not be localized near the boundary. In [34], the authors get around this problem by varying a quotient of determinants.

(b) Various modifications and generalizations of the Scott–Wojciechowski formula are to be expected, in particular for the ζ-determinant over a partitioned manifold (i.e., a reduction formula to the hypersurface and a true pasting formula). Even so such results are not yet obtained, Scott's and Wojciechowski's formula provides an illuminating example of the meaning of the geometry of Cauchy data spaces for the study of spectral invariants.

8.3. AN ADIABATIC PASTING FORMULA

At the end of this exposition let us present a recent adiabatic splitting formula for the determinant proved by Park and Wojciechowski ([29]):

Theorem 8.3. *Let $R \in \mathbb{R}$ be positive, let M^R denote the stretched partitioned manifold $M^R = M_0 \cup_\Sigma [-R, 0] \times \Sigma \cup_\Sigma [0, R] \times \Sigma \cup_\Sigma M_1$, and let A_R, $A_{0,R}$, $A_{1,R}$ denote the corresponding Dirac operators. We assume that the tangential operator $\mathfrak{B}$ is invertible. Then*

$$\lim_{R \to \infty} \frac{\det_z A_R^2}{\left(\det_z (A_{0,R})_{\mathrm{Id}-\Pi_>}^2\right) \cdot \left(\det_z (A_{1,R})_{\Pi_>}^2\right)} = 2^{-\zeta_{\mathfrak{B}^2}(0)}.$$

References

[1]   Arnold, V.I., Characteristic class entering in quantization conditions, *Funkcional. Anal. i Priložen.* **1** (1967), 1–14. (Russian; English translation *Functional Anal. Appl.* **1** (1967), 1–13; French translation Complément 1 to Maslov, V.P., "*Théorie des perturbations et méthodes asymptotiques*", Dunod, Gauthier-Villars, Paris 1972, pp. 341–361).

[2]   Atiyah, M.F., Patodi, V.K., and Singer I.M., Spectral asymmetry and Riemannian geometry. I, *Math. Proc. Cambridge Phil. Soc.* **77** (1975), 43–69.

[3]   Bär, C., On nodal sets for Dirac and Laplace operators, *Comm. Math. Phys.* **188** (1997), 709–721.

[4]   Bojarski, B., The abstract linear conjugation problem and Fredholm pairs of subspaces, in '*In Memoriam I.N. Vekua*', Tbilisi Univ, Tbilisi 1979, pp. 45–60 (Russian).

[5]   Booß–Bavnbek, B., The unique continuation property for Dirac operators - revisited, in *Geometry and Topology: Aarhus*, K. Grove, I.H. Madsen, and E. Pedersen (eds.), Proceedings (Aarhus 1998), Amer. Math. Soc. Series Contemporary Mathematics, vol. 258, Providence, R.I. 2000, pp. 21–32.

[6]   ———, Boundary reduction of spectral invariants - results and puzzles, in *Stochastic Processes, Physics and Geometry: New Interplays. A Volume in Honor of Sergio Albeverio*, F. Gesztesy, H. Holden, and S. Paycha (eds.), Canadian Mathematical Society, Toronto 2000, pp. xxx–xxx (in print).

[7]   Booss–Bavnbek, B. and Bleecker, D.D., *Index theory with applications to mathematics and physics*, (in preparation).

[8]   Booss–Bavnbek, B. and Furutani, K., The Maslov index — a functional analytical definition and the spectral flow formula, *Tokyo J. Math.* **21** (1998), 1–34.

[9]   ———, Symplectic functional analysis and spectral invariants, in *Geometric Aspects of Partial Differential Equations*, B. Booss-Bavnbek and K.P. Wojciechowski (eds.), Amer. Math. Soc. Series Contemporary Mathematics, vol. 242, Providence, R.I. 1999, pp. 53–83.

[10] Booss–Bavnbek, B., Furutani, K., and Otsuki, N., Criss–cross reduction of the Maslov index and a proof of the Yoshida–Nicolaescu theorem, *Tokyo J. Math.* (in print).

[11] Booss–Bavnbek, B., Morchio, G., Strocchi, F., and Wojciechowski, K.P., Grassmannian and chiral anomaly, *J. Geom. Phys.* **22** (1997), 219–244.

[12] Booss–Bavnbek, B., Scott, S.G., and Wojciechowski, K.P., The ζ–determinant and $\mathscr{C}$–determinant on the Grassmannian in dimension one, *Letters in Math. Phys.* **45** (1998), 353–362.

[13] Booss–Bavnbek, B. and Wojciechowski, K.P., Desuspension of splitting elliptic symbols, Part I, *Ann. Glob. Analysis and Geometry* **3** (1985), 337–383. Part II, *Ann. Glob. Analysis and Geometry* **4** (1986), 349–400. Addendum, *Ann. Glob. Analysis and Geometry* **5** (1987), 87.

[14] _____, *Elliptic boundary problems for Dirac operators*, Birkhäuser, Boston 1993.

[15] Brüning, J. and Lesch, M., Spectral theory of boundary value problems for Dirac type operators, in *Geometric Aspects of Partial Differential Equations*, B. Booss-Bavnbek and K.P. Wojciechowski (eds.), Amer. Math. Soc. Series Contemporary Mathematics, vol. 242, Providence, R.I., 1999, pp. 203–215.

[16] Calderón, A.P., Boundary value problems for elliptic equations, in *Outlines of the Joint Soviet–American Symposium on Partial Differential Equations*, Novosibirsk 1963, pp. 303–304.

[17] Cappell, S.E., Lee, R., and Miller, E.Y., On the Maslov index, *Comm. Pure Appl. Math.* **47** (1994), 121–186.

[18] _____, Selfadjoint elliptic operators and manifold decompositions Part I: Low eigenmodes and stretching, *Comm. Pure Appl. Math.* **49** (1996), 825–866. Part II: Spectral flow and Maslov index, *Comm. Pure Appl. Math.* **49** (1996), 869–909. Part III: Determinant line bundles and Lagrangian intersection, *Comm. Pure Appl. Math.* **52** (1999), 543–611.

[19] _____, *Surgery formulae for analytical invariants of manifolds*, Preprint, 2000.

[20] Daniel, M. and Kirk, P., with an appendix by K.P. Wojciechowski, A general splitting formula for the spectral flow, *Michigan Math. Journal* **46** (1999), 589–617.

[21] de Gosson, M., Maslov classes, metaplectic representation and Lagrangian quantization, *Mathematical Research*, vol. 95, Akademie Verlag, Berlin 1997.

[22] Grubb, G., Trace expansions for pseudodifferential boundary problems for Dirac-type operators and more general systems, *Ark. Mat.* **37** (1999), 45–86.

[23] Hawking, S.W., Quantum cosmology, in *Relativité, groupes et topologie II*, École d'été de physique théorique (Les Houches, 1983), B.S. DeWitt and R. Stora (eds.), North–Holland, Amsterdam 1984, pp. 333–379.

[24] Hörmander, L., Pseudo-differential operators and non-elliptic boundary problems, *Ann. of Math.* **83** (1966), 129–209.

[25] Lawruk, B., Śniatycki, J., and Tulczyjew, W.M., Special symplectic spaces, *J. Differential Equations* **17** (1975), 477–497.

[26] Leray, J., *Analyse Lagrangiénne et mécanique quantique: Une structure mathématique apparentée aux développements asymptotiques et à l'indice de Maslov*, Série Math. Pure et Appl., I.R.M.P., Strasbourg 1978 (English translation 1981, MIT Press).

[27] Lesch, M. and Tolksdorf, J., On the determinant of one-dimensional elliptic boundary value problems, *Comm. Math. Phys.* **193** (1998), 643–660.

[28] Nicolaescu, L., The Maslov index, the spectral flow, and decomposition of manifolds, *Duke Math. J.* **80** (1995), 485–533.

[29] Park, J. and Wojciechowski, K.P., *Relative ζ-determinant and adiabatic decomposition of the ζ–determinant of the Dirac Laplacian*, Preprint, 2000.

[30] Phillips, J., Self-adjoint Fredholm operators and spectral flow, *Canad. Math. Bull.* **39** (1996), 460–467.

[31] Schulze, B.W., An algebra of boundary value problems not requiring Shapiro–Lopatinskij conditions, *J. Funct. Anal.* (2000), (in print).

[32] Scott, S.G., Determinants of Dirac boundary value problems over odd-dimensional manifolds, *Comm. Math. Phys.* **173** (1995), 43–76.

[33] Scott, S.G. and Wojciechowski, K.P., ζ-determinant and the Quillen determinant on the Grassmannian of elliptic self-adjoint boundary conditions, *C. R. Acad. Sci., Serie I* **328** (1999), 139–144.

[34] ———, *The ζ-determinant and Quillen determinant for a Dirac operator on a manifold with boundary*, GAFA 2000, (in print).

[35] Seeley, R.T., Singular integrals and boundary value problems, *Amer. J. Math.* **88** (1966), 781–809.

[36] ———, *Complex powers of an elliptic operator*, AMS Proc. Symp. Pure Math. X. AMS Providence 1967, pp. 288–307.

[37] ———, Topics in pseudo-differential operators, in *CIME Conference on Pseudo–Differential Operators (Stresa 1968)*. Ed. Cremonese, Rome 1969, pp. 167–305.

[38] Swanson, R.C., Fredholm intersection theory and elliptic boundary deformation problems, I, *J. Differential Equations* **28** (1978), 189–201.

[39] Taylor, M.E., *Partial differential equations I - basic theory*, Springer, New York 1996.

[40] Wojciechowski, K.P., The ζ–determinant and the additivity of the η–invariant on the smooth, self-adjoint Grassmannian, *Comm. Math. Phys.* **201** (1999), 423–444.

[41] Yoshida, T., Floer homology and splittings of manifolds, *Ann. of Math.* **134** (1991), 277–323.

ON THE CAUCHY PROBLEM FOR KIRCHHOFF EQUATIONS OF p-LAPLACIAN TYPE

KUNIHIKO KAJITANI

In memory of Jean Leray

Contents

1   Introduction 357
2   Proof of theorem 357

M. de Gosson (ed.), Jean Leray '99 Conference Proceedings, 355-362.
© *2003 Kluwer Academic Publishers.*

1. Introduction

We shall obtain the local existence theorem in Gevrey class of solutions to the Cauchy problem for Kirchhoff equations of p-Laplacian type. Let consider the following Cauchy problem

$$\partial_t^2 u(t,x) - A\left(\|\nabla_x u(t)\|_{L^p(\mathbb{R}^n)}^p\right) \Delta_p u(t,x) = f(t,x),$$

$$t \in (0,T),\ x \in \mathbb{R}^n, \tag{1.1}$$

$$u(0,x) = u_0(x), \qquad u_t(0,x) = u_1(x), \quad x \in \mathbb{R}^n, \tag{1.2}$$

where A is a positive and two times continuously differentiable function and Δ_p is defined by

$$\Delta_p u = \nabla_x \left(|\nabla_x u|^{p-2} \nabla_x u\right). \tag{1.3}$$

We introduce a functional space as follows

$$H_{\rho,s}^l = \left\{u \in L^2\left(\mathbb{R}^n\right);\ \langle\xi\rangle^l e^{\rho\langle\xi\rangle^{1/s}} \hat{u}(\xi) \in L^2\right\},$$

where $\hat{u}$ stands for a Fourier transform of u and $l \in \mathbb{R}^1, s \geq 1$ and $\rho > 0$ and denote by H^l the usual Sobolev space. Now we can state our main theorem.

Theorem 1.1. *Assume p is an even positive integer, $A(\eta) \geq a_0 > 0$ is in $C^2([0,\infty))$ and there are a positive integer $l, \rho_0 > 0, 1 < s < 4/3$ such that the initial data u_0, u_1 belong to $H_{\rho_0,s}^l$ and $f \in C^0([0,T_0];\ H_{\rho_0,s}^l)$. Then there is $T_0 > T > 0, \rho_1 > 0$ such that the Cauchy problem (1.1) and (1.2) has a unique solution u in $C^2([0,T];\ H_{\rho_1,s}^l)$.*

It is well known that when $p = 2$ and $s = 1$ the Cauchy problem (1.1) and (1.2) has a global real analytic solution. For example see the article of Bernstein [1] and of Pohozaev [4]. When $A(\eta) \equiv 1$, we can take Gevrey index s such that $1 < s \leq 2$, (cf. [2]).

2. Proof of theorem

For $a(t) \in C^1([0,T])$ we consider

$$\partial_t^2 u(t,x) - a(t)\Delta_p u(t,x) = f(t,x), \quad t \in (0,T),\ x \in \mathbb{R}^n. \tag{2.1}$$

Denote

$$B_{T,M} = \left\{a(t) \in C^1([0,T]);\ a(t) \geq a_0,\ |a|_{C^1([0,T])} \leq M\right\}, \tag{2.2}$$

where $|a|_{C^1([0,T])} = \sup_{k \leq 1, t \in [0,T]} |\frac{d^k}{dt^k} a(t)|$ and a_0 is a positive constant. Then it follows from [3] that we can prove

Proposition 2.1. Assume p is an even positive integer and there are a positive integer l, $\rho_0 > 0$, $1 < s < 4/3$ such that the initial data $u_0, u_1 \in H^l_{\rho_0,s}$ and $f \in C^0([0, T_0]; H^l_{\rho_0,s})$. Then there are $T_0 \geq T > 0$, $M > 0$ and ρ_1 such that for any $a(t) \in B_{T,M}$ the Cauchy problem (2.1), (1.2) has a unique solution $u \in C^2([0, T]; H^l_{\rho_1,s})$ satisfying

$$\|\partial_t u(t)\|_{H^l_{\rho_1,s}} + \|u(t)\|_{H^l_{\rho_1,s}} \leq C(M), \quad t \in [0, T]. \tag{2.3}$$

Consider the following linear hyperbolic equation

$$\partial_t^2 u(t, x) - a(t, x, D)u(t, x) - b(t, x, D)u(t, x) = f(t, x),$$
$$t \in (0, T), \ x \in \mathbb{R}^n, \tag{2.4}$$
$$u(0, x) = u_t(0, x) = 0, \quad x \in \mathbb{R}^n. \tag{2.5}$$

We assume that $a(t, x, \xi) \in C^1([0, T]; \gamma^d S^2)$, $b(t, x, \xi) \in C^0([0, T]; \gamma^d S^1)$ and $a(t, x, \xi) \geq 0$ in $[0, T] \times \mathbb{R}^{2n}$ that assures the hyperbolicity of the equation of (2.4), where $a \in \gamma^s S^m$ means that there are $C_a > 0$, $\rho_a > 0$ such that

$$\left|\partial_\xi^\alpha \partial_x^\beta a(x, \xi)\right| \leq C_a \rho^{-|\alpha+\beta|} |\alpha + \beta|!^s \langle\xi\rangle^{m-|\alpha|}, \quad x, \xi \in \mathbb{R}^n.$$

Then we have from [3, Theorem 4.4].

Proposition 2.2. Assume that $1 < d < 4/3$ and the above conditions are valid. Then we have a priori estimate below for the solution u of (2.4) and (2.5)

$$\|u_t(t)\|_{H^l_{\rho(t),s}} + \|u(t)\|_{H^l_{\rho(t),s}} \leq Ce^{ct} \int_0^t \|f(\tau)\|_{H^l_{\rho(\tau),s}} d\tau, \quad t \in [0, T], \tag{2.6}$$

where $C, c > 0$ and $\rho(t) = c_1 - c_2 t$ (c_1, c_2 positive constants).

Let u be a solution in $C^2([0, T]; H^l_{\rho_1,s})$ of (2.1)–(1.2). Put $v(t, x) = u(t, x) - (u_0(x) + u_1(x)t)$. Then $v \in C^2([0, T]; H^l_{\rho_1,s})$ satisfies a linear hyperbolic equation below

$$\partial_t^2 v(t, x) - a(t) \left(|\nabla_x u|^{p-2} \Delta v + \sum_{j,k=1}^n u_{x_j} u_{x_k} v_{x_j x_k} \right)$$

$$= f(t, x) + a(t) \left(|\nabla_x u|^{p-2} \Delta (u_0 + tu_1) + \sum_{j,k=1}^n u_{x_j} u_{x_k} (u_0 + tu_1)_{x_j x_k} \right)$$

$$v(0, x) = v_t(0, x) = 0.$$

Applying Proposition 2.2 to the above equation we obtain

$$\|v_t(t)\|_{H^l_{\rho_1,s}} + \|v(t)\|_{H^l_{\rho_1,s}} \leq C(M)T, \quad t \in [0, T].$$

Therefore we obtain from (2.3) and (2.6)

$$\|u_t(t)\|_{H^l_{\rho_1,s}} + \|u(t)\|_{H^l_{\rho_1,s}} \le C, \tag{2.7}$$

if we take $T > 0$ such that $C(M)T \le 1$, where C is independent of M.

For $a \in B_{T,M}$ we denote by $u_a(t, x)$ the solution of (2.1)–(1.2) and define

$$\Psi(a)(t) = A\left(\|\nabla_x u_a(t, \cdot)\|^p_{L^p}\right). \tag{2.8}$$

Then we can obtain:

Proposition 2.3. Assume p is an even positive integer and there are a positive integer l, $\rho_0 > 0$, $1 < s < 4/3$ such that the initial data $u_0, u_1 \in H^l_{\rho_0,s}$ and $f \in C^0([0, T_0]; H^l_{\rho_1,s})$. Then there are $T_0 > T > 0$, $M > 0$ such that Ψ is a contraction mapping in $B_{T,M}$.

This proposition implies our Theorem 1.1.

To prove this proposition we need a lemma. Define an energy for the equation (2.1)–(1.2) as follows

$$e(t)^2 = \frac{1}{2}\|u(t)\|^2_{L^2} + \frac{a(t)}{p}\|\nabla_x u(t)\|^p_{L^p}. \tag{2.9}$$

Then we have the following lemma.

Lemma 2.4. Let a be in $B_{T,M}$ and u in $\cap^2_{k=0} C^k([0, T]; H^{2-k}(\mathbb{R}^n))$ a solution of (2.1)–(1.2) with $f \in C^0([0, T]; L^2(\mathbb{R}^n))$. Then

$$e(t) \le e^{Mt/2a_0}\left[e(0) + \int_0^t \frac{1}{2}\|f(s)\|_{L^2}\right], \quad t \in (0, T). \tag{2.10}$$

Proof. Denote by $e'(t)$ the derivative of a function $e(t)$. From (2.1)

$$2e(t)e'(t) = \frac{a'(t)}{p}\|\nabla_x u(t)\|^p_{L^p} + \Re(u_t(t), f(t))_{L^2}$$

$$\le \frac{M}{a_0}e(t)^2 + e(t)\|f(t)\|_{L^2},$$

which implies (2.10). □

Proof of Proposition 2.1. In brief we denote

$$\eta_a(t) = \|\nabla_x u_a(t)\|^p_{L^p}.$$

It follows from Lemma 2.4

$$\eta_a(t) \le p/a_0 e^{Mt/2a_0}\left[e(0) + \frac{1}{2}\int_0^t \|f(s)\|_{L^2}\right] \tag{2.11}$$

$$\le Ce^{CMT}(1 + T) \le C_1, \quad t \in [0, T],$$

where C_1 is independent of M, if we choose T such that $TM \leq 1$. Moreover by virtue of Sobolev's lemma we can estimate

$$|\eta_a'(t)| = p \left| \int |\nabla_x u_a(t, x)|^{p-2} \nabla_x u_a \cdot \nabla_x u_{at}(t, x) dx \right|$$
$$\leq C(n, p) \|\nabla_x u(t)\|_{H^{n/2+1}}^{p-2} \|\nabla_x u_{at}\|_{L^2} \|\nabla_x u_a(t)\|_{L^2}.$$

Hence we can see by use of (2.7)

$$|\eta_a'(t)| \leq C_1, \quad t \in [0, T],$$

if we choose $T > 0$ such that $C(M)T \leq 1$, where C_1 is independent of M. Therefore when we take $M > 0$ such that $M \geq C_1 |A|_{C^2([0,C_1])}$, $\Phi(a)$ belongs to $B_{T,M}$ for any $a \in B_{T,M}$. Next we shall prove that there are $M > 0, T > 0$ such that

$$|\Phi(a) - \Phi(a')|_{C^1([0,T])} \leq \frac{1}{2}|a - a'|_{C^0([0,T])}, \quad a, a' \in B_{T,M}. \quad (2.12)$$

Put $w(t, x) = u_a(t, x) - u_{a'}(t, x)$. Since we have

$$|\eta_a(t) - \eta_{a'}(t)| \leq (\|\nabla_x u_a(t)\|_{L^p} + \|\nabla_x u_{a'}(t)\|_{L^p})^{p-1} \|\nabla_x w(t)\|_{L^p}, \quad (2.13)$$

we can see from (2.7) and (2.13)

$$|(\Phi(a) - \Phi(a'))(t)| = \left| \int_0^1 A'(\eta_{a'} + \theta(\eta_a - \eta_{a'})) d\theta (\eta_a - \eta_{a'})(t) \right| \quad (2.14)$$
$$\leq |A|_{C^1([0,C_1])} C \|\nabla_x w\|_{L^p}.$$

Decompose

$$\frac{d\Phi(a)(t)}{dt} - \frac{d\Phi(a')(t)}{dt}$$
$$= A'(\eta_a(t)) \eta_a'(t) - A'(\eta_{a'}(t)) \eta_{a'}'(t)$$
$$= \int_0^1 A''(\eta_a(t) + \theta(\eta_a - \eta_{a'})(t)) d\theta (\eta_a - \eta_{a'})(t) \eta_a'(t) \quad (2.15)$$
$$+ A'(\eta_{a'}(t)) (\eta_a' - \eta_{a'}')(t).$$

Moreover Sobolev's lemma implies

$$
\begin{aligned}
\left|\eta_a'(t) - \eta_{a'}'(t)\right| &= p \left| \int |\nabla_x u_a(t,x)|^{p-2} \nabla_x u_a \cdot \nabla_x u_{at}(t,x) \right. \\
&\quad \left. - |\nabla_x u_{a'}(t,x)|^{p-2} \nabla_x u_{a'} \cdot \nabla_x u_{a't}(t,x) dx \right| \\
&\leq 2p \left(\|\nabla_x u_a(t)\|_{H^{n/2+1}} + \|\nabla_x u_a(t)\|_{H^{n/2+1}} \right)^{p-2} \\
&\quad \times \|\nabla_x w_a(t)\|_{L^2} \|\nabla_x u_{at}(t)\|_{L^2} + p \|\nabla_x u_{a'}(t)\|_{H^{n/2+1}}^{p-2} \\
&\quad \times \|\nabla_x u_{a'}(t)\|_{L^2} \|\nabla_x w_t(t)\|_{L^2} .
\end{aligned}
\tag{2.16}
$$

Hence we obtain from (2.7), (2.13) and (2.15)

$$
\begin{aligned}
\left| \frac{d\Phi(a)(t)}{dt} - \frac{d\Phi(a')(t)}{dt} \right| &\leq C_3 \left(\|\nabla_x w(t)\|_{L^p} + \|\nabla_x w(t)\|_{L^2} \right. \\
&\quad \left. + \|\nabla_x w_t(t)\|_{L^2} \right), \quad t \in [0,T],
\end{aligned}
\tag{2.17}
$$

where C_1 is independent of M. On the other hand $w = u_a - u_{a'}$ satisfies the following hyperbolic linear equation

$$
\begin{aligned}
&\partial_t^2 w - a(t) \left(|\nabla_x u_a|^{p-2} \Delta w + p |\nabla_x u_a|^{p-4} \sum_{j,k=1}^n u_{ax_j} u_{a'x_k} \partial_{x_j} \partial_{x_k} w \right) \\
&- \sum_j^n b_j (u_a, u_{a'}) \partial_{x_j} w = (a - a')(t) \Delta_p u_{a'}, \qquad w(0) = w_t(0) = 0.
\end{aligned}
$$

Therefore it follows from Proposition 2.2 that w satisfies

$$
\begin{aligned}
\|w_t\|_{H^l_{\rho(t),s}} + \|w\|_{H^l_{\rho(t),s}} &\leq C e^{ct} \int_0^t \left| (a-a')(\tau) \right| \left\| \Delta_p u_{a'}(\tau) \right\|_{H^l_{\rho(\tau),s}} d\tau, \quad t \in [0,T].
\end{aligned}
\tag{2.18}
$$

Noting

$$
\|\nabla_x w\|_{L^p} \leq \|\nabla_x w\|_{H^{n/2+1}}^{(p-2)/p} \|\nabla_x w\|_{L^2}^{2/p} \leq \|\nabla_x w\|_{H^l_{\rho_1,s}},
$$

and for $\rho(t) \leq 1/2\rho_1$

$$
\left\| \Delta_p u_{a'}(\tau) \right\|_{H^l_{\rho(\tau),s}} \leq C \|u_{a'}(\tau)\|_{H^{l+2}_{\rho(\tau),s}}^{p-2} \leq C \|u_{a'}(\tau)\|_{H^l_{\rho_1,s}}^{p-2}
$$

we can estimate by vitrue of (2.14), (2.17) and (2.18)

$$
\left| (\Phi(a) - \Phi(a')) \right|_{C^1([0,T])} \leq C(M) T |a - a'|_{C^0([0,T])},
$$

which implies (2.12), if we take T such that $C(M)T \leq 1/2$. Thus we have proved Proposition 2.2. $\qquad \square$

References

[1] Bernstein, S., Sur une classe d'équations fonctionnelles aux dérivèes partielles, *Izv. Akad. Nauk SSSR, Ser. Mat.* **4** (1940), 17–26.

[2] Kajitani, K., Local solution of Cauchy problem for nonlinear hyperbolic systems in Gevrey class, *Hokkaido Math. J.* **12** (1983), 434–460.

[3] Kajitani, K., The Cauchy problem for nonlinear hyperbolic systems, *Bull. Sci. Math.* **110** (1986), 3–48.

[4] Pohozaev, S.I., On a class of quasilinear hyperbolic equations, *Math. USSR. Sb.* **96** (1975), 152–166.

A REMARK ON SURGERY IN INDEX THEORY OF ELLIPTIC OPERATORS

V.E. NAZAIKINSKII AND B.Y. STERNIN

Contents

1 Introduction 365
2 Bottleneck spaces and a general relative index theorem 366
3 Applications to index formulas for operators with symmetry conditions on manifolds with singularities 369
    3.1   The index theorem for elliptic pseudo-differential operators 369
    3.2   The index theorem for Fourier integral operators 370

M. de Gosson (ed.), Jean Leray '99 Conference Proceedings, 363-372.
© *2003 Kluwer Academic Publishers.*

Abstract. *We suggest a general version of the relative index formula related to surgery and apply it to obtain index formulas for elliptic pseudo-differential operators and Fourier integral operators on manifolds with conical singularities.*

1. Introduction

Recently, a number of index theorems for elliptic operators with symmetry conditions on manifolds with singularities have been obtained. These results, in particular, include index theorems for elliptic differential and pseudo-differential operators ([11], [12] and [27]) as well as an index theorem for quantized contact transformations [21]. The main tool in all these papers is essentially glueing the second copy of the (stretched) manifold and passing to operators on the double of the original manifold. The double is a smooth closed manifold, where the classical index theory (for the case of pseudo-differential operators) or the recent result owed to Epstein and Melrose [10] (for the case of quantized canonical transformations) can be applied. The problem is then to express the difference between the index of the original operator and half the index of the new operator on the double via invariants of the conormal symbol, which is performed in the papers cited above with the help of various techniques. Glueing the second copy and passing to operators on the double (or, more generally, on a manifold containing the original manifold as a part) is in fact a very old technique. It was successfully applied to boundary value problems already in early papers of the Soviet school (e.g., see [1], [7], [9], [30] and the survey [2]) and later was used extensively by various authors. We only mention the papers [17], [18] and [28], where glueing via an orientation-reversing automorphism of the boundary was use to handle the index of the singnature operator, and the papers [14] and [15]; of course, this list is by no means complete. The usefulness of passing to the double in index problems is in fact due to a certain 'relative index principle', which can be informally stated as follows: if we modify an elliptic operator on some part of the manifold where it is defined (a modification may involve surgery), then the resulting increment in the index is independent of the structure of the operator on any other part of the manifold that has an empty intersection with this part. This relative index principle (which is obvious for smooth compact manifolds in view of local index formulas; e.g., see [13]) was proved in a number of special cases. We mention the paper [16], where a relative index theorem for Dirac operators on complete non-compact Riemannian manifolds was proved, the paper [3], where this theorem was generalized to arbitrary essentially self-adjoint supersymmetric Fredholm elliptic operators on the first order, and also the papers [6], [8], [29] and the book [5]. Again, it would be hard to give a complete list.

We suggest a general version of the relative index theorem and apply it to derive index theorems, generalizing those known previously, for elliptic pseudo-differential operators and Fourier integral operators on manifolds with singularities. This general version involves an abstract notion of elliptic operator. We note in passing that the notion of general elliptic operators in Hilbert $C(X)$-modules introduced by Atiyah [4] is not sufficient for our aims: by definition, Atiyah's operators have compact commutators with operators of multiplication by functions, and so this definition does not cover Fourier integral operators considered in our second index theorem. On the other hand, Atiyah's operators satisfy our definition.

In section 1 we give the necessary definitions and the general relative index theorem, and in section 2 we consider applications. Proofs are omitted. They can be found in the complete version [25] of this paper, which appeared while this volume was in preparation.

The authors are grateful to A.Yu. Savin and V.E. Shatalov for valuable discussions and to B.-W. Schulze for his attention and support.

The work is supported by Institut für Mathematik, Universität Potsdam, and RFBR grants Nos. 00-01-00161, 99-01-01100, 99-01-01254, 02-01-00118.

2. Bottleneck spaces and a general relative index theorem

Definition. (a) A *bottleneck space* is a separable Hilbert space H equipped with the structure of a module over the commutative topological algebra $C^\infty([-1, 1])$ (the action is continuous, and the unit function $1 \in C^\infty([-1, 1])$ acts as the identity operator on H).

(b) The *support of an element* $h \in H$ is the closed set

$$\text{supp}\, h = \bigcap \varphi^{-1}(0) \subset [-1, 1], \tag{2.1}$$

where $\varphi^{-1}(0)$ is the full preimage of the point 0 and the intersection is taken over all elements $\varphi \in C^\infty([-1, 1])$ such that $\varphi h = 0$. For an arbitrary subset $F \subset [-1, 1]$, by $H(F)$ we denote the closure in H of the set of elements $h \in H$ supported in F.

Since for any locally finite open cover there exists a subordinate partition of unity in $C^\infty([-1, 1])$, it follows that the notion of support thus introduced possesses natural properties. In particular, if $F_1, F_2 \subset [-1, 1]$ are subsets such that $\overline{F_1} \cap \overline{F_2} = \emptyset$, then $H(F_1 \cup F_2) = H(F_1) \oplus H(F_2)$ (the sum is direct but not necessarily orthogonal).

Example 2.1. Let M be a manifold (possibly, non compact or with singularities), and let $\chi : M \to [-1, 1]$ be a continuous function that is smooth in the smooth part of M and satisfies the following condition: the *bottleneck $B =$* $\chi^{-1}((-1, 1))$ is a compact set lying in the smooth part of M. The Sobolev spaces $H^s(M)$ (with arbitrary weights near singular points and at infinity) can be equipped with the structure of a bottleneck space as follows: for $h \in H^s(M)$ and $\varphi \in C^\infty([-1, 1])$, we set

$$(\varphi h)(x) = \varphi(\chi(x))h(x), \quad x \in M. \tag{2.2}$$

The construction described in this example is the main construction used in applications.

Definition. Let H_1, H_2 be bottleneck spaces, and let $F \subset [-1, 1]$ be a given subset. We say that H_1 *coincides with H_2 on F* and write $H_1 \overset{F}{=} H_2$, if we are given an isomorphism (not necessarily isometric) $H_1(F) \approx H_2(F)$. In this case we also say that H_1 and H_2 are *modifications* of each other on $[-1, 1]\backslash F$ and write $H_1 \overset{[-1,1]\backslash F}{\longleftrightarrow} H_2$. A square

$$
\begin{array}{ccc}
H_1 & \overset{B}{\longleftrightarrow} & H_2 \\
A \updownarrow & & \updownarrow C \\
H_3 & \overset{D}{\longleftrightarrow} & H_4
\end{array}
$$

of modifications is said to commute if the corresponding diagram

$$
\begin{array}{ccc}
H_1(F) & \approx & H_2(F) \\
\wr\wr & & \wr\wr \qquad\quad F = [-1, 1]\backslash(A \cup B \cup C \cup D) \\
H_3(F) & \approx & H_4(F),
\end{array}
$$

of isomorphisms commutes.

Definition. (a) A *proper operator* in bottleneck spaces H and G is a continuous (in the uniform operator topology) family of bounded linear operators

$$D_\delta : H \longrightarrow G, \quad \delta > 0, \tag{2.3}$$

such that for any $\varepsilon > 0$ there exists a $\delta_0 > 0$ such that

$$\operatorname{supp} D_\delta h \subset U_\varepsilon(\operatorname{supp} h), \quad \forall h \in H \tag{2.4}$$

for any $\delta < \delta_0$. Here $U_\varepsilon(F)$ is the ε-neighborhood of F.

(b) An *elliptic operator* in bottleneck spaces H and G is a proper operator (2.3) such that D_δ is Fredholm for each δ and has an almost inverse $D_\delta^{[-1]}$ such that the family $D_\delta^{[-1]}$ is a proper operator.

We usually omit the parameter δ in the notation of a proper operator.

One can readily see that the product of proper operators (if it is defined) is a proper operator. In bottleneck spaces obtained by the construction of Example 2.1 on compact manifolds with singularities, elliptic pseudo-differential operators are elliptic in the sense of the above definition (or, more precisely, can be included in a family depending on the parameter δ and satisfying the definition). The same pertains to Fourier integral operators on manifolds with conical singularities if the bottleneck lies in a neighborhood of a conical point (or conical points) and has the form $B = \Omega \times [r_1, r_2]$ with sufficiently small $r_1, r_2 > 0$.

Now let

$$A_i : H_i \longrightarrow G_i, \quad i = 1, 2,$$

be proper operators in bottleneck spaces, and let $F \subset [-1, 1]$ be an open subset (in the topology of $[-1, 1]$).

Definition. We say that A_1 coincides with A_2 on F and write $A_1 \overset{F}{=} A_2$ if $H_1 \overset{F}{=} H_2$, $G_1 \overset{F}{=} G_2$, and for each compact set $K \subset F$ the following condition holds: there exists a $\delta_0 > 0$ such that

$$A_{1\delta}h = A_{2\delta}h, \quad \text{for } \delta < \delta_0 \text{ and } \operatorname{supp} h \subset K. \tag{2.5}$$

(This is well defined, since $h \in H_1(F) \approx H_2(F)$ and with regard for the fact that F is open, $A_{1\delta}h$, $A_{2\delta}h \in G_1(F) \approx G_2(F)$ for small δ.) In this case we also say that A_1 is obtained from A_2 by a *modification* on $[-1, 1]\backslash F$ and write $A_1 \overset{[-1,1]\backslash F}{\longleftrightarrow} A_2$. The notion of a *commutative square* of modifications is obvious.

Lemma 2.2. *Let $A_1 \overset{F}{=} A_2$ be elliptic operators in bottleneck spaces. Then $A_1^{[-1]} \overset{F}{=} A_2^{[-1]} + K$, where K is a proper compact operator. In particular, for $A_1 = A_2 = A$ we find that any two almost-inverses of A differ by a proper compact operator.*

Lemma 2.3. *Let $[-1, 1] = \bigcup_j F_j$ be an open cover, and let $A_1 \overset{F_j}{=} A_2$ for all j, where A_1 and A_2 are proper operators. Then $A_{1\delta} = A_{2\delta}$ for sufficiently small δ.*

Now we are in a position to state our main result.

Theorem 2.4. *Let the following diagram of modifications of elliptic operators in bottleneck spaces commute:*

$$
\begin{array}{ccc}
D & \overset{\{-1\}}{\longleftrightarrow} & D_- \\
{\scriptstyle\{1\}}\big\uparrow\big\downarrow & & \big\uparrow\big\downarrow{\scriptstyle\{1\}} \\
D_+ & \overset{\{-1\}}{\longleftrightarrow} & D_{+-}
\end{array}
$$

Then

$$\text{ind}(D) - \text{ind}(D_-) = \text{ind}(D_+) - \text{ind}(D_{+-}).$$

3. Applications to index formulas for operators with symmetry conditions on manifolds with singularities

In this section we present applications of the general relative index theorem to operators on manifolds with conical singularities. For definitions and further references concerning operators on manifolds with conical singularities, we refer the reader to [26].

3.1. THE INDEX THEOREM FOR ELLIPTIC PSEUDO-DIFFERENTIAL OPERATORS

Let N be a compact manifold with conical singularities. Without loss of generality, we assume that there is only one conical point α; the base Ω of the corresponding cone is not assumed to be connected. Let

$$\widehat{D} : H^s(N, E) \longrightarrow H^{s-m}(N, F)$$

be an elliptic pseudo-differential operator in Sobolev spaces on N with conormal symbol $D_0(p)$. (We assume, without loss of generality, that the weight exponent is zero and omit it altogether in the notation of Sobolev spaces.)

Next, let $g : \Omega \to \Omega$ be a diffeomorphism, and let

$$\mu_E : E|_\Omega \longrightarrow g^*(E|_\Omega), \qquad \mu_F : F|_\Omega \longrightarrow g^*(F|_\Omega)$$

be bundle isomorphisms. Suppose that the following condition is satisfied.

Condition A. *The conormal symbols*

$$D_0(p) \qquad and \qquad \mu_F^{-1} g^* D_0(-p) \left(g^*\right)^{-1} \mu_E$$

are homotopic in the class of elliptic conormal symbols.

We denote the homotopy by D_{0t},

$$D_{00}(p) = D_0(p), \qquad D_{01}(p) = \mu_F^{-1} g^* D_0(-p) \left(g^*\right)^{-1} \mu_E.$$

We construct a closed manifold $\mathcal{N}$ and bundles $\mathcal{E}, \mathcal{F}$ over it as follows. Deleting a small neighborhood of the conical point from N, we obtain a manifold $\widetilde{N}$ with boundary $\partial \widetilde{N} = \Omega$. We attach two copies of $\widetilde{N}$ to two faces of the cylinder $\Omega \times [0, 1]$ using the indentity mapping on the left face and the mapping g on the right face for glueing. We extend the bundles $E|_\Omega$ and $F|_\Omega$ to the cylinder $\Omega \times [0, 1]$ in a natural way and join them with the corresponding bundles on two copies of $\widetilde{N}$ identically on the left face and using the isomorphisms μ_E, μ_F on the right face. The bundles thus obtained over $\mathcal{N}$ will be denoted by $\mathcal{E}$ and $\mathcal{F}$.

Theorem 3.1. *Under the above-mentioned conditions, the index formula*

$$\operatorname{ind} \widehat{D} = \frac{1}{2} \{\operatorname{ind} \mathscr{D} + \operatorname{sf} D_{0t}\}$$

holds, where

$$\mathscr{D} : H^s(\mathcal{N}, \mathscr{E}) \longrightarrow H^{s-m}(\mathcal{N}, \mathscr{F})$$

is the elliptic operator on the closed manifold $\mathcal{N}$ whose symbol is obtained by the natural glueing of two copies of the principal symbol of $\widehat{D}$ via the principal symbol of the homotopy D_{0t} and sf is the spectral flow of the family of conormal symbols, introduced in [22].

3.2. THE INDEX THEOREM FOR FOURIER INTEGRAL OPERATORS

Let us now state the index theorem for Fourier integral operators. For the definitions of homogeneous canonical transformations and Fourier integral operators on manifolds with singularities we refer the reader to [21] (see also [20]).

Let N_1 and N_2 be compact manifolds with conical singularities. The bases of the corresponding cones will be denoted by Ω_1 and Ω_2, respectively. Let

$$g : T^*N_1 \setminus \{0\} \longrightarrow T^*N_2 \setminus \{0\}$$

be a homogeneous canonical transformation of the corresponding (compressed) cotangent bundles without the zero sections. Let

$$\widehat{T} = T(g, a) : H^s(N_1, E) \longrightarrow H^{s-m}(N_2, F)$$

be an elliptic Fourier integral operator with amplitude a of some order m associated with g. We suppose that the conormal symbol $T_0(p)$ of the operator $\widehat{T}$ satisfies the following symmetry condition.

Let $g_j : \Omega_j \to \Omega_j$ be diffeomorphisms, and let

$$\mu_E : E|_{\Omega_1} \longrightarrow g_1^*(E|_{\Omega_1}), \qquad \mu_F : F|_{\Omega_2} \longrightarrow g_2^*(F|_{\Omega_2})$$

be bundle isomorphisms.

Condition B. *The conormal symbols*

$$T_0(p) \qquad and \qquad \mu_F^{-1} g_2^* D_0(-p) \left(g_1^*\right)^{-1} \mu_E$$

are homotopic in the class of elliptic conormal symbols of Fourier integral operators.

By T_{0t} we denote the corresponding homotopy. Under these assumptions, the following theorem, which is completely similar to the index theorem for pseudo-differential operators stated in the preceding subsection, is valid.

Theorem 3.2. *Under the above-mentioned conditions, the index formula*

$$\text{ind}\,\widehat{T} = \frac{1}{2}\,\{\text{ind}\,\mathcal{T} + \text{sf}\,T_{0t}\}$$

holds, where $\mathcal{T} : H^s(\mathcal{N}_\infty, \mathcal{E}) \rightarrow H^{s-m}(\mathcal{N}_\epsilon, \mathcal{F})$ *is the elliptic Fourier integral operator whose symbol is obtained by the natural glueing of two copies of the principal symbol of* $\widehat{T}$ *via the principal symbol of the homotopy* T_{0t} *and* sf *is the spectral flow of the family of conormal symbols.*

Along with the spectral flow, this index formula involves the index of a Fourier integral operator on a closed manifold, which can be computed by the formulas given (for quantized contact transformations between one and the same manifold) in [10] and for general Fourier integral operators in [19].

The results of present paper were partially published in [23] and [24].

References

[1] Agranovich, M., Elliptic singular integro-differential operators, *Uspekhi Matem. Nauk (5)* **20** (1965), 3–20. [Russian].

[2] ———, Elliptic boundary problems, In *Partial Differential Equations IX. Elliptic Boundary Value Problems, number 79 in Encyclopaedia of Mathematical Sciences*, M. Agranovich, Yu.V. Egorov, and M.A. Shubin (eds.), Berlin–Heidelberg, Springer–Verlag 1997, pp. 1–144.

[3] Anghel, N., An abstract index theorem on non-compact Riemannian manifolds, *Houston J. of Math.* **19** (1993), 223–237.

[4] Atiyah, M.F., Global theory of elliptic operators, In *Proc. of the Int. Symposium on Functional Analysis*, Tokyo, University of Tokyo Press 1969, pp. 21–30.

[5] Booß–Bavnbek, B. and Wojciechowski, K., *Elliptic boundary problems for Dirac operators*, Birkhäuser, Boston–Basel–Berlin 1993.

[6] Bunke, U., Relative index theory, *J. Funct. Anal.* **105** (1992), 63–76.

[7] Dezin, A.A., *Invariant differential operators and boundary value problems*, vol. 68 of Trudy MIAN. AN SSSR, Moscow 1962. [Russian].

[8] Donnelly, H, Essential spectrum and the heat kernel, *J. Funct. Anal.* **75** (1987), 362–381.

[9] Dynin, A.S., Multidimensional elliptic boundary problems with one unknown function, *DAN USSR* **141** (1961), 285–287. [Russian].

[10] Epstein, C. and Melrose, R., Contact degree and the index of Fourier integral operators, *Math. Res. Lett. (3)* **5** (1998), 363–381.

[11] Fedosov, B.V., Schulze, B.-W., and Tarkhanov, N., The index of higher order operators on singular surfaces, *Pacific J. of Math. (1)* **191** (1999), 25–48.

[12] ———, *A remark on the index of symmetric operators*, Univ. Potsdam, Institut für Mathematik, Potsdam, February 1998, Preprint N 98/4.

[13] Gilkey, P.B., *Invariance theory, the heat equation and the Atiyah–Singer index theorem*, Publish of Perish. Inc., Wilmington Delawaere 1984.

[14]   Gilkey, P.B. and Smith, L., The η invariant for a class of elliptic boundary value problems, *Comm. Pure Appl. Math.* **36** (1983), 85–132.

[15]   _____, The twisted index problem for manifolds with boundary, *J. Diff. Geometry (3)* **18** (1983), 393–444.

[16]   Gromov, M. and Lawson Jr, H.B., Positive scalar curvature and the Dirac operator on complete Riemannian manifolds, *Publ. Math. IHES* **58** (1983), 295–408.

[17]   Hsiung, Ch.-Ch., The signature and G-signarute of manifolds with boundary, *J. Diff. Geometry* **6** (1972), 595–598.

[18]   _____, A remark on cobordism of manifolds with boundary, *Arch. Math.* **XXVII** (1976), 551–555.

[19]   Leichtnam, E., Nest, R., and Tsygan, B., *Local formula for the index of a Fourier integral operator*, preprint math.DG/0004022, 2000.

[20]   Melrose, R., Transformation of boundary problems, *Acta Math.* **147** (1981), 149–236.

[21]   Nazaikinskii, V., Schulze, B.-W., and Sternin, B., *The index of quantized contact transformations on manifolds with conical singularities*, Univ. Potsdam, Institut für Mathematik, Potsdam, August 1998, Preprint N 98/16.

[22]   Nazaikinskii, V. and Sternin, B., Localization and surgery in index theory of elliptic operators. In *Conference: Operator Algebras and Asymptotics on Manifolds with Singularities*, Warsaw, Stefan Banach International Mathematical Center, Universität Potsdam, Institut für Mathematik 1999, pp. 27–28.

[23]   _____, Localization and surgery in the index theory to elliptic operators, *Russian Math. Dokl. (1)* **370** (2000), 19–23.

[24]   _____, *Surgery of manifolds and the relative index to elliptic operators*, Univ. Potsdam, Institut für Mathematik, Potsdam, Juli 1999, Preprint N 99/17.

[25]   _____, *The index locality principle in elliptic theory, Funktsionalayi Ancliz i Prilozhen. (2)* **35** (2001), 37–52. [Russian].

[26]   Schulze, B.-W., Sternin, B., and Shatalov, V., *Differential equations on singular manifolds. Semiclassical theory and operator algebras*, vol. 15 of Mathematics Topics, Wiley–VCH Verlag, Berlin–New York 1998.

[27]   _____, On the index of differential operators on manifolds with conical singularities, *Annals of Global Analysis and Geometry (2)* **16** (1998), 141–172.

[28]   Stong, R.E., Manifolds with reflecting boundary, *J. Diff. Geometry* **9** (1974), 465–474.

[29]   Teleman, N., The index of signature operators on Lipschitz manifolds, *Publ. Math. IHES* **58** (1984), 39–78.

[30]   Volpert, A.I., On index and normal solvability to boundary value problems for elliptic systems of differential equations on the plane, *Trudy Mosk. Matem. Ob-va* **10** (1961), 41–87. [Russian].

THE η INVARIANT AND ELLIPTIC OPERATORS IN SUBSPACES

A. SAVIN, B.-W. SCHULZE AND B. STERNIN

Contents

1   Introduction.. 375
2   The η invariant and the dimension of subspaces 376
    2.1   The η invariant.. 376
    2.2   Pseudo-differential subspaces 378
3   Index theory in subspaces and the η invariant 379
    3.1   Elliptic operators in subspaces........................... 379
    3.2   Application to η invariants 380
4   Index theory modulo n 383
5   A formula for the fractional part of invariants 385

M. de Gosson (ed.), Jean Leray '99 Conference Proceedings, 373-387.
© *2003 Kluwer Academic Publishers.*

1. Introduction

P. Gilkey [1] observed that for differential operators satisfying the parity condition

$$\text{ord } A + \dim M \equiv 1 \pmod 2, \tag{1.1}$$

the spectral Atiyah–Patodi–Singer η invariant is rigid: for a one-parameter operator family, the corresponding family of η invariants is a piecewise constant function. In particular, the fractional part of the spectral η invariant of a self-adjoint elliptic operator is in fact a homotopy invariant depending on the principal symbol of the operator alone. Thus, we arrive at the problem of computing the fractional part of the η invariant. This fractional homotopy invariant was successfully applied in several problems of topology and differential geometry (e.g., see [2], [3], [4] and [5]).

Our approach to this problem is based on the following observation: the η invariant of an operator A satisfying condition (1.1) is completely determined by the non-negative spectral subspace $\widehat{L}_+(A)$ of this operator. This subspace is generated by the eigenvectors of A corresponding to non-negative eigenvalues. The fractional part of the invariant, in turn, is determined by the so-called *symbol of the subspace*. This is a vector bundle on the cospheres S^*M; it is generated by the positive eigenspaces of the self-adjoint symbol $\sigma(A)$. First, this allows us to identify the η invariant of self-adjoint elliptic operators with a dimension-type functional on the corresponding spectral subspaces [6]. Second, we can apply the index formula for elliptic operators in subspaces [6] and [7]. The index formula gives an expression for the fractional part of the η invariant for the case in which the operator A defines a trivial element in the group $K_c^1(T^*M)$. In the general case, however, the index formula in subspaces reduces the computation of the fractional part to the 'index modulo n' problem for operators in subspaces. The term 'modulo n' refers to the fact that in this case the index of an elliptic operator in subspaces, being reduced modulo n, becomes an invariant of the principal symbol of the operator. It turns out that such elliptic operators on a closed manifold define the K-theory with coefficients in $\mathbf{Z}_n$. In particular, the index of an operator is computed modulo n by the direct image in this K-theory.

Complete proofs of the results stated in the present paper are contained in [8].

We conclude this introduction with the following table, which indicates some analogies between the APS theory [9] and elliptic theory in subspaces.

self-adjoint operator A	pseudo-differential subspace $\widehat{L}$
η invariant $\eta(A)$	dimension functional $d(\widehat{L})$
spectral flow	index of elliptic operators in subspaces
index theorem for flat bundles	fractional part of d, Theorem 5.2

The work is supported by Institut für Mathematik, Universität Potsdam, and RFBR grants Nos. 00-01-00161, 02-01-00118, 02-01-06515.

2. The η invariant and the dimension of subspaces

2.1. THE η INVARIANT

Let $A : C^\infty(M, E) \to C^\infty(M, E)$ be an elliptic self-adjoint pseudo-differential operator of positive order on a closed manifold M. Consider the spectral η-function of A

$$\eta(A, s) = \sum_{\lambda_n \neq 0} |\lambda_n|^{-s} \operatorname{sgn} \lambda_n, \quad \operatorname{Re} s \gg 0,$$

where the sum is taken over the eigenvalues λ_n of A. The Atiyah–Patodi–Singer η invariant of A [9] is defined via the analytic continuation to zero:

$$\eta(A) = \frac{\dim \ker A + \eta(A, 0)}{2}.$$

This is well defined on odd-dimensional manifolds according to [9]; for even-dimensional manifolds, this was proved in [10].

Example 2.1. Consider the elliptic self-adjoint first-order differential operator

$$A = -i\frac{d}{dt} + c, \quad c \in \mathbb{R},$$

on the unit circle S^1 with coordinate t. The eigenvalues are $\lambda_n = n + c$. A computation shows that

$$\eta(A) = \frac{1}{2} - \{c\},$$

where $\{\}$ denotes the fractional part. For example, at $c = 1/2$ the eigenvalues have symmetry with respect to origin and the η invariant is zero. Because of this the η invariant is often referred to as a measure of spectral asymmetry. The η invariant defines a piecewise smooth function of the parameter c, and the jumps occur at those values of c for which an eigenvalue changes its sign.

The η invariant has similar properties in the general case as well. The following result was obtained in [9].

Theorem 2.2. *Consider a smooth family of elliptic self-adjoint pseudo-differential operators A_t, ord $A_t > 0$. Then*

- *the family of fractional parts $\{\eta(A_t)\}$ is a smooth function of the parameter t;*

- *the jump at an isolated discontinuity point t' is equal to the number of eigenvalues of operators A_t that change sign from negative to positive as the parameter t passes the value t' minus the number of eigenvalues changing sign from positive to negative. This number is known as the spectral flow of the family A_t. Thus*

$$\eta\left(A_{t'+0}\right) - \eta\left(A_{t'-0}\right) = \text{sf } \{A_t\}_{t \in [t'-\varepsilon, t'+\varepsilon]} . \tag{2.1}$$

Example 2.1 shows that the η invariant in an operator family can take arbitrary real values. However, in [1] and [11] the following rigidity result was obtained.

Theorem 2.3. *Let A_t be an operator family as in Theorem 2.2. Suppose in addition that the A_t are differential operators. Then*

$$\frac{d}{dt} \{\eta\left(A_t\right)\} = 0$$

provided that

$$\text{ord } A + \dim M \equiv 1 \quad (\text{mod } 2). \tag{2.2}$$

Remark. It is possible to extend (see [1]) this rigidity result to a class of pseudo-differential operators. In the present exposition, we omit this generalization.

Corollary. 1. $\{\eta(A)\}$ is a homotopy invariant of principal symbol of A;

2. $\eta(A)$ is invariant under homotopies of A for which the spectral flow in (2.1) is zero.

This conditional homotopy invariance admits an interesting interpretation.
Let $\widehat{L}_+(A) \subset C^\infty(M, E)$ be the subspace generated by the eigenvectors of A corresponding to non-negative eigenvalues. In general position, the zero spectral flow condition for a homotopy of A means that the spectral subspace $\widehat{L}_+(A)$ varies continuously. Therefore, in this case a) $\eta(A)$ is a homotopy invariant of the subspace $\widehat{L}_+(A)$, while (2.1) implies that this homotopy invariant is a dimension-type functional of such subspaces; b) the difference of the η invariants for two subspaces that differ by a finite-dimensional space is equal to the dimension of this space.

These two properties a), b) were investigated in [6] and [7], where dimension-type functionals were studied in the framework of general *pseudo-differential subspaces*.

2.2. PSEUDO-DIFFERENTIAL SUBSPACES

Definition. A subspace $\widehat{L} \subset C^\infty(M, E)$ is said to be *pseudo-differential* if $\widehat{L} = \operatorname{Im} P$ for some pseudo-differential projection $P : C^\infty(M, E) \to C^\infty(M, E)$ of order zero.

Definition. The *symbol of* $\widehat{L}$ is the sub-bundle $L = \operatorname{Im} \sigma(P) \subset \pi^* E$, $\pi : S^* M \to M$, determined by the principal symbol of P.

The following proposition is based on the results due to R. Seeley [12].

Proposition 2.4. Let A be an elliptic self-adjoint operator of positive order on M. Then the spectral subspace $\widehat{L}_+(A)$ is pseudo-differential, and the symbol $L_+(A)$ is equal to the non-negative spectral sub-bundle for the principal symbol $\sigma(A)$:

$$L_+(A) = L_+(\sigma(A)).$$

The principal symbols of differential operators are invariant (anti-invariant) with respect to antipodal involution

$$\alpha : T^* M \longrightarrow T^* M, \qquad \alpha(x, \xi) = (x, -\xi),$$

of the cotangent bundle according to the parity of the order of the operator:

$$\alpha^* \sigma(A) = \pm \sigma(A).$$

This motivates the following definition.

Definition. A pseudo-differential subspace $\widehat{L} \subset C^\infty(M, E)$ is said to be *even*, (respectively, *odd*) if its symbol is invariant: $L = \alpha^* L$, (respectively, anti-invariant: $L \oplus \alpha^* L = \pi^* E$) with respect to the involution.

The spectral subspaces of differential self-adjoint operators are, of course, even or odd according to the parity of the order of the operator. The semigroups of even (odd) subspaces are denoted by $\widehat{\operatorname{Even}}(M)$ and $\widehat{\operatorname{Odd}}(M)$.

The parity condition (2.2) has a counterpart for pseudo-differential subspaces: we require that the parities of the subspace and of the dimension of the manifold be opposite. In [6] and [7] it is shown that these subspaces possess a unique homotopy invariant similar to the (co)dimension of finite-dimensional vector spaces.

Theorem 2.5. *There exists a unique homotopy invariant functional*

$$d : \widehat{\operatorname{Even}}\left(M^{\mathrm{odd}}\right) \longrightarrow \mathbf{Z}\left[\frac{1}{2}\right] \qquad or \qquad d : \widehat{\operatorname{Odd}}\left(M^{\mathrm{ev}}\right) \longrightarrow \mathbf{Z}\left[\frac{1}{2}\right]$$

with the following properties:

1. (*relative dimension*) *For a pair of subspaces* $\widehat{L} + L_0, \widehat{L}$ *which differ by a finite-dimensional space* L_0, *one has*

$$d\left(\widehat{L} + L_0\right) - d\left(\widehat{L}\right) = \dim L_0;$$

2. (*complement*)

$$d\left(\widehat{L}\right) + d\left(\widehat{L}^{\perp}\right) = 0,$$

where $\widehat{L}^{\perp}$ *is the orthogonal complement of* $\widehat{L}$.

In the theorem, by $\mathbf{Z}[1/2] \subset \mathbb{Q}$ we denote the ring of dyadic rationals.

Remark. In some sense, the functional d measures the deviation of $\widehat{L}$ from the space of sections of a vector bundle, since the complement property implies

$$d\left(C^{\infty}(M, E)\right) = 0. \tag{2.3}$$

Of course, for spectral subspaces of differential operators this functional coincides with the η invariant.

Theorem 2.6 ([6] and [7]). *For an elliptic self-adjoint differential operator* A, ord $A > 0$, *one has*

$$d\left(\widehat{L}_+(A)\right) = \eta(A). \tag{2.4}$$

The dimension functional d can be expressed in terms of the index of an elliptic operator. However, we have to consider operators acting in pseudo-differential subspaces rather than in spaces of sections of vector bundles. Now we describe briefly the basics of elliptic theory in subspaces.

3. Index theory in subspaces and the η invariant

3.1. ELLIPTIC OPERATORS IN SUBSPACES

Let us consider two pseudo-differential subspaces $\widehat{L}_{1,2} \subset C^{\infty}(M, E_{1,2})$ and a pseudo-differential operator

$$D : C^{\infty}(M, E_1) \longrightarrow C^{\infty}(M, E_2).$$

Suppose that $D\widehat{L}_1 \subset \widehat{L}_2$.

Definition. The restriction

$$D : \widehat{L}_1 \longrightarrow \widehat{L}_2 \tag{3.1}$$

is called an *operator in subspaces*.

Definition. *The principal symbol of the operator in subspaces (3.1) is the* homomorphism

$$\sigma(D) : L_1 \longrightarrow L_2$$

of vector bundles over S^*M.

Theorem 3.1. *The closure of the operator (3.1) with respect to the Sobolev norms has the Fredholm property if and only if the principal symbol $\sigma(D)$ is elliptic, i.e.,*

$$\sigma(D) : L_1 \longrightarrow L_2$$

is an isomorphism.

Suppose that the operator (3.1) acts in subspaces of the same parity. This symmetry permits us to define a classical elliptic operator, which we denote by $\tilde{D}$. In the case of odd subspaces, this operator

$$\tilde{D} : C^\infty(M, E_1) \longrightarrow C^\infty(M, E_2)$$

has the principal symbol

$$\sigma\left(\tilde{D}\right) = \sigma(D) \oplus \alpha^*\sigma(D) : L_1 \oplus \alpha^* L_1 \longrightarrow L_2 \oplus \alpha^* L_2.$$

In the case of even subspaces, the operator

$$\tilde{D} : C^\infty(M, E_1) \longrightarrow C^\infty(M, E_1)$$

has the principal symbol

$$\sigma\left(\tilde{D}\right) = \left[\alpha^*\sigma(D)\right]^{-1}\sigma(D) \oplus 1 : L_1 \oplus L_1^\perp \longrightarrow L_1 \oplus L_1^\perp.$$

Theorem 3.2. *(the index formula for operators in subspaces) Suppose that the operator (3.1) acts in subspaces of the same parity, opposite to the parity of the dimension of the manifold:*

$$\widehat{L}_{1,2} \in \widehat{\text{Even}}\left(M^{\text{odd}}\right) \quad \text{or} \quad \widehat{\text{Odd}}\left(M^{\text{ev}}\right). \tag{3.2}$$

Then

$$\text{ind}\left(D : \widehat{L}_1 \longrightarrow \widehat{L}_2\right) = \frac{1}{2}\text{ind}\,\tilde{D} + d\left(\widehat{L}_1\right) - d\left(\widehat{L}_2\right). \tag{3.3}$$

3.2. APPLICATION TO η INVARIANTS

The index formula in subspaces gives an integrality theorem for the dimension functional d. More precisely, suppose that one of the subspaces in it is the space $C^\infty(M, F)$ of sections of a vector bundle. Then the index formula (3.3) implies that the fractional part of the dimension d of the subspace is either 0 or $1/2$.

Proposition 3.3. Let

$$D : \widehat{L} \longrightarrow C^\infty(M, F) \tag{3.4}$$

be an elliptic operator in subspaces. Suppose that the subspace satisfies the parity conditions (3.2); then

$$\{d\,(\widehat{L})\} = \left\{ \frac{1}{2} \operatorname{ind} \widetilde{D} \right\}.$$

For a given subspace $\widehat{L}$, there exists an elliptic operator of the form (3.4) if and only if the symbol $L \in \operatorname{Vect}(S^*M)$ is the pullback of some bundle from the base:

$$L \in \pi^* \operatorname{Vect}(M), \quad \pi : S^*M \longrightarrow M.$$

Let us rewrite this condition in terms of K-theory. In [9] it is shown that the coboundary map δ in the exact sequence

$$\longrightarrow K\,(B^*M) \longrightarrow K\,(S^*M) \xrightarrow{\ \delta\ } K_c^1\,(T^*M) \longrightarrow \tag{3.5}$$

induces an isomorphism $K\,(S^*M)/K(M) \overset{\delta}{\simeq} K_c^1(T^*M)$. Therefore, Proposition 3.3 applies when

$$[L] = 0 \in K_c^1\,(T^*M) \simeq K\,(S^*M)\,/K(M).$$

Example 3.4. On a closed odd-dimensional manifold M, consider the space of closed 1-forms

$$\widehat{L} = \ker d \subset C^\infty\left(M, \Lambda^1(M)\right).$$

This subspace is even, since it is the non-negative spectral subspace of the second order self-adjoint operator $d\delta - \delta d$:

$$\widehat{L} = \widehat{L}_+(d\delta - \delta d).$$

The exterior differential d defines an elliptic operator in these subspaces:

$$d : C^\infty(M) \longrightarrow \widehat{L}.$$

Its index is equal to the difference of the corresponding Betti numbers,

$$\operatorname{ind}\left(d : C^\infty(M) \longrightarrow \widehat{L}\right) = \dim H^0(M) - \dim H^1(M),$$

by de Rham's theorem. On the other hand, an explicit computation shows that the symbol of the corresponding classical elliptic operator in (3.3) is homotopic to a constant symbol. Thus, the index formula (3.3) yields (cf. [1])

$$d\left(\widehat{L}\right) = \eta(d\delta - \delta d) = \dim H^0(M) - \dim H^1(M), \quad \{\eta(d\delta - \delta d)\} = 0.$$

Example 3.5. Consider the self-adjoint Euler operator

$$d + \delta : C^\infty(M, \Lambda(M)) \longrightarrow C^\infty(M, \Lambda(M))$$

on a closed even-dimensional manifold. Its non-negative spectral subspace $\widehat{L}_+(d + \delta)$ is odd, since $d + \delta$ is a first-order differential operator. The projection $\Lambda(M) \to \Lambda^{\mathrm{odd}}(M)$ induces an elliptic operator in subspaces,

$$\widehat{L}_+(d + \delta) \xrightarrow{P} C^\infty\left(M, \Lambda^{\mathrm{odd}}(M)\right).$$

Hence, Proposition 3.3 gives the following formula for fractional part of the η invariant:

$$\{\eta(d + \delta)\} = \left\{\frac{1}{2} \operatorname{ind} \widetilde{P}\right\}.$$

A direct computation shows that the symbol $\sigma(\widetilde{P})$ is homotopic to the symbol of the Euler operator

$$d + \delta : C^\infty\left(M, \Lambda^{\mathrm{ev}}(M)\right) \longrightarrow C^\infty\left(M, \Lambda^{\mathrm{odd}}(M)\right).$$

Hence

$$\{\eta(d + \delta)\} = \left\{\frac{1}{2}\chi(M)\right\},$$

where χ is the Euler characteristic. A non-trivial fractional part occurs, for instance, on the projective space $\mathbf{RP}^{2n} : \chi(\mathbf{RP}^{2n}) = 1$.

In the general case, the symbol $[L] \in K_c^1(T^*M)$ is non-trivial. However, under the parity condition it is a 2-torsion element.

Theorem 3.6. *The induced involution α^* in K-theory of the cotangent bundle T^*M modulo 2-torsion coincides with involution $(-1)^{\dim M}$:*

$$\alpha^* : K_c^*(T^*M) \otimes \mathbf{Z}\left[\frac{1}{2}\right] \longrightarrow K_c^*(T^*M) \otimes \mathbf{Z}\left[\frac{1}{2}\right], \quad \alpha^* = (-1)^{\dim M}.$$

This property is based on the Mayer-Vietoris principle and a computation at a point $x \in M$:

$$\alpha_x^* : K_c^*(T_x^*M) \otimes \mathbf{Z}\left[\frac{1}{2}\right] \longrightarrow K_c^*(T_x^*M) \otimes \mathbf{Z}\left[\frac{1}{2}\right].$$

An explicit formula for the Bott generator of this group $K_c^*(\mathbf{R}^{\dim M}) \simeq \mathbf{Z}$ shows that $\alpha_x^* = (-1)^{\dim M}$.

A subspace $\widehat{L}$ under the parity conditions according to Theorem 3.6 admits an elliptic operator

$$D : 2^N \widehat{L} \xrightarrow{\quad} C^\infty(M, F) \tag{3.6}$$

(for some $N \geq 0$ and some vector bundle $F \in \text{Vect}(M)$). The index formula (3.3) gives

$$\{d\,(\widehat{L})\} = -\left\{\frac{1}{2^{N+1}} \text{ ind } \widetilde{D}\right\} + \left\{\frac{1}{2^N} \text{ ind } D\right\}. \tag{3.7}$$

However, for $N > 0$ the index of the operator D in subspaces cannot be neglected. Let us analyze this term.

The obvious inclusion of cyclic groups $\mathbf{Z}_{2^N} \subset \mathbf{Z}[1/2]/\mathbf{Z}$ gives the equality

$$\left\{\frac{1}{2^N} \text{ ind } D\right\} = \frac{1}{2^N} \quad \text{mod } 2^N - \text{ind } D,$$

where mod 2^N-ind $D \in \mathbf{Z}_{2^N}$ is the residue class of the index of D.

The logarithmic property of the index implies the following important observation.

Lemma 3.7. *The mod 2^N-index of the elliptic operator (3.6) is determined by the principal symbol of this operator.*

This mod 2^N-index is computed in the next section via a technique that generalizes [9].

4. Index theory modulo n

Let n be a positive integer.

Definition. *An elliptic operator modulo n is an elliptic operator in subspaces of the form*

$$D = n\widehat{L}_1 \oplus C^\infty(M, E_1) \xrightarrow{\quad} n\widehat{L}_2 \oplus C^\infty(M, F_1), \tag{4.1}$$

$$\widehat{L}_1 \subset C^\infty(M, E), \qquad \widehat{L}_2 \subset C^\infty(M, F). \tag{4.2}$$

Elliptic direct sums of the form

$$n\left(\widehat{L}_1 \oplus C^\infty(M, E_1)\right) \xrightarrow{nD} n\left(\widehat{L}_2 \oplus C^\infty(M, F_1)\right) \tag{4.3}$$

are called *trivial operators modulo n*. The group of stable homotopy classes of elliptic operators modulo n is denoted by $\text{Ell}(M, \mathbf{Z}_n)$. Recall that stable homotopy classes of classical elliptic operators $\text{Ell}(M)$ define the K-groups $K_c(T^*M)$ (see [13]). Similarly, the group $\text{Ell}(M, \mathbf{Z}_n)$ defines the K-theory with coefficients $\mathbf{Z}_n$.

Theorem 4.1 ([8]). *There is an isomorphism of groups*

$$\text{Ell}(M, \mathbf{Z}_n) \xrightarrow{\chi_n} K_c\left(T^*M, \mathbf{Z}_n\right).$$

Let us give an explicit formula for the isomorphism χ_n. Recall that the K-theory with coefficients is defined by means of a Moore space, which we denote by M_n. This is a topological space with K-groups equal to

$$\overline{K}^0\left(M_n\right) = \mathbf{Z}_n, \qquad K^1\left(M_n\right) = 0.$$

For instance, for finite cyclic groups $\mathbf{Z}_n$ we can take the two-dimensional complex obtained from the disc D^2 with the following identification of points on its boundary:

$$M_n = \left\{ D^2 \subset C \,\middle|\, |z| \le 1 \right\} \Big/ \left\{ e^{i\varphi} \sim e^{i\left(\varphi + \frac{2\pi k}{n}\right)} \right\}.$$

The generator of the reduced group $\overline{K}^0\left(M_n\right) = \mathbf{Z}_n$ is denoted by γ_n. We also fix a trivialization β of the direct sum $n\gamma_n$:

$$n\gamma_n \xrightarrow{\beta} C^n.$$

For a topological space X, the K-groups with coefficients $\mathbf{Z}_n$ are defined by the formula

$$K^*\left(X, \mathbf{Z}_n\right) = K^*\left(X \times M_n, X \times pt\right). \tag{4.4}$$

Let us now define the map χ_n.

An arbitrary operator of the form (4.1) is stably homotopic to an elliptic operator

$$n\widehat{L} \xrightarrow{D} C^\infty(M, F). \tag{4.5}$$

For this operator we define the following family of classical elliptic symbols:

$$\chi_n[D] = \left[\pi^*F \xrightarrow{\sigma^{-1}(D)} nL \xrightarrow{\beta^{-1} \otimes 1_L} \gamma_n \otimes nL \xrightarrow{1_\gamma \otimes \sigma(D)} \gamma_n \otimes \pi^*F \right] \tag{4.6}$$

$$\in K_c\left(T^*M \times M_n, T^*M \times pt\right).$$

The symbols act on the manifold M, while the parameter space is M_n (the element in the square brackets is understood in the sense of the difference construction for elliptic families [14]).

Theorem 4.2. *(mod n-index formula) The following triangle is commutative*

$$\text{Ell}\,(M, \mathbf{Z}_n) \xrightarrow{\ \text{mod}\,n-\text{ind}\ } \mathbf{Z}_n$$

with χ_n going down to $K_c(T^*M, \mathbf{Z}_n)$ and $p_!$ going up to $\mathbf{Z}_n$.

where $p_! : K(T^*M, \mathbf{Z}_n) \to \mathbf{Z}_n$ is the direct image map in K-theory with coefficients.

Remark. In terms of elliptic families, this theorem expresses the mod n-index of an operator in subspaces via the index of a family of classical elliptic operators parametrized by the Moore space M_n.

Let us also mention an exact sequence which relates elliptic operators modulo n to the usual elliptic operators. To this end, we introduce the 'odd' elliptic group $\text{Ell}_1(M)$ (cf. [9]) that is generated by pseudo-differential subspaces. More precisely, two pseudo-differential subspaces are called *stably homotopic* if they are homotopic modulo finite-dimensional spaces and spaces of sections of vector bundles. Now $\text{Ell}_1(M)$ is the group of stable homotopy classes of pseudo-differential subspaces. The symbol map establishes an isomorphism with the corresponding K-group (see (3.5)):

$$\text{Ell}_1(M) \xrightarrow{\ \chi_1\ } K_c^1(T^*M), \qquad \widehat{L} \longmapsto \delta[L].$$

The following diagram is commutative:

$$
\begin{array}{ccccccccc}
\text{Ell}(M) & \xrightarrow{\times n} & \text{Ell}(M) & \xrightarrow{\ i\ } & \text{Ell}\,(M, \mathbf{Z}_n) & \xrightarrow{\ j\ } & \text{Ell}_1(M) & \xrightarrow{\times n} & \text{Ell}_1(M) \\
\downarrow{\scriptstyle \chi_0} & & \downarrow{\scriptstyle \chi_0} & & \downarrow{\scriptstyle \chi_n} & & \downarrow{\scriptstyle \chi_1} & & \downarrow{\scriptstyle \chi_1} \\
K_c(T^*M) & \xrightarrow{\times n} & K_c(T^*M) & \xrightarrow{\ i'\ } & K_c(T^*M, \mathbf{Z}_n) & \xrightarrow{\ j'\ } & K_c^1(T^*M) & \xrightarrow{\times n} & K_c^1(T^*M)
\end{array}
$$

$$(4.7)$$

here the map $\times n$ takes an element x of an abelian group to the multiple nx, the forgetful map i is induced by the embedding of classical operators in the set of operators modulo n, and the boundary map j is defined by the formula

$$j\left[n\widehat{L}_1 \oplus C^\infty(M, E_1) \xrightarrow{\ D\ } n\widehat{L}_2 \oplus C^\infty(M, F_1) \right] = \left[\widehat{L}_1 \right] - \left[\widehat{L}_2 \right].$$

5. A formula for the fractional part of invariants

When we apply the index modulo n formula to the expression for the fractional part of the dimension functional (3.7) in section 3, the following result is obtained.

Theorem 5.1. *For a subspace $\widehat{L}$ with parity conditions (3.2) and an isomorphism $\sigma : 2^N L \to \pi^* F$ (see (3.6)), the fractional part of the dimension functional is computed by the formula*

$$\{d\left(\widehat{L}\right)\} = \frac{1}{2^{N+1}} p_! \left[(1 \pm \alpha^*)\sigma\right], \quad \left[(1 \pm \alpha^*)\sigma\right] \in K_c\left(T^*M, \mathbf{Z}_{2^{N+1}}\right),$$

where $p_! : K(T^*M, \mathbf{Z}_{2^{N+1}}) \to \mathbf{Z}_{2^{N+1}}$ is the direct image map and the sign coincides with the parity of the subspace.

Even though the construction of the element $[(1 \pm \alpha^*)\sigma]$ depends on the number N and the isomorphism σ, in the limit as $N \to \infty$ the result becomes in a certain sense unique. Let us introduce the corresponding notation.

Consider the increasing sequence of groups

$$\mathbf{Z}_2 \subset \mathbf{Z}_4 \subset \cdots \subset \mathbf{Z}_{2^{N'}} \subset \cdots$$

with the fractional parts of dyadic numbers as its injective limit:

$$\varinjlim_{N' \to \infty} \mathbf{Z}_{2^{N'}} = \mathbf{Z}\left[\frac{1}{2}\right] \Big/ \mathbf{Z}.$$

Consider also the corresponding sequence in K-theory:

$$K_c\left(T^*M, \mathbf{Z}\left[\frac{1}{2}\right] \Big/ \mathbf{Z}\right) = \varinjlim_{N' \to \infty} K_c\left(T^*M, \mathbf{Z}_{2^{N'}}\right). \tag{5.1}$$

Theorem 5.2. *In the notation of the previous theorem, the element*

$$[L] = i_*\left[(1 \pm \alpha^*)\sigma\right] \in K_c\left(T^*M, \mathbf{Z}\left[\frac{1}{2}\right] \Big/ \mathbf{Z}\right), \quad i : \mathbf{Z}_{2^{N+1}} \subset \mathbf{Z}\left[\frac{1}{2}\right] \Big/ \mathbf{Z}$$

is well defined (i.e., it does not depend on the choice of σ) and

$$\{d\left(\widehat{L}\right)\} = p_![L], \quad p_! : K_c\left(T^*M, \mathbf{Z}\left[\frac{1}{2}\right] \Big/ \mathbf{Z}\right) \longrightarrow \mathbf{Z}\left[\frac{1}{2}\right] \Big/ \mathbf{Z}.$$

Let us consider an application of these results.

M. Karoubi [15] showed that on an orientable manifold M the antipodal involution α induces in K-theory the map $(-1)^{\dim M}$:

$$\alpha^* = (-1)^{\dim M} : K_c\left(T^*M, \mathbf{Z}_{2^N}\right) \longrightarrow K_c\left(T^*M, \mathbf{Z}_{2^N}\right).$$

Consequently,

$$2[L] = i_*\left[2(1 \pm \alpha^*)\sigma\right] = i_*(1 \pm \alpha^*)[2\sigma] = i_*[0] = 0.$$

Thus, we obtain the following corollary (cf. [11]).

Corollary. The η invariant of elliptic differential operators with parity conditions (2.2) on orientable manifolds has at most 2 in its denominator:

$$\{2\eta(A)\} = 0.$$

References

[1] Gilkey, P.B., The eta invariant of even order operators, *Lecture Notes in Mathematics* **1410** (1989), 202–211.

[2] Stolz, S., Exotic structures on 4-manifolds detected by spectral invariants, *Invent. Math.* **94** (1988), 147–162.

[3] Bahri, A. and Gilkey, P., The eta invariant, Pin^c bordism, and equivariant $Spin^c$ bordism for cyclic 2-groups. *Pacific Jour. Math. (1)* **128** (1987), 1–24.

[4] Gilkey, P.B., *The eta invariant of Pin manifolds with cyclic fundamental groups*, *Periodica Math. Hungarica* **36** (1998), 139–170.

[5] Liu, K. and Zhang, W., Elliptic genus and η-invariant, *Int. Math. Research Notices* **8** (1994), 319–328.

[6] Savin, A.Yu. and Sternin, B., Elliptic operators in even subspaces, *Matem. sbornik (8)* **190** (1999), 125–160. English transl.: *Sbornik: Mathematics (8)* **190** (1999), 1195–1228; math.DG/9907027.

[7] _____, Elliptic operators in odd subspaces, *Sbornik: Mathematics, (8)* **191** (2000), 1191-1213, math.DG/9907039.

[8] Savin, A.Yu., Schulze, B.-W, and Sternin, B., appears as "The Eta Invariant and Elliptic operators in subspaces", vol. 26, in k-Theory Journal, no. 3, 2001, math.DG/9907047.

[9] Atiyah, M., Patodi, V., and Singer, I., Spectral asymmetry and Riemannian geometry III, *Math. Proc. Cambridge Philos. Soc.* **79** (1976), 71–99.

[10] Gilkey, P.B., The residue of the global eta function at the origin, *Adv. in Math.* **40** (1981), 290–307.

[11] _____, The eta invariant for even dimensional Pin^c manifolds, *Advances in Mathematics* **58** (1985), 243–284.

[12] Seeley, R.T., Complex powers of an elliptic operator, *Proc. Sympos. Pure Math.* **10** (1967), 288–307.

[13] Atiyah, M. and Singer, I.M., The index of elliptic operators I, *Ann. of Math.* **87** (1968), 484–530.

[14] _____, The index of elliptic operators IV, *Ann. Math.* **93** (1971), 119–138.

[15] Karoubi, M., *Seminaire Heidelberg - Saarbruecken - Strasbourg sur la K-theorie*, vol. 36 of Lecture Notes in Math. 1970.

REGULARISATION OF MIXED BOUNDARY PROBLEMS

B.-W. SCHULZE, A. SHLAPUNOV AND N. TARKHANOV

Contents

1 Introduction 391
2 A crack problem 395
3 Adjoint operator 398
4 Iterations 403
5 Cauchy problem 405
6 The Zaremba problem 406

M. de Gosson (ed.), Jean Leray '99 Conference Proceedings, 389-410.
© *2003 Kluwer Academic Publishers.*

Abstract. *We show an application of the spectral theorem in constructing approximate solutions of mixed boundary value problems for elliptic equations.*[1]

1. Introduction

When studying boundary value problems for solutions of an elliptic differential equation $Pu = f$, P being of type $E \to F$ for some vector bundles E and F, one uses any left parametrix of P to reduce the problem to the boundary. A powerful tool for such a reduction is Green's formula for P which brings together the values of Pu in a domain $\mathcal{D}$ and the Cauchy data $t(u)$ of u on the boundary of $\mathcal{D}$ to present u in $\mathcal{D}$ modulo smoothing operators.

In case P has a left fundamental solution Φ, Green's formula reads $u = \mathcal{P}_{\mathrm{dl}} t(u) + \mathcal{P}_{\mathrm{v}} Pu$ in $\mathcal{D}$, where $\mathcal{P}_{\mathrm{dl}} u_0$ and $\mathcal{P}_{\mathrm{v}} f$ are the double layer and volume potentials corresponding to Φ. If, moreover, Φ is a right fundamental solution, then the potential $\mathcal{P}_{\mathrm{dl}} u_0$ satisfies the homogeneous equation $Pu = 0$ in $\mathcal{D}$, and so Green's formula provides us with a soft version of the Hodge decomposition.

However, right fundamental solutions are available only for those P which are determined, i.e., bear as many scalar equations as the number of unknown functions. For over determined elliptic operators P, the construction of kernels Φ with the property that $P \mathcal{P}_{\mathrm{dl}} u_0 = 0$ in $\mathcal{D}$, for each density u_0, is a significant problem. In complex analysis such Green's formulas are known as Cauchy–Fantappiè formulas. They give rise to explicit solutions of the inhomogeneous Cauchy–Riemann system $\bar{\partial} u = f$ on strongly pseudo-convex manifolds or domains in $\mathbb{C}^N$.

Consider the iterations $\mathcal{P}_{\mathrm{dl}}^N t(u)$, for $N = 1, 2, \ldots$, in any function space H invariant under $\mathcal{P}_{\mathrm{dl}}$. If Φ is a right fundamental solution of P, then these iterations stabilize because $P\mathcal{P}_{\mathrm{dl}} t(u) = 0$ in $\mathcal{D}$ and so $\mathcal{P}_{\mathrm{dl}}^N t(u) = \mathcal{P}_{\mathrm{dl}} t(u)$ for all N. In general, since $\mathcal{P}_{\mathrm{dl}}$ is the identity operator on solutions of $Pu = 0$, one may conjecture that the iterations converge to a projection of H onto the subspace V_1 of H consisting of u satisfying $Pu = 0$ in $\mathcal{D}$. Were such the case, the equality

$$u = \pi_{V_1} u + \sum_{\nu=0}^{\infty} \mathcal{P}_{\mathrm{dl}}^{\nu} \mathcal{P}_{\mathrm{v}} Pu, \tag{1.1}$$

for $u \in H$, would give us a substitute for the Hodge decomposition.

This idea goes back to the work of Romanov [6] who studied the iterations of the Martinelli–Bochner integral in $\mathbb{C}^N$, $N > 1$. His results were extended in a beautiful way to arbitrary over determined elliptic systems by Nacinovich and Shlapunov [3].

---

[1] *AMS subject classification*: primary: 35J70; secondary: 35Cxx, 35M10.

To handle the iterations of $\mathcal{P}_{\mathrm{dl}}$, the idea is to present this integral as a self-adjoint operator on a Hilbert space H. While H is specified within the Sobolev space H^p in $\mathcal{D}$, p being the order of P, the Hermitian structure of H is different from that induced by H^p. If moreover $0 \leq \mathcal{P}_{\mathrm{dl}} \leq I$, then the iterations of $\mathcal{P}_{\mathrm{dl}}$ converge, by the spectral theorem, to the projection of H onto the eigenspace of $\mathcal{P}_{\mathrm{dl}}$ corresponding to the eigenvalue 1. This eigenspace in turn coincides with the null space of $\mathcal{P}_{\mathrm{v}} P$ in H, and hence with the null space of P if $\mathcal{P}_{\mathrm{v}}$ is identified with the adjoint of P in H.

The choice of H is actually suggested by the restriction of Φ to the closed domain $\overline{\mathcal{D}}$. If $\mathcal{P}_{\mathrm{v}} f$ meets automatically some conditions on the boundary of $\mathcal{D}$, for each $f \in L^2(\mathcal{D}, F)$, these should be incorporated in the definition of H.

To introduce the relevant Hermitian structure in H, we choose any domain $\mathcal{D}'$ with C^∞ boundary, such that $\overline{\mathcal{D}} \subset \mathcal{D}'$, and we fix an extension operator $e : H \to H^p(\mathcal{D}', E)$ with the properties that $t(e(u)) = 0$ on $\partial \mathcal{D}'$, for each $u \in H$, and $e(\mathcal{P}_{\mathrm{v}} f) = \mathcal{P}_{\mathrm{v}} f$, for each $f \in L^2(\mathcal{D}, F)$. In this way we actually reduce the problem to that on a compact surrounding manifold. As but one example of $e(u)$ we show the solution of the Dirichlet problem for the Laplacian $\Delta = P^* P$ in $\mathcal{D}' \setminus \overline{\mathcal{D}}$ with data $t(u)$ on $\partial \mathcal{D}$ and 0 on $\partial \mathcal{D}'$. Define a scalar product in H by

$$h(u, v) = (Pe(u), Pe(v))_{L^2(\mathcal{D}', F)}, \qquad (1.2)$$

then the identity $h(\mathcal{P}_{\mathrm{v}} f, v) = (f, Pv)_{L^2(\mathcal{D}, F)}$, for every $v \in H$, is immediate. It in turn implies the self-adjointness of $\mathcal{P}_{\mathrm{dl}}$ with respect to (1.2).

In [3], H is the whole space $H^p(\mathcal{D}, E)$, and $\mathcal{P}_{\mathrm{v}} f = \mathcal{G} P^*(\chi_{\mathcal{D}} f)$ where $\mathcal{G}$ is Green's function of the Dirichlet problem for Δ in $\mathcal{D}'$ and $\chi_{\mathcal{D}}$ is the characteristic function of $\mathcal{D}$. In this setting, the orthogonal decomposition (1.1) applies to constructing approximate solutions for the inhomogeneous equation $Pu = f$ and the Dirichlet system for the Laplacian Δ in $\mathcal{D}$.

This work was intended as an attempt to develop the approach of [3] to derive approximate solutions of the generalized Zaremba problem in $\mathcal{D}$. It consists of finding a function $u \in H^p(\mathcal{D}, E)$ satisfying $\Delta u = f$ in $\mathcal{D}$, whose 'tangential part' $t(u)$ takes prescribed values on a part σ of $\partial \mathcal{D}$ while the 'normal part' $n(Pu)$ takes prescribed values on the complementary part $\partial \mathcal{D} \setminus \bar{\sigma}$. Since the Dirichlet problem in $\mathcal{D}$ is elliptic, we may assume without loss of generality that $t(u) = 0$ on σ. This causes H to be a subspace of $H^p(\mathcal{D}, E)$ whose elements satisfy $t(u) = 0$ on σ. Moreover, $\mathcal{G}$ should be Green's function of the Dirichlet problem in the domain with crack $\mathcal{D}' \setminus \bar{\sigma}$.

Thus, the study of the Zaremba problem leads to the Dirichlet problem for the generalized Laplace operator Δ in a domain with cracks. Note that both mixed boundary value problems and crack problems can be treated in the framework of any calculus on manifolds with edges of codimension 1 on the boundary. At present, several calculi on such singular configurations are known, namely those by Vishik and Eskin (cf. [2, section 24]), Maz'ya and Plamenevskii (cf. [4, Chapter 8]), Schulze [7], Duduchava and Wendland [1], etc. Either of these calculi is applicable.

The important point to note here is the function spaces used as domains for pseudo-differential operators involved. These are scales of weighted Sobolev spaces $H^{s,\gamma}(\mathcal{D}, E)$ parametrised by $s, \gamma \in \mathbb{R}$, where s specifies the smoothness and γ stands for the weight. While the spaces may differ from each other in various calculi, they usually coincide with $L^2(\mathcal{D}, E)$, for $s, \gamma = 0$. As weight functions, one uses the powers of the regularised distance to the set of singularities, here to the edge $\partial\sigma$. Moreover, we may always specify the weight exponent γ in such a way that admissible operators of order m map $H^{s,\gamma}(\mathcal{D}, E)$ to $H^{s-m,\gamma-m}(\mathcal{D}, F)$.

Let us dwell upon the contents of the paper. In section 2 we formulate the Dirichlet problem for the generalised Laplacian in a domain with a crack along a hypersurface with C^∞ boundary. When compared to general crack problems, the Dirichlet problem has the peculiarity of being formally self-adjoint. Moreover, any solution of the corresponding homogeneous problem is trivial unless it is rather singular. Hence it follows that the Dirichlet problem is uniquely solvable provided it is Fredholm. Since the Dirichlet problem is elliptic in the usual sense on the smooth part of the boundary, i.e., away from $\partial\sigma$, the Fredholm property of this problem is equivalent to the bijectivity of an operator-valued symbol called the *edge symbol*. This symbol lives in the cotangent bundle to $\partial\sigma$ with the zero section removed. In fact, the edge symbol is the Dirichlet problem for a differential operator of special form in the plane with a cut along the non-negative semiaxis. The operator is obtained from P^*P by freezing the coefficients at any point $y \in \partial\sigma$ and substituting the covariable η for the derivatives along the edge. It is supplied by the Dirichlet boundary conditions from both sides of the cut. Using polar coordinates in the normal plane to the edge at y, we see that the conormal symbol of the edge symbol is $(\sigma^P(P))^*\sigma^P(P)$, the principal symbol of P being evaluated at $\eta = 0$ and

$$\xi_1 = \cos\varphi D_r - (1/r)\sin\varphi D_\varphi,$$
$$\xi_2 = \sin\varphi D_r + (1/r)\cos\varphi D_\varphi,$$

ξ_1, ξ_2 standing for the conormal variables. Thus, if supplied by the Dirichlet conditions at the endpoints of $[0, 2\pi]$, the conormal symbol belongs to the same class of boundary value problems, now on a segment. The condition of bijectivity of the conormal symbol gives us a discrete set Σ of 'prohibited' weights γ, independent of η. For $\gamma \notin \Sigma$, the edge symbol is a Fredholm operator in Sobolev spaces of weight γ over $\mathbb{R}^2 \setminus \overline{\mathbb{R}}_+$. It is to be expected that the null space of the edge symbol is trivial if $\gamma \gg 0$ is large enough. By duality, if $-\gamma \gg 0$ is large enough, then the cokernel of the edge symbol is trivial. It is now the property of P, t and the singular configuration whether or not, given a γ, the edge symbol is an isomorphism for all $\eta \neq 0$. For the Dirichlet problem in a domain with a smooth edge, the edge symbol is known to be an isomorphism for γ in an interval around p (see [4, sections 6.1.3 and 8.4.2]). It follows that the Dirichlet problem for the Laplacian $\Delta = P^*P$ is solvable in $H^{s,p}(\mathcal{D}, E)$, $\mathcal{D}$ being a domain with smooth edges. Then, we make use of a familiar classical scheme to construct Green's function of the problem.

Section 3 contains a construction of a scalar product $h(u, v)$ in the space $H^{p,p}(\mathcal{D}, E)$, such that the norm $\sqrt{h(u, u)}$ is equivalent to the original one. If restricted to the subspace H of $H^{p,p}(\mathcal{D}, E)$ which consists of the functions with vanishing Dirichlet data on σ, this scalar product makes self-adjoint the double layer potential related to Green's function of a larger domain $\mathcal{D}'$ with a crack along σ.

Section 4 provides a detailed exposition of iteration of the double layer and volume potentials. Let A be a non-negative selfadjoint bounded operator in a Hilbert space H, satisfying $\|A\| \leq 1$. By the spectral theorem, we have $A^N = \int_{0-}^{1+0} \lambda^N \, dI(\lambda)$ for any $N = 1, 2, \ldots$, where $I(\lambda)$ is a resolution of the identity in H corresponding to A. Hence it follows that the iterations of A converge to the orthogonal projection of H onto the null space of the complementary operator $I - A$. As but one application of this we show the orthogonal decomposition

$$I = \lim_{N \to \infty} A^N + \sum_{v=0}^{\infty} A^v (I - A)$$

in the space H. Obviously, (1.1) is a very particular case of this latter formula.

In section 5 we indicate how (1.1) may be used to yield approximate solutions to the Cauchy problem for $Pu = f$ in $\mathcal{D}$ with data on σ. Write π_{V_0} for the orthogonal projection of H onto the null space of A. Consider the operator $\mathcal{R} = \sum_{v=0}^{\infty} (I - A)^v$ in H with a domain Dom $\mathcal{R}$ consisting of those $f \in H$ for which this series converges. If $f \in \text{Dom}\,\mathcal{R}$, then $A\mathcal{R}f = f - \pi_{V_0}f$. If moreover f is orthogonal to the null space of A, then $A\mathcal{R}f = f$. Conversely, if $f = Au$, for some $u \in H$, then $f \in \text{Dom}\,\mathcal{R}$ and $f \perp \ker A$. On the other hand, the equation $Bu = f$, for a non-necessarily self-adjoint operator $B \in \mathcal{L}(H, H)$, reduces to $B^*Bu = B^*f$ under the additional condition $f \perp \ker B^*$. Assuming $\sigma \neq \emptyset$, we apply this approach to the operator $B = P$ whose adjoint is $\mathcal{P}_v$.

In section 6 we derive in a similar way an interesting formula for solutions of the generalised Zaremba problem in $\mathcal{D}$, the Dirichlet data being given on $\sigma \neq \emptyset$.

Using explicit formulas for Green's function of $\mathcal{D}' \backslash \bar{\sigma}$ we are able to construct approximate solutions for the Cauchy and Zaremba problems in $\mathcal{D}$ with data on σ, thus recovering the results of [3].

2. A crack problem

Let X be an open set in $\mathbb{R}^n$ and $P \in \mathrm{Diff}^p(X; E, F)$ an elliptic differential operator of order p on X, where $E = X \times \mathbb{C}^k$, $F = X \times \mathbb{C}^l$.

Having fixed Hermitian metrics on E and F, denote by $P^* \in \mathrm{Diff}^p(X; F, E)$ the formal adjoint for P. Then, $\Delta = P^*P$ is an elliptic differential operator of order $2p$ and type $E \to E$.

Assume that Δ bears the uniqueness property for the Cauchy problem in the small on X. Then it has a two sided fundamental solution G on X.

Let $\mathscr{D}$ be a relatively compact domain in X. We emphasise that the boundary of $\mathscr{D}$ need not be smooth. Denote by $\mathscr{S}_P(\mathscr{D})$ the space of all solutions of the system $Pu = 0$ in $\mathscr{D}$.

In section 3 we construct a special scalar product $h(\cdot, \cdot)$ on the weighted Sobolev space $H^{p,p}(\mathscr{D}, E)$. This is obtained by using a fundamental solution of Δ having special properties at the boundary of a larger domain $\mathscr{O} \Subset X$ with singular boundary.

Throughout this section we will assume that the boundary of $\mathscr{D}$ consists of two smooth pieces, namely $\partial \mathscr{D} = \sigma \cup (\partial \mathscr{D} \setminus \sigma)$ where σ is an open connected subset of $\partial \mathscr{D}$ with smooth boundary. Fix a domain $\mathscr{D}'$ with C^∞ boundary, such that $\mathscr{D} \Subset \mathscr{D}' \Subset X$.

Let we be given a Dirichlet system $\{B_j\}_{j=0}^{p-1}$ of order $(p-1)$ on the smooth part of $\partial \mathscr{D} \cup \partial \mathscr{D}'$, with B_j being of type $E \to F_j$ and order m_j. Denote by $\{C_j\}_{j=0}^{p-1}$ the Dirichlet system adjoint to $\{B_j\}$ with respect to Green's formula for P (cf. [9, 9.2.1]). As is well known, C_j is of type $F^* \to F_j^*$ and order $p-1-m_j$. The only singularities of the coefficients of $\{C_j\}$ lie on the interface $\partial \sigma \subset \partial \mathscr{D}$.

By the above, there is a Green's operator G_P for P whose restriction to the smooth part of $\partial \mathscr{D} \cup \partial \mathscr{D}'$ is $G_P(*_F g, u) = (t(u), n(g))_y\, ds$, where ds is the area form of $\partial(\mathscr{D}' \setminus \overline{\mathscr{D}})$ and

$$t(u) = \bigoplus_{j=0}^{p-1} B_j u,$$

$$n(g) = \bigoplus_{j=0}^{p-1} *_{F_j}^{-1} C_j *_F g,$$

for $u \in C^\infty(E)$ and $g \in C^\infty(F)$. Moreover, $*_F$ and $*_{F_j}$ are sesquilinear bundle isomorphisms $F \to F^*$ and $F_j \to F_j^*$ induced by the Hermitian metrics on F and F_j, respectively.

Recall that $H^{s,\gamma}(\mathscr{D}, E)$ stands for the weighted Sobolev space of sections of E over $\mathscr{D}$ of smoothness s and weight γ, both s and γ being real numbers.

For $s \in \mathbb{Z}_+$, it coincides with the completion of the C^∞ sections of E vanishing near $\partial\sigma$, with respect to the norm

$$\|u\|_{H^{s,\gamma}(\mathscr{D},E)} = \left(\int_{\mathscr{D}} \sum_{|\alpha| \leq s} \text{dist}(x, \partial\sigma)^{2(|\alpha|-\gamma)} \left| D^\alpha u(x) \right|^2 dx \right)^{1/2}.$$

We denote by $H^{s-\frac{1}{2},\gamma-\frac{1}{2}}(\partial\mathscr{D}, E)$ the space consisting of the traces of functions $u \in H^{s,\gamma}(\mathscr{D}, E)$ on the smooth part of $\partial\mathscr{D}$. As usual, we endow this space by the quotient norm; an inner description is available, too.

In his edge calculus Schulze [7] uses another scale of weighted Sobolev spaces, denoted by $W^{s,\gamma}(\mathscr{D}, E)$. While $H^{s,\gamma}(\mathscr{D}, E)$ and $W^{s,\gamma}(\mathscr{D}, E)$ are quite different from each other, for arbitrary s and γ, they coincide on the diagonal $s = \gamma$, i.e., $H^{s,s}(\mathscr{D}, E) = W^{s,s}(\mathscr{D}, E)$ for all $s \in \mathbb{Z}_+$ (cf. Schulze [8, Proposition 3.1.5]). This is just the case in the present paper where we deal, for the most part, with the spaces $H^{p,p}(\mathscr{D}, E)$.

Lemma 2.1. *Let* $\gamma \in \mathbb{R}$. *For each* $u \in H^{p,\gamma}(\mathscr{D}, E)$ *and* $g \in H^{p,p-\gamma}(\mathscr{D}, F)$, *we have*

$$\int_{\partial\mathscr{D}} (t(u), n(g))_x \, ds = \int_{\mathscr{D}} \left((Pu, g)_x - (u, P^*g)_x \right) dx.$$

Proof. Indeed, if $g = g_\nu$ vanishes near $\partial\sigma$, then the equality is simply a Green's formula for P (cf. [9, 9.2.1]). Since

$$\left| \int_{\partial\mathscr{D}} (t(u), n(g_\nu))_x \, ds \right| \leq c \, \|u\|_{H^{p-1,\gamma-\frac{1}{2}}(\partial\mathscr{D},E)} \|g_\nu\|_{H^{p-1,p-\gamma-\frac{1}{2}}(\partial\mathscr{D},F)}$$

and

$$\left| \int_{\mathscr{D}} (Pu, g_\nu)_x \, dx \right| \leq c \|u\|_{H^{p,\gamma}(\mathscr{D},E)} \|g_\nu\|_{H^{0,p-\gamma}(\mathscr{D},F)},$$

$$\left| \int_{\mathscr{D}} (u, P^*g_\nu)_x \, dx \right| \leq c \|u\|_{H^{0,\gamma}(\mathscr{D},E)} \|g_\nu\|_{H^{p,p-\gamma}(\mathscr{D},F)},$$

with c a constant independent of u and g_ν, the proof is completed by a passage to the limit in the formula for g_ν vanishing near $\partial\sigma$ and approximating an arbitrary $g \in H^{p,p-\gamma}(\mathscr{D}, F)$. $\square$

Consider the following Dirichlet problem for the Laplacian $\Delta = P^*P$ in the domain $\mathbb{O} = \mathscr{D}' \setminus \bar{\sigma}$ having a crack along $\bar{\sigma}$. Let $s \in \mathbb{Z}_+$ satisfy $s \geq p$, and let $\gamma \in \mathbb{R}$. Given any $u_0 \in \oplus H^{s-m_j-\frac{1}{2},\gamma-m_j-\frac{1}{2}}(\partial\mathbb{O}, F_j)$, find a $u \in H^{s,\gamma}(\mathbb{O}, E)$ such that

$$\begin{cases} \Delta u = 0 & \text{in } \mathbb{O}, \\ t(u) = u_0 & \text{on } \partial\mathbb{O}. \end{cases} \tag{2.1}$$

Since $\{B_j\}_{j=0}^{p-1}$ is a Dirichlet system of order $(p-1)$, we may solve (2.1) first in the class of weighted Sobolev spaces, thus reducing the problem to that of finding a $\tilde{u} \in H^{s,\gamma}(\mathbb{O}, E)$ satisfying $\Delta\tilde{u} = f$ in $\mathbb{O}$ and $t(\tilde{u}) = 0$ on $\partial\mathbb{O}$, for given $f \in H^{s-2p,\gamma-2p}(\mathbb{O}, E)$. The advantage of using this reduction lies in the fact that the latter problem can be posed as a weak one. Namely, given any $f \in H^{s-2p,\gamma-2p}(\mathbb{O}, E)$, $\gamma \geq p$, find a $\tilde{u} \in H^{s,\gamma}(\mathbb{O}, E)$ such that $t(\tilde{u}) = 0$ on $\partial\mathbb{O}$ and

$$\int_{\mathbb{O}} (P\tilde{u}, Pv)_x \, dx = \int_{\mathbb{O}} (f, v)_x \, dx \qquad (2.2)$$

for all $v \in H^{s,\gamma}(\mathbb{O}, E)$ satisfying $t(v) = 0$ on $\partial\mathbb{O}$. Note that the restriction $\gamma \geq p$ is necessary because otherwise the pairings in (2.2) are not defined.

If $s = \gamma = p$, then (2.2) is uniquely solvable for all $f \in H^{-p,-p}(\mathbb{O}, E)$. Indeed, denote by H the subspace of $H^{p,p}(\mathbb{O}, E)$ consisting of all u satisfying $t(u) = 0$ on $\partial\mathbb{O}$. The sesquilinear form $(u, v)_H = \int_{\mathbb{O}}(Pu, Pv)_x \, dx$ is easily seen to define a scalar product on H. Moreover, the norm $u \mapsto \sqrt{(u, u)_H}$ on H is equivalent to that induced by $H^{p,p}(\mathbb{O}, E)$, hence H is Hilbert. For every $f \in H^{-p,-p}(\mathbb{O}, E)$, the functional $v \mapsto \int_{\mathbb{O}} (f, v)_x \, dx$ is continuous on H. By Riesz' theorem, this functional can be written in the form (2.2) with some $u \in H$, as desired.

That we have chosen $s = p$ is not important at all, for the Dirichlet problem meets the Lopatinskii condition on the smooth part of $\partial\mathbb{O}$. However, the exponent $\gamma = p$ is of exceptional character. More precisely, the following is true.

Lemma 2.2. *There is a closed set $\Sigma \subset \mathbb{R}$ with the property that, if $\gamma \notin \Sigma$, then the Dirichlet problem (2.1) has a unique solution $u \in H^{s,\gamma}(\mathbb{O}, E)$, for each data $u_0 \in \oplus H^{s-m_j-\frac{1}{2},\gamma-m_j-\frac{1}{2}}(\partial\mathbb{O}, F_j)$. Moreover, $p \notin \Sigma$.*

Proof. Indeed, (2.1) is an elliptic boundary value problem in the domain $\mathcal{D}'$ with a crack along the smooth surface σ. It can be treated in the framework of analysis on manifolds with boundary and edges of codimension 1 on the boundary. The fact that $p \notin \Sigma$ is a consequence of the *formal self-adjointness* of the Dirichlet problem (see [4, sections 6.1.3 and 8.4.2]). $\qquad \Box$

In particular, we are able to invoke the classical scheme of constructing Green's function of an elliptic boundary value problem by improving a two-sided fundamental solution, the domain being $\mathbb{O} = \mathcal{D}' \setminus \bar{\sigma}$.

Proposition 2.3. Suppose $1 \notin \Sigma$. Then, there exists a Green's function $\mathcal{G}(x, y)$ of problem (2.1), i.e., a two sided fundamental solution of Δ in $\mathbb{O}$, such that:

1) $\mathcal{G}$ extends to a smooth matrix-valued function away from the diagonal in $(\overline{\mathcal{D}'} \setminus \partial\sigma) \times (\overline{\mathcal{D}'} \setminus \partial\sigma)$, the smoothness near $\sigma \times \sigma$ being understood sidewise;

2) $t(\mathcal{G}(\cdot, y)) = 0$ on $\partial\mathcal{D}' \cup \sigma$, for each $y \in \mathcal{O}$, the operator t acting in the variable x.

Proof. It is easy to verify that, given any fixed $y \in \mathcal{O}$, the function $u_0 = t(G(\cdot, y))$ belongs to $\oplus H^{\infty, \gamma - m_j - \frac{1}{2}}(\partial\mathcal{D}' \cup \sigma, F_j \otimes E_y^*)$ for all $\gamma < 1$. Denote by $R(\cdot, y)$ the unique solution of problem (2.1) corresponding to this u_0. Then $\mathcal{G}(x, y) = G(x, y) - R(x, y)$ bears all the desired properties. $\qquad\square$

If multiplied by an excision function of y, Green's function $\mathcal{G}(\cdot, y)$ actually belongs to $H^{\infty, p}(\mathcal{O}, E \otimes E_y^*)$, for each $y \in \mathcal{O}$, as is seen from the proof of Proposition 2.3. For more details we refer the reader to [4, Chapter 8, section 6].

When compared to explicit formulas, the proof highlights the structure of $\mathcal{G}$. Indeed, $\mathcal{G}$ can be constructed within the pseudo-differential calculus on $\mathcal{O}$ developed in Schulze [7], where $\mathcal{O}$ is thought of as a manifold with edge $\partial\sigma$ on the boundary. Thus, $\mathcal{G}$ inherits the structure of operators in this calculus, cf. also [8].

Remark 2.4. Since the problem (2.1) is formally self-adjoint, Green's function $\mathcal{G}$ actually satisfies $\mathcal{G}(x, y)^* = \mathcal{G}(y, x)$ on $\mathcal{O} \times \mathcal{O}$. However, we will not use this symmetry.

3. Adjoint operator

Let $\mathscr{S}_\Delta(\widehat{\mathcal{D}}' \setminus \overline{\mathcal{D}})$ be the set of all solutions to the equation $\Delta u = 0$ in $\mathcal{D}' \setminus \overline{\mathcal{D}}$ such that u has finite order of growth near $\partial\mathcal{D}'$ and $t(u) = 0$ on $\partial\mathcal{D}'$.

Using Lemma 2.2 for $\mathcal{D}' \setminus \overline{\mathcal{D}}$ instead of $\mathcal{D}' \setminus \bar{\sigma}$, we obtain readily a linear isomorphism

$$H^{p, \gamma}(\mathcal{D}' \setminus \overline{\mathcal{D}}, E) \cap \mathscr{S}_\Delta(\widehat{\mathcal{D}}' \setminus \overline{\mathcal{D}}) \xrightarrow{t_+} \oplus H^{p - m_j - \frac{1}{2}, \gamma - m_j - \frac{1}{2}}(\partial\mathcal{D}, F_j)$$

given by $u \mapsto t(u)|_{\partial\mathcal{D}}$. Finally, composing the inverse t_+^{-1} with the trace operator

$$H^{p, \gamma}(\mathcal{D}, E) \xrightarrow{t_-} \oplus H^{p - m_j - \frac{1}{2}, \gamma - m_j - \frac{1}{2}}(\partial\mathcal{D}, F_j)$$

yields a continuous linear mapping

$$H^{p, \gamma}(\mathcal{D}, E) \xrightarrow{\mathscr{E}} H^{p, \gamma}(\mathcal{D}' \setminus \overline{\mathcal{D}}, E) \cap \mathscr{S}_\Delta(\widehat{\mathcal{D}}' \setminus \overline{\mathcal{D}}). \qquad (3.1)$$

Suppose $\gamma \geq p$. Then we have $Pu \in L^2(\mathcal{D}, F)$ and $P\mathscr{E}(u) \in L^2(\mathcal{D}' \setminus \overline{\mathcal{D}}, E)$ for each $u \in H^{p, \gamma}(\mathcal{D}, E)$. Consider the Hermitian form

$$h(u, v) = \int_{\mathcal{D}} (Pu, Pv)_x \, dx + \int_{\mathcal{D}' \setminus \overline{\mathcal{D}}} (P\mathscr{E}(u), P\mathscr{E}(v))_x \, dx$$

defined for $u, v \in H^{p, \gamma}(\mathcal{D}, E)$.

Proposition 3.1. The Hermitian form $h(\cdot, \cdot)$ defines a scalar product in $H^{p,\gamma}$ $(\mathcal{D}, E)$.

Proof. It suffices to prove that $h(u, u) = 0$ implies $u \equiv 0$ in D. To do this, pick a $u \in H^{p,\gamma}(\mathcal{D}, E)$ with $h(u, u) = 0$. Then, $u \in \mathcal{S}_P(\mathcal{D})$ and $\mathcal{E}(u) \in \mathcal{S}_P(\mathcal{D}' \setminus \overline{\mathcal{D}})$. By the construction of $\mathcal{E}$, we have $t(u) = t(\mathcal{E}(u))$ on the smooth part of the boundary of $\mathcal{D}$. Hence there is a solution $u \in \mathcal{S}_P(\mathcal{D}' \setminus \partial\sigma)$ such that $u = u$ in $\mathcal{D}$ and $u = \mathcal{E}(u)$ in $\mathcal{D}' \setminus \overline{cD}$. In particular, $t(u) = 0$ on $\partial\mathcal{D}'$. Since the uniqueness property holds true for the Cauchy problem in the small for the Laplacian P^*P, it does so for P. This gives $u \equiv 0$ on all of $\mathcal{D}' \setminus \partial\sigma$ (cf., for instance, [9, 10.3.1]). In particular, $u = 0$ in $\mathcal{D}$, as desired. $\qquad\square$

Set

$$\Phi(x, y) = *_F P(y, D) *_E^{-1} \mathcal{G}(x, y),$$

for $(x, y) \in (\mathcal{D}' \setminus \bar{\sigma}) \times (\mathcal{D}' \setminus \bar{\sigma})$. Moreover, introduce the double layer and volume potentials by

$$-\mathcal{P}_{dl}u(x) = \int_{\partial\mathcal{D}} \left(t(u), n \left(*^{-1}\Phi(x, \cdot) \right) \right)_y ds, \quad x \in \mathcal{D}' \setminus \partial\mathcal{D};$$

$$\mathcal{P}_v f(x) = \int_{\mathcal{D}} \left(f, *^{-1}\Phi(x, \cdot) \right)_y dy, \quad x \in \mathcal{D}',$$

for $u \in H^{p,\gamma}(\mathcal{D}, E)$ and $f \in H^{0,\gamma-p}(\mathcal{D}, F)$. By the above, the integrals make sense provided $\gamma \geq p$.

Since $\mathcal{P}_{dl}u$ does depend only on the Cauchy data $t(u)$ of u on $\partial\mathcal{D}$, the designation $\mathcal{P}_{dl}u_0$ still makes sense for any $u_0 \in \oplus H^{p-m_j-\frac{1}{2}, p-m_j-\frac{1}{2}}(\partial\mathcal{D}, F_j)$.

Let $(\mathcal{P}_v f)^-$ and $(\mathcal{P}_v f)^+$ be the restrictions of $\mathcal{P}_v f$ to $\mathcal{D}$ and $\mathcal{D}' \setminus \overline{\mathcal{D}}$, respectively.

Lemma 3.2. *There exists a constant $c > 0$ with the property that, for every $f \in H^{0,\gamma-p}(\mathcal{D}, F)$, we have*

$$\| (\mathcal{P}_v f)^- \|_{H^{p,\gamma}(\mathcal{D},E)} \leq c\|f\|_{H^{0,\gamma-p}(\mathcal{D},F)},$$

$$\|(\mathcal{P}_v f)^+\|_{H^{p,\gamma}(\mathcal{D}'\setminus\overline{\mathcal{D}},E)} \leq c\|f\|_{H^{0,\gamma-p}(\mathcal{D},F)}.$$

Proof. This follows from the boundedness of edge pseudo-differential operators in the weighted Sobolev spaces. $\qquad\square$

This lemma gives us a hint that the proper choice of the weight exponent γ should be $\gamma = p$.

Lemma 3.3. *For every* $u \in H^{p,\gamma}(\mathcal{D}, E)$,

$$\int_{\mathcal{D}} \left(Pu, *^{-1}\Phi(x, \cdot) \right)_y dy + \int_{\mathcal{D}' \setminus \overline{\mathcal{D}}} \left(P\mathscr{E}(u), *^{-1}\Phi(x, \cdot) \right)_y dy$$

$$= \begin{cases} u(x), & x \in \mathcal{D}, \\ \mathscr{E}(u)(x), & x \in \mathcal{D}' \setminus \overline{\mathcal{D}}. \end{cases}$$

Proof. Since $\mathscr{G}(x, y)$ differs from $G(x, y)$ by a kernel $R(x, y)$ which satisfies $\Delta'(y, D)R(x, y) = 0$ over $\mathcal{O} \times \mathcal{O}$, it follows that

$$\Phi P = \mathscr{G} P^* P = (G - R)\Delta = \mathrm{Id}$$

on compactly supported sections of $E \mid_{\mathcal{O}}$. In other words, Φ is a left fundamental solution of P on $\mathcal{O}$. By Green's formula,

$$-\int_{\partial(\mathcal{D}' \setminus \overline{\mathcal{D}})} \left(t(E(u)), n(*^{-1}\Phi(x, \cdot)) \right)_y ds$$

$$+\int_{\mathcal{D}' \setminus \overline{\mathcal{D}}} \left(P\mathscr{E}(u), *^{-1}\Phi(x, \cdot) \right)_y dy = \begin{cases} 0, & x \in \mathcal{D}, \\ \mathscr{E}(u)(x), & x \in \mathcal{D}' \setminus \overline{\mathcal{D}}. \end{cases}$$

We now make use of the property that $t(\mathscr{E}(u)) = t(u)$ on $\partial \mathcal{D}$ and $t(\mathscr{E}(u)) = 0$ on $\partial \mathcal{D}'$. Hence we deduce that the integral over $\partial(\mathcal{D}' \setminus \overline{\mathcal{D}})$ on the left hand side is equal to $-\mathscr{P}_{\mathrm{dl}} u(x)$, for all $x \in \mathcal{D}' \setminus \partial \mathcal{D}$. This latter potential can in turn be expressed from Green's formula for u over the domain $\mathcal{D}$, thus implying the desired equality. $\qquad \square$

If $\gamma \geq 1$, then the elements of $H^{p,\gamma}(\mathcal{D}, E)$ vanish on the boundary of σ, to which refers the weight exponent γ. Denote by $H^{p,\gamma}_{[\sigma]}(\mathcal{D}, E)$ the closure in $H^{p,\gamma}(\mathcal{D}, E)$ of the C^∞ sections of E vanishing near $\bar{\sigma}$.

Proposition 3.4. *Suppose* $u \in H^{p,\gamma}(\mathcal{D}, E)$ *and* $f \in H^{0,\gamma-p}(\mathcal{D}, F)$. *For each* $v \in H^{p,2p-\gamma}_{[\sigma]}(\mathcal{D}, E)$, *it follows that*

$$h\left(\mathscr{P}_v f, v\right) = \int_{\mathcal{D}} (f, Pv)_x \, dx,$$

$$h\left(\mathscr{P}_{\mathrm{dl}} u, v\right) = \int_{\mathcal{D}' \setminus \overline{\mathcal{D}}} (P\mathscr{E}(u), P\mathscr{E}(v))_x \, dx.$$

Proof. Indeed, if $f \in H^{p,\gamma-p}(\mathcal{D}, F)$ and $v \in H^{p,2p-\gamma}(\mathcal{D}, E)$, then integration by parts gives

$$\int_{\mathcal{D}} (f, Pv)_x \, dx - \int_D \left(P^* f, v \right)_x dx = \int_{\partial \mathcal{D}} (n(f), t(v))_x \, ds. \qquad (3.2)$$

Analogously,

$$\int_{\mathcal{D}'\backslash\mathcal{D}} (P\mathcal{E}(u), P\mathcal{E}(v))_x \, dx = - \int_{\partial\mathcal{D}} (n(P\mathcal{E}(u)), t(v))_x \, ds \qquad (3.3)$$

for all $u \in H^{2p,\gamma}(\mathcal{D}, E)$ and $v \in H^{p,2p-\gamma}(\mathcal{D}, E)$, because $t(\mathcal{E}(v)) = t(v)$ on $\partial\mathcal{D}$ and $t(\mathcal{E}(v)) = 0$ on $\partial\mathcal{D}'$.

Fix $u \in H^{2p,\gamma}(\mathcal{D}, E)$ and $v \in H^{p,2p-\gamma}(\mathcal{D}, E)$, and apply formula (3.2) for $f = Pu$. Using (3.2) and (3.3), we obtain

$$h(u, v) = \int_{\partial\mathcal{D}} (n(Pu) - n(P\mathcal{E}(u)), t(v))_x \, ds + \int_{\mathcal{D}} (\Delta u, v)_x \, dx.$$

Let us take $f \in C^{\infty}_{\text{comp}}(\mathcal{D}, F)$. In this case integrating by parts shows readily that

$$\mathcal{P}_v f(x) = \int_{\mathcal{D}} \left(P^* f, *_E^{-1} \mathcal{G}(x, \cdot) \right)_y \, dy$$

for all $x \notin \partial\mathcal{D}$. From Proposition 2.3 we deduce that $\mathcal{P}_v f$ is a C^{∞} section of E over $\mathcal{D}' \backslash \bar{\sigma}$, such that $t(\mathcal{P}_v f)$ vanishes on both σ and $\partial\mathcal{D}'$. Moreover, it satisfies $\Delta\mathcal{P}_v f = 0$ on $\mathcal{D}' \backslash \overline{\mathcal{D}}$. Hence it follows that $\mathcal{E}(\mathcal{P}_v f)^-$ is simply equal to the restriction of $\mathcal{P}_v f$ to $\mathcal{D}' \backslash \overline{\mathcal{D}}$, for $t(\mathcal{P}_v f)^+ = t(\mathcal{P}_v f)^-$ on $\partial\mathcal{D} \backslash \bar{\sigma}$. Now we can substitute $(\mathcal{P}_v f)^-$ for u in the formula above, to obtain

$$h(\mathcal{P}_v f, v) = \int_{\partial\mathcal{D}} \left(n\left(P\,(\mathcal{P}_v f)^-\right) - n\left(P\,(\mathcal{P}_v f)^+\right), t(v) \right)_x \, ds$$
$$+ \int_{\mathcal{D}} (\Delta\mathcal{P}_v f, v)_x \, dx.$$

Since $\mathcal{P}_v f$ is smooth away from $\bar{\sigma}$ in $\mathcal{D}'$, the jump of $n(P\mathcal{P}_v f)$ across $\partial\mathcal{D} \backslash \bar{\sigma}$ is zero. On the other hand, this jump has nothing to do on the rest part of the boundary of $\mathcal{D}$, provided that $t(v) = 0$ on σ. Thus, the first summand on the right hand side of the last equality is equal to zero, if v varies over $H^{p,2p-\gamma}_{[\sigma]}(\mathcal{D}, E)$. Taking into account that $\Delta\mathcal{P}_v f(x) = P^* f(x)$ for all $x \in \mathcal{D}$, we obtain

$$h(\mathcal{P}_v f, v) = \int_{\mathcal{D}} \left(P^* f, v \right)_x \, dx = \int_{\mathcal{D}} (f, Pv)_x \, dx.$$

Since $C^{\infty}_{\text{comp}}(\mathcal{D}, F)$ is dense in $H^{0,\gamma-p}(\mathcal{D}, F)$, this formula holds for every $f \in H^{0,\gamma-p}(\mathcal{D}, F)$ and $v \in H^{p,2p-\gamma}_{[\sigma]}(\mathcal{D}, E)$, as desired.

Finally, we can express the double layer potential $\mathcal{P}_{\text{dl}} u$ from Green's formula for P in $\mathcal{D}$ as $\mathcal{P}_{\text{dl}} u = u - \mathcal{P}_v Pu$. Hence

$$h(\mathcal{P}_{\text{dl}} u, v) = h(u, v) - h(\mathcal{P}_v Pu, v)$$
$$= \int_{\mathcal{D}'\backslash\overline{\mathcal{D}}} (P\mathcal{E}(u), P\mathcal{E}(v))_x \, dx$$

for any $u \in H^{p,\gamma}(\mathfrak{D}, u)$ and $v \in H^{p,2p-\gamma}_{[\sigma]}(\mathfrak{D}, E)$, as is clear from what has already been proved. This establishes the formula. $\square$

Not only does Proposition 3.4 specify the adjoint operator for P under the scalar product $h(u, v)$, but does also show the self-adjointness of the double layer potential.

Lemma 3.5. *The mappings t_-, t_+ induce topological isomorphisms of Hilbert spaces, namely*

$$H^{p,\gamma}(\mathfrak{D}, E) \cap \mathscr{S}_\Delta(\mathfrak{D}) \xrightarrow{t_-} \oplus H^{p-m_j-\frac{1}{2},\gamma-m_j-\frac{1}{2}}(\partial\mathfrak{D}, F_j),$$

$$H^{p,\gamma}(\mathfrak{D}' \setminus \overline{\mathfrak{D}}, E) \cap \mathscr{S}_\Delta(\widehat{\mathfrak{D}}' \setminus \mathfrak{D}) \xrightarrow{t_+} \oplus H^{p-m_j-\frac{1}{2},\gamma-m_j-\frac{1}{2}}(\partial\mathfrak{D}, F_j).$$

Proof. This follows from Lemma 2.2 as is explained at the very beginning of this section. $\square$

Proposition 3.6. *The topologies induced in $H^{p,p}(\mathfrak{D}, E)$ by $h(\cdot, \cdot)$ and by the original scalar product are equivalent.*

Proof. As the mapping (3.1) is continuous, we see, for any $u \in H^{p,p}(\mathfrak{D}, E)$, that

$$h(u, u) = \|Pu\|^2_{L^2(\mathfrak{D}, F)} + \|P\mathscr{E}(u)\|^2_{L^2(\mathfrak{D}'\setminus\overline{\mathfrak{D}}, F)}$$

$$\leq c \left(\|u\|^2_{H^{p,p}(\mathfrak{D}, E)} + \|\mathscr{E}(u)\|^2_{H^{p,p}(\mathfrak{D}'\setminus\overline{\mathfrak{D}}, E)} \right)$$

$$\leq c (1 + c') \|u\|^2_{H^{p,p}(\mathfrak{D}, E)},$$

the constants c and c' being independent of u. Here, we use the identification $L^2(\mathfrak{D}, F) = H^{0,0}(\mathfrak{D}, F)$ and the fact that P acts continuously from $H^{s,\gamma}(\mathfrak{D}, E)$ to $H^{s-p,\gamma-p}(\mathfrak{D}, F)$.

Conversely, Lemmas 3.2 and 3.3 imply that

$$\frac{1}{2} \|u\|^2_{H^{p,p}(\mathfrak{D}, E)} \leq \|\mathscr{P}_v Pu\|^2_{H^{p,p}(\mathfrak{D}, E)} + \|\mathscr{P}_v P\mathscr{E}(u)\|^2_{H^{p,p}(\mathfrak{D}, E)}$$

$$\leq c\|Pu\|^2_{L^2(\mathfrak{D}, F)} + c\|P\mathscr{E}(u)\|^2_{L^2(\mathfrak{D}'\setminus\overline{\mathfrak{D}}, F)}$$

$$= ch(u, u),$$

with c a constant independent of u. This completes the proof. $\square$

As mentioned above, Proposition 3.4 allows one to specify the adjoint operator of P with respect to the scalar product $h(u, v)$ on $H^{p,p}_{[\sigma]}(\mathfrak{D}, E)$ as the volume potential $\mathscr{P}_v$.

Theorem 3.7. *In the Hilbert space $H^{p,p}_{[\sigma]}(\mathfrak{D}, E)$ endowed with the scalar product $h(\cdot, \cdot)$, we have $\|P\| \leq 1$ and $P^* = \mathscr{P}_v$.*

Proof. This is an immediate consequence of Propositions 3.1, 3.4 and 3.6. $\square$

4. Iterations

The following theorem can actually be formulated in the abstract context of selfadjoint operators in Hilbert spaces.

Theorem 4.1. *In the strong operator topology on the space* $\mathcal{L}(H_{[\sigma]}^{p,p}(\mathcal{D}, E))$, *we have*

$$\lim_{N \to \infty} \mathcal{P}_{dl}^N = \pi_{V_1},$$

$$\lim_{N \to \infty} (\mathcal{P}_v P)^N = \pi_{V_0},$$

where

$$V_0 = \left\{ u \in H_{[\sigma]}^{p,p}(\mathcal{D}, E) : P\mathcal{E}(u) = 0 \text{ in } \mathcal{D}' \setminus \overline{\mathcal{D}} \right\},$$

$$V_1 = \left\{ u \in H_{[\sigma]}^{p,p}(\mathcal{D}, E) : Pu = 0 \text{ in } \mathcal{D} \right\}.$$

Proof. First, Theorem 3.7 implies that the integrals $\mathcal{P}_{dl}$ and $\mathcal{P}_v P$ define bounded selfadjoint non-negative operators in the Hilbert space $H_{[\sigma]}^{p,p}(\mathcal{D}, E)$ with the scalar product $h(\cdot, \cdot)$. Moreover, their norms do not exceed 1.

By the spectral theorem for bounded selfadjoint operators we conclude readily that

$$\mathcal{P}_{dl}^N = \int_{0-}^{1+0} \lambda^N dI(\lambda), \tag{4.1}$$

where $(I(\lambda))_{0 \leq \lambda \leq 1}$ is a resolution of the identity in the Hilbert space $H_{[\sigma]}^{p,p}(\mathcal{D}, E)$ corresponding to the self-adjoint operator $0 \leq \mathcal{P}_{dl} \leq I$ (see, for instance, [10, Chapter XI, sections 5, 6]).

Passing to the limit in (4.1) yields

$$\lim_{N \to \infty} \mathcal{P}_{dl}^N = I(1+0) - I(1-0),$$

the operator on the right side being an orthogonal projection of $H_{[\sigma]}^{p,p}(\mathcal{D}, E)$ onto a (closed) subspace $V(1)$. Obviously,

$$(I - \mathcal{P}_{dl}) \lim_{N \to 0} \mathcal{P}_{dl}^N u = 0$$

for all $u \in H_{[\sigma]}^{p,p}(\mathcal{D}, E)$, i.e., $V(1) \subset \ker(I - \mathcal{P}_{dl})$. Finally, if $u \in \ker(I - \mathcal{P}_{dl})$ then

$$u = \mathcal{P}_{dl} u + (I - \mathcal{P}_{dl}) u = \mathcal{P}_{dl} u = \mathcal{P}_{dl}^N u,$$

for every $N \geq 0$. Hence it follows that

$$u = \lim_{N \to \infty} \mathcal{P}_{\mathrm{dl}}^N u,$$

and so $V(1) = \ker(I - \mathcal{P}_{\mathrm{dl}})$ whence

$$\lim_{N \to \infty} \mathcal{P}_{\mathrm{dl}}^N = \pi_{\ker \mathcal{P}_{\mathrm{v}} P}.$$

Arguing in the same way we obtain

$$\lim_{N \to \infty} (\mathcal{P}_{\mathrm{v}} P)^N = \pi_{\ker \mathcal{P}_{\mathrm{dl}}}.$$

Now Proposition 3.4 implies that the null space of $\mathcal{P}_{\mathrm{v}} P$ in $H^{p,p}_{[\sigma]}(\mathscr{D}, E)$ coincides with $H^{p,p}_{[\sigma]}(\mathscr{D}, E) \cap \mathscr{S}_P(\mathscr{D})$.

Finally, Proposition 3.4 and Lemma 3.3, if combined with Green's formula for P, imply that $\mathcal{P}_{\mathrm{dl}} u = 0$ if and only if $P\mathscr{E}(u) = 0$ in $\mathscr{D}' \setminus \overline{\mathscr{D}}$. $\qquad \square$

Recall that the operator P bears the uniqueness property for the Cauchy problem in the small on X. Hence, if the boundary of $\mathscr{D}$ is connected, [9, Theorem 10.3.5] implies that V_0 coincides with the space of all $u \in H^{p,p}(\mathscr{D}, E)$ such that $t(u) = 0$ on $\partial \mathscr{D}$.

On the other hand, if $\sigma \neq \emptyset$, then the uniqueness theorem for P gives $V_1 \equiv 0$.

Corollary 4.2. In the strong operator topology of the space $\mathscr{L}(H^{p,p}_{[\sigma]}(\mathscr{D}, E))$, we have

$$I = \lim_{N \to \infty} \mathcal{P}_{\mathrm{dl}}^N + \sum_{v=0}^{\infty} \mathcal{P}_{\mathrm{dl}}^v \mathcal{P}_{\mathrm{v}} P, \qquad (4.2)$$

$$I = \lim_{N \to \infty} (\mathcal{P}_{\mathrm{v}} P)^N + \sum_{v=0}^{\infty} (\mathcal{P}_{\mathrm{v}} P)^v \mathcal{P}_{\mathrm{dl}}. \qquad (4.3)$$

Proof. Indeed, Green's formula $\mathcal{P}_{\mathrm{dl}} + \mathcal{P}_{\mathrm{v}} P = I$ in $\mathscr{D}$ implies

$$I = \mathcal{P}_{\mathrm{dl}}^N + \sum_{v=0}^{N-1} \mathcal{P}_{\mathrm{dl}}^v \mathcal{P}_{\mathrm{v}} P$$

$$= (\mathcal{P}_{\mathrm{v}} P)^N + \sum_{v=0}^{N-1} (\mathcal{P}_{\mathrm{v}} P)^v \mathcal{P}_{\mathrm{dl}},$$

for every $N \in \mathbb{N}$. Now using Theorem 4.1 we can pass to the limit for $N \to \infty$, thus obtaining (4.3) and (4.2). $\qquad \square$

For applications to the Cauchy problem for solutions of $Pu = 0$ in $\mathscr{D}$ with data on σ, we also need the following result.

Corollary 4.3. Suppose $\sigma \neq \emptyset$. Then, in the strong operator topology of $\mathscr{L}(L^2(\mathscr{D}, F))$, we have

$$I = \pi_{\ker \mathscr{P}_v} + \sum_{v=0}^{\infty} P \mathscr{P}_{dl}^v \mathscr{P}_v.$$

Proof. Arguing as in the proof of Corollary 4.2, we obtain

$$I = \lim_{N \to \infty} (I - P\mathscr{P}_v)^N + \sum_{v=0}^{\infty} (I - P\mathscr{P}_v)^v P\mathscr{P}_v$$

$$= \pi_{\ker P\mathscr{P}_v} + \sum_{v=0}^{\infty} (I - P\mathscr{P}_v)^v P\mathscr{P}_v,$$

I being the identity operator on $L^2(\mathscr{D}, F)$.

Assume that $f \in L^2(\mathscr{D}, F)$ and $P\mathscr{P}_v f = 0$ in $\mathscr{D}$. As $t(\mathscr{P}_v f) = 0$ on σ, it follows, by the uniqueness property for the Cauchy problem in the small, that $\mathscr{P}_v f = 0$ in $\mathscr{D}$. Consequently, the kernel of $P\mathscr{P}_v$ on $L^2(\mathscr{D}, F)$ coincides with that of $\mathscr{P}_v$.

To complete the proof, it suffices to observe that

$$(I - P\mathscr{P}_v)^v P\mathscr{P}_v = P (I - \mathscr{P}_v P)^v \mathscr{P}_v = P \mathscr{P}_{dl}^v \mathscr{P}_v,$$

for any $v \in \mathbb{Z}_+$, which establishes the formula. $\qquad \square$

5. Cauchy problem

Consider the following Cauchy problem, for the operator P and the Dirichlet system $\{B_j\}_{j=0}^{p-1}$. Given $f \in L^2(\mathscr{D}, F)$ and $u_0 \in \oplus H^{p-m_j-\frac{1}{2}, p-m_j-\frac{1}{2}}(\sigma, F_j)$, find a section $u \in H^{p,p}(\mathscr{D}, E)$ satisfying

$$\begin{cases} Pu = f & \text{in } \mathscr{D}, \\ t(u) = u_0 & \text{on } \sigma. \end{cases} \tag{5.1}$$

This problem is well known to be ill posed unless $\sigma = \partial \mathscr{D}$. Using Corollary 4.2 we obtain easily approximate solutions to the problem. To this end, set

$$\mathscr{P}_{dl} u_0(x) = -\int_{\sigma} \left(u_0, n \left(*^{-1} \Phi(x, \cdot) \right) \right)_y ds, \quad x \in \mathscr{D}. \tag{5.2}$$

Note that the integral on the right hand side of formula (5.2) is well defined because of the properties of $\Phi(x, \cdot)$ and u_0.

Theorem 5.1. *The problem (5.1) has at most one solution. It is solvable if and only if $f \perp \ker \mathscr{P}_v$ and the series*

$$u = \mathscr{P}_{dl} u_0 + \sum_{v=0}^{\infty} \mathscr{P}_{dl}^v \mathscr{P}_v (f - P\mathscr{P}_{dl} u_0) \qquad (5.3)$$

converges in $H^{p,p}(\mathscr{D}, E)$. Moreover, the series, if converges, is the solution to this problem.

Proof. For the uniqueness theorem for solutions of (5.1) we refer the reader to [9, 10.3.1].

Lemma 2.2 and Proposition 2.3 imply that the potential $\mathscr{P}_{dl} u_0$ belongs to $H^{p,p}(\mathscr{D}, E) \cap \mathscr{S}_{\Lambda}(\mathscr{D})$ and satisfies $t(\mathscr{P}_{dl} u_0) = u_0$ on σ. Thus, by setting $u = \mathscr{P}_{dl} u_0 + u$ we reduce the problem (5.1) to the problem of finding a function $u \in H^{p,p}(\mathscr{D}, E)$ such that

$$\begin{cases} Pu = f & \text{in } \mathscr{D}, \\ t(u) = 0 & \text{on } \sigma, \end{cases} \qquad (5.4)$$

where $f = f - P\mathscr{P}_{dl} u_0$. It is a simple matter to see, by Theorem 3.7, that $f \perp \ker \mathscr{P}_v$ if and only if $f \perp \ker \mathscr{P}_v$.

Let the problem (5.4) be solvable. Then $\tilde{f} \perp \ker \mathscr{P}_v$. Moreover, equality (4.2) implies that the series $\tilde{u} = \sum_{v=0}^{\infty} \mathscr{P}_{dl}^v \mathscr{P}_v \tilde{f}$ converges and gives a solution to this problem. Therefore, the problem (5.1) is solvable, too, and its solution is presented by (5.3).

Conversely, let $f \perp \ker \mathscr{P}_v$ and the series (5.3) converge in $H^{p,p}(\mathscr{D}, E)$. Then Corollary 4.3 implies that

$$Pu = \sum_{v=0}^{\infty} P\mathscr{P}_{dl}^v \mathscr{P}_v f = f - \pi_{\ker \mathscr{P}_v} f = f,$$

as desired.   □

The theorem gains in interest if we realise that the null space of $\mathscr{P}_v$ consists of all $g \in L^2(\mathscr{D}, F)$ satisfying $P^* g = 0$ in $\mathscr{D}$ and $n(g) = 0$ on $\partial \mathscr{D}$ (cf. [9, Lemma 10.2.20]). Hence, $f \perp \ker \mathscr{P}_v$ implies, in particular, that $P_1 f = 0$, for each differential operator P_1 over $\mathscr{D}$ satisfying $P_1 P = 0$.

6. The Zaremba problem

Consider the following generalised Zaremba problem. Given $f \in H^{-p,-p}(\mathscr{D}, E)$ and

$$u_0 \in \oplus H^{p-m_j-\frac{1}{2}, p-m_j-\frac{1}{2}}(\sigma, F_j),$$

$$u_1 \in \oplus H^{-p+m_j+\frac{1}{2}, -p+m_j+\frac{1}{2}}(\partial \mathscr{D} \setminus \bar{\sigma}, F_j),$$

find a section $u \in H^{p,p}(\mathcal{D}, E)$ such that

$$\begin{cases} \Delta u = f & \text{in } \mathcal{D}, \\ t(u) = u_0 & \text{on } \sigma, \\ n(Pu) = u_1 & \text{on } \partial\mathcal{D} \setminus \bar{\sigma}. \end{cases} \tag{6.1}$$

It is worth pointing out that the trace of $n(Pu)$ on $\partial\mathcal{D} \setminus \bar{\sigma}$ is not defined for any $u \in H^{p,p}(\mathcal{D}, E)$, for the order of $n \circ P$ is $2p-1$. To cope with this, a familiar way is to assign an operator L with a dense domain Dom $L \hookrightarrow H^{p,p}(\mathcal{D}, E)$ to (6.1), such that both $t(u)$ and $n(Pu)$ are well defined for all $u \in$ Dom L. More precisely, Dom L is defined to be the completion of $C_{\text{comp}}^{\infty}(\overline{\mathcal{D}} \setminus \partial\sigma, E)$ with respect to the graph norm of $u \mapsto (u, t(u), n(Pu))$ in $H^{p,p}(\mathcal{D}, E) \oplus \mathcal{D} \oplus \mathfrak{N}$, where

$$\mathcal{D} = \oplus H^{p-m_j-\frac{1}{2}, p-m_j-\frac{1}{2}}(\sigma, F_j),$$
$$\mathfrak{N} = \oplus H^{-p+m_j+\frac{1}{2}, -p+m_j+\frac{1}{2}}(\partial\mathcal{D} \setminus \bar{\sigma}, F_j).$$

For more details, see Roitberg [5] and elsewhere. Then, (6.1) defines a continuous operator Dom $L \to H^{-p,-p}(\mathcal{D}, E) \oplus \mathcal{D} \oplus \mathfrak{N}$ by $Lu = (\Delta u, t(u), n(Pu))$.

If P is the gradient operator in $\mathbb{R}^n$, then (6.1) is just the classical Zaremba problem in $\mathcal{D}$.

As the Dirichlet problem is elliptic, it is easy to reduce the mixed problem (6.1) to that with $u_0 = 0$. To this end, set $u = \mathcal{P}_{\text{dl}}u_0 + \tilde{u}$, the integral $\mathcal{P}_{\text{dl}}u_0$ being given by (5.2). Then the data in (6.1) transform to $\tilde{f} = f$ and $\tilde{u}_0 = 0$, $\tilde{u}_1 = u_1 - n(P\mathcal{P}_{\text{dl}}u_0)$.

The advantage of studying the problem (6.1) with $u_0 = 0$ lies in the fact that it can be thought of in the following weak sense. For any f and u_1, as above, find a $u \in H_{[\sigma]}^{p,p}(\mathcal{D}, E)$ such that

$$\int_{\mathcal{D}} (Pu, Pv)_x \, dx = \int_{\mathcal{D}} (f, v)_x \, dx + \int_{\partial\mathcal{D}} (u_1, t(v))_x \, ds \tag{6.2}$$

for all $v \in H_{[\sigma]}^{p,p}(\mathcal{D}, E)$.

For Zaremba data f and u_1, we introduce the simple layer and volume potentials by

$$\mathcal{P}_{\text{sl}}u_1(x) = \int_{\partial\mathcal{D} \setminus \bar{\sigma}} \left(u_1, t\left(*^{-1}\mathcal{G}(x, \cdot) \right) \right)_y \, ds, \quad x \in \mathcal{D};$$
$$\mathcal{G}(\chi_{\mathcal{D}}f)(x) = \int_{\mathcal{D}} \left(f, *^{-1}\mathcal{G}(x, \cdot) \right)_y \, dy, \quad x \in \mathcal{D}, \tag{6.3}$$

where $\chi_{\mathcal{D}}$ is the characteristic function of $\mathcal{D}$.

Recall, cf. Remark 2.4, that $t(*^{-1}\mathcal{G}(x, \cdot))$ vanishes on σ, for each fixed $x \in \mathcal{D}' \setminus \bar{\sigma}$, hence the first integral in (6.3) is actually over all of $\partial\mathcal{D}$.

Theorem 6.1. *If $\sigma \neq \emptyset$, then the problem (6.1) has no more than one solution. It is solvable if and only if the series*

$$u = \mathscr{P}_{dl}u_0 + \sum_{\nu=0}^{\infty} \mathscr{P}_{dl}^{\nu} \left(\mathscr{G}(\chi_{\mathscr{D}} f) + \mathscr{P}_{sl}(u_1 - n(P\mathscr{P}_{dl}u_0))) \right) \qquad (6.4)$$

converges in $H^{p,p}(\mathscr{D}, E)$. Moreover, the series, when it converges, is the solution to this problem.

Proof. The uniqueness theorem for the problem (6.1) follows from the weak setting (6.2) immediately. Indeed, let $u \in H_{[\sigma]}^{p,p}(\mathscr{D}, E)$ be any solution of the problem corresponding to the zero data. Then $\int_{\mathscr{D}}(Pu, Pv)_x \, dx = 0$ for all $v \in H_{[\sigma]}^{p,p}(\mathscr{D}, E)$. In particular, substituting $v = u$ we conclude that $u \in H_{[\sigma]}^{p,p}(\mathscr{D}, E) \cap \mathscr{S}_P(\mathscr{D})$. Since P bears the uniqueness property for the Cauchy problem in the small on X we see that $u \equiv 0$, which is our claim.

Since $t(*^{-1}\mathscr{G}(x, \cdot)) = 0$ on σ, for each $x \in \mathscr{D}$, it follows from Green's formula for P that

$$\mathscr{P}_{v}Pu = \mathscr{G}(\chi_{\mathscr{D}} \Delta u) + \mathscr{P}_{sl}n(Pu)$$

for all $u \in H^{p,p}(\mathscr{D}, E)$. Hence, formula (4.2) shows that any solution u to the problem (6.1) with $u_0 = 0$ can be presented by the formula

$$u = \sum_{\nu=0}^{\infty} \mathscr{P}_{dl}^{\nu} \left(\mathscr{G}(\chi_{\mathscr{D}} f) + \mathscr{P}_{sl} u_1 \right). \qquad (6.5)$$

It follows that any solution to the general problem (6.1) can be written by formula (6.4).

Conversely, let the series (6.4) converge to a function $u \in H^{p,p}(\mathscr{D}, E)$. Then the series (6.5) converges, too, to the function $u = u - \mathscr{P}_{dl}u_0$ lying in $H_{[\sigma]}^{p,p}(\mathscr{D}, E)$. Since

$$\mathscr{P}_{v}Pu = (I - \mathscr{P}_{dl}) u$$

$$= \sum_{\nu=0}^{\infty} \mathscr{P}_{dl}^{\nu} \left(\mathscr{G}(\chi_{\mathscr{D}} f) + \mathscr{P}_{sl}u_1 \right) - \sum_{\nu=1}^{\infty} \mathscr{P}_{dl}^{\nu} \left(\mathscr{G}(\chi_{\mathscr{D}} f) + \mathscr{P}_{sl}u_1 \right)$$

$$= \mathscr{G}(\chi_{\mathscr{D}} f) + \mathscr{P}_{sl}u_1,$$

we conclude that

$$\int_{\mathscr{D}} (Pu, Pv)_x \, dx = h(\mathscr{P}_{v}Pu, v) = h(\mathscr{G}(\chi_{\mathscr{D}} f) + \mathscr{P}_{sl} u_1, v)$$

for any $v \in H_{[\sigma]}^{p,p}(\mathscr{D}, E)$, the first equality being due to Proposition 3.4.

Set $U = \mathcal{G}(\chi_{\mathcal{D}} f) + \mathcal{P}_{sl} u_1$. Using the fact that $\mathcal{G}(x, y)$ is a two-sided fundamental solution for Δ in $\mathcal{D}' \setminus \bar{\sigma}$, we obtain

$$\Delta U^- = f, \qquad \mathcal{C}\left(U^-\right) = U^+,$$

where U^- and U^+ stand for the restrictions of U to $\mathcal{D}$ and $\mathcal{D}' \setminus \overline{\mathcal{D}}$, respectively. Therefore, if v is a C^∞ section of E vanishing in a neighbourhood of $\bar{\sigma}$, we have

$$\int_{\mathcal{D}} (Pu, Pv)_x \, dx = \int_{\mathcal{D}} (f, v)_x \, dx + \int_{\partial \mathcal{D}} \left(n \left(PU^- \right) - n \left(P \mathcal{C} \left(U^- \right) \right), t(v) \right)_x \, ds$$

$$= \int_{\mathcal{D}} (f, v)_x \, dx + \int_{\partial \mathcal{D}} \left(n \left(PU^- \right) - n \left(PU^+ \right), t(v) \right)_x \, ds.$$

By a jump theorem for the simple layer potential (cf. [9, Theorem 10.1.5]), the difference $n(PU^-) - n(PU^+)$ is equal to u_1 on $\partial \mathcal{D} \setminus \bar{\sigma}$. Therefore,

$$\int_{\mathcal{D}} (Pu, Pv)_x \, dx = \int_{\mathcal{D}} (f, v)_x \, dx + \int_{\partial \mathcal{D}} (u_1, t(v))_x \, ds,$$

which gives (6.2) if combined with the fact that the functions $v \in C_{loc}^\infty(X, E)$ vanishing near $\bar{\sigma}$ are dense in $H_{[\sigma]}^{p,p}(\mathcal{D}, E)$. Thus, the problem (6.1) is solvable, which completes the proof. $\qquad \square$

Were the problem (6.1) first reduced to the particular case corresponding to $f = 0$ and $u_0 = 0$, we would obtain yet another formula for approximate solutions of this problem.

Theorem 6.1 just amounts to saying that the operator L bears the following two properties: 1) L is injective, and 2) $(f, u_0, u_1) \in \operatorname{Ran} L$ is equivalent to the convergence of the series (6.4) in $H^{p,p}(\mathcal{D}, E)$. Moreover, since the Zaremba problem is formally self-adjoint with respect to Green's formula for Δ, it follows that the range of L is dense in $H^{-p,-p}(\mathcal{D}, E) \oplus \mathcal{D} \oplus \mathfrak{N}$.

Remark 6.2. If (6.1) is Fredholm, the series (6.4) converges in $H^{p,p}(\mathcal{D}, E)$, for each data (f, u_0, u_1).

References

[1] Duduchava, R. and Wendland, W., The Wiener–Hopf method for systems of pseudo-differential equations with an application to crack problems, *Integral Equ. and Operator Theory* **23** (1995), 294–335.

[2] Eskin, G., *Boundary value problems for elliptic pseudo-differential operators*, Nauka, Moscow 1973.

[3] Nacinovich, M. and Shlapunov, A. A., On iterations of the Green integrals and their applications to elliptic differential complexes, *Math. Nachr.* **180** (1996), 243–286.

[4] Nazarov, S.-A. and Plamenevskii, B.-A., *Elliptic boundary value problems in domains with piecewise smooth boundary*, Nauka, Moscow 1991.

[5] Roitberg, Ya.-A., *Elliptic boundary value problems in generalized functions*, Kluwer Academic Publishers, Dordrecht NL 2000.

[6] Romanov, A.-V., Convergence of iterations of Bochner–Martinelli operator, and the Cauchy–Riemann system, *Dokl. Akad. Nauk SSSR (4)* **242** (1978), 780–783.

[7] Schulze, B.-W., Crack problems in the edge pseudo-differential calculus, *Applic. Analysis* **45** (1992), 333–360.

[8] _____, *Pseudo-differential calculus and applications to non-smooth configurations*, Preprint 99/16, Univ. Potsdam, July 1999, p. 135.

[9] Tarkhanov, N.-N., *The Cauchy problem for solutions of elliptic equations*, Akademie–Verlag, Berlin 1995.

[10] Yosida, K., *Functional analysis*, Springer–Verlag, Berlin et al. 1965.

PART V

MATHEMATICAL PHYSICS

COVARIANT METHOD FOR SOLUTION OF CAUCHY'S PROBLEM BASED ON LIE GROUP ANALYSIS AND LERAY'S FORM

NAIL H. IBRAGIMOV

Contents

1 Preliminaries 415
    1.1   Invariance principle 415
    1.2   Preliminaries on transformations of distributions, Leray's form
           and auxiliary differential equations 416
2 Application to the wave equation 418
    2.1   The initial-value problem for the fundamental
           solution and its symmetries 419
    2.2   Derivation of the fundamental solution 420

M. de Gosson (ed.), Jean Leray '99 Conference Proceedings, 413-421.
© *2003 Kluwer Academic Publishers.*

Abstract. *The Lie group theory and Leray's form provide a covariant (i.e., independent of a coordinate system) method for the calculation of fundamental solutions for linear hyperbolic equations, and hence for the solution of Cauchy's problem. Here, the method is illustrated by the classical wave equation.*

1. Preliminaries

Lie group methods are usually regarded to be not particularly useful for solving the Cauchy problem because arbitrary initial conditions break the symmetry group of a differential equation in question. It seems that this argument is irrefutable. However, the author noted recently [1](i) that the fundamental solutions of the classical equations of mathematical physics are in fact group-invariant solutions. This observation, amplified by the formulation of what is called the *invariance principle* [1](ii), led to the development of a systematic group theoretical method suited for solving the Cauchy problem [1](iii)–(iv). In this new approach, the philosophy of Lie symmetries is combined with the theory of distributions.

The new approach provides a simple way for constructing fundamental solutions of linear differential equations. The method is applicable not only to equations with constant coefficients but also to numerous equations with variable coefficients provided that the latter admit 'sufficiently wide' Lie groups. By its very nature Lie symmetry analysis provides tools for obtaining results independent on a choice of coordinate systems. In consequence, the group theoretic method together with the invariant definition of the Dirac measure via Leray's form [2] provide a covariant method, unlike the accustomed Fourier transform method, for calculating fundamental solutions of the Cauchy problem for linear hyperbolic differential equations.

1.1. INVARIANCE PRINCIPLE

Recall that a differential equation is said to be *invariant* with respect to a Lie group G of invertible transformations of the independent and dependent variables, $x^1, \ldots, x^n$ and u, involved in the differential equation, if the solutions of the differential equation are merely permuted among themselves by every transformation of the group G. The group G is also called a *symmetry group* or a group *admitted* by the differential equation in question.

It is evident that any subgroup $H \subset G$ of a symmetry group G is also a symmetry group, and hence H maps every solution of the differential equation in question again into solutions of the same equation. It may happen that some of solutions are individually unaltered under the action of the subgroup H. These solutions are termed H-invariant solutions or, for brevity, *invariant solutions*. In other words, every transformation from H maps an invariant solution into itself. The invariant solutions can be expressed, under certain regularity conditions, via invariants $J(x^1, \ldots, x^n, u)$ of the subgroup H. Furthermore, the invariants $J(x^1, \ldots, x^n, u)$ of the subgroup H are determined by the system of homogeneous linear partial differential equations $X_i J = 0$, where X_i are basic infinitesimal generators of H. In consequence, the search of this particular type of solutions reduces the number of the independent variables of the differential equation. For example, if H is a one-parameter subgroup, then the number of the independent variables reduces by one. The further reduction is achieved by considering invariant solutions under multi-parameter subgroups.

Let us illustrate the notion of invariant solutions by considering the classical wave equation $u_{tt} = \Delta u$. Here $\Delta u \equiv u_{xx} + u_{yy} + u_{zz}$ is the Laplacian in three variables x, y, and z; the subscripts denote the partial differentiation. The wave equation is invariant under the 15-parameter conformal group G of the Minkowski space–time. The time-translations $\bar{t} = t + a$ form a one-parameter (with the group parameter a) subgroup $H \subset G$. The invariants of H are the independent variables x, y, z and the dependent variable u. Hence, the invariant solutions are stationary solutions $u = v(x, y, z)$. They are determined by the Laplace equation $\Delta u = 0$, i.e., the number of the independent variables is reduced by one. To further reduce the number of the independent variables, let us take, e.g., the subgroup $H \subset G$ consisting of time-translations and rotations in the space x, y, z. This subgroup has two invariants, $r = \sqrt{x^2 + y^2 + z^2}$ and u. Consequently, the invariants solutions are spherically symmetric stationary solutions $u = w(r)$. Substituting the latter expression in the wave equation, one arrives at the ordinary differential equation $(r^2 w')' = 0$, whence $w = C_1 r^{-1} + C_2$. Hence, the invariance under the subgroup H, comprising time-translations and spatial rotations, reduces the number of the independent variables by three.

Invariance principle: Consider a boundary-value problem (in particular, the Cauchy problem) for a differential equation admitting a Lie group G. If the boundary-value problem is invariant under a subgroup $H \subset G$, then one should seek the solution of the problem among H-invariant solutions of the differential equation. Here, the invariance of the boundary-value problem means that *the differential equation, the manifold where the data are given, and the data themselves are invariant under H.*

1.2. PRELIMINARIES ON TRANSFORMATIONS OF DISTRIBUTIONS, LERAY'S FORM AND AUXILIARY DIFFERENTIAL EQUATIONS

Consider a one-parameter group of transformations $\bar{x} = g(x, a)$ in $\mathbb{R}^n$, where $x = (x^1, \ldots, x^n)$. The infinitesimal transformation of this group is written

$$\bar{x}^i \approx x^i + a\xi^i(x), \quad i = 1, \ldots, n. \tag{1.1}$$

Given a distribution $f(x)$, its image $\bar{f}$ under the infinitesimal transformation (1.1) is given by (see, e.g., [1](ii)):

$$\bar{f} \approx f - a D_i \left(\xi^i \right) f, \quad \text{where } D_i \left(\xi^i \right) = \sum_{i=1}^{n} \frac{\partial \xi^i}{\partial x^i}. \tag{1.2}$$

In particular, let $f(x)$ be the Dirac measure $\delta(x)$. Then in view of the well known formula from the theory of distributions:

$$\phi(x)\delta(x) = \phi(0)\delta(x) \tag{1.3}$$

the transformation law (1.2) becomes:

$$\bar{\delta} \approx \delta - a D_i \left(\xi^i \right) \Big|_{x=0} \delta. \tag{1.4}$$

Consider a hypersurface in $\mathbb{R}^n$ given by $P(x) = 0$ with a continuously differentiable function $P(x)$ such that $\nabla P \neq 0$ on the hypersurface. *Leray's form* for this hypersurface is an $(n-1)$-differential form ω such that

$$dP \wedge \omega = dx^1 \wedge \cdots \wedge dx^n.$$

It can be represented in the form (see Leray [2] (1953), Ch. IV, section 1)

$$\omega = (-1)^{i-1} \frac{dx^1 \wedge \cdots \wedge dx^{i-1} \wedge dx^{i+1} \wedge \cdots \wedge dx^n}{P_i}$$

for any fixed i such that $P_i \equiv \partial P(x)/\partial x^i \neq 0$.

The *Heaviside function* $\theta(P)$ on the surface $P(x) = 0$ is defined by

$$\theta(P) = \begin{cases} 1, & P \geq 0, \\ 0, & P < 0. \end{cases}$$

It can be identified with the distribution

$$\langle \theta(P), \varphi \rangle = \int_{P \geq 0} \varphi(x)\, dx.$$

The *Dirac measure* $\delta(P)$ on the surface $P(x) = 0$ is defined by

$$\langle \delta(P), \varphi \rangle = \int_{P=0} \varphi\omega,$$

where ω is Leray's form. These two distributions are related by $\theta'(P) = \delta(P)$.

The above distributions are involved in the solutions of the following first-order *auxiliary differential equations* useful for our purposes. The simplest of these equations is the equation

$$xf' = 0 \tag{1.5}$$

with one independent variable x. Its classical solution is $f = $ const. Now, taking in equation (1.3) $\phi(x) = x$, one has the identity $x\delta(x) = 0$. Invoking that $\theta'(x) = \delta(x)$, where $\theta(x)$ is the Heaviside function with one variable x, we conclude that equation (1.5) has, in distributions, the solution $f = \theta(x)$ different from $f = $ const. Since equation (1.5) is linear, the linear combination

$$f = C_1\theta(x) + C_2 \tag{1.6}$$

provides a distribution-solution involving two arbitrary constants, C_1 and C_2. It is known in the theory of distributions that (1.6) furnishes the general solution of equation (1.5).

The equations (1.3) and (1.5) generalized to many variables x^i have the forms

$$P\delta(P) = 0 \tag{1.7}$$

and

$$Pf'(P) = 0, \tag{1.8}$$

respectively. The general solution of the later equation in distributions is similar to (1.6) and has the form

$$f = C_1\theta(P) + C_2, \quad C_1, C_2 = \text{const.} \tag{1.9}$$

Furthermore, differentiating (1.7) successively, one arrives at the following:

$$P\delta^{(m)}(P) + m\delta^{(m-1)}(P) = 0, \quad m = 1, 2, \ldots, \tag{1.10}$$

where $\delta^{(m)}$ is the mth derivative of $\delta(P)$ with respect to P. It means that the first-order differential equation

$$Pf'(P) + mf(P) = 0 \tag{1.11}$$

has, along with the classical solution $f = P^{-m}$, also a non-classical solution given by the distribution $f = \delta^{(m-1)}(P)$. Their linear combination with arbitrary constant coefficients C_1 and C_2,

$$f = C_1\delta^{(m-1)}(P) + C_2 P^{-m}, \tag{1.12}$$

furnishes the general solution to (1.11) in distributions.

2. Application to the wave equation

The key point of the success of the group theoretic approach to fundamental solutions is that the specific initial-value problem defining fundamental solutions, unlike the general Cauchy problem with arbitrary initial conditions, inherits certain symmetries of the differential equation. That's why one can simplify the calculation of fundamental solutions by searching them as invariant solutions under the inherited symmetry group.

For those differential equation whose fundamental solutions are usual functions, one can use the classical Lie theory to obtain the fundamental solutions in terms of usual invariants of the inherited symmetry group. This is the case, e.g., for the Laplace and heat equations [1](i).

It is well known, however, that the fundamental solution of the wave equation is not a usual function. Rather, it is a distribution. Therefore, it was suggested in [1](ii) to tackle the problem as follows.

As mentioned in section 1.1, the classical wave equation is invariant under the conformal group of the Minkowski space–time. We first use a subgroup of the conformal group, namely the group of isometric motions (the Lorentz group) to find two classical invariants. Then we relate these invariants by a differential equation obtained from the invariance test under a proper conformal subgroup, namely a one-parameter group of dilations. The general solution in distributions of the latter equation gives us the fundamental solution. Note that this approach does need the wave equation at all. It suffices to use only the symmetry group, i.e., the conformal group. Hence, the propagation of a wave is uniquely determined by the initial data and the principle of conformal symmetry!

2.1. THE INITIAL-VALUE PROBLEM FOR THE FUNDAMENTAL SOLUTION AND ITS SYMMETRIES

Recall that the Cauchy problem for the wave equation

$$u_{tt} - u_{xx} - u_{yy} - u_{zz} = 0 \tag{2.1}$$

with arbitrary initial conditions $u|_{t=0} = u_0(x, y, z)$, $u_t|_{t=0} = u_1(x, y, z)$ can be reduced to the special Cauchy problem

$$u_{tt} - u_{xx} - u_{yy} - u_{zz} = 0 \quad (t > 0), \quad u|_{t=0} = 0, \quad u_t|_{t=0} = h(x, y, z). \tag{2.2}$$

Furthermore, the solution of the latter problem is given by an explicit integral formula (convolution) provided that the fundamental solution is known defined as follows. The distribution $u = E(t, x, y, z)$ is called the fundamental solution of the Cauchy problem (2.2) if it solves the problem (2.2) with $h(x, y, z)$ replaced by Dirac's δ-function, i.e.,

$$u_{tt} - u_{xx} - u_{yy} - u_{zz} = 0 \quad (t > 0), \quad u|_{t=0} = 0, \quad u_t|_{t=0} = \delta(x, y, z). \tag{2.3}$$

Let us find the symmetries of the problem (2.3). It is well known that the wave equation (2.1) is conformally invariant. It means that the wave equation admits the isometric motions, i.e., the 10-parameter Lorentz group, dilations and proper conformal transformations. Their generators are:

$$X_0 = \frac{\partial}{\partial t}, \qquad X_i = \frac{\partial}{\partial x^i}, \qquad X_{ij} = x^j \frac{\partial}{\partial x^i} - x^i \frac{\partial}{\partial x^j},$$

$$X_{0i} = t \frac{\partial}{\partial x^i} + x^i \frac{\partial}{\partial t}, \qquad Z_1 = t \frac{\partial}{\partial t} + x^i \frac{\partial}{\partial x^i}, \qquad Z_2 = u \frac{\partial}{\partial u},$$

$$Y_0 = \left(t^2 + |x|^2\right)\frac{\partial}{\partial t} + 2tx^i \frac{\partial}{\partial x^i} - 2tu \frac{\partial}{\partial u},$$

$$Y_i = 2tx^i \frac{\partial}{\partial t} + \left(2x^i x^j + \left(t^2 - |x|^2\right)\delta^{ij}\right)\frac{\partial}{\partial x^j} - 2x^i u \frac{\partial}{\partial u}, \quad i, j = 1, 2, 3.$$

(2.4)

Here we ignore the trivial infinitesimal symmetries of the form $X = \tau \partial/\partial u$ with $\tau = \tau(t, x, y, z)$ an arbitrary solution of the wave equation. According to the definition of the invariance of an initial value problem given in section 1.1, we will now restrict the above symmetries by imposing to the invariance of the initial manifold given by $t = 0$ and of the initial data of the problem (2.3) under the linear span of the operators (2.4). Since the equation $t = 0$ is invariant under all the operators (2.4) except X_0, the invariance condition of the initial manifold cuts of the latter generator. Now, let us note that the invariance of the initial data in (2.3) requires, in particular, that the support of $\delta(x, y, z)$, i.e., the point $x = 0$, $y = 0$, $z = 0$, be invariant. This further reduces the operators (2.4) by the translation generators X_i. Hence, the symmetries of the wave equations are restricted now to the rotations, the Lorentz boosts, dilations and the proper conformal transformations, i.e., to the following generators from (2.4): $X_{ij}, X_{0i}, Z_1, Z_2, Y_0, Y_i$. The latter operators leave invariant the first initial condition in (2.3), i.e., the equation $u|_{t=0} = 0$. Imposing the invariance of the second initial condition in (2.3), invoking the transformation (1.4), one concludes that the initial-value problem (2.3) admits the homogeneous Lorentz group (the rotations and Lorents boosts), a dilation and conformal transformations generated by

$$X_{ij}, X_{0i}, Z = Z_1 - 2Z_2, Y_0, Y_i, \quad i, j = 1, 2, 3. \tag{2.5}$$

2.2. DERIVATION OF THE FUNDAMENTAL SOLUTION

As mentioned in the preamble to section 2, we calculate the fundamental solution by means of the invariance principle in two steps. At first, we find a basis of invariants of the homogeneous Lorentz group with the generators X_{ij} and X_{0i}. These invariants are $\Gamma = t^2 - |x|^2$ and u. The second step is to satisfy the invariance under the dilation Z from (2.5). The restriction of the latter to functions of two variable Γ and u has the form

$$Z = 2\Gamma \frac{\partial}{\partial \Gamma} - 2u \frac{\partial}{\partial u}. \tag{2.6}$$

Now, looking for the fundamental solution in the form $u = f(\Gamma)$, one obtains the invariance test under the dilation generator Z in the form of equation (1.11) with $m = 1$:

$$\Gamma f'(\Gamma) + f(\Gamma) = 0.$$

Hence, by (1.12),

$$u = C_1 \delta(\Gamma) + C_2 \Gamma^{-1}. \tag{2.7}$$

Substituting (2.7) in the initial conditions from (2.3) and invoking the well known equation $\lim_{t \to 0} \delta(\Gamma) = 0$, one obtains $C_1 = 1/(2\pi)$, $C_2 = 0$. Thus, one ultimately arrives at the following fundamental solution:

$$u = \frac{1}{2\pi} \delta(\Gamma). \tag{2.8}$$

The proper conformal transformations were not used here. One can verify that the generators Y_0 and Y_i of the conformal transformations are also admitted by the distribution (2.8) and hence provide *excess symmetries* of the fundamental solution.

Acknowledgements

I thank the National Research Foundation of South Africa for its continuing support of my research. I am grateful to the University of Karlskrona/Ronneby and to the organizers of the Leray conference, in particular to Maurice and Charlyne de Gosson, for the warm hospitality during the conference.

References

[1]   Ibragimov, N.H., (i) *Primer of the group analysis*, Moscow: Znanie, no. 8, 1989.
      (ii) Group analysis of ordinary differential equations and the invariance principle in mathematical physics (to the 150th anniversary of Sophus Lie), *Uspekhi Mat. Nauk* (1992). Engl. transl. in *Russian Math. Surveys* **47**(4) (1992), 89–156.
      (iii) Small effects in physics hinted by the Lie group philosophy: Are they observable? I. From Galilean principle to heat diffusion, Proceedings of Int. Conf. on Modern Group Analysis V, Johannesburg, South Africa, January 16–22, 1994, published in *Lie Groups and Their Applications*, vol. 1(1), 1994, 113–123.
      (iv) Differential equations with distributions: Group theoretic treatment of fundamental solutions, Chapter 3 in *CRC Handbook of Lie Group Analysis of Differential Equations*, N.H. Ibragimov (ed.), vol. 3, Boca Raton, Florida: CRC Press 1996, 69–90.
[2]   Leray, J., *Hyperbolic differential equations*, Mimeographed, The Institute for Advanced Study, Princeton 1953. Published in Russian translation by Ibragimov, N.H, Moscow: Nauka 1984.

LIOUVILLE FORMS, PARALLELISMS AND CARTAN CONNECTIONS

PAULETTE LIBERMANN

Cet article est dédié à la mémoire de Jean Leray,
en témoignage d'admiration

Contents

1   Introduction 425
2   On semi-basic and vertical forms 425
3   Connections and parallelism [8] 427
4   Parallelism and momentum maps 429
5   Principal connections and Cartan connections 431
6   Cartan connections on homogeneous spaces 434
7   Connections on frame bundles 435

M. de Gosson (ed.), Jean Leray '99 Conference Proceedings, 423-436.
© 2003 *Kluwer Academic Publishers.*

1. Introduction

This paper is a survey (without proofs) of subjects connected with the title. It does not deal directly with Mechanics but investigates tools utilized in Analytical Mechanics. For a direct treatment see the contribution of C. Marle to this publication [12].

The first subject was investigated in a recent article [10] whose source was a paper due to D. Alekseevsky, J. Grabowski, G. Marmo and P.W. Michor [2].

The investigations concerning parallelisms and Cartan connections summarize a part of the results of a previous paper [8]. For the parallelism we formulate in terms of jets and groupoids certain ideas of M. Lazard [7]. The theory of Cartan connections was introduced and developed by C. Ehresmann in his famous paper on 'Connexions infinitésimales' [4]. See also Kobayashi [5]. A Cartan connection gives rise to a special parallelism.

We explain links between parallelisms and momentum maps. We refer to [11] and [12] for momentum maps.

All manifolds and maps are supposed to be C^∞. The projections $TM \to N$, $T^*N \to N$, $TT^*N \to TN$ will be denoted p, q, Tq for any manifold N.

2. On semi-basic and vertical forms

Let $\pi : E \to M$ be a locally trivial fibration, $VE = \ker T\pi$ be the vertical bundle. From the inclusion $i : VE \to TE$, we deduce the projection $j : T^*E \to V^*E$ whose kernel is the annihilator $(VE)^0$ of VE; this kernel may be identified with the sub-bundle $\pi^*T^*M = E \times_M T^*M$ of T^*E.

A section $\eta : E \to E \times_M T^*M$ is called a *semi-basic form* on E. This section induces a morphism $f = \text{pr}_2 \circ \eta$ from E to T^*M (where pr_2 is the projection $E \times_M T^*M \to T^*M$). In particular for the fibration $q : T^*M \to M$, the natural Liouville form θ_M on T^*M corresponds to the identity mapping of T^*M. The form η on E is the pull-back $f^*\theta_M$. Conversely for any morphism $f : E \to T^*M$, the pull-back $f^*\theta_M$ is semi-basic. By means of adapted coordinates $(x^1, \ldots, x^n, y^1, \ldots, y^k)$ in $\pi^{-1}(U)$ and $(x^1, \ldots, x^n, p_1, \ldots, p_n)$ in $q^{-1}(U)$, the forms η and θ_M are written

$$\eta = \sum_{i=1}^{n} a_i \left(x^1, \ldots, x^n, y^1, \ldots, y^k \right) dx^i, \qquad \theta_M = \sum_{i=1}^{n} p_i \, dx^i,$$

and f is represented by $p_i = a_i$ $(i = 1, \ldots, n)$.

A section $\mu : E \to V^*E$ is called a *vertical form*; it acts only on vertical vectors. This vertical form may be represented in $\pi^{-1}(U)$ by

$$\mu = \sum_{j=1}^{k} b_j \left(x^1, \ldots, x^n, y^1, \ldots, y^k \right) dy^j.$$

If λ is any form on E, λ is written

$$\lambda = \sum_{i=1}^{n} A_i \, dx^i + \sum_{j=1}^{k} B_j \, dy^j,$$

then

$$j\lambda = \sum_{j=1}^{k} B_j \, dy^j.$$

We define a *generalized Liouville form* with respect to the locally trivial fibration $\pi : E \to M$ as a *semi-basic form η such that $d\eta$ is symplectic*. Then the fibers are isotropic submanifolds of E; so each fiber is of dimension $k \le n$. For $k = n$, we obtain a *Liouville form* in the sense of [2]. This is the case of the natural Liouville form θ_M on T^*M. The fibers are Lagrangian submanifolds and the mapping f such that $\eta = f^*\theta_M$ (and hence $d\eta = f^*d\theta_M$) is locally a diffeomorphism; so η may be written $\eta = \sum_{i=1}^{n} y^i \, dx^i$.

In order to obtain a local model for generalized Liouville forms, we prescribe the extra conditions (which are fulfilled for any Liouville form)

1°) the foliation $\mathscr{F}$ induced by the fibers of $\pi : E \to M$ is *symplectically complete*, i.e., the Poisson bracket (defined by the dual Λ of $d\eta$) of two first integrals of $\mathscr{F}$ is also a first integral;

2°) the foliation is *homogeneous* with respect to the vector field Z (such that $i(Z)d\eta = \eta$), i.e., the Lie derivative $\mathscr{L}(Z)h$ is a first integral whenever h is a first integral.

As proved in [9], the distribution which is orthogonal to $\mathscr{F}$ is also completely integrable and defines a homogeneous symplectically complete foliation $\mathscr{F}^{\perp}$.

When the dimension k of the fibers is given, we only have one model. We can find local coordinates $x^1, \ldots, x^{2s}, x^{2s+1}, \ldots, x^n, z^1, \ldots, z^k$ (with $2s = n - k$), where $x^1, \ldots, x^{2s}, x^{2s+1}, \ldots, x^n$ are first integrals of $\mathscr{F}$, $x^{2s+1}, \ldots, x^n$ are also first integrals of $\mathscr{F}^{\perp}$ in such a way that η may be written $\eta = \eta_1 + \eta_2$ with

$$\eta_1 = x^1 \, dx^2 + \cdots + x^{2s-1} \, dx^{2s}, \qquad \eta_2 = z^1 \, dx^{2s+1} + \cdots + z^k \, dx^n.$$

A generalized Liouville form with $k = 1$ occurs in Mechanics. Let N be a manifold such that T^*N is endowed with a time-dependent Hamiltonian H_t, i.e., there exists a function $H : T^*N \times \mathbb{R} \to \mathbb{R}$. Let us consider the manifold $Q = N \times \mathbb{R}$ (configuration space–time). We obtain

$$T^*Q = T^*N \times T^*\mathbb{R} = T^*N \times \mathbb{R} \times \mathbb{R}^*.$$

The natural Liouville form θ_Q is written

$$\theta_Q = \theta_N + p_0\, dt,$$

where p_0 and t are coordinates on $\mathbb{R}^*$ and $\mathbb{R}$. For the trivial fibration $\pi : T^*Q \to T^*N \times \mathbb{R}$, the form θ_Q is a generalized Liouville form η with $\eta_1 = \theta_N$ and $\eta_2 = p_0\, dt$.

Let φ be the injection: $T^*N \times \mathbb{R} \to T^*N \times \mathbb{R} \times \mathbb{R}^*$ defined by $p_0 + H = 0$ (energy equation). Then on $T^*N \times \mathbb{R}$ the Poincaré–Cartan integral invariant

$$\omega = \theta_N - H\, dt$$

is the pull-back $\varphi^*\theta_Q$.

3. Connections and parallelism [8]

let $\pi : E \to M$ be a locally trivial fibration, $J_1 E$ be the set of all 1-jets of local sections with the projection $\delta_1 : J_1 E \to E$.

A *connection* C on E is a lifting

$$C : E \longrightarrow J_1 E.$$

It is a first order differential system. A solution of C is a section $\sigma : U \subset M \to E$ such that for any $x \in U$ the jet $j_x^1\sigma$ belongs to $C(E)$. The connection is said to be *integrable* if for any $y \in E$, there exists a solution σ such that $\sigma(\pi(y)) = y$. According to Frobenius' theorem, C is integrable if and only if the composed map $J_1 C \circ C$ from E to $J_1 J_1 E$ takes its values in $J_2 E$ (set of 2-jets of local sections). The obstruction to integrability is the *curvature*. It is a lifting $\rho : E \to L^2_{E,a}(\pi^*TM; VE)$, set of alternate bilinear maps from $\pi^*TM \times \pi^*TM$ to VE.

As any $z \in J_1 E$ may be considered as an injective linear map from $T_{\pi(y)}E$ to $T_y E$ (with $y = \delta_1(z)$), the connection induces a splitting

$$TE = VE \oplus \mathcal{H}$$

and conversely. According to a second version of Frobenius theorem, the connection is integrable if and only if the distribution $\mathcal{H}$ is completely integrable.

A manifold M is said to be *parallelizable* if there exists a mapping

$$\omega : TM \longrightarrow \mathbb{A},$$

(where $\mathbb{A}$ is a vector space) such that for any $x \in M$, the restriction ω_x of ω to $T_x M$ is an isomorphism onto $\mathbb{A}$.

It follows a trivialization

$$\phi : TM \longrightarrow M \times \mathbb{A}$$

defined for any $x \in M$ and any $v \in T_x M$ by

$$\phi(v) = (x, \omega_x v).$$

With each mapping $f : U \subset M \to M$ we associate its *reduced differential*

$$D_{\text{red}} f : U \longrightarrow L(\mathbb{A}, \mathbb{A})$$

defined by

$$D_{\text{red}} f(x) = \omega_{f(x)} \circ T_x f \circ \omega_x^{-1}.$$

The *reduced expression* of the vector field $X : M \to TM$ is the map $\omega \circ X : M \to \mathbb{A}$. The vector field X is said to be *invariant* if $\omega \circ X$ is constant.

For any pair $(x, x') \in M \times M$, the isomorphism $\omega_{x'}^{-1} \circ \omega_x$ is an isomorphism from $T_x M$ onto $T_{x'} M$; it may be considered as the 1-jet $j_x^1 f$ of a diffeomorphism $f : U \ni x \to M$ such that $x' = f(x)$. So the parallelism ω induces a lifting

$$C : M \times M \longrightarrow \Gamma^1(M),$$

where $\Gamma^1(M)$ is the groupoid of all 1-jets of local diffeomorphisms on M.

As a diffeomorphism $f : U \subset M \to M$ induces a section $x \mapsto (x, f(x))$ of the fibration $\text{pr}_1 : M \times M \to M$, the *lifting C is a connection.*

A solution of the differential system C is a diffeomorphism $f : U \subset M \to M$ such that for any $x \in U$, the jet $j_x^1 f$ is equal to $C(x, f(x))$. Such a solution is called a *translation*; identifying $j_x^1 f$ and $T_x f$, we check that $D_{\text{red}} f(x)$ is the identity map of.

The connection C lifts every vector field X tangent to M into a horizontal vector field on $M \times M$; it is the vector field $Y(x, x') = (X(x), \omega_{x'}^{-1} \circ \omega_x X(x))$. A vector field X on M is invariant if and only if its natural lift on $M \times M$ defined by $(x, x') \mapsto (X(x), X(x'))$ is horizontal.

We deduce that the *parallelism* (considered as a connection C) *is integrable if and only if the bracket of two invariant vector fields is an invariant vector field.* It follows a *Lie algebra* structure on $\mathbb{A}$.

Examples. (i) A Lie group admits integrable parallelisms. The left and right translations as well as the left and right invariant vector fields are translations and invariant vector fields in our sense.

(ii) Let M be a simply connected manifold endowed with an integrable parallelism such that all translations are defined on the whole of M. If we fix a point on M, we obtain a Lie group structure on M.

(iii) The sphere S_7 admits a parallelism defined by means of the Cayley numbers. This parallelism is not integrable.

Remark. The usual point of view for introducing connections linked with a parallelism (not necessarily integrable) is the following. The choice of a basis of the tangent space $T_{x_0} M$ at a point x_0 determines n linearly independent invariant vector fields, hence a section of the frame bundle $H(M)$ (or $\{e\}$-structure); it follows a principal connection on $H(M)$ with null curvature. The trajectories of the invariant vector fields are the geodesics of this connection.

4. Parallelism and momentum maps

For any manifold M, the natural Liouville form θ_M is the form on T^*M which associates with any $v \in TT^*M$ the scalar

$$\langle p(v), Tq(v) \rangle.$$

It is known (see [10]) that any vector field X on M is lifted on T^*M to an infinitesimal automorphism $\tilde{X}$ of the form θ_M; it is the Hamiltonian vector field defined as follows. The vector field X induces a section $\overline{X}$ of the bundle $T^*M \times_M TM \to T^*M$; as θ_M is a section of $T^*M \times_M T^*M \to T^*M$, we may define $h = \langle \theta_M, X \rangle$ by $h = \langle \theta_M, \overline{X} \rangle$. Then $\tilde{X}$ is the Hamiltonian vector field such that $i(\tilde{X})d\theta_M = -dh$. It can be checked that $h = \langle \theta_M, \tilde{X} \rangle$.

Suppose now that M is parallelizable. From the map $\omega : TM \to \mathbb{A}$, we obtain by duality the mapping

$$\mu : T^*M \longrightarrow \mathbb{A}^*$$

such that for any $x \in M$, the mapping $\mu_x : T^*M \to \mathbb{A}$ is the contragredient of ω_x.

It follows a trivialization

$$\psi : T^*M \longrightarrow M \times \mathbb{A}^*$$

such that for any $x \in M$ and any $\eta \in T_x^*M$, we have

$$\psi(\eta) = (x, \mu_x \eta).$$

Any morphism f from $E \to M$ to $T^*M \to M$ induces a map $E \to \mathbb{A}^*$ and conversely.

We have the notion of invariant 1-form; a form η on M is said to be *invariant* if $\mu \circ \eta$ is constant. This condition is equivalent to the following

$$\left(\omega_x^{-1}\right)^* \eta(x) = \left(\omega_{x'}^{-1}\right)^* \eta(x') \quad \text{for any pair } (x, x') \in M \times M.$$

The Liouville form θ_M may be defined as the form associating with $v \in TT^*M$ the scalar

$$\langle \mu \circ p(v), \omega \circ Tq(v) \rangle.$$

Let X^a be the invariant vector field on M whose reduced expression is a. Then its lift $\widetilde{X}^a$ to T^*M is the vector field defined by the condition: for any $y \in T^*M$,

$$i\left(\widetilde{X}^a\right) d\theta_M|_y = -\langle \mu(y), a \rangle.$$

The bracket $[X, Y]$ of two vector fields X and Y on M is lifted to the bracket $[\widetilde{X}, \widetilde{Y}]$. So when the parallelism is integrable, the lifts of the invariant vector fields constitute a Lie algebra. We have the *Hamiltonian action of a Lie algebra on T^*M* in the sense of [11, chapter 4]. These considerations do not imply that the vector fields are complete.

The Hamiltonian action of a Lie group G on its cotangent bundle T^*G is studied in [11] and [12].

Let L_g (resp., R_g) be the left (resp., the right) translation $s \mapsto gs$ (resp., $s \mapsto sg$). The left parallelism on G corresponds to the maps

$$\omega^L : TG \longrightarrow \mathcal{G}, \qquad \mu^L : T^*G \longrightarrow \mathcal{G}^*$$

such that for any $g \in G$

$$\omega_g^L = T_g L_{g^{-1}}, \qquad \mu_g^L = {}^t\left(\omega_g^L\right)^{-1} = {}^t\left(T_e L_g\right).$$

We define ω^R and μ^R similarly.

We recall that a symplectic action of a Lie group G on a symplectic manifold M is said to be Hamiltonian if there exists a map $J : M \to \mathcal{G}^*$ (called *momentum map*) such that for any $c \in \mathcal{G}$ the associated fundamental vector field Y^c is Hamiltonian and admits as Hamiltonian the function f defined by $f(x) = \langle J(x), c \rangle$.

This is the case of the left action or the right action of G on T^*G, with momentum maps

$$J_L = \mu^R, \qquad J_R = \mu^L.$$

The exchange between R and L comes from the fact that the fundamental vector field Y^c corresponding to a left action of G is a right invariant vector field.

It is also proved that the left (resp., right) orbits of the left (resp., right) action of G are the level sets of J_R (resp., J_L). These orbits are orthogonal with respect of the symplectic form $d\theta_G$. The connected components of these orbits constitute symplectically complete foliations in the sense of section 2. When G is connected, it follows from the theory of symplectically complete foliations (see [9]) that there exists on $\mathcal{G}^*$ an unique Poisson structure such that J_R is a Poisson map. The map J_L is a Poisson map for the opposite Poisson structure on $\mathcal{G}^*$. We recover the 'Kirillov–Kostant–Souriau' Poisson structures on $\mathcal{G}^*$ (see [12]).

These properties have led M. Condevaux, P. Dazord and P. Molino to introduce the notion of 'generalized momentum map' [3]. The authors consider a symplectically complete foliation; an atlas of 'local slidings' along the leaves of the orthogonal foliation constitutes the generalized momentum map.

Remark. C. Albert and P. Dazord [1] have exhibited a symplectic groupoid structure on T^*G as follows. The set of unities is $T_e^*G = \mathcal{G}^*$ with projections μ^L and μ^R from T^*G onto $\mathcal{G}^*$ (in our notations). The product $u \circ v$ of $u \in T_g^*G$ and $v \in T_s^*G$ is defined if and only if

$$\mu_g^L(u) = \mu_s^R(v) = b.$$

Then

$$u \circ v = \left(\mu_g^L\right)^{-1} \left(\mu_s^R\right)^{-1} b = \left(\mu_g^L\right)^{-1} v.$$

By means of the left trivialization $\psi^L : T^*G \to G \times \mathcal{G}^*$ (with $\psi^L(\eta) = (g, \mu_g^L\eta)$ for any $g \in G$ and $\eta \in T^*G$), u and v may be written, respectively, $u = (g, \xi)$, $v = (s, \lambda)$. Then $(g, \xi) \circ (s, \lambda) = (gs, \lambda)$ if $\xi = \mathrm{Ad}_s^* \lambda$.

5. Principal connections and cartan connections

In this section, as we shall have to consider the left parallelism on a Lie group and a parallelism on a principal bundle P, we shall denote by α the left invariant form $\omega^L : TG \to \mathcal{G}$. This form α (called *Maurer–Cartan form of* G) satisfies the relations

$$L_g^* \alpha = \alpha, \qquad R_g^* \alpha = \mathrm{Ad}\left(g^{-1}\right)\alpha, \qquad d\alpha + \frac{1}{2}[\alpha, \alpha] = 0.$$

Let $\pi : P \to M$ be a principal G-bundle (where M is not assumed to be parallelizable). The action of G on P by right translations $z \mapsto zg$ being free and regular, any $z \in P$ determines a diffeomorphism from $P_{\pi(z)}$ onto G which maps z on e. Hence we obtain an isomorphism ϖ_z from $T_z P_{\pi(z)}$ onto $\mathcal{G}$. We deduce a map

$$\varpi : VP \longrightarrow \mathcal{G}$$

called the *vertical parallelism* in the sense of [8]. We have

$$\varpi_{zg} = \mathrm{Ad}\left(g^{-1}\right)\varpi_z \quad \text{for } g \in G.$$

Considering the contragredient ${}^t\varpi_z^{-1}$ of ϖ_z, we obtain a map

$$\kappa : VP \longrightarrow \mathcal{G}^*.$$

The right action of G on P (whose orbits are the fibers) lifts to a Hamiltonian action on T^*P. It is proved in [10] that the corresponding momentum map is

$$J = \kappa \circ j : T^*P \longrightarrow \mathcal{G}^*,$$

where j is the natural projection $T^*P \to V^*P$.

So $J^{-1}(0)$, kernel of J, is the kernel of the projection j; according to the first section, $J^{-1}(0)$ is the set $\widetilde{P} = P \times_M T^*M$ of semi-basic forms on P.

Let $\beta : TP \to \mathcal{G}$ be a connection form on P inducing a principal connection. We recall that $\beta_{zg} = \text{Ad}(g^{-1})\beta_z$ for $z \in P$, $g \in G$. The restriction of β to VP is the form ϖ defined above.

In the splitting (cf. section 3)

$$TP = VP \oplus \mathcal{H},$$

the horizontal bundle $\mathcal{H} = \ker \beta$ is G-invariant.

By duality we obtain

$$T^*P = V^*P \oplus \mathcal{H}^* = \mathcal{H}^0 \oplus \widetilde{P},$$

identifying $\mathcal{H}^*$ with the annihilator $(VP)^0$ of VP in TP, i.e., with $\widetilde{P}$ and identifying V^*P with $\mathcal{H}^0$.

We have proved in [10] that any $\varphi \in T_z^*P$ may be written $\varphi = \varphi_1 + \varphi_2$ with $\varphi_1 = \varphi - \beta_z^* J(\varphi)$ belonging to $\widetilde{P}$ and $\varphi_2 = \beta_z^* J(\varphi)$ vanishing on $\ker \beta$.

Let $\widehat{G}$ be a Lie group (with Lie algebra $\widehat{\mathcal{G}}$) such that G is a closed subgroup of $\widehat{G}$ and P be a G-principal bundle.

A *Cartan connection* on P is a form

$$\omega : TP \longrightarrow \widehat{\mathcal{G}}$$

satisfying the conditions:

1) The restriction of ω to VP is the form $\varpi : VP \to \mathcal{G}$ defined above;
2) $\omega_{zg} = \text{ad}(g^{-1})\omega_z$ for $g \in G$;
3) ω defines a parallelism on P.

It follows that $\dim \widehat{G} = \dim G + \dim M$. A Cartan connection is *not* a connection in the usual sense because ω takes its values in $\widehat{\mathcal{G}}$. But it induces a principal connection on a principal $\widehat{G}$-bundle $\widehat{P}$ obtained by enlarging the structure group of G to $\widehat{G}$. This bundle is

$$\widehat{P} = P \times_G \widehat{G},$$

quotient of $P \times \widehat{G}$ by the diagonal right action of G on $P \times \widehat{G}$, i.e., the bundle $\widehat{P}$ is the quotient of $P \times \widehat{G}$ by the equivalence relation $(z, s) \sim (zg, sg)$ for any $z \in P$, $s \in \widehat{G}$, $g \in G$. Let Λ be the projection $P \times \widehat{G} \to \widehat{P}$.

Let us consider the form $\omega - \alpha$ on $P \times \widehat{G}$ (where α is the Maurer–Cartan form on $\widehat{G}$). As

$$\omega_{zg} = \mathrm{Ad}\left(g^{-1}\right)\omega_z, \qquad \alpha_{sg} = \mathrm{Ad}\left(g^{-1}\right)\alpha_s$$

we have

$$\omega_{zg} - \alpha_{sg} = \mathrm{Ad}\left(g^{-1}\right)(\omega_z - \alpha_s)$$

and the form $\omega - \alpha$ is the pullback $\Lambda^*\widehat{\beta}$ of a *connection form* $\widehat{\beta}$ on $\widehat{P}$. The restriction to P of the form $\widehat{\beta}$ is the form ω.

It can be checked (see [4] and [8]) that if X and Y are vector fields on P and $\widehat{G}$ whose reduced expression is the same (i.e., there exists $a \in \mathcal{G}$ such that $\omega(X_z) = \alpha(Y_s) = a$ for any pair $(z, s) \in P \times \widehat{G}$), then the pair (X, Y) is projectable by Λ onto a vector field Z tangent to the horizontal distribution $\widehat{\mathcal{H}} = \ker \widehat{\beta}$. We shall say that the *Cartan connection ω is integrable* if the corresponding parallelism is *integrable* in the sense of section 3. As the parallelism on $\widehat{G}$ is integrable, *the Cartan connection ω is integrable if and only if the connection $\widehat{\beta}$ on $\widehat{P}$ is integrable.*

The obstruction to the integrability is the curvature Ω defined by

$$\Omega = d\omega + \frac{1}{2}[\omega, \omega].$$

Let $T_P\widehat{P}, V_P\widehat{P}, T_P^*\widehat{P}, V_P^*\widehat{P}$ be the restriction of $T\widehat{P}, V\widehat{P}, T^*\widehat{P}, V^*\widehat{P}$ to P. The vertical parallelism $\widehat{\varpi} : V\widehat{P} \to \mathcal{G}$ and the Cartan connection $\omega : TP \to \mathcal{G}$ induce an isomorphism φ from $V_P\widehat{P}$ onto TP; for $z \in P$, the restriction φ_z of φ is the isomorphism $\omega_z^{-1} \circ \widehat{\varpi}_z$ from $V_z\widehat{P}$ onto T_zP.

Let us define $\widehat{\kappa} : V^*\widehat{P} \to \mathcal{G}^*$ and $\widehat{j} : T^*\widehat{P} \to V^*\widehat{P}$ as were defined κ and j for T^*P and V^*P. The momentum map $\widehat{J}$ of the symplectic action of $\widehat{G}$ on $T^*\widehat{P}$ is

$$\widehat{J} = \widehat{\kappa} \circ \widehat{j} : T^*\widehat{P} \longrightarrow \mathcal{G}^*.$$

Then the restriction $\widehat{J}_P$ of $\widehat{J}$ to $T_P^*\widehat{P}$ may be written

$$\widehat{J}_P = \mu \circ \mathcal{P},$$

where $\mathcal{P}$ is the projection $T_P^*\widehat{P} \to T^*P$ and

$$\mu : T^*P \longrightarrow \mathcal{G}^*$$

is defined by the trivialization $T^*P \approx P \times \mathcal{G}^*$ associated with the parallelism ω (see section 4).

The momentum map $J : T^*P \to \mathcal{G}^*$ may be written

$$J = \gamma \circ \mu,$$

where γ is the projection $\widehat{\mathcal{G}}^* \to \mathcal{G}^*$.

Remarks. (i) The restriction $\widehat{\mathcal{H}}_P$ to P of the horizontal distribution $\widehat{\mathcal{H}} = \ker \widehat{\beta}$ is transversal to both vector bundles $V_P \widetilde{P}$ and TP. So if the Cartan connection (and hence $\widehat{\mathcal{H}}$) is *integrable*, then any $z \in P$ admits a neighborhood U in $\widehat{P}$ such that the trace on U of every horizontal leaf (if not empty) intersects P and the fiber $\widehat{P}_{\widehat{\pi}(z)}$ in one point. It follows a local diffeomorphism of P onto $\widehat{P}_{\widehat{\pi}(z)}$ (hence on $\widehat{G}$).

(ii) According to section 3, the parallelism on P induces a lifting $C : P \times P \to \Gamma^1(P)$ (groupoid of the 1-jets of all local diffeomorphisms on P). In the case of a Cartan connection the result may be improved. We shall sum up the investigations of [8].

The lifting C takes its values in the groupoid $\Theta^1 \subset \Gamma^1(P)$, set of the 1-jets of all local automorphisms of P; such a local automorphism φ is defined in an open set $\pi^{-1}(U)$ (where U is an open subset of M) and satisfies the relation $\varphi(zg) = \varphi(z)g$ for any $g \in G$.

Let us consider the gauge groupoid Φ of P, i.e., the quotient of $P \times P$ by the diagonal right action of G. Let π_1 and π_2 be the projections of Φ onto M. An invertible π_1-section $\sigma_1 : U \to \Phi$ (with $\pi_1 \circ \sigma_1 = \mathrm{id}_U$) satisfies the condition: the map $\pi_2 \circ \sigma_1$ is a diffeomorphism f inducing the π_2-section $\sigma_2 = \sigma_1 \circ f^{-1}$. There exists a (1-1) correspondence between the set of all local automorphsims of P and the set of all invertible π_1-sections of Φ. The set $\Phi^{(1)}$ of 1-jets of all invertible π_1-sections is a groupoid and the groupoid Θ^1 may be identified with the fibered product $P \times_M \Phi^{(1)}$; then the lifting C induces a lifting

$$C' : \Phi \longrightarrow \Phi^{(1)},$$

which is a groupoid morphism. Conversely such a lifting C' induces a Cartan connection on P.

It would be interesting to study a groupoid Φ which admits a lifting $C' : \Phi \to \Phi^{(1)}$ but not being a gauge groupoid.

6. Cartan connections on homogeneous spaces

Let $\widehat{G}$ be a Lie group and G a closed subgroup. The right action of G on $\widehat{G}$ defines a principal G-bundle structure on $\widehat{G}$, whose base M is the homogeneous space $\widehat{G}/G$. We shall set $P = \widehat{G}$ when considering $\widehat{G}$ as a principal G-bundle.

The Maurer–Cartan form α on $\widehat{G}$ is an integrable Cartan connection form. It induces a principal connection on the $\widehat{G}$-bundle $\widehat{P} = P \times_G \widehat{G}$, as we have seen in section 5. This bundle $\widehat{P} = P \times_G P$ is the gauge groupoid of P. As is shown in [8], $\widehat{P}$ is diffeomorphic to $P \times \widehat{G}/G$ (i.e., $\widehat{G} \times \widehat{G}/G$) with projections π_1 and π_2 defined by $\pi_1(g, x) = (e, x)$, $\pi_2(g, x) = (e, gx)$.

Remark. Let $\mathcal{J}_L$ be the momentum map of the left action of $\widehat{G}$ on $T^*\widehat{G}$. If $b \in \widehat{\mathcal{G}}^*$ is invariant for the coadjoint action of G on $\widehat{\mathcal{G}}^*$, then the submanifold $\mathcal{J}_L^{-1}(b)$ is invariant under the left action of G on T^*G as it is shown in [12] and the quotient manifold $\mathcal{J}_L^{-1}(b)/G$ (i.e., the set of orbits of the left action of G on $\mathcal{J}_L^{-1}(b)$) has a reduced symplectic structure.

We are in a special case of the situation studied in [6] as follows; $\pi : P \to M$ is a principal G-bundle and the Lie group G is a subgroup of a Lie group $\widehat{G}$ acting on P.

7. Connections on frame bundles

Among the homogeneous spaces let us consider $G = L_n = \mathrm{GL}(n, \mathbb{R})$ and $\widehat{G} = L_n \times \mathbb{R}^n$ (semi-direct product); the group $\widehat{G}$ is the affine group $\mathcal{A}(n, \mathbb{R})$.

If we restrain L_n to the orthogonal group $\mathrm{SO}(n)$, then $\widehat{G}$ is the group of Euclidian displacements and $\widehat{G}/G$ is endowed with an Euclidian structure. For $n = 3$, C. Marle [12] has considered this situation when studying the motion of the rigid body.

For a n-dimensional manifold M, the frame bundle $\pi : H(M) \to M$ is a principal L_n-bundle. An L_n-connection on $H(M)$ is called a linear connection. It is known that there exists on $H(M)$ an equivariant form η with values in $\mathbb{R}^n$. It is the form defined by

$$\eta(v) = h^{-1}(\pi(v)) \quad \text{for } v \in T_h H(M).$$

As $\ker \eta = VH$, with any L_n-connection form β on $H(M)$ is associated a form $\omega = \beta + \eta$ which is a Cartan connection. So *every frame bundle is parallelizable*. The form ω is the restriction of a $\mathcal{A}(n)$-connection form on the bundle of affine frames. The parallelism associated with ω is integrable if the corresponding $\mathcal{A}(n)$-connection has a null curvature. This is equivalent to the nullity of the curvature and torsion of the linear connection β.

Every sub-bundle of the frame bundle is also parallelizable. This is in particular true for Riemannian structures.

Instead of L_n, we could consider the group L_n^r (whose elements are r-jets of local diffeomorphisms of $\mathbb{R}^n$ with source and target 0). Then $H^r(M)$ is the set of frames of order r. In particular the projective and conformal structures are structures of order 2. See [4] and [5].

In [2] are considered principal bundles for which there exists a form η with values in $\mathbb{R}^n$ such that η is equivariant.

Other examples of parallelizable principal bundles are the SU_2-bundle $S_7 \to S_4$ and the S_1-bundle $S_7 \to \mathbb{P}_3(\mathbb{C})$.

References

[1] Albert, C. and Dazord, P., *Groupoïdes de Lie et groupoïdes symplectiques,* M.S.R.I. Berkeley Publications, vol. 20, 1989; Publications dépt. Math. Lyon, Nouvelle série 1990.

[2] Alekseevsky, D., Grabowski, J., Marmo, G., and Michor, P.W., Poisson structures on the cotangent bundle of a Lie group or a principal bundle and their reductions, *J. Math. Phys.* **35** (1994), 4909–4924.

[3] Condevaux, M., Dazord, P., and Molino, P., *Géométrie du moment,* Publications dépt. math. Lyon, Nouvelle série 1983, 131–160.

[4] Ehresmann, Ch., *Les connexions infinitésimales,* Colloq. Topologie (Bruxelles 1950). Œvres complètes, Partie I 2, 179–205.

[5] Kobayashi, S., *Transformations groups in differential geometry,* Springer–Verlag, Berlin, Heidelberg, New York 1972.

[6] Kummer, M., On the construction of the reduced phase of a Hamiltonian system with symmetry, *Indiana University math. J.* **30** (1981), 281–291.

[7] Lazard, M., *Leçons de calcul différentiel et intégral dans les variétés banachiques,* unpublished 1970.

[8] Libermann, P., Parallélismes, *Journ. of Diff. Geom. (4)* **8** (1973), 511–539.

[9] _____, *Problèmes d'équivalence et géométrie symplectique,* Proc. IIIe rencontre de Geom. Schnepfenried, Astérisque 107–108 1983, 43–68. *Cartan-Darboux theorems for Pfaffian forms on foliated manifolds,* Proc. VIth Int. Colloq. Diff. Geom. Santiago 1989, 125–144.

[10] Libermann, P. and Marle, C.-M., *On Liouville forms,* Banach Center publications, vol. 51, 2000, Institute of Mathematics, Polish Academy of Sciences, Warszawa.

[11] _____, *Symplectic geometry and analytical mechanics,* D. Reidel Publishing Company, Dordrecht 1987.

[12] Marle, C.-M., *On mechanical systems with a Lie group as configuration space.*

[13] Weinstein, A., A universal phase space for particles in Yang–Mills field, *Letters Math. Phys.* **2** (1978), 417–420.

A TWO-DIMENSIONAL NON-LINEAR SHELL MODEL
OF KOITER'S TYPE

PHILIPPE G. CIARLET

Contents

1 Geometrical preliminaries 439
2 A two-dimensional non-linear shell model proposed by W.T. Koiter 440
3 A two-dimensional non-linear shell model of Koiter's type 444
4 Conclusions 448

M. de Gosson (ed.), Jean Leray '99 Conference Proceedings, 437-449.
© *2003 Kluwer Academic Publishers.*

Abstract. *A two-dimensional non-linear shell model, based on* a priori *assumptions of geometrical and mechanical nature, has been proposed by W.T. Koiter [8]. However, the 'exact' change of curvature tensor appearing in this model possesses a denominator that may vanish for some displacements.*

The purpose of this paper is to propose a related non-linear shell model, where the exact change of curvature tensor is replaced by an ad hoc 'modified' tensor, the form of which is suggested by recent works of V. Lods and B. Miara, then of A. Roquefort. The advantage of this tensor is that it no longer has a possibly vanishing denominator; in fact, the covariant components of this tensor turn out to be simply polynomials of degree ≤ 3 with respect to the covariant components of the unknown displacement field and of their partial derivatives.

It will be shown elsewhere, in a joint work with A. Roquefort, how this two-dimensional non-linear shell model 'of Koiter's type' can be fully justified for all kinds of geometries of the middle surface and of boundary conditions, by means of a formal asymptotic analysis of its solution with the thickness as the 'small' parameter.

1. Geometrical preliminaries

Greek indices and exponents (except ε, and ν in the notation ∂_ν) belong to the set $\{1, 2\}$, Latin indices and exponents belong to the set $\{1, 2, 3\}$, and the summation convention with respect to repeated indices and exponents is systematically used. The Euclidean scalar product and the exterior product of $a, b \in \mathbb{R}^3$ are noted $a \cdot b$ and $a \wedge b$ and the Euclidean norm of $a \in \mathbb{R}^3$ is noted $|a|$.

Let ω be an open, bounded, connected subset of $\mathbb{R}^2$, whose boundary γ is Lipschitz–continuous, the set ω being locally on one side of γ. Let $y = (y_\alpha)$ denote a generic point in the set $\overline{\omega}$, and let $\partial_\alpha := \partial/\partial y_\alpha$. Let there be given an injective mapping $\theta \in \mathscr{C}^2(\overline{\omega}; \mathbb{R}^3)$ such that the two vectors

$$a_\alpha(y) := \partial_\alpha \theta(y)$$

are linearly independent at all points $y \in \overline{\omega}$. These two vectors thus span the tangent plane to the *surface*

$$S := \theta(\overline{\omega})$$

at the point $\theta(y)$, and the unit vector

$$a_3(y) := \frac{a_1(y) \wedge a_2(y)}{|a_1(y) \wedge a_2(y)|}$$

is normal to S at the point $\boldsymbol{\theta}(y)$. The three vectors $\boldsymbol{a}_i(y)$ constitute the *covariant basis at the point* $\boldsymbol{\theta}(y)$, while the three vectors $\boldsymbol{a}^i(y)$ defined by the relations

$$\boldsymbol{a}^i(y) \cdot \boldsymbol{a}_j(y) = \delta^i_j,$$

where δ^i_j is the Kronecker symbol, constitute the *contravariant basis at the point* $\boldsymbol{\theta}(y) \in S$ (note that $\boldsymbol{a}^3(y) = \boldsymbol{a}_3(y)$ and that the vectors $\boldsymbol{a}^\alpha(y)$ are also in the tangent plane to S at $\boldsymbol{\theta}(y)$).

The covariant and contravariant components $a_{\alpha\beta}$ and $a^{\alpha\beta}$ of the *metric tensor of* S, the *Christoffel symbols* $\Gamma^\sigma_{\alpha\beta}$, and the covariant and mixed components $b_{\alpha\beta}$ and b^β_α of the *curvature tensor of* S are then defined by letting (the explicit dependence on the variable $y \in \overline{\omega}$ is henceforth dropped):

$$a_{\alpha\beta} := \boldsymbol{a}_\alpha \cdot \boldsymbol{a}_\beta, \qquad a^{\alpha\beta} := \boldsymbol{a}^\alpha \cdot \boldsymbol{a}^\beta, \qquad \Gamma^\sigma_{\alpha\beta} := \boldsymbol{a}^\sigma \cdot \partial_\beta \boldsymbol{a}_\alpha,$$

$$b_{\alpha\beta} := \boldsymbol{a}^3 \cdot \partial_\beta \boldsymbol{a}_\alpha, \qquad b^\beta_\alpha := a^{\beta\sigma} b_{\sigma\alpha}.$$

The *area* element along S is $\sqrt{a}\, y$, where

$$a := \det(a_{\alpha\beta}).$$

Note that $\sqrt{a} = |\boldsymbol{a}_1 \wedge \boldsymbol{a}_2|$.

Remark. All details needed here about the differential geometry of surfaces in $\mathbb{R}^3$ are provided in, e.g., Ciarlet [3, Chap. 2].

2. A two-dimensional non-linear shell model proposed by W.T. Koiter

Consider an *elastic shell* with *middle surface* $S = \boldsymbol{\theta}(\overline{\omega})$ and *thickness* $2\varepsilon > 0$, i.e., an elastic body whose *reference configuration* is the set $\boldsymbol{\Theta}(\overline{\omega} \times [-\varepsilon, \varepsilon])$, where

$$\boldsymbol{\Theta}\left(y, x^\varepsilon_3\right) := \boldsymbol{\theta}(y) + x^\varepsilon_3 \boldsymbol{a}_3(y) \quad \text{for all } \left(y, x^\varepsilon_3\right) \in \overline{\omega} \times [-\varepsilon, \varepsilon].$$

Let $\Omega^\varepsilon := \omega \times] - \varepsilon, \varepsilon[$, let $x^\varepsilon = (x^\varepsilon_i)$ denote a generic point in the set $\overline{\Omega}^\varepsilon$ (hence $x^\varepsilon_\alpha = y_\alpha$), and let $\partial^\varepsilon_i := \partial/\partial x^\varepsilon_i$. Then, for $\varepsilon > 0$ small enough, the mapping $\boldsymbol{\Theta} : \overline{\Omega}^\varepsilon \to \mathbb{R}^3$ is injective and the three vectors

$$\boldsymbol{g}^\varepsilon_i\left(x^\varepsilon\right) := \partial^\varepsilon_i \boldsymbol{\Theta}\left(x^\varepsilon\right)$$

are linearly independent at all points $x^\varepsilon \in \overline{\Omega}^\varepsilon$ (Ciarlet [3, Thm. 3.1-1]). The three vectors $\boldsymbol{g}^\varepsilon_i(x^\varepsilon)$ constitute the *covariant basis at the point* $\boldsymbol{\Theta}(x^\varepsilon)$.

Let γ_0 be a subset of $\gamma = \partial\omega$ satisfying

$$\text{length } \gamma_0 > 0.$$

The shell is subjected to a *homogeneous boundary condition of place* along the portion $\Theta(\gamma_0 \times [-\varepsilon, \varepsilon])$ of its lateral face $\Theta(\gamma \times [-\varepsilon, \varepsilon])$; this means that its displacement field vanishes on the set $\Theta(\gamma_0 \times [-\varepsilon, \varepsilon])$.

The shell is subjected to *applied body forces* in its interior $\Theta(\Omega^\varepsilon)$ and to *applied surface forces* on its 'upper' and 'lower' faces $\Theta(\Gamma_+^\varepsilon)$ and $\Theta(\Gamma_-^\varepsilon)$, where $\Gamma_\pm^\varepsilon := \omega \times \{\pm\varepsilon\}$. The applied forces are given by the contravariant components (i.e., over the vectors g_i^ε of the covariant basis) $f^{i,\varepsilon} \in L^2(\Omega^\varepsilon)$ and $h^{i,\varepsilon} \in L^2(\Gamma_+^\varepsilon \cup \Gamma_-^\varepsilon)$ of their densities per unit volume and per unit area, respectively. We then define functions $p^{i,\varepsilon} \in L^2(\omega)$ by letting

$$p^{i,\varepsilon} := \int_{-\varepsilon}^\varepsilon f^{i,\varepsilon} x_3^\varepsilon + h_+^{i,\varepsilon} + h_-^{i,\varepsilon} \quad \text{where } h_\pm^{i,\varepsilon} := h^{i,\varepsilon}(\cdot, \pm\varepsilon).$$

Remark. The exponents ε appearing in the above notations (and other ones defined later) may be deemed superfluous since no asymptotic analysis as ε approaches zero is performed here. They are nevertheless kept for the sake of coherence with the notations used in Ciarlet [3] and in Ciarlet and Roquefort [5].

Finally, the elastic material constituting the shell is assumed to be *homogeneous* and *isotropic* and its reference configuration is assumed to be a *natural state*. Hence the material is characterized by two *Lamé constants* $\lambda > 0$ and $\mu > 0$ (see, e.g., Ciarlet [2, Chap. 3]). The functions

$$a^{\alpha\beta\sigma\tau} := \frac{4\lambda\mu}{\lambda + 2\mu} a^{\alpha\beta} a^{\sigma\tau} + 2\mu \left(a^{\alpha\sigma} a^{\beta\tau} + a^{\alpha\tau} a^{\beta\sigma}\right)$$

denote the contravariant components of the *two-dimensional elasticity tensor* of the shell.

Remark. The specific form of the functions $a^{\alpha\beta\sigma\tau}$ can be fully justified, in both the linear and non-linear cases, by asymptotic analysis of the solution of the three-dimensional equations as the thickness 2ε approaches zero; see Ciarlet [3]

In the two-dimensional shell equations proposed by Koiter [8], the *unknown* is the vector field $\boldsymbol{\zeta}^\varepsilon = (\zeta_i^\varepsilon) : \overline{\omega} \to \mathbb{R}^3$, where the three functions $\zeta_i^\varepsilon : \overline{\omega} \to \mathbb{R}$ denote the covariant components of the displacement field of the middle surface $S = \boldsymbol{\theta}(\overline{\omega})$ of the shell. This means that the displacement vector of the point $\boldsymbol{\theta}(y) \in S$ is $\zeta_i^\varepsilon(y) \boldsymbol{a}^i(y)$ for each $y \in \overline{\omega}$.

Given a vector field $\boldsymbol{\eta} = (\eta_i) : \overline{\omega} \to \mathbb{R}^3$ with smooth enough components $\eta_i : \overline{\omega} \to \mathbb{R}$, let

$$a_{\alpha\beta}(\boldsymbol{\eta}) := \boldsymbol{a}_\alpha(\boldsymbol{\eta}) \cdot \boldsymbol{a}_\beta(\boldsymbol{\eta}), \quad \text{where } \boldsymbol{a}_\alpha(\boldsymbol{\eta}) := \partial_\alpha \left(\boldsymbol{\theta} + \eta_i \boldsymbol{a}^i\right),$$

denote the covariant components of the metric tensor of the 'deformed' surface $(\boldsymbol{\theta} + \eta_i \boldsymbol{a}^i)(\overline{\omega})$. Then

$$G_{\alpha\beta}(\boldsymbol{\eta}) := \frac{1}{2} \left(a_{\alpha\beta}(\boldsymbol{\eta}) - a_{\alpha\beta} \right)$$

denote the covariant components of the *change of metric tensor associated with the displacement field* $\eta_i \boldsymbol{a}^i$ of S. An easy computation then shows that (see, e.g., Ciarlet [3, Thm. 9.1-1]):

$$G_{\alpha\beta}(\boldsymbol{\eta}) = \frac{1}{2} \left(\eta_{\alpha\|\beta} + \eta_{\beta\|\alpha} + a^{mn} \eta_{m\|\alpha} \eta_{n\|\beta} \right),$$

where
$$a^{\alpha 3} = a^{3\alpha} = 0 \qquad \text{and} \qquad a^{33} = 1$$

(otherwise the functions $a^{\alpha\beta}$ are the contravariant components of the metric tensor of S; cf. section 1) and

$$\eta_{\alpha\|\beta} = \partial_\beta \eta_\alpha - \Gamma^\sigma_{\alpha\beta} \eta_\sigma - b_{\alpha\beta} \eta_3 \qquad \text{and} \qquad \eta_{3\|\beta} = \partial_\beta \eta_3 + b^\sigma_\beta \eta_\sigma.$$

If the two vectors $a_\alpha(\boldsymbol{\eta})$ are linearly independent at all points of ω, let

$$b_{\alpha\beta}(\boldsymbol{\eta}) := \frac{1}{\sqrt{a(\boldsymbol{\eta})}} \partial_{\alpha\beta} \left(\boldsymbol{\theta} + \eta_i \boldsymbol{a}^i \right) \cdot \{ \boldsymbol{a}_1(\boldsymbol{\eta}) \wedge \boldsymbol{a}_2(\boldsymbol{\eta}) \},$$

where
$$a(\boldsymbol{\eta}) := \det \left(a_{\alpha\beta}(\boldsymbol{\eta}) \right),$$

denote the covariant components of the curvature tensor of the 'deformed' surface $(\boldsymbol{\theta} + \eta_i \boldsymbol{a}^i)(\overline{\omega})$. Then

$$R_{\alpha\beta}(\boldsymbol{\eta}) := b_{\alpha\beta}(\boldsymbol{\eta}) - b_{\alpha\beta}$$

denote the covariant components of the *change of curvature tensor associated with the displacement field* $\eta_i \boldsymbol{a}^i$ of S. Note that $\sqrt{a(\boldsymbol{\eta})} = |\boldsymbol{a}_1(\boldsymbol{\eta}) \wedge \boldsymbol{a}_2(\boldsymbol{\eta})|$.

The two-dimensional equations proposed by Koiter [8] for modeling a *non-linearly elastic shell* are derived from three-dimensional elasticity on the basis of two *a priori* assumptions: One assumption, of a *geometrical* nature, is the *Kirchhoff–Love assumption*; it asserts that any point situated on a normal to the middle surface remains on the normal to the deformed middle surface after the deformation has taken place and that, in addition, the distance between such a point and the middle surface remains constant. The other assumption, of a *mechanical* nature, asserts that the state of stress inside the shell is planar and parallel to the middle surface (this second assumption is itself based on delicate *a priori* estimates due to John [7]).

Taking these *a priori* assumptions into account, W.T. Koiter then reaches the conclusion that the unknown vector field $\zeta^\varepsilon = (\zeta_i^\varepsilon)$ should be a *minimizer* of the *energy* j_K^ε defined by (cf. Koiter [8, eqs. (4.2), (8.1) and (8.3)]):

$$j_K^\varepsilon(\eta) = \frac{\varepsilon}{2} \int_\omega a^{\alpha\beta\sigma\tau} G_{\sigma\tau}(\eta) G_{\alpha\beta}(\eta) \sqrt{a}\, y$$

$$+ \frac{\varepsilon^3}{3} \int_\omega a^{\alpha\beta\sigma\tau} R_{\sigma\tau}(\eta) R_{\alpha\beta}(\eta) \sqrt{a}\, y - \int_\omega p^{i,\varepsilon} \eta_i \sqrt{a}\, y.$$

One virtue of this expression is that a formal linearization immediately reduces it to the energy proposed by Koiter [9] for modeling a *linearly* elastic shell. More specifically, let $[\cdots]^{\text{lin}}$ denote the linear part with respect to $\eta = (\eta_i)$ in any smooth enough function $[\cdots]$ of η vanishing for $\eta = \mathbf{0}$. Then (cf. Ciarlet [3, Thms. 2.4-1 and 2.5-1])

$$[G_{\alpha\beta}(\eta)]^{\text{lin}} = \gamma_{\alpha\beta}(\eta) \qquad \text{and} \qquad [R_{\alpha\beta}(\eta)]^{\text{lin}} = \rho_{\alpha\beta}(\eta),$$

where

$$\gamma_{\alpha\beta}(\eta) := \frac{1}{2} \left(\partial_\beta \widetilde{\eta} \cdot a_\alpha + \partial_\alpha \widetilde{\eta} \cdot a_\beta \right),$$

$$\rho_{\alpha\beta}(\eta) := \left(\partial_{\alpha\beta} \widetilde{\eta} - \Gamma_{\alpha\beta}^\sigma \partial_\sigma \widetilde{\eta} \right) \cdot a_3,$$

respectively, denote the covariant components of the *linearized change of metric*, and *linearized change of curvature*, tensors associated with the displacement field $\widetilde{\eta} := \eta_i a^i$ of S.

Remark. The above expressions of the functions $\gamma_{\alpha\beta}(\eta)$ and $\rho_{\alpha\beta}(\eta)$, which are due to Blouza and Le Dret [1], are substantially shorter than the more 'classical' ones used by Koiter [8] (and by many other authors since then), *viz.*,

$$\gamma_{\alpha\beta}(\eta) = \frac{1}{2} \left(\eta_\alpha|_\beta + \eta_\beta|_\alpha \right) - b_{\alpha\beta} \eta_3,$$

$$\rho_{\alpha\beta}(\eta) = \eta_3|_{\alpha\beta} - b_\alpha^\sigma b_{\sigma\beta} \eta_3 + b_\alpha^\sigma \eta_\sigma|_\beta + b_\beta^\tau \eta_\tau|_\alpha + b_\beta^\tau|_\alpha \eta_\tau,$$

where the covariant derivatives $\eta_\alpha|_\beta$, $\eta_3|_{\alpha\beta}$, and $b_\beta^\tau|_\alpha$ are given by

$$\eta_\alpha|_\beta := \partial_\beta \eta_\alpha - \Gamma_{\alpha\beta}^\sigma \eta_\sigma, \qquad \eta_3|_{\alpha\beta} := \partial_{\alpha\beta} \eta_3 - \Gamma_{\alpha\beta}^\sigma \partial_\sigma \eta_3,$$

$$b_\beta^\tau|_\alpha := \partial_\alpha b_\beta^\tau + \Gamma_{\alpha\sigma}^\tau b_\beta^\sigma - \Gamma_{\alpha\beta}^\sigma b_\sigma^\tau.$$

But the functions $b_{\alpha\beta}(\eta)$, and thus the functions $R_{\alpha\beta}(\eta)$, are not defined at those points of $\overline{\omega}$ where the two vectors $a_\alpha(\eta)$ are linearly dependent, since the denominator $\sqrt{a(\eta)}$ in $b_{\alpha\beta}(\eta)$ vanishes at those points: For this reason, the problem of finding the minimizers of the energy j_K^ε is not well posed.

The objective of this paper is to propose one way to circumvent this difficulty.

Remark. This difficulty is not encountered in the *linearized* theory, since the functions $\rho_{\alpha\beta}(\eta)$ are well defined functions in the space $L^2(\omega)$ for *all* $\eta = (\eta_i) \in H^1(\omega) \times H^1(\omega) \times H^2(\omega)$.

3. A two-dimensional non-linear shell model of Koiter's type

To begin with we observe that the *strain energy*

$$\eta \longrightarrow \frac{\varepsilon}{2} \int_\omega a^{\alpha\beta\sigma\tau} G_{\sigma\tau}(\eta) G_{\alpha\beta}(\eta) \sqrt{a}\, y + \frac{\varepsilon^3}{6} \int_\omega a^{\alpha\beta\sigma\tau} R_{\sigma\tau}(\eta) R_{\alpha\beta}(\eta) \sqrt{a}\, y$$

found in the energy j_K^ε proposed by W.T. Koiter is the *sum* of the *strain energy*

$$\eta \longrightarrow \frac{\varepsilon}{2} \int_\omega a^{\alpha\beta\sigma\tau} G_{\sigma\tau}(\eta) G_{\alpha\beta}(\eta) \sqrt{a}\, y$$

of a non-linearly elastic 'membrane' shell and of the *strain energy*

$$\eta \longrightarrow \frac{\varepsilon^3}{6} \int_\omega a^{\alpha\beta\sigma\tau} R_{\sigma\tau}(\eta) R_{\alpha\beta}(\eta) \sqrt{a}\, y$$

of a non-linearly elastic 'flexural' shell recently justified by a *formal asymptotic analysis*, with the thickness 2ε as the 'small' parameter, *of the three-dimensional displacement field* by Miara [12] and Lods and Miara [11], respectively.

We recall that, in this approach, a non-linearly elastic shell is called a *'membrane'* shell if the manifold

$$\left\{ \eta \in \mathbf{W}^{1,4}(\omega);\ \eta = \mathbf{0} \text{ on } \gamma_0,\ a_{\alpha\beta}(\eta) - a_{\alpha\beta} = 0 \text{ in } \omega \right\}$$

contains only the field $\eta = \mathbf{0}$, while it is called a *'flexural'* shell if the manifold (∂_ν denotes the outer normal derivative operator along the boundary γ of ω)

$$\mathcal{M}_F(\omega) := \left\{ \eta \in \mathbf{H}^2(\omega);\ \eta = \partial_\nu \eta = \mathbf{0} \text{ on } \gamma_0,\ a_{\alpha\beta}(\eta) - a_{\alpha\beta} = 0 \text{ in } \omega \right\}$$

contains *non-zero* fields η and if the tangent space at each $\eta \in \mathcal{M}_F(\omega)$ contains *non-zero* elements (see Miara [12], Lods and Miara [11] and Ciarlet [3]).

In this approach, the *energy*

$$\eta \longrightarrow \frac{\varepsilon}{2} \int_\omega a^{\alpha\beta\sigma\tau} G_{\sigma\tau}(\eta) G_{\alpha\beta}(\eta) \sqrt{a}\, y - \int_\omega p^{i,\varepsilon} \eta_i \sqrt{a}\, y$$

of a 'membrane' shell is to be minimized over the *vector space*

$$\mathbf{W}_M(\omega) := \left\{ \eta \in \mathbf{W}^{1,4}(\omega);\ \eta = \mathbf{0} \text{ on } \gamma_0 \right\},$$

while the *energy*

$$\eta \longrightarrow \frac{\varepsilon^3}{6} \int_\omega a^{\alpha\beta\sigma\tau} R_{\sigma\tau}(\eta) R_{\alpha\beta}(\eta) \sqrt{a}\, y - \int_\omega p^{i,\varepsilon} \eta_i \sqrt{a}\, y$$

of a *'flexural' shell* is to be minimized over the manifold $\mathcal{M}_F(\omega)$ defined above. Thanks to the relations $a_{\alpha\beta}(\eta) - a_{\alpha\beta} = 0$ in ω satisfied by the fields $\eta \in \mathcal{M}_F(\omega)$, the functions $R_{\alpha\beta}(\eta)$ are well defined as functions of $L^2(\omega)$ for all fields $\eta \in \mathcal{M}_F(\omega)$ (for a proof, see Ciarlet and Coutand [4]).

Remark. (1) While the problem of finding a minimizer to the energy of a 'membrane' shell is open, that of finding a minimizer to the energy of a 'flexural' shell has been solved by Ciarlet and Coutand [4].

(2) The relations $a_{\alpha\beta}(\eta) - a_{\alpha\beta} = 0$ found in the definition of the manifold $\mathcal{M}_F(\omega)$ mean that *the metric of the surface* $S = \theta(\overline{\omega})$ *and the metric of the 'deformed' surface* $(\theta + \eta_i a^i)(\overline{\omega})$ *are the same.* This is why the fields $\eta_i a^i$ associated with the elements $\eta = (\eta_i) \in \mathcal{M}_F(\omega)$ are called *inextensional displacement fields* of the surface S.

(3) *Another* two-dimensional minimization problem has been derived from three-dimensional non-linear elasticity by Le Dret and Raoult [10]. Using Γ–*convergence theory*, they have shown that the minimizers of the three-dimensional energy weakly converge in an *ad hoc* space $\mathbf{W}^{1,p}(\Omega)$ as the thickness of the shell approaches zero, thus providing the first (and only one to this date) justification of a two-dimensional non-linear shell model by a *convergence theorem*.

The limit *'Le Dret–Raoult stored energy function'* is again that of a *non-linearly elastic 'membrane' shell*, in the sense that it is a function of the change of metric tensor only. For a St Venant–Kirchhoff material, it does *not* coincide, however, with the stored energy function found *via* the method of formal asymptotic expansions, save when the singular values of the limit deformation gradients belong to a specific subset of the plane (Genevey [6]).

The origin of such differences may lie in that the latter approach models shells made with *'soft'* elastic materials (like the sails of a sailboat or a balloon), while the former models shells made with *'rigid'* elastic materials (like the hull of a ship). But this affirmation is yet to be substantiated (more comments about the comparison between these two 'membrane' shell models are given in Ciarlet [3, sect. 9.5]).

A close scrutiny of the formal asymptotic analysis that leads to the equations of a non-linearly elastic 'flexural' shell shows that the energy to be minimized over the manifold $\mathcal{M}_F(\omega)$ is in fact of the form (Lods and Miara [11, Thm. 2])

$$\eta \longrightarrow \frac{\varepsilon^3}{6} \int_\omega a^{\alpha\beta\sigma\tau} R^\flat_{\sigma\tau}(\eta) R^\flat_{\alpha\beta}(\eta) \sqrt{a}\, y - \int_\omega p^{i,\varepsilon} \eta_i \sqrt{a}\, y,$$

where the functions $R_{\alpha\beta}^{\flat}(\eta)$ are given by

$$R_{\alpha\beta}^{\flat}(\eta) := F_{\alpha\|\beta}(\eta, \psi(\eta)) - \frac{1}{2} g^{\sigma\tau,1} \eta_{\sigma\|\alpha} \eta_{\tau\|\beta}$$
$$+ \left(\Gamma_{\alpha\beta}^{p,1} + \frac{1}{2} a^{mn} \left\{ \Gamma_{m\alpha}^{p,1} \eta_{n\|\beta} + \Gamma_{n\beta}^{p,1} \eta_{m\|\alpha} \right\} \right) \eta_p,$$

where the functions $g^{\sigma\tau,1}$ and $\Gamma_{m\alpha}^{p,1}$ are known functions of $y \in \overline{\omega}$ (they depend only on the mapping $\boldsymbol{\theta}$),

$$F_{\alpha\|\beta}(\eta, \psi) := \frac{1}{2} \left(\psi_{\alpha\|\beta} + \psi_{\beta\|\alpha} + a^{mn} \left\{ \eta_{m\|\alpha} \psi_{n\|\beta} + \eta_{n\|\beta} \psi_{m\|\alpha} \right\} \right)$$

and the field $\psi(\eta) = (\psi_i(\eta))$ is defined by

$$\psi_1(\eta) := b_1^{\alpha} \eta_{\alpha} + \left(1 + a^{\alpha 2} \eta_{\alpha\|2} \right) \eta_{3\|1} - a^{\alpha 2} \eta_{\alpha\|1} \eta_{3\|2},$$
$$\psi_2(\eta) := b_2^{\alpha} \eta_{\alpha} + \left(1 + a^{\alpha 1} \eta_{\alpha\|1} \right) \eta_{3\|2} - a^{\alpha 1} \eta_{\alpha\|2} \eta_{3\|1},$$
$$\psi_3(\eta) := -a^{\alpha\beta} \eta_{\alpha\|\beta} - a^{\alpha 1} a^{\beta 2} \left(\eta_{\alpha\|1} \eta_{\beta\|2} - \eta_{\alpha\|2} \eta_{\beta\|1} \right).$$

As is easily verified, these functions $R_{\alpha\beta}^{\flat}(\eta)$ are polynomials of degree ≤ 3 in the 'variables' η_i, $\partial_{\alpha}\eta_j$ and $\partial_{\beta\sigma}\eta_k$, and they are well defined functions in $L^2(\omega)$ for *all* fields $\eta = (\eta_i)$ in the vector space $\mathbf{W}^{2,p}(\omega)$, for any $p > 2$.

A second virtue of these functions is that they satisfy the relations (Lods and Miara [11, Thm. 2])

$$R_{\alpha\beta}^{\flat}(\eta) = R_{\alpha\beta}(\eta) \quad \text{for all } \eta \in \mathbf{W}^{2,p}(\omega) \text{ such that } a_{\alpha\beta}(\eta) - a_{\alpha\beta} = 0 \text{ in } \omega,$$

which explain why the energy of a flexural shell may be indifferently expressed in terms of either the functions $R_{\alpha\beta}^{\flat}(\eta)$ or the functions $R_{\alpha\beta}(\eta)$, and that it may then be minimized over the manifold $\mathcal{M}_F(\omega)$ (as already noted, when $a_{\alpha\beta}(\eta) - a_{\alpha\beta} = 0$ in ω, it suffices that $\eta \in \mathbf{H}^2(\omega)$ in order that $R_{\alpha\beta}^{\flat}(\eta)$ be well defined functions in $L^2(\omega)$).

As shown by Roquefort [13], the functions $R_{\alpha\beta}^{\flat}(\eta)$ take a remarkably simpler form when they are expressed in terms of the *deformation* $(\boldsymbol{\theta} + \eta_i \mathbf{a}^i)$ of the middle surface associated with a field $\eta = (\eta_i)$, viz.,

$$R_{\alpha\beta}^{\flat}(\eta) = R_{\alpha\beta}^{\sharp}(\eta) \quad \text{for all } \eta \in \mathbf{W}^{2,p}(\omega), \text{ for any } p > 2,$$

where

$$R_{\alpha\beta}^{\sharp}(\eta) := \frac{1}{\sqrt{a}} \partial_{\alpha\beta} \left(\boldsymbol{\theta} + \eta_i \mathbf{a}^i \right) \cdot \{ \mathbf{a}_1(\eta) \wedge \mathbf{a}_2(\eta) \} - b_{\alpha\beta}.$$

Remark. These alternate expressions for the functions $R^{\flat}_{\alpha\beta}(\eta)$ again show (this time immediately) that they coincide with the functions $R_{\alpha\beta}(\eta)$ when η belongs to the manifold $\mathcal{M}_F(\omega)$.

As a consequence of the above observations, it is proposed here to consider the following *two-dimensional non-linear shell model of Koiter's type* (which was first hinted at in Ciarlet [3, sect. 11.1]): The unknown $\zeta^{\varepsilon} = (\zeta_i)$ should be a *minimizer* of the *energy* j^{ε} defined by

$$j^{\varepsilon}(\eta) := \frac{\varepsilon}{2} \int_{\omega} a^{\alpha\beta\sigma\tau} G_{\sigma\tau}(\eta) G_{\alpha\beta}(\eta) \sqrt{a}\, y$$

$$+ \frac{\varepsilon^3}{6} \int_{\omega} a^{\alpha\beta\sigma\tau} R^{\sharp}_{\sigma\tau}(\eta) R^{\sharp}_{\alpha\beta}(\eta) \sqrt{a}\, y - \int_{\omega} p^{i,\varepsilon} \eta_i \sqrt{a}\, y$$

over a vector space of smooth enough vector fields $\eta = (\eta_i)$ satisfying *boundary conditions of 'strong clamping'* $\eta = \partial_{\nu}\eta = 0$ on γ_0 (cf. Ciarlet [3, sect. 10.5]).

Besides the fact that the functions $R^{\sharp}_{\alpha\beta}(\eta)$ are defined for *all* fields in a *vector space* (by contrast, the functions $R_{\alpha\beta}(\eta)$ are not necessarily well defined outside the manifold $\mathcal{M}_F(\omega)$), a major virtue of this model (established elsewhere; cf. Ciarlet and Roquefort [5]) is the following:

According to two mutually exclusive sets of assumptions bearing on the associated manifold of inextensional displacements, which turn out to be exactly the same as those leading to the distinction between 'membrane' and 'flexural' shells (cf. sect. 2), *the leading term of a formal asymptotic expansion* (with the thickness as the 'small' parameter) *of the solution of this two-dimensional model satisfies either the two-dimensional equations of a non-linearly elastic 'membrane' shell or those of a non-linearly elastic 'flexural' shell* (recalled at the beginning of this section).

These conclusions being identical to those reached by Miara [12] and Lods and Miara [11] about the leading term of the formal asymptotic expansion of the solution of the *three-dimensional equations* (again with the thickness as the 'small' parameter), *the non-linear shell model of Koiter's type proposed here is thus justified* (at least formally).

Finally, it is to be emphasized that these two-dimensional equations model *any* non-linearly elastic shell, i.e., irrespectively of the 'geometry' of its middle surface $\theta(\overline{\omega})$ and of the set $\theta(\gamma_0)$ (the middle line of that portion of the lateral surface of the shell where boundary conditions are imposed). *Remarks.* (1) Interestingly, *exactly* the same functions $R^{\sharp}_{\alpha\beta}(\eta)$ are also found in Koiter [8, eq. (4.11)], under the appellation of covariant components of a '*modified change of curvature tensor*'. However, W.T. Koiter does not provide much comment about the raison d'être of these functions, which he most likely derived out of completely different considerations!

(2) In view of the above observations, the functions $R^\sharp_{\alpha\beta}(\eta)$ may be aptly called the covariant components of the '*Koiter–Lods–Miara–Roquefort modified change of curvature tensor*'.

(3) W.T. Koiter proposes, again without any hint, yet *another* '*modified change of curvature tensor*', whose components $R^*_{\alpha\beta}(\eta)$ are defined as

$$R^*_{\alpha\beta}(\eta) = R_{\alpha\beta}(\eta) - \frac{1}{2}b^\sigma_\alpha G_{\sigma\beta}(\eta) - \frac{1}{2}b^\tau_\beta G_{\alpha\tau}(\eta).$$

The functions $R^*_{\alpha\beta}(\eta)$ are thus subjected to the same limitations as the functions $R_{\alpha\beta}(\eta)$.

(4) The proposed model may be viewed as the 'non-linear counterpart' of the famed *linear Koiter model* (proposed by Koiter [9]), whose solution likewise behaves like the solution of the three-dimensional equations of linearized elasticity when the thickness approaches zero (see Ciarlet [3, Chap. 7]).

(5) By contrast with the functions $R_{\alpha\beta}(\eta)$, the functions $R^\sharp_{\alpha\beta}(\eta)$ do not reduce by formal linearization to the components $\rho_{\alpha\beta}(\eta)$ of the linearized change of curvature tensor. As is easily checked, they instead satisfy

$$\left[R^\sharp_{\alpha\beta}(\eta)\right]^{\mathrm{lin}} = \rho_{\alpha\beta}(\eta) + b_{\alpha\beta}a^{\sigma\tau}\gamma_{\sigma\tau}(\eta),$$

where the functions $\gamma_{\sigma\tau}(\eta)$ are the covariant components of the linearized change of metric tensor. This property is, however, of no consequence on the soundness of the model, since, as already noted, its solution behaves appropriately for 'small' values of ε.

4. Conclusions

On the one hand, *the solution of the proposed model thus adequately 'captures' the 'limit' behavior*, i.e., when the thickness is 'small enough', *of a non-linearly elastic shell*, whether it be that of a '*flexural*' shell or that of a '*membrane*' shell.

On the other, *its stored energy function remains of a 'reasonable' computational complexity*, in that it consists of *polynomials with respect to the unknowns*, i.e., the covariant components of the displacement field and their partial derivatives.

For these reasons, it is hoped that this model can be successfully used in *numerical simulations of non-linearly elastic shells*, notably those that are expected to undergo 'large' displacements.

References

[1] Blouza, A. and Le Dret, H., Existence and uniqueness for the linear Koiter model for shells with little regularity, *Quart. Appl. Math.* **57** (1999), 317–337.

[2] Ciarlet, P.G., Mathematical elasticity, Volume I: *Three-Dimensional Elasticity*, North–Holland, Amsterdam 1988.

[3] _____, Mathematical elasticity, Volume III: *Theory of Shells*, North–Holland, Amsterdam 2000.

[4] Ciarlet, P.G. and Coutand, D., An existence theorem for non-linearly elastic 'flexural' shells, *J. Elasticity* **50** (1998), 261–277.

[5] Ciarlet, P.G. and Roquefort, A., Justification of a two-dimensional non-linear shell model of Koiter's type, to appear, (2000).

[6] Genevey, K., Remarks on non-linear membrane shell problems, *Math. Mech. Solids* **2** (1997), 215–237.

[7] John, F., Estimates for the derivatives of the stresses in a thin shell and interior shell equations, *Comm. Pure Appl. Math.* **18** (1965), 235–267.

[8] Koiter, W.T., On the non-linear theory of thin elastic shells, *Proc. Kon. Ned. Akad. Wetensch.* **B69** (1966), 1–54.

[9] _____, On the foundations of the linear theory of thin elastic shells, *Proc. Kon. Ned. Akad. Wetensch.* **B73** (1970), 169–195.

[10] Le Dret, H. and Raoult, A., The membrane shell model in non-linear elasticity: A variational asymptotic derivation, *J. Nonlinear Sci.* **6** (1996), 59–84.

[11] Lods, V. and Miara, B., Nonlinearly elastic shell models. II. The flexural model, *Arch. Rational Mech. Anal.* **142** (1998), 355–374.

[12] Miara, B., Nonlinearly elastic shell models. I. The membrane model, *Arch. Rational Mech. Anal.* **142** (1998), 331–353.

[13] Roquefort, A., *Sur quelques questions liées aux modéles non linéaires de coques minces*, Doctoral Dissertation, Universit é Pierre et Marie Curie, Paris 2000.

A MODEL OF THE PROCESS OF THINKING BASED ON THE DYNAMICS OF BUNDLES OF BRANCHES AND SETS OF BUNDLES OF p-ADIC TREES

ANDREI KHRENNIKOV

This research was supported by the grant 'Strategic Investigations' of the University of Växjö and a visiting professor fellowship at the Science University of Tokyo

Contents

1 Introduction 453
2 m-adic hierarchic chains for coding of information 453
3 Dynamical evolution of information 456
4 m-adic representation for information states 457
5 Semi-metric spaces of sets 459
6 Dynamics preserving the order of associations 465

M. de Gosson (ed.), Jean Leray '99 Conference Proceedings, 451-467.

1. Introduction

The system of p-adic numbers $\mathbf{Q}_p$, constructed by K. Hensel in the 1890s, was the first example of an infinite number field (i.e., a system of numbers where the operations of addition, subtraction, multiplication and division are well defined) which was different from a subfield of the fields of real and complex numbers. During much of the last 100 years p-adic numbers were considered only in pure mathematics, but in recent years they have been extensively used in theoretical physics (see, for example, the books [7] and [11]), the theory of probability [7] and investigations of chaos in dynamical systems [7]. In [1], [4] and [8] p-adic dynamical systems were applied to the simulation of functioning of complex information systems (in particular, cognitive systems). In this paper we continue these investigations. We study the collective dynamics of information states. We found that such a dynamics has some advantages compared to the dynamics of individual information states. First of all, the use of collections of sets (instead of single points) as primary information (in particular, cognitive) units extremely extends the ability of an information system to operate with large volumes of information. Another advantage is that (in the opposite to the dynamics of single states) the collective dynamics is essentially more regular. As we have seen [1] and [8] discrete dynamical systems over fields of p-adic numbers have the large spectrum of non-attracting behaviours. Starting with the initial point $x_0 \in \mathbf{Q}_p$ iterations need not arrive to an attractor. In particular, there are numerous cycles (and cyclic behaviour depends cruicially on the prime number p) as well as Siegel disks. In our information model attractors are considered as solutions of problems (coded by initial information states $x_0 \in \mathbf{Q}_p$)[1]. The absence of an attractor implies that in such a model the problem x_0 could not be solved. In the opposite to dynamics of single information states (p-adic numbers) collective dynamics practically always have attractors (at least for the dynamical systems which have been studied in [1] and [8]). So here each problem has the definite solution[2].

There are no physical reasons to use only prime numbers $p > 1$ as the basis for the description of a physical or information model. Therefore, we use systems of so called m-adic numbers, where $m > 1$ is an arbitrary natural number, see, for example, [10].

2. m-adic hierarchic chains for coding of information

The abbreviation 'I' will be used for information. The symbol τ will be used to denote an I-system.

---

[1] This is more or less the standard viewpoint for models based on neural networks, see, for example, [2].

[2] Of course, the construction of this solution needs time and memory resources. An information system may have or may not have such resources.

We shall use neurophysiologic terminology: elementary units for I-processing are called neurons, a 'thinking device' of τ is called brain. In our model it is supposed that each neuron n has $m > 1$ levels of excitement, $\alpha = 0, 1, \ldots, m - 1$. Individual neurons has no I-meaning in this model. Information is represented by chains of neurons, $\mathcal{N} = (n_0, n_1, \ldots, n_M)$. Each chain of neurons $\mathcal{N}$ can (in principle) perform m^M different I-states

$$x = (\alpha_0, \alpha_1, \ldots, \alpha_{M-1}), \quad \alpha \in \{0, 1, \ldots, m - 1\}, \tag{2.1}$$

corresponding to different levels of excitement for neurons in $\mathcal{N}$. Denote the set of all possible I-states by the symbol X_I.

In our model each chain of neurons $\mathcal{N}$ has a *hierarchic structure*: neuron n_0 is the most important, neuron n_1 is less important than neuron $n_0, \ldots$, neuron n_j is less important neurons than $n_0, \ldots, n_{j-1}$. This hierarchy is based on the possibility of a neuron to ignite other neurons in this chain: n_0 can ignite all neurons $n_1, \ldots, n_k, \ldots, n_M$, n_1 can ignite all neurons $n_2, \ldots, n_k, \ldots, n_M$, and so on; but the neuron n_j cannot ignite any of the previous neurons $n_0, \ldots, n_{j-1}$. Moreover, the process of igniting has the following structure. If n_j has the highest level of excitement, $\alpha_j = m - 1$, then increasing of α_j to one unit induces the complete relaxation of the neuron n_j, $\alpha_j \to \alpha'_j = 0$, and increasing to one unit of the level of excitement α_{j+1} of the next neuron in the chain,

$$\alpha_{j+1} \longrightarrow \alpha'_{j+1} = \alpha_{j+1} + 1. \tag{2.2}$$

If neuron n_{j+1} already was maximally exited, $\alpha_{j+1} = m - 1$, then transformation (2.2) will automatically imply the change to one unit of the state of neuron n_{j+2} (and the complete relaxation of the neuron n_{j+1}) and so on[3].

We shall use the abbreviation HCN for *hierarchic chain of neurons*. This hierarchy is called a *horizontal hierarchy* in the I-performance in brain.

HCNs provide the basis for forming *associations*. Of course, a single HCN is not able to form associations. Such an ability is a feature of an ensemble B^τ of HCNs of τ. Let $s \in \{0, 1, \ldots, m - 1\}$. A set

$$A_s = \{x = (\alpha_0, \ldots, \alpha_M) \in X_I : \alpha_0 = s\} \subset X_I$$

is called an association of the order 1. This association is represented by a collection B_s^τ of all HCNs $\mathcal{N} = (n_0, n_1, \ldots, n_M)$ which have the state $\alpha_0 = s$ for neuron n_0. Thus any association A_s of the order 1 is represented in the brain of τ by some set B_s^τ of HCNs. Of course, if the set B_s^τ is empty the association A_s does not present in the brain (at this instance of time). Associations of higher orders are defined in the same way. Let $s_0, \ldots, s_{l-1} \in \{0, 1, \ldots, m-1\}$, $l \leq M$. The set

$$A_{s_0 \cdots s_l} = \{x = (\alpha_0, \ldots, \alpha_M) \in X_I : \alpha_0 = s_0, \ldots, \alpha_{l-1} = s_{l-1}\}$$

---

[3] In fact, transformation (2.2) is the addition with respect to mod m.

is called an association of the order l. Such an association is represented by a set $B^\tau_{s_0 \cdots s_l} \subset B^\tau$ of HCN. We remark that associations of the order M coincide with I-states for HCN. We shall demonstrate that an I-system τ obtains large advantages by working with associations of orders $l \ll M$.

Denote the set of all associations of order l by the symbol $X_{A,l}$. We set $X_A = \cup_l X_{A,l}$. This is the set of all possible associations which can be formed on the basis of the I-space X_I.

Sets of associations $J \subset X_A$ also have a cognitive meaning. Such sets of associations will be called *ideas* of τ (of the order 1). Denote the set of all ideas by the symbol X_{ID}[4]. *Homogeneous ideas* are ideas which are formed by associations of the same order. For example, ideas $J = \{A_s, \ldots, A_q\}, s, \ldots, q \in \{0, 1, \ldots, m-1\}$, or $J = \{A_{s_1 s_2}, \ldots, A_{q_1 q_2}\}, s_i, \ldots, q_i \in \{0, 1, \ldots, m-1\}$ are homogeneous. An idea $J = \{A_s, A_{s_1 s_2}, \ldots, A_{q_1 q_2 \cdots q_l}\}$ is not homogeneous. Denote the space of all ideas formed by associations of the fixed order l by the symbol $X_{ID,l}$ (these ideas are homogeneous). Denote the space of all ideas formed by associations of orders less or equal to L (where L is the fixed number) by the symbol X^L_{ID}.

The hierarchy I-state $\rightarrow$ association $\rightarrow$ idea is called a *vertical hierarchy* in the I-performance in the brain.

Remark. (Associations, ideas and complexity of cognitive behaviour) One of the main features of our model is that not only I-states $x \in X_I$, but also associations $A \in X_A$ and ideas $J \in X_{ID}$ have a cognitive meaning. One of the reasons to use such a model is that complex cognitive behaviour can be demonstrated not only by living organisms τ which are able to perform in 'brains' large amounts of 'pure information' (I-states), but also by some living organisms with negligibly small 'brains'. It is well known that some primitive organisms τ_{pr} having (approximately) $N = 300$ nervous cells can demonstrate rather complex cognitive behaviour: ability for learning, complex sexual (even homosexual) behaviour. Suppose, for example, that the basis m of the coding system of τ_{pr} is equal to 2. Here each nervous cell n can yield two states: 0, non-firing, and 1, firing. Non-hierarchic coding of information gives the possibility to perform in the brain (at each instance of time) 300 bits of information. In the process of 'thinking' (see section 5) τ_{pr} transforms these 300 bits into another 300 bits. It seems that such 300 bits I-dynamics could not give a complex cognitive behaviour. We now suppose that τ_{pr} has the ability to create hierarchic chains of nervous cells (horizontal hierarchy). Let, for example, such $HCNs$ have the length $L = 5$. Thus τ_{pr} has $N = 60$ $HCNs$ (so the set $B^{\tau_{pr}}$ has 60 elements). The total number of I-states, $x = (\alpha_0, \alpha_1, \alpha_2, \alpha_3, \alpha_4), \alpha_j = 0, 1$, which can be performed by $HCNs$ of the length $L = 5$ is equal to $N_I = 2^5 = 32$.

---

[4] In principle, it is possible to consider sets of ideas of the order 1 as new cognitive objects (ideas of the order 2) and so on. However, we restrict our attention to dynamics of ideas of order 1.

Thus brain's hardware $B^{\tau_{\mathrm{pr}}}$ can perform all I-states (since $N_I < N$). We assume that all I-states are performed by the brain at each instant of time. We suppose that τ_{pr} is able to use the vertical hierarchy in the I-performance. The τ_{pr} have $N_a = 2^k$ associations of order $k = 1, 2, \ldots, 5$. The number of homogeneous ideas of τ_{pr} is equal

$$N_{ID} = \left(2^2 - 1\right) + \left(2^{2^2} - 1\right) + \left(2^{2^3} - 1\right) + \left(2^{2^4} - 1\right) + \left(2^{2^5} - 1\right)$$
$$= 4295033103 \gg 300$$

(each term contains -1, because empty sets of associations are not considered as ideas). Hence τ_{pr} works with more than 4295033103 'ideas' (having at the same time only $N_I = 32$ I-strings in his brain).

In our model 'hardware' of the brain of τ is given by an ensemble B^τ of HCNs. For an HCN $\mathcal{N} \in B^\tau$, we set $i(\mathcal{N}) = x$, where x is the I-state of $\mathcal{N}$. The map $i : B^\tau \to X_I$ gives the correspondence between states of brain and states of mind[5]. In general it may be that $i(\mathcal{N}_1) = i(\mathcal{N}_2)$ for $\mathcal{N}_1 \neq \mathcal{N}_2$. It is natural to assume that in general the map i depends on the time parameter $t : i = i_t$. In particular, if t is discrete, we obtain a sequence of maps $i_t : t = 0, 1, 2, \ldots$.

Let O be some subset of X_I. The space of associations which are composed by I-states x belonging to the set O is denoted by the symbol $X_A(O)$. The corresponding space of ideas is denoted by the symbol $X_{ID}(O)$.

In the spatial domain model each HCN $\mathcal{N}$ is a chain of physical neurons which are connected by axons and dendrites, see, for example, [5]. In principle, such a chain of neurons can be observed (as a spatial structure in the Euclidean space $\mathbf{R}^3$), compare with [3]. In the frequency domain model [6] digits $\alpha_j \in \{0, 1, \ldots, m - 1\}$ can be considered as (discretized) frequencies of oscillations for neurons $n_j, j = 0, 1, 2, \ldots$, which form a 'frequency HCN' $\mathcal{N}$. Here $\mathcal{N}$ need not be imagine as a connected spatial structure. It may have a dust-like structure in $\mathbf{R}^3$.

3. Dynamical evolution of information

In this section shall we study the simplest dynamics of I-states, associations and ideas. Such I-dynamics is 'ruled' by a function $f : X_I \to X_I$ which does not depend on time and random fluctuations. This 'process of thinking' has no memory: the previous I-state x determines a new I-state y via the transformation $y = f(x)$. In this model time is discrete, $t = 0, 1, 2, \ldots, n, \ldots, K$. Set

$$U_0^\tau = i_o\left(B^\tau\right), \qquad U_1^\tau = i_1\left(B^\tau\right), \qquad \ldots, \qquad U_n^\tau\left(B^\tau\right), \ldots. \quad (3.1)$$

---

[5] In fact, the map i provides the connection between the material and the mental worlds.

A set U_n^τ of I-states is called an I-universe of τ. This is the set of all I-states which are generated by the ensemble B^τ of HCNs of τ at the instant of the time $t = n$. We suppose that sets $\{U_n^\tau\}_{n=0}^\infty$ of I-states can be obtained by iterations of one fixed map $f : X_I \to X_I$. Thus dynamics (3.1) of I-universe of τ is induced by pointwise iterations:

$$x_{n+1} = f(x_n). \tag{3.2}$$

If $x \in U_n^\tau$, then $y = f(x) \in U_{n+1}^\tau$. Each $x_0 \in U_0^\tau$ evolves via in I-trajectory: $x_0, x_1 = f(x_0), x_2 = f(x_1) = f^2(x_0), \ldots, x_{n+1} = f(x_n) = f^n(x_0), \ldots$. Here the symbol f^n denotes n th iteration of f.

Suppose that, for each association A, its image $B = f(A) = \{y = f(x) : x \in A\}$ is again an association. Denote the class of all such maps f by the symbol $\mathscr{A}(X_I)$. If $f \in \mathscr{A}(X_I)$, then dynamics (3.2) of I-states of τ induces dynamics of associations

$$A_{n+1} = f(A_n). \tag{3.3}$$

Starting with an association A_0 (which is a subset of I-universe U_0^τ) τ obtains a sequence of associations: $A_0, A_1 = f(A_0), \ldots, A_{n+1} = f(A_n), \ldots$. Dynamics of associations (3.3) induces dynamics of ideas: $J' = f(J) = \{B^\tau = f(A) : A \in J\}$. Thus each idea evolves by iterations:

$$J_{n+1} = f(J_n). \tag{3.4}$$

Starting with an idea J_0 τ obtains a sequence of ideas: $J_0, J_1 = f(J_0), \ldots, J_{n+1} = f(J_n), \ldots$. In particular, by choosing $J_0 = U_0^\tau$ we obtain dynamics of I-universe (which is also an idea of τ).

We are interested in attractors of dynamical system (3.4) (these are ideas–solutions). To define attractors in the space of ideas X_{ID}, we have to define a convergence in this space. This will be done in section 5.

4. m-adic representation for information states

It is surprising that number systems which provide the adequate mathematical description of HCN were developed long time ago by purely number theoretical reasons. These are systems of so called m-adic numbers, $m > 1$ is a natural numbers. First we note that in mathematical model it would be useful to consider infinite I-states:

$$x = (\alpha_0, \alpha_1, \ldots, \alpha_M, \ldots), \qquad \alpha_j = 0, 1, \ldots, m - 1. \tag{4.1}$$

Such an I-state x can be generated by an ideal infinite HCN $\mathcal{N}$. Denote the set of all vectors (4.1) by the symbol $\mathbf{Z}_m$. This is an ideal I-space, $X_I = \mathbf{Z}_m$.

On this space we introduce a metric ρ_m corresponding to the hierarchic structure between neurons in chain $\mathcal{N}$ having an I-state x: two I-states x and y are close with respect to ρ_m if initial (sufficiently long) segments of x and y coincides. If $x = (\alpha_0, \ldots, \alpha_M, \ldots)$, $y = (\beta_0, \ldots, \beta_M, \ldots)$, and $\alpha_0 = \beta_0, \ldots, \alpha_{k-1} = \beta_{k-1}$, but $\alpha_k \neq \beta_k$, then $\rho_m(x, y) = 1/m^k$. Such a metric is well know in number theory. This is an ultra-metric: the strong triangle inequality

$$\rho_m(x, y) \leq \max\left[\rho_m(x, z), \rho_m(x, y)\right] \tag{4.2}$$

holds true. This inequality has the simple I-meaning. Let $\mathcal{N}, \mathcal{M}, \mathcal{L}$ be HCNs having I-states x, y, z respectively. Denote by $k(\mathcal{N}, \mathcal{M})$ ($k(\mathcal{N}, \mathcal{L})$ and $k(\mathcal{M}, \mathcal{L})$) length of an initial segment in chains $\mathcal{N}$ and $\mathcal{M}$ ($\mathcal{N}$ and $\mathcal{L}$, $\mathcal{M}$ and $\mathcal{L}$) such that neurons in $\mathcal{N}$ and $\mathcal{M}$ have the same level of exiting. Then is evident that

$$k(\mathcal{N}, \mathcal{M}) \geq \min\left[k(\mathcal{N}, \mathcal{L}), k(\mathcal{L}, \mathcal{M})\right]. \tag{4.3}$$

But this gives inequality (4.2). As in every metric space, in $(\mathbf{Z}_m, \rho_m)$ we can introduce balls, $U_r(a) = \{x \in \mathbf{Z}_m : \rho_m(a, x) \leq r\}$ and spheres $S_r(a) = \{x \in \mathbf{Z}_m : \rho_m(a, x) = r\}$ (with center at $a \in \mathbf{Z}_m$ of radius $r > 0$). There is one to one correspondence between balls and associations. Let $r = 1/p^l$ and $a = (a_0, a_1, \ldots, a_{l-1}, \ldots)$. The $U_r(a) = \{x = (x_0, x_1, \ldots, x_{l-1}, \ldots) : x_0 = a_0, x_1 = a_1, \ldots, x_{l-1} = a_{l-1}\} = A_{a_0 a_1 \cdots a_{l-1}}$. The space of associations X_A coincides with the space of all balls. The space of ideas X_{ID} is the space which elements are families of balls. Geometrically $\mathbf{Z}_m$ can be represented as a homogeneous tree.

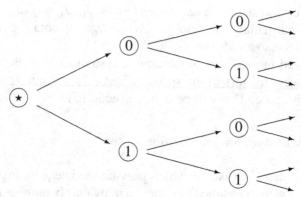

Figure 1. The 2-adic tree.

Associations are represented as bundles of branches on the m-adic tree. Ideas are represented as sets of bundles. So dynamics (3.2), (3.3), (3.4) are, respectively, dynamics of branches, bundles and sets of bundles on the m-adic tree.

I-dynamics on $\mathbf{Z}_m$ is generated by maps $f : \mathbf{Z}_m \to \mathbf{Z}_m$. We are interested in maps which belong to the class $\mathcal{A}(\mathcal{O})$, where $\mathcal{O}$ is some subset of $\mathbf{Z}_m$. They map a ball onto a ball: $f(U_r(a)) = U_{r'}(a')$. To give examples of such maps, we use the standard algebraic structure on $\mathbf{Z}_m$.

Each sequence $x = (\alpha_0, \alpha_1, \ldots, \alpha_M, \ldots)$ is identified with an m-adic number

$$x = \sum_{j=0}^{\infty} \alpha_j m^j = \alpha_0 + \alpha_1 \cdot m + \alpha_2 \cdot m^2 + \cdots + \alpha_n \cdot m^n + \cdots. \qquad (4.4)$$

It is possible to work with such series as with ordinary numbers. Addition, subtraction and multiplication are well defined. Analysis is much simpler for prime numbers $m = p > 1$. Therefore we study mathematical models for p-adic numbers. Let $f_n(x) = x^n$ (n-times multiplication of x), $n = 2, 3, 4, \ldots$. We shall prove in section that f_n belongs to the class $\mathcal{A}(Z_m^*)$, where $\mathbf{Z}_m^* = \mathbf{Z}_m \setminus \{0\}$. Hence here associations are transformed into associations.

m-adic balls $U_r(a)$ have an interesting property which will be used in our cognitive model. Each point $b \in U_r(a)$ can be chosen as a center of this ball: $U_r(a) \equiv U_r(b)$. Thus each I-state x belonging to an association A can be chosen as a *base* of this association. m-adic balls have another interesting property which will by used in our cognitive model. Let $U_r(a)$ and $U_s(b)$ be two balls. If the intersection of these balls is not empty, then one of the balls is contained in another.

5. Semi-metric spaces of sets

Let X be a set. A function $\rho : X \times X \to \mathbb{R}_+$ is said to be a metric[6] if it has the following properties: 1) $\rho(x, y) = 0$ iff $x = y$; 2) $\rho(x, y) = \rho(y, x)$; 3) $\rho(x, y) \leq \rho(x, z) + \rho(z, y)$ (the triangle inequality). The pair (X, ρ) is called a metric space. The set $U_r(a) = \{x \in X : \rho(x, a) \leq r\}, a \in X, r > 0$, is a ball in X. This set is closed in the metric space (X, ρ).

A metric ρ on X is called an *ultra-metric* if the so called *strong triangle inequality*

$$\rho(a, b) \leq \max[\rho(a, c), \rho(c, b)], \quad a, b, c \in X,$$

holds true; in such a case (X, ρ) is called an ultra-metric space.

A distance between a point $a \in X$ and a subset B of X is defined as

$$\rho(a, B) = \inf_{b \in B} \rho(a, b)$$

(if B is a finite set, then $\rho(a, B) = \min_{b \in B} \rho(a, b)$).

---

[6] The symbol $\mathbb{R}_+$ denotes the set of non-negative real numbers.

Denote by $\mathrm{Sub}(X)$ the system of all subsets of X. The *Hausdorf* distance between two sets A and B belonging to $\mathrm{Sub}(X)$ is defined as

$$\rho(A, B) = \sup_{a \in A} \rho(a, B) = \sup_{a \in A} \inf_{b \in B} \rho(a, b). \tag{5.1}$$

If A and B are finite sets, then

$$\rho(A, B) = \max_{a \in A} \rho(a, B) = \max_{a \in A} \min_{b \in B} \rho(a, b).$$

The Hausdorf distance ρ is not a metric on the set $Y = \mathrm{Sub}(X)$. In particular, $\rho(A, B) = 0$ does not imply that $A = B$. For instance, let A be a subset of B. Then, for each $a \in A$, $\rho(a, B) = \inf_{b \in B} \rho(a, b) = \rho(a, a) = 0$. So $\rho(A, B) = 0$. However, in general $\rho(A, B) = 0$ does not imply $A \subset B$[7]. Moreover, the Hausdorf distance is not symmetric: in general $\rho(A, B) \neq \rho(B, A)$[8].

We shall use the following simple fact. Let B be a closed subset in the metric space X[9]. Then $\rho(A, B) = 0$ iff $A \subset B$. In particular, this holds true for finite sets.

The triangle inequality

$$\rho(A, B) \leq \rho(A, C) + \rho(C, B), \quad A, B, C \in Y,$$

holds true for the Hausdorf distance.

Let T be a set. A function $\rho : T \times T \to \mathbb{R}_+$ for that the triangle inequality holds true is called a *pseudo-metric* on T; (T, ρ) is called a pseudo-metric space. So the Hausdorf distance is a pseudo-metric on the space Y of all subsets of the metric space X; (Y, ρ) is a pseudo-metric space.

The strong triangle inequality

$$\rho(A, B) \leq \max[\rho(A, C), \rho(C, B)], \quad A, B, C \in Y,$$

holds true for the Hausdorf distance corresponding to an ultra-metric ρ on X. In this case the Hausdorf distance ρ is an *ultra-pseudo-metric* on the set $Y = \mathrm{Sub}(X)$.

We can repeat the previous considerations starting with the Hausdorf pseudo-metric on Y. We set $Z = \mathrm{Sub}(Y)$ and define the Hausdorf pseudo-metric on Z via (5.1). As $\rho : Y \times Y \to \mathbb{R}_+$ is not a metric (and only a pseudo-metric) the Hausdorf pseudo-metric $\rho : Z \times Z \to \mathbb{R}_+$ does not have the same properties

---

[7] Let B be a non-closed subset in the metric space X and let A be the closure of B. Thus B is a proper subset of A. Here, for each $a \in A$, $\rho(a, B) = \inf_{b \in B} \rho(a, b) = 0$. Hence $\rho(A, B) = 0$.

[8] Let $A \subset B$ and let $\rho(b, A) \neq 0$ at least for one point $b \in B$. Then $\rho(A, B) = 0$. But $\rho(B, A) \geq \rho(b, A) > 0$.

[9] A closed set B can be defined as a set having the property: for each $x \in X$, $\rho(x, B) = 0$ implies that $x \in B$.

as $\rho : Y \times Y \rightarrow \mathbb{R}_+$. In particular, even if $A, B \in Z = \mathrm{Sub}(Y)$ are finite sets, $\rho(A, B) = 0$ does not imply that A is a subset of B. For example, let $A = \{u\}$ and $B = \{v\}$ are single-point sets $(u, v \in Y = \mathrm{Sub}(X))$ and let $u \subset v$ (as subsets of X). Then $\rho(u, v) = 0$. If u is a proper subset of v, then A is not a subset of B (in the space Y).

Proposition 5.1. Let $A, B \in Z = \mathrm{Sub}(Y)$ be finite sets and let elements of B be closed subsets of X. If $\rho(A, B) = 0$, then, for each $u \in A$, there exists $v \in B$ such that $u \subset v$.

Proof. As $\rho(A, B) = 0$, then, for each $u \in A$, $\rho(u, B) = \min_{b \in B} \rho(u, b) = 0$. Thus, for each $u \in A$, there exists $v \in B$ such that $\rho(u, v) = 0$. As v is a closed subset of X, this implies that $u \subset v$. $\qquad\square$

Let $A, B \in Z$ and let, for each $u \in A$, there exists $v \in B$ such that $u \subset v$. Such a relation between sets A and B is denoted by the symbol $A \subset\subset B$ (in particular, $A \subset B$ implies that $A \subset\subset B$). We remark that $A \subset\subset B$ and $B \subset\subset A$ do not imply $A = B$. For instance, let $A = \{u_1, u_2\}$ and let $B = \{u_2\}$, where $u_1 \subset u_2$. We also remark that $A_1 \subset\subset B_1$ and $A_2 \subset\subset B_2$ implies that $A_1 \cup A_2 \subset\subset B_1 \cup B_2$.

Let $f : T \rightarrow T$, where $T = Y$ or $T = Z$, be a map. Let H be a fixed point of f, $f(H) = H$. A basin of attraction of H is the set $AT(H) = \{J \in T : \lim_{n\to\infty} \rho(f^n(J), H) = 0\}$. We remark that $J \in AT(H)$ means that iterations $f^n(J)$ of the set J are (approximately) absorbed by the set H. The H is said to be an attractor for the point $J \in Z$ if, for any fixed point H' of f such that $\lim_{n\to\infty} \rho(f^n(J), H') = 0$ (so $J \in AT(H')$), we have $H \subset H'$. Thus an attractor for the set $J (\in \mathrm{Sub}(Y))$ is the minimal set that attracts J. The attractor is uniquely defined.

Let $T = Z = \mathrm{Sub}(Y)$, $Y = \mathrm{Sub}(X)$. The H is said to be an $\subset\subset$-attractor for the point $J \in Z$ if, for any fixed point H' of f such that $\lim_{n\to\infty} \rho(f^n(J), H') = 0$ (so $J \in AT(H')$), we have $H \subset\subset H'$. $\subset\subset$-attractor is not uniquely defined. For example, let $J = \{u\}$, $u \in Y$, $f(u) = u$. Here the set J is an $\subset\subset$-attractor (for itself) as well as any refinement of $J : A = \{u, v_1, \ldots, v_N\}$, where $v_j \subset u$.

All previous considerations can be repeated if, instead of the spaces $Y = \mathrm{Sub}(X)$ and $Z = \mathrm{Sub}(Y)$ of all subsets, we consider some families of subsets: $U \subset \mathrm{Sub}(X)$ and $V = \mathrm{Sub}(U)$. We obtain pseudo-metric spaces (U, ρ) and (V, ρ).

Let $f : U \rightarrow U$ be a map. For $u \in U$, we set

$$O_{+,k}(u) = \left\{ f^l(u) : l \geq k \right\}, \quad k = 0, 1, 2, \ldots,$$

$$\text{and } O_\infty(u) = \cap_{k=0}^\infty O_{+,k}(u).$$

For a set $J \in V$, we set $O_{+,k}(J) = \cup_{u \in J} O_{+,k}(u)$ and $O_\infty(J) = \cup_{u \in J} O_\infty(u)$.

Lemma 5.2. *Let the space U be finite. Then, for each $J \in V$, J is attracted by the set $O_\infty(J)$.*

Proof. First we remark that, as $O_\infty(u) \subset \cdots \subset O_{+,k+1}(u) \subset O_{+,k}(u) \cdots \subset O_{+,0}(u)$, and $O_{+,0}(u)$ is finite, we obtain that $O_\infty(u) \equiv O_{+,k}(u)$ for $k \geq N(u)$ (where $N(u)$ is sufficiently large).

We prove that, for each $u \in J$, the set $O_\infty(u)$ is f-invariant and

$$\lim_{k \to \infty} \rho \left(f^n(u), O_\infty(u) \right) = 0.$$

As $O_\infty(u) \equiv O_{+,k}(u), k \geq N(u)$, and $f(O_{+,k}(u)) = O_{+,k+1}(u)$, we obtain that $f(O_\infty(u)) = O_\infty(u)$. If $k \geq N(u)$, then $f^k(u) \in O_{+,k}(u) = O_\infty(u)$. Thus $\rho(f^k(u), O_\infty(u)) = 0$. We have

$$f\left(O_\infty(J)\right) = \cup_{u \in J} f\left(O_\infty(u)\right) = \cup_{u \in J} O_\infty(u) = O_\infty(J).$$

So $O_\infty(J)$ is invariant. Let $N(J) = \max_{u \in J} N(u)$. If $k \geq N(J)$, then, for each $u \in J$, $\rho(f^k(u), O_\infty(J)) \leq \rho_p(f^k(u), O_\infty(u)) = 0$. So $J \in AT(O_\infty(J))$.

A pseudo-metric ρ (on some space) is called *bounded from below* if

$$\delta = \inf\{q = \rho(a, b) \neq 0\} > 0. \tag{5.2}$$

If ρ is a metric, then (5.2) is equivalent to the condition

$$\delta = \inf\{q = \rho(a, b) : a \neq b\} > 0. \qquad \square$$

Theorem 5.3. *Let the space U be finite and let the Hausdorf distance on the space U be a metric which is bounded from below. Then each set $J \in V$ has an attractor, namely the set $O_\infty(J)$.*

Proof. By Lemma 5.2 we have that $J \in AT(O_\infty(J))$. We need to prove that if, for some set $A \in V$,

$$\lim_{k \to \infty} \rho \left(f^k(u), A \right) = 0, \tag{5.3}$$

then $O_\infty(u) \subset A$. Let $\rho(f^l(u), A) < \delta$ for $l \geq k \geq N(u)$ (here δ is defined by condition (5.2)). As A is a finite set (so $\rho(d, A) = \min_{a \in A} \rho(d, a)$), we obtain that

$$\rho \left(f^l(u), a \right) = 0 \tag{5.4}$$

for some $a = a(u, l) \in A$. Hence

$$f^l(u) = a(u, l) \in A, \quad l \geq k. \tag{5.5}$$

Thus $O_\infty(u) = O_{+,k}(u) \subset A$. Let

$$\lim_{k \to \infty} \rho\left(f^k(J), A\right) = 0. \qquad (5.6)$$

As U is finite (and so J is also finite), (5.6) holds true iff (5.3) holds true for all $u \in J$. Thus $O_\infty(u) \subset A$ for each $u \in J$. So $O_\infty(J) \subset A$. □

If the Hausdorf distance is not a metric on U (and only a pseudo-metric), then (in general) the set $O_\infty(J)$ is not an attractor for the set J. However, we have the following result:

Theorem 5.4. *Let the space U be finite and let all elements of the space $U \subset$ Sub(X) be closed subsets of the metric space (X, ρ). Let the Hausdorf pseudo-metric on the space U be bounded from below. Then each set $J \in V$ has an $\subset\subset$-attractor, namely the set $O_\infty(J)$.*

Proof. By Lemma 5.2 we again have that $J \in AT(O_\infty(J))$. We need to prove that if, for some set $A \in V$, (5.3) holds true, then $O_\infty(u) \subset\subset A$. We again obtain condition (5.4). However, as ρ is just a pseudo-metric, this condition does not imply (5.5). We apply Proposition 5.1 and obtain that $f^l(u) \subset a(u, l)$. As $O_\infty(u) = O_{+,k}(u)$ for sufficiently large k, we obtain that, for each $w \in O_\infty(u)(w = f^l(u), l \geq k)$, there exists $a \in A$ such that $w \subset a$. Thus $O_\infty(u) \subset\subset A$. □

In applications to the I-processing we shall use the following construction.
Let (X, ρ) be an ultra-metric space. We choose $U \subset$ Sub(X) as the set of all balls $U_r(a)$. The Hausdorf distance is an ultra-pseudo-metric on the space of balls U. As balls are closed, $\rho(U_r(a), U_s(b)) = 0$ implies $U_r(a) \subset U_s(b)$. In particular, $\rho(U_r(a), U_r(b)) = 0$ implies $U_r(a) = U_r(b)$.

Proposition 5.5. *Let $U_r(a) \cap U_s(b) = \emptyset$. Then $\rho(U_r(a), U_s(b)) = \rho(a, b)$.*

Proof. We have $\rho(U_r(a), U_s(b)) \geq \rho(a, U_s(b))$. If $y \in U_s(b)$ then $\rho(a, b) > s \geq \rho(b, y)$. Thus $\rho(a, y) = \rho(a, b)$ and, consequently, $\rho(a, b) \leq \rho(U_r(a), U_s(b))$. On the other hand, for each $x \in U_r(a)$, $\rho(x, U_s(b)) \leq \rho(x, b) = \rho(a, b)$. Hence $\sup_{x \in U_r(a)} \rho(x, U_s(b)) \leq \rho(a, b)$.

We choose $V = $ Sub(U), the space of all subsets of the space of balls.
Let $X = \mathbf{Z}_m$ and $\rho = \rho_m$. The space of associations X_A can be identified with the space of balls U. Here $\rho_m(A, B) = 0$ iff A is a sub-association of B : $A \subset B$. Thus $\rho_m(A_{\alpha_0 \cdots \alpha_l}, A_{\beta_0 \cdots \beta_m}) = 0$ iff $l \geq s$ and $\alpha_0 = \beta_0, \ldots, \alpha_s = \beta_s$. In particular, if $A, B \in X_{A,l}$ (associations of the same order l), then $\rho_m(A, B) = 0$ iff $A = B$. □

The space of ideas X_{ID} can be identified with the space $V = \text{Sub}(U)$ (of collections of balls). In such a way we introduce the Hausdorf ultra-pseudo-metric on the space of ideas. In further constructions we shall also choose some subspaces of the space of associations X_A and the space of ideas X_{ID} as spaces U and V, respectively.

In particular, the space $U = X_{A,l}$ of associations of the order l can be identified with the space of all balls having radius $r = 1/p^l$. The Hausdorf distance ρ_m is the **metric** on the space $U = X_{A,l}$. This metric is bounded from below with $\delta = 1/p^l$. So $(X_{A,l}, \rho_m)$ is a finite metric space with the metric (the Hausdorf distance) which is bounded from below. Theorem 5.1 can be applied for spaces $U = X_{A,l}$ and $V = X_{ID,l} = \text{Sub}(X_{A,l})$ (homogeneous ideas consisting of associations of the order l).

Theorem 5.6. *Let $f : X_{ID,l} \to X_{ID,l}$ be a map induced by some map $f : X_{A,l} \to X_{A,l}$. Each idea $J \in X_{ID,l}$ has an attractor, namely the set $O_\infty(J) \in X_{ID,l}$.*

In fact, the proof of Theorem 5.3 gives the algorithm for construction of the attractor $H = O_\infty(J)$. The brain of a cognitive system τ produces iterations $J, J_1 = f(J), \ldots, J_n = f(J_{n-1}), \ldots$ until the first coincidence of a new idea J_s with one of the previous ideas: $J_s = J_n$. As $J_{n+j} = J_{s+j}$, $O_{+,n}(J) = \{J_n, \ldots, J_{s-1}\} = O_\infty(J)$ is the attractor.

Remark. In fact, attractors in the space of ideas are given by cycles of associations. Dynamical systems over p-adic trees have a large number of cycles for I-states as well as for associations. This is one of the main disadvantages of the process of thinking in the domain of I-states and associations: starting with the initial I-state x_0 (or the association A_0) the brain of τ will often obtain no definite solution. However, by developing the ability to work with collections of associations (ideas) cognitive systems transferred this disadvantage into the great advantage: richness of cyclic behaviour on the level of associations implies richness of the set of possible ideas-solutions.

Let $U = \cup_{l=1}^{L} X_{A,l}$. This is the collection of all associations of orders less or equal to L (all balls $U_{1/p^l}(a), a \in \mathbf{Z}_m, l \leq L$). Let $V = X_{ID}^L = \text{Sub}(U)$. The Hausdorf distance is not a metric on the U. It is just a pseudo-metric: if $U_{1/p^l}(a) \subset U_{1/p^k}(b)$, then $\rho_m(U_{1/p^l}(a), U_{1/p^k}(b)) = 0$. However, the Hausdorf distance is bounded from below. By Proposition 5.5 if $\rho_m(U_{1/p^l}(a), U_{1/p^k}(b)) \neq 0$, then $\rho_m(U_{1/p^l}(a), U_{1/p^k}(b)) = \rho_m(a, b) \geq 1/p^L$. Thus we can apply Theorem 5.6 and obtain:

Theorem 5.7. *Let $f : X_A \to X_A$ be a map and let, for some M, the induced map $f : X_{ID}^L \to X_{ID}^L$. Then each idea $J \in X_{ID}^M$ has an CC-attractor, namely the set $O_\infty(J) \in X_{ID}^M$.*

As it was already been noted, $\subset\subset$-attractor is not unique. It seems that the brain of τ could have problems to determine uniquely the solution of a problem J. However, it would be natural for τ to produce the solution of J as 'algorithmically' determined attractor $G(J)$[10].

6. Dynamics preserving the order of associations

We set $\mathbb{O} = S_1(0)$ (the unit sphere in $\mathbf{Z}_p$ with the center at zero). In this section we will present a large class of maps $f : \mathbb{O} \to \mathbb{O}$ which produce dynamics of associations with the property $f : X_{A,m}(\mathbb{O}) \to X_{A,m}(\mathbb{O})$ for all m (associations of the order m are transformed into associations of the same order m).

We consider the map $f : \mathbf{Z}_p \to \mathbf{Z}_p$, $f = f_n(x) = x^n (n = 2, 3, \dots)$. The sphere $\mathbb{O} = S_1(0)$ is an invariant subset of this map. We shall study dynamics generated by f in the I-space $X_I = \mathbb{O}$ and corresponding dynamics in spaces of associations $X_A(\mathbb{O})$ and $X_{ID}(\mathbb{O})$. We start with the following mathematical result:

Theorem 6.1. *The f_n-image of any ball in $\mathbf{Z}_p^*$ is again a ball in $\mathbf{Z}_p^*$.*

Proof. Let $U_r(a) \subset \mathbf{Z}_p^*, r = 1/p^m$. As $0 \notin U_r(a)$, we have $|a|_p > r$. We shall prove that $f(U_r(a)) = U_s(b)$, where $b = a^n$ and $s = r|n|_p|a|_p^{n-1}$. First we prove that $f(U_r(a)) \subset U_s(b)$. Here we use the following result: $\qquad\square$

Lemma 6.2. *Let $a, \xi \in \mathbf{Z}_p$ and let $|a|_p > |\xi|_p$. Then*

$$\left|(a + \xi)^n - a^n\right|_p \leq |n|_p|a|_p^{n-1}|\xi| \qquad (6.1)$$

for every natural number n, where equality holds for $p > 2$.

To prove Lemma 6.2, we use the following result [4]:

Lemma 6.3. *Let $\gamma \in \mathbb{O}$ and $u \in \mathbf{Z}_p$, $|u|_p \leq 1/p$. Then $|(\gamma + u)^n - \gamma^n|_p \leq |n|_p|u|_p$ for every $n \in \mathbb{N}$, where equality holds for $p > 2$.*

By using (6.1) we obtain that $f(U_r(a)) \subset U_s(b)$. We prove that $f(U_r(a)) = U_s(b)$. Let $y = a^n + \beta$, where $|\beta|_p \leq s$. We must find $\xi, |\xi|_p \leq r$, such that $(a+\xi)^n = a^n + \beta$ or $(1 + \xi/a)^n = 1 + \beta/a^n$. Formally $1 + \xi/a = (1 + \beta/a^n)^{1/n}$.

---

[10] We repeat again that $f : \mathbf{Z}_m \to \mathbf{Z}_m$ is not a recursive function. So we a use more general viewpoint of the notion of an algorithm: a recursive functions which works with non-recursive blocks f. In any case we do not accept Church's thesis.

The p-adic binom $(1+\lambda)^{1/n}$ is analytic for $|\lambda|_p \leq |n|_p/p$. We have $|\beta/a^n|_p \leq r|n|_p/|a|_p \leq |n|_p/p$. So $\xi = a[(1+\beta/a^n)^{1/n} - 1] \in \mathbf{Z}_p$. We have to prove that $|\xi|_p \leq r$. We have

$$|\xi| \leq |a|_p \max_{1 \leq j < \infty} \frac{|\beta|_p^j}{|n|_p^j |a^n|_p^j |j!|_p}$$

$$\leq r \max_{1 \leq j < \infty} \left(\frac{r}{|a|_p}\right)^{j-1} \frac{1}{|j!|_p}.$$

By using the inequality $1/|j!|_p \leq p^{(j-1)/(p-1)}$, we obtain

$$|\xi| \leq r \max_{1 \leq j < \infty} \left(\frac{rp^{1/(p-1)}}{|a|_p}\right)^{j-1} \leq r.$$

In particular, this theorem implies that:

If n is not divisible by p, the f_n-image of each ball $U_{1/p^m}(a) \subset \mathbb{O}$ is a ball $U_{1/p^m}(b) \subset \mathbb{O}$.

In this case $f_n : X_{A,m}(\mathbb{O}) \to X_{A,m}(\mathbb{O})$ for all m. Hence we can apply Theorems 5.6, 5.7. Each problem $J \in X_{ID,m}(\mathbb{O})$ has the solution $O_\infty(J) \in X_{ID,m}(\mathbb{O})$ which is the attractor (in the space $X_{ID,m}(\mathbb{O})$) for J. Each problem $J \in X_{ID}^M(\mathbb{O})$ has the solution $O_\infty(J) \in X_{ID}^M(\mathbb{O})$ which is a $\subset\subset$-attractor (in the space $X_{ID}(\mathbb{O})$) for J. Moreover, the construction of the solution $O_\infty(J)$ can be reduced to purely arithmetical computations.

We set $R_{p^m} = \{1, 2, \ldots, p^m - 1\}$. We consider mod p^m multiplication on R_{p^m} (this is the ring of mod p^m residue classes). The metric ρ_p on R_{p^m} is induced from $\mathbf{Z}_p$. This metric is bounded from below with $\delta = 1/p^m$. We denote by the symbol $R_{p^m}^\star$ the subset of R_{p^m} consisting of all j which are not divisible by p. We introduce the function $f_{n,(m)} : R_{p^m} \to R_{p^m}$ by setting $f_{n,(m)}(x) = x^n \bmod p^m$. We remark that $f_{n,(m)}$ maps the set $R_{p^m}^\star$ into itself.

Let $a \in R_{p^m}^\star$. Here set $O_{+,k}(a) = \{a^{n^l} : l \geq k\}, k = 0, 1, 2, \ldots$, and (as usual) $O_\infty(a) = \cap_{k=1}^\infty O_{+,k}(a)$ and $O_\infty(D) = \cup_{d \in D} O_\infty(d)$ for $D \subset R_{p^m}^\star$. Let $J \in X_{ID,m}(\mathbb{O})$. So $J = \{U_{1/p^m}(d)\}_{d \in D}$, where $D \subset R_{p^m}^\star$. Thus, instead of f_n-dynamics of homogeneous ideas $J \in X_{ID,m}(\mathbb{O})$ τ can use $f_{n,(m)}$-dynamics of collections of points $d \in R_{p^m}^\star$. It is performed via mod p^m arithmetics for natural numbers. In particular, the attractor $O_\infty(J) = \{U_{1/p^m}(t) : t \in O_\infty(D)\}$. Therefore, the solution $O_\infty(J)$ of the problem J can constructed purely mod p^m-arithmetically.

Conjecture. The process of thinking (at least its essential part) is based on mod p^m arithmetics.

The same considerations can be used for non-homogeneous ideas $J \in X_{ID}^M(\mathbb{O})$. Here $J = \{J_m\}$, where $J_m \in X_{ID,m}(\mathbb{O})$. Due to properties of the map f_n all homogeneous ideas J_m proceed independently.

References

[1]   Albeverio, S., Khrennikov, A.Yu., and Kloeden, P., Memory retrieval as a p-adic dynamical system, *Biosystems* **49** (1999), 105–115.

[2]   Amit, D.J., *Modeling of brain functions*, Cambridge University Press, Cambridge 1989.

[3]   Cohen, J.D., Perlstein, W.M., Braver, T.S., Nystrom, L.E., Noll, D.C., Jonides, J. and Smith, E.E., Temporal dynamics of brain activation during working memory task, *Nature* **386** (1997), April 10, 604–608.

[4]   Dubischar, D., Gundlach, V.M., Steinkamp, O., and Khrennikov, A.Yu., A p-adic model for the process of thinking disturbed by physiological and information noise, *J. Theor. Biology* **197** (1999), 451–467.

[5]   Eccles, J.C., *The understanding of the brain*, McGraw-Hill Book Company, New York 1974.

[6]   Hoppensteadt, F.C., *An introduction to the mathematics of neurons: modeling in the frequency domain*, 2nd. edition, Cambridge University Press, New York 1997.

[7]   Khrennikov, A.Yu., *p-adic valued distributions in mathematical physics*, Kluwer Academic Publishers, Dordrecht 1994.

[8]   _____, *Non-archimedean analysis: quantum paradoxes, dynamical systems and biological models*, Kluwer Academic Publ., Dordrecht 1997.

[9]   _____, Human subconscious as a p-adic dynamical system, *J. Theor. Biol.* **193** (1998), 179–196.

[10] Mahler, K., *p-adic numbers and their functions*, Cambridge tracts in math., vol. 76, Cambridge Univ. Press, Cambridge 1980.

[11] Vladimirov, V.S., Volovich, I.V., and Zelenov, E.I., *p-adic analysis and mathematical physics*, World Scientific Publ., Singapore 1994.

GLOBAL WAVE MAPS ON BLACK HOLES

YVONNE CHOQUET–BRUHAT

Contents

1   Introduction 471
2   Definitions 471
3   Cauchy problem 472
4   Wave maps outside a Schwarzchild black hole 474
5   First energy estimate 476
6   Second energy estimates 478

M. de Gosson (ed.), Jean Leray '99 Conference Proceedings, 469-482.
© *2003 Kluwer Academic Publishers.*

1. Introduction

Wave maps are the generalization of the usual wave equation to mappings from a pseudo-Riemannian manifold of hyperbolic (Lorentzian) signature, called the source, into another pseudo-Riemannian manifold, called the target, which can be of arbitrary signature. Wave maps appear in various areas of physics, with properly Riemannian or Lorentzian targets. The first ones to be considered were the σ-models, mappings from Minkowski space–time into a sphere.

Wave maps are the counterpart in hyperbolic signature of the source of the harmonic maps between properly Riemannian manifolds, which have a long mathematical history. The existence of harmonic maps is an elliptic problem, it has only a global meaning and depends fundamentally on the geometrical and topological properties of the source and target manifolds.

The natural problem for wave maps is the Cauchy problem: determination of a wave map from its Cauchy data on a space-like submanifold of the source. It splits, as for other non-linear hyperbolic equations, into a local problem, that is existence for a small time interval, and a global problem, existence on the whole source manifold. The local problem is quite easy to solve, in spite of the non-linearity of the equations and the fact that the unknown is not vector-valued, though some subtlety occurs in the lowering of the regularity required from the data. The global problem has yet received only partial answers, of a positive or negative character. The global existence on an arbitrary regularly hyperbolic source manifold has been proved only when this manifold is of dimension 2 and the target is a complete Riemannian manifold (cf. Choquet–Bruhat [6]). This result was proved on Minkowski space–time M_2 by Gu Chaohao [4] (Riemann method of characteristics), Ginibre and Velo [5] (energy method).

The global existence of spherically symmetric wave maps on 4-dimensional Minkowski space–time into Riemannian targets has been established by Christodoulou and Talvilar–Zadeh [9]: it is again a problem with one space variable, but the proof is delicate owing to the singularity which appears in the equation at the center of symmetry.

We will in this article prove global existence of spherically symmetric wave maps from the outside of a Schwarzhild black hole. We will not encounter difficulties with the center of symmetry, which does not belong to the source manifold, but we will be confronted with the problem that the outside of a Schwarzhchild black hole is a globally, but not regularly, hyperbolic manifold, we will admit some classes of targets which are not Riemannian but pseudo-Riemannian of Lorentzian signature.

2. Definitions

Let (V, g) and (M, h) be two smooth pseudo-Riemanian manifolds of arbitrary signature and dimensions $n + 1$ and d. Let u be a mapping from V into M.

In an open set ω of V with local coordinates (x^α), with ω sufficiently small for the mapping u to take its value in a coordinate chart (y^A) of M, the mapping u is represented by d functions u^A of the $n+1$ variables x^α. The mapping u is differentiable at $x \in V$ if the functions u^A are differentiable. The differential $du(x)$ at x is a linear map between the tangent space at x to V and the tangent space to M at $u(x)$, it is therefore an element of the tensor product of the cotangent space to V at x by the tangent space to M at $u(x)$. The differential itself is a section of the vector bundle E with base V and fiber E_x at x this tensor product. The metrics g on V and h on M together with the mapping u endow E_x with a scalar product

$$G(x) \equiv g(x) \otimes h(u(x)).$$

We denote by ∇ the torsion-free linear connexion on the vector bundle E which leaves invariant this scalar product. If f is a section of E represented in local coordinates by the $(n+1) \times d$ functions f_α^a we have:

$$\nabla_\alpha f_\beta^a(x) \equiv \partial_\alpha f_\beta^a(x) - \Gamma_{\alpha\beta}^\mu(x) f_\mu^a(x) + \partial_\alpha u^b(x) \Gamma_{bc}^a(u(x)) f_\beta^c(x),$$

where $\Gamma_{\alpha\beta}^\mu$ and Γ_{bc}^a denote, respectively, the components of the Riemannian connections of g and h. Analogous formulas give the covariant derivatives of sections of bundles over V with fiber $\otimes^p T_x^* V \otimes T_{u(x)} M$.

Definition. A mapping $u : (V, g) \to (M, h)$ is called a wave map if it satisfies the following second order partial differential equation, taking its values in TM:

$$g \cdot \nabla^2 u = 0.$$

This equation, invariant under isometries of V and M, reads in local coordinates on V and M as:

$$g^{\alpha\beta} \nabla_\alpha \partial_\beta u^a \equiv g^{\alpha\beta} \left(\partial_{\alpha\beta}^2 u^a - \Gamma_{\alpha\beta}^\lambda \partial_\lambda u^a + \Gamma_{bc}^a \partial_\alpha u^b \partial_\beta u^c \right) = 0.$$

Wave maps are critical points of the Dirichlet integral:

$$\mathscr{E}(u) \equiv \int_V g^{\alpha\beta} h_{ab} \partial_\alpha u^a \partial_\beta u^b \mu_g,$$

with μ_g the volume element of g.

Notice that the Dirichlet integral is not a positive functional of u if g and h are not both properly Riemannian, it is not to be called the energy of the mapping u.

3. Cauchy problem

The wave map equation in local coordinates on V and M reads as a system of semilinear, quasi-diagonal, second order equations with diagonal principal part: the wave operator of the metric $\mathbf{g}$. the natural problem to solve is the Cauchy problem, i.e., the construction of a wave map taking together with its first derivative given values on a given space-like submanifold S_0 of V:

$$u_{S_0} = \varphi,$$

where φ is a map from S_0 into M and

$$(\partial_t u)_{S_0}(x) = \psi(x) \in T_{\varphi(x)}M, \quad \text{with } x \in S_0.$$

To apply to this problem the results local in time known for scalar-valued systems we will use the following definition and lemma.

Definition. The manifold (M, h) is said to be regularly embedded in the pseudo-Euclidean manifold (R^{N+1}, Q) if it is defined by P smooth scalar equations $F^{(p)} = 0$ on R^{N+1}, of rank P on M, and h is the pullback of the metric Q under this embedding.

We denote by $v^{(p)}$ the normal to M in R^{N+1} defined by the gradient of $F^{(p)}$, supposed to be non-isotropic on M. We set $v_{(p)} = m_{pq}v^{(q)}$ where m_{pq} is the inverse, supposed to exist, of the matrix with elements $(v^{(p)}, v^{(q)})_Q$, scalar product in the metric Q.

Lemma 3.1. *Let (M, h) be regularly embedded in (R^{N+1}, Q). Then $u : (V, \mathbf{g})$ $\to (M, h)$ is a wave map if and only if the mapping $U = i \circ u : V \to R^{N+1}$ satisfies the system of $N + 1$ scalar semilinear equations on (V, g)*

$$g^{\alpha\beta}\left\{\nabla_\beta\partial_\alpha U^A + \partial_\alpha U^B \partial_\beta U^C Q^{AD}v_{(I)D}\partial_B v_C^{(I)}\right\} = 0$$

and the initial data for U are such that $U^A{}_{|S_0} \equiv \Phi^A$ take their values in M while $(\partial_t U^A)_{|S_0} = \Psi^A$ take their values in the tangent space to M.

The proof (Choquet–Bruhat [6], see for special cases Ginibre and Velo [5]) is for the 'if' part, by applying the variational calculus with constraints on the Dirichlet integral after remarking that $\mathscr{E}(U)$, constructed with $\mathbf{g}$, Q and U is equal to $\mathscr{E}(u)$ when $h = i^*Q$ and $U = i \circ u$. The 'only if' converse results

The equation for $U \equiv i \circ u$ reads as a semilinear quasi-diagonal hyperbolic system of $N + 1$ scalar second order equations on V. The Leray theory, as completed by Dionne [2], Choquet–Bruhat–Christodoulou–Francaviglia [3], gives a local existence and uniqueness theorem which will follow, modulo the following definition:

Definition. The manifold $(V, \mathbf{g})$ is said to be regularly hyperbolic if:

1) It is globally hyperbolic, hence V is of the type $S \times R$, with S an n-dimensional oriented smooth manifold, the past of any compact subset of V intersects any $S_t \equiv S \times \{t\}$ along a compact set.

2) The metric **g**, assumed here for simplicity to be smooth, can be written

$$\mathbf{g} = -N^2 dt^2 + g_{ij}\theta^i\theta^j, \qquad \theta^i \equiv dx^i + \beta^i dt,$$

with $0 < B_1 \le N \le B_2$ on V while the metrics $g_t = g_{ij}dx^i dx^j$, induced by **g** on S_t, are uniformly equivalent to a given smooth Riemannian metric e on S which is Sobolev regular, i.e., such that the usual embedding and multiplication properties hold for the Sobolev spaces W_s^p of tensors on S defined through the metric e.

Theorem 3.2. *Let $(V, \mathbf{g})$ be a smooth regularly hyperbolic manifold. Let (M, h) be a Lorentzian manifold regularly embedded by i in a pseudo-Euclidean space (R^{N+1}, Q). Let $\varphi : S \to M$ and $\psi : S \to TM$ be Cauchy data for u. Suppose that the $2(N+1)$ functions $\Phi = i(\varphi)$ and $\Psi = i'\psi$ are respectively in H_s and H_{s-1}, with $s > (1/2)n + 1$, then there exists a number $T > 0$ and a wave map $u : (S \times [0, T]) \to (M, h)$ taking the given data. The interval T of existence for any given s is equal to the interval corresponding to s_0, smallest integer greater than $(1/2)n + 1$. The solution is unique and depends continuously on the data.*

4. Wave maps outside a Schwarzschild black hole

We consider as source the manifold $(V, \mathbf{g})$ with $V \equiv \Omega \times R$, Ω the exterior $2m$ of a ball in R^3 and **g** the spherically symmetric Schwarzhild metric:

$$\mathbf{g} \equiv -\left(1 - \frac{2m}{r}\right) dt^2 + \left(1 - \frac{2m}{r}\right)^{-1} dr^2 + r^2 \left(d\theta^2 + \sin^2\theta d\varphi^2\right).$$

This manifold is globally hyperbolic, but not regularly hyperbolic in the slicing defined by the t coordinate owing to the behaviour of the metric at the horizon $r = 2m$.

We assume that the target manifold (M, h) is regularly embedded in a pseudo-Euclidean space (R^{N+1}, Q). We denote by i the embedding. If u is a spherically symmetric wave map from $(V, \mathbf{g})$ into (M, h) then the set of scalar functions $U \equiv i \circ u \equiv (U^A)$, $A = 1, \ldots, N+1$ depends only on t and r. They satisfy a system of quasi-diagonal semilinear second order hyperbolic equations on V which reads:

$$-\partial_t\left\{\left(1 - \frac{2m}{r}\right)^{-1}\partial_t U^A\right\} + r^{-2}\partial_r\left\{r^2\left(1 - \frac{2m}{r}\right)\partial_r U^A\right\}$$

$$+ Q^{AD}v_{(p)D}\partial_B v_C^{(p)}\left\{-\left(1 - \frac{2m}{r}\right)^{-1}\partial_t U^B\partial_t U^C\right.$$

$$\left. + \left(1 - \frac{2m}{r}\right)\partial_r U^B\partial_r U^C\right\} = 0.$$

To eliminate the singularity at the horizon we introduce on V the Regge–Wheeler coordinate:

$$\rho \equiv r + 2m \log(r - 2m),$$

hence

$$\frac{dr}{d\rho} = 1 - 2mr^{-1}.$$

With this new coordinate the Schwarzchild metric reads:

$$\mathbf{g} = \left(1 - \frac{2m}{r}\right)\left(-dt^2 + d\rho^2\right) + r^2\left(d\theta^2 + \sin^2\theta \, d\varphi^2\right),$$

with $-\infty < \rho < +\infty$, r is now a C^∞ function of ρ, increasing from $2m$ to $+\infty$.

A straightforward calculation shows that the equation satisfied by U reads in these coordinates, after multiplication by $(1 - 2mr^{-1})$:

$$-\frac{\partial^2 U^A}{\partial t^2} + \frac{\partial^2 U^A}{\partial \rho^2} + \frac{2}{r}\left(1 - \frac{2m}{r}\right)\frac{\partial U^A}{\partial \rho}$$
$$+ Q^{AD} v_{(p)D} \partial_B v_C^{(p)}\left(-\frac{\partial U^B}{\partial t}\frac{\partial U^C}{\partial t} + \frac{\partial U^B}{\partial \rho}\frac{\partial U^C}{\partial \rho}\right) = 0. \tag{4.1}$$

We see on this formula that $U \equiv (U^A)$ satisfies a system of quasi-diagonal semilinear second order hyperbolic equations on a regularly hyperbolic manifold, the Minkowski space–time (R^2, η), with smooth coefficients (recall that we have $r > 2m$ when $-\infty < \rho < \infty$ and that the v's are smooth functions of U by hypothesis).

It results from the Leray theory that this system admits a solution, local in time, for Cauchy data $\rho \mapsto (\Phi(\rho), \Psi(\rho))$ if $\Phi \in C_b^0$; $\partial_\rho \Phi$ and $\Psi \in H_1$. By the theorem of the previous section and the Leray uniqueness theorem this solution satisfies the constraints $F^{(p)} = 0$, hence defines a wave map $u : (V, \mathbf{g}) \to (M, h)$, if the initial data satisfy these constraints and their differentiated equations.

Remark. The equation that we have written for $U \equiv i \circ u$ in the t, ρ coordinates is the equation satisfied by a mapping $U = i \circ u$ with u a wave map from (R^2, η) into (M, h), with an added linear term $\ell_\rho \partial U/\partial \rho$ where

$$\ell_\rho \equiv 2mr^{-1}\left(1 - 2mr^{-1}\right).$$

We will call it the quasi wave map equation.

We have prove in Choquet–Bruhat 1987 the global existence of u for properly Riemannian targets by establishing *a priori* bounds of the H_1 norms of the 't-Cauchy data' $U(\cdot, t)$ and $\partial_t U(\cdot, t)$ through energy estimates using the fact that U satisfies the constraints. We will show how this result extends to some classes of Lorentzian targets: we will suppose from now on that (M, h) is regularly embedded into the Minkowski space (R^{N+1}, η_{N+1}) that is the metric Q of the embedding space is

$$Q \equiv \eta_{N+1} \equiv \eta_{AB} dX^A dX^B \equiv - \left(dX^0 \right)^2 + \delta_{IJ} dX^I dX^J,$$

$$\text{with } I, J = 1, \ldots, N.$$

5. First energy estimate

When the metric of the target is not properly Riemannian we cannot use it directly to define a stress energy tensor with adequate properties for the mapping U. We will define the stress energy tensor of $U : (R^2, \eta) \to (R^{N+1}, Q)$ as the covariant 2-tensor on R^2 given by, with δ the Euclidean metric on R^{N+1},

$$T_{\alpha\beta} \equiv \partial_\alpha U \cdot \partial_\beta U - \frac{1}{2} \eta_{\alpha\beta} \partial_\lambda U \cdot \partial^\lambda U$$

$$\equiv \delta_{AB} \left(\partial_\alpha U^A \partial_\beta U^B - \frac{1}{2} \eta_{\alpha\beta} \partial_\lambda U^A \partial^\lambda U^B \right).$$

Greek indices are raised with the metric η of R^2, a dot denotes the scalar product in the Euclidean metric δ of R^{N+1}.

We will use the following definition relating the Lorentzian structures of (M, h) and (R^{N+1}, η_{N+1}).

Definition. A Lorentzian manifold (M, h) regularly embedded in a Minkowski space (R^{N+1}, η) is said to be *properly embedded* if the equations $F^{(p)}(U) = 0$ which define the embedding depend only on the space components of U in R^{N+1}, i.e., $F^{(p)}(U) \equiv F^{(p)}(U^I)$.

Lemma 5.1. *Let (M, h) be a Lorentzian manifold regularly and properly embedded in a Minkowski space. Then if U satisfies the quasi wave map equation and takes its values in M the divergence of its stress energy tensor is given by:*

$$\partial_\alpha T_\lambda^\alpha = -\partial_\lambda U \cdot \ell_\rho \partial_\rho U \equiv -\delta_{AB} \partial_\lambda U^A \ell_\rho \partial_\rho U^B.$$

Proof. A straightforward calculation gives:

$$\partial_\alpha T_\lambda^\alpha \equiv \delta_{AE} \partial_\lambda U^E \eta^{\alpha\beta} \partial_\alpha \partial_\beta U^A$$

hence (to simplify the writing we suppose there is only one equation to define M as a submanifold of R^{N+1})

$$\partial_\alpha T_\lambda^\alpha = -\delta_{AE} \partial_\lambda U^E \left\{ \ell_\rho \partial_\rho U^A + \eta^{AD} \nu_D \partial_B \nu_C \left(-\frac{\partial U^B}{\partial t} \frac{\partial U^C}{\partial t} + \frac{\partial U^B}{\partial \rho} \frac{\partial U^C}{\partial \rho} \right) \right\}.$$

If the set of unknowns U^A satisfy the equation $F(U) = 0$ it holds also:

$$\partial_\lambda F(U) \equiv \partial_\lambda U^A v_A = 0, \qquad \text{since} \qquad v_A = \partial_{U^A} F.$$

If F depends only on the U^I then

$$v_0 = 0,$$

therefore

$$\eta^{AD} v_D = v_A$$

and

$$\delta_{AE} \partial_\lambda U^E \eta^{AD} v_D = \partial_\lambda U^A v_A = 0. \qquad \square$$

Definition. The energy–momentum vector of U with respect to a vector X on (R^2, η) is:

$$\mathcal{P}^\alpha \equiv T^\alpha_\beta X^\beta.$$

We choose for X the time-like Killing field of (R^2, η) with components: $X_1 = 0$, and $X_0 = 1$. The corresponding energy–momentum vector is time-like or null and future directed. Its divergence is given by

$$\partial_\alpha \mathcal{P}^\alpha = \ell_\rho \frac{\partial U}{\partial \rho} \cdot \frac{\partial U}{\partial t}.$$

Definition. 1) The quantity $\mathcal{P}^\alpha X_\alpha = \mathcal{P}^0$ is by definition the energy density of U at time t with respect to the chosen X. We have:

$$\mathcal{P}^0 \equiv \frac{1}{2} \delta_{AB} \left\{ \partial_\rho U^A \partial_\rho U^B + \partial_t U^A \partial_t U^B \right\} \geq 0.$$

2) The energy $E_t(U)$ of U at time t is the integral of the energy density

$$E_t(U) \equiv \frac{1}{2} \int_{-\infty}^\infty \left\{ \partial_t U \cdot \partial_t U + \partial_\rho U \cdot \partial_\rho U \right\} \partial \rho.$$

Theorem 5.2. *The energy at time t is bounded as follows by the energy at time 0:*

$$E_t(U) \leq E_0(U) \exp\left(\frac{t}{4m}\right).$$

This bound gives a bound of the L^2 norms of $\partial_\rho U$ and $\partial_t U$ at time t.

Proof. The integration of the divergence equation for $\mathcal{P}$ gives for a mapping U with square integrable differential the energy equality:

$$
\begin{aligned}
E_t(U) &\equiv \frac{1}{2} \int_{-\infty}^{\infty} \{\partial_t U \cdot \partial_t U + \partial_\rho U \cdot \partial_\rho U\} \, \partial\rho \\
&= E_0(u) + \int_0^t \int_{-\infty}^{\infty} \partial_\tau U \cdot \ell_\rho \partial_\rho U d\rho \, d\tau,
\end{aligned}
$$

which equality implies, since $0 < \ell_\rho \equiv 2r^{-1}(1 - 2mr^{-1}) \leq 1/4m$ the inequality:

$$
E_t(U) \leq E_0(U) + \frac{1}{4m} \int_0^t E_\tau(U) \, d\tau.
$$

We deduce from the Gromwall lemma that $E_t(U)$ is bounded for all finite t by a solution of the corresponding equality, hence the result of the lemma. $\qquad\square$

6. Second energy estimates

To estimate the H_1 norms we set $U' \equiv \partial_\rho U$ and define for U' quantities analogous to the ones introduced for U, that is with X the Killing vector used for the first energy:

$$
\mathcal{P}'^\alpha \equiv T'^{\alpha\beta} X_\beta
$$

and

$$
T'_{\alpha\beta} \equiv \partial_\alpha U' \cdot \partial_\beta U' - \frac{1}{2} \eta_{\alpha\beta} \partial_\lambda U' \cdot \partial^\lambda U'.
$$

Lemma 6.1. *The divergence of the energy–momentum vector, $J \equiv \partial_\alpha \mathcal{P}'^\alpha$, is of the form:*

$$
J = \{\partial_t U' \cdot \{\ell_\rho \partial_\rho U' + \partial_\rho \ell_\rho U'\} + \mathcal{M}(\partial U', \partial U)\},
$$

where $\mathcal{M}$ is linear in $\partial U'$ and cubic in ∂U, with coefficients which are uniformly bounded if it is so of the second fundamental form K of M as submanifold of (R^{N+1}, η_{N+1}) and the value on M of the derivative of K: we will say in this case that (M, h) is C^2 uniformly embedded in (R^{N+1}, η_{N+1}).

Proof. We have by elementary calculus

$$
\partial_\alpha \mathcal{P}'^\alpha \equiv J = \partial_t U' \cdot \partial^\alpha \partial_\alpha U' \equiv \delta_{AE} \partial_t U'^E \partial^\alpha \partial_\alpha U'^A.
$$

We deduce the equation satisfied by U' by derivation with respect to ρ of the quasi wave map equation for U. We find

$$
\partial^\alpha \partial_\alpha U'^A \equiv -\frac{\partial^2 U'^A}{\partial t^2} + \frac{\partial^2 U'^A}{\partial \rho^2} = -f^A,
$$

with

$$f^A \equiv \ell_\rho \frac{\partial U'^A}{\partial \rho} + \partial_\rho \ell_\rho U'^A + 2\eta^{AD} v_D \partial_B v_C \left(-\frac{\partial U'^B}{\partial t}\frac{\partial U^C}{\partial t} + \frac{\partial U'^B}{\partial \rho}\frac{\partial U^C}{\partial \rho} \right)$$

$$+ \eta^{AD} \left\{ \partial_E v_D \partial_B v_C + v_D \partial^2_{EB} v_C^{(I)} \right\}$$

$$\times \left(-\frac{\partial U^B}{\partial t}\frac{\partial U^C}{\partial t} + \frac{\partial U^B}{\partial \rho}\frac{\partial U^C}{\partial \rho} \right)\frac{\partial U^E}{\partial \rho} = 0$$

we have

$$\partial_\rho \ell_\rho \equiv 2r^{-2}\left(1 - 2mr^{-1}\right)\left(-1 + 4mr^{-1}\right),$$

hence the function $\partial_\rho \ell_\rho$ is uniformly bounded on $r > 2m > 0$ by

$$\partial_\rho \ell_\rho \leq m^{-2}C, \quad C = \left(9 + \sqrt{17}\right)^3 \left(3 + \sqrt{17}\right) 2^{-9}.$$

To estimate the non-linear terms we proceed as follows.

If U satisfies the constraints $F = 0$ then we have by deriving it with respect to t and ρ:

$$\partial_\rho \partial_t F(U) \equiv \partial_t U'^A v_A + \partial_t U^A \partial_\rho U^B \partial_B v_A = 0.$$

If F depends only on the U^I then $v_0 = 0$ and

$$\delta_{AE}\partial_t U'^E \eta^{AD} v_D = \partial_t U'^A v_A = -\partial_t U^A \partial_\rho U^D \partial_D v_A.$$

The result follows because the hypothesis on the second fundamental form K of M embedded in (R^{N+1}, Q) is equivalent to the uniform boundedness on M of $\partial_A v_B$ and $\partial^2_{AB} v_D$. $\qquad\square$

Theorem 6.2. *Under the hypothesis of the above lemma the L^2 norms on $R \times \{t\}$ of $\partial_\rho U'$ and $\partial_t U'$ are continuous for $t \in R$ and bounded for any finite t.*

Proof. We have

$$\mathscr{P}'^0 \equiv \frac{1}{2}\left\{ \partial_t U' \cdot \partial_t U' + \partial_\rho U' \cdot \partial_\rho U' \right\}.$$

We set:

$$E'_t(U) = \int_{-\infty}^{\infty} \mathscr{P}'^0 d\rho \equiv \frac{1}{2}\Sigma_{A=1,\dots,N+1} \left\{ \left\| \partial_\rho U'^A \right\|^2 + \left\| \partial_t U'^A \right\|^2 \right\}.$$

The integration of the divergence of $\mathcal{P}'$ leads to an inequality of the form:

$$E'_t(U) \le E'_0(U) + C \int_0^t \left\{ \left[\int_{-\infty}^\infty \left(|\partial_t U|^6 + |\partial_\rho U|^6 \, d\rho \right)^{1/2} \right. \right.$$

$$\times \left\{ E'_\tau(U) \right\}^{1/2} d\tau + \frac{1}{4} m^{-1} \int_0^t E_t(U) d\tau$$

$$\left. + C m^{-2} \int_0^t \left\{ E_t(U) \right\}^{1/2} \left\{ E'_t(U) \right\}^{1/2} d\tau \right\}.$$

We estimate the terms in $|\partial U|^6$ by using Sobolev inequalities on R. We set, with ∂U denoting some $\partial_t U^A$ or $\partial_\rho U^A$:

$$f \equiv |\partial U|^2,$$

then

$$\|\partial U\|^2_{L^6} \equiv \|f\|_{L^3}$$

whilst

$$\partial_\rho f = 2 \partial U \partial_\rho \partial U.$$

Hence by the Cauchy–Schwarz inequality:

$$\|\partial_\rho f\|_{L^1} \le 2 \|\partial U\|_{L^2} \|\partial_\rho \partial U\|_{L^2}.$$

This inequality implies *a fortiori*:

$$\|\partial_\rho f\|_{L^1} \le E_t(U)^{1/2} E'_t(U)^{1/2}.$$

By the Sobolev embedding theorem on R there exists a constant C such that:

$$\|f\|_{C^0_b} \le C \left\{ \|f\|_{L^1} + \|\partial_\rho f\|_{L^1} \right\}$$

hence:

$$\left\{ \int_{-\infty}^\infty |\partial U|^6 d\rho \right\}^{1/2} = \|f\|_{L^3}^{3/2} \le \|f\|_{C^0_b} \|f\|_{L^1}^{1/2}$$

$$\le C E_t(U) \left\{ E_t(U)^{1/2} + E'_t(U)^{1/2} \right\}.$$

We deduce from this inequality that $E'_t(U)$ satisfies a linear integral inequality whose coefficients are continuous functions of $t \in R$, bounded for all finite t. Therefore $E'_t(U)$ enjoys the same continuity and boundedness properties. We have proved the following theorem, uniqueness and domain of dependence properties are standard for hyperbolic equations.   □

Theorem 6.3. *Let (M, h) be a Lorentzian manifold C^2 regularly and properly embedded into a Minkowski space. If the Cauchy data Φ is continuous and takes its values in M while $\partial_\rho \Phi$ and $\Psi \in H_1$, with Ψ taking its values in $T_\Phi M$, then there exists a global spherically symmetric wave map from the exterior of a Schwarzchild black hole into (M, h) taking these Cauchy data. The solution is unique. The solution in a compact set depends only on the data in the intersection of the initial line with the past of this set.*

Remark. The condition imposed on Φ and Ψ in the coordinate ρ is equivalent to the condition on these data in the coordinate r given by $\partial_r \Phi \in H^{Sch}_{1, 1/2}$, $\Psi \in H^{Sch}_{1, -1/2}$ with the definition of the weighted Sobolev spaces $H^{Sch}_{1, \delta}$ on $(2m, \infty)$:

$$\| f \|_{H^{Sch}_{1;\delta}} = \left\{ \int_{2m}^{\infty} \left[\left(1 - 2mr^{-1} \right)^{2\delta} |f|^2 + \left(1 - 2mr^{-1} \right)^{2\delta+2} |\partial_r f|^2 \right] dr \right\}^{1/2}.$$

Corollary. The global existence theorem still holds if $\partial_\rho \Phi$ and Ψ are only supposed to be in H_1 in any bounded interval of R.

Proof. It is standard for hyperbolic equations when global existence is not restricted to small data. It goes as follows. Consider at an arbitrary time T some bounded interval ω_T of R. The past of its compact closure intersects the initial line $t = 0$ along a compact set ω_0. There exist initial data which are in H_1 on this line and coincide with the given data on ω_0. To these initial data correspond a global wave map. The covering of the line $t = T$ by bounded intervals, and the uniqueness theorem which permits the matching of the obtained wave maps proves the corollary. $\qquad\Box$

Acknowledgment

I thank J. Isenberg for calling my attention to the interest of Lorentzian targets and for discussions on the subject.

References

[1] Leray, J., *Hyperbolic differential equations*, I.A.S. Princeton 1952.

[2] Dionne, P., Problèmes hyperboliques bien posés, *Journal D'Analyse Mathematique* n° (1962), 1–56.

[3] Christodoulou, D. and Francaviglia, M., Cauchy data on a manifold, *Ann. Inst. Poincaré XXXI* n° **4** (1979), 399–414.

[4] Chaohao, Gu., On the Cauchy problem for harmonic maps on two dimensional Minkowski space, *Comm. Pure and App; Maths.* **33** (1980), 727–737.

[5] Ginibre, J. and Velo, G., The Cauchy problem for the O(N), CP(N-1) and GC(N,p) models, *Ann. of Phys.(2)* **142** (1982), 393–415.

[6] Choquet–Bruhat, Y., Hyperbolic harmonic maps, *Ann Inst. Poincaré 46* n° **1** (1987), 97–111.

[7] Shatah, J., Weak solutions and development of singularities in the SU(2) σ-model *Comm. Pure Appl. Math* **41** (1988), 459–469.

[8] Choquet–Bruhat, Y. and DeWitt–Morette, C., *Analysis, manifolds and physics II*, North Holland 1989.

[9] Christodoulou, D. and Talvilar–Zadeh, A., On the regularity of spherically symmetric wave maps, *Comm. Pure Ap. Math.* **46** (1993), 1041–1091.

[10] Nicolas, J.P., Non-linear Klein Gordon equations on Schwarzchild like metrics, *J. Math. Pures App.* **74** (1995), 35–58.

[11] Choquet–Bruhat, Y., *Wave maps on the outside of a Schwarzchild black hole*, Proceedings of Mogran 7, N. Ibragimov (ed.), publications of the University of Trondheim.

ENTANGLEMENT, PARATAXY, AND COSMOLOGY

ERNST BINZ AND WALTER SCHEMPP

Contents

1 Introduction.. 486
2 The crystalline micro system 491
3 The planetary macro system 523
4 The projective unification 534
5 Concluding remarks.. 537

M. de Gosson (ed.), Jean Leray '99 Conference Proceedings, 483-542.
© *2003 Kluwer Academic Publishers.*

Ich habe hundertmal so viel über Quantenprobleme nachgedacht wie über die allgemeine Relativitätstheorie. Das Erstaunlichste an der Welt aber ist, daß man sie verstehen kann.
Albert Einstein

Es erscheint hart, dem Herrgott in die Karten zu gucken. Aber daß er würfelt und sich telepathischer Mittel bedient wie es ihm von der gegenwärtigen Quantentheorie zugemutet wird, kann ich keinen Augenblick glauben.
Albert Einstein

Eine relativ zu einem Bezugssystem mit der Geschwindigkeit v gleichförmig bewegte Uhr geht von diesem Bezugssystem beurteilt im Verhältnis $1 : \sqrt{1-(v^2/c^2)}$ langsamer als die nämliche Uhr, falls sie relativ zu jenem Bezugssystem ruht.
Albert Einstein

Matter tells space how to curve, and space tells matter how to move.
John A. Wheeler

Sommerfeld *waxed* Kepplerian. As Keppler *had solved the problems of planetary motion by speaking the "language of the spheres," one could solve the problems of quantum spectroscopy by learning to speak the "language of the spectra." This language, wrote* Sommerfeld, *is "an atomic music of the spheres, a harmonizing of whole number relationships."*
David C. Cassidy

Heisenberg *sagte gern: "Die Mathematik ist klüger als wir." Er meinte, sie enthält und enthüllt bei richtigem Gebrauch Strukturen, die zu erfassen unser Anschauungsvermögen noch zu schwach ausgebildet war.*
Carl Friedrich von Weizsäcker

Theorien kommen zustande durch ein vom empirischen Material inspiriertes Verstehen, welches am besten im Anschluß an Plato *als zur Deckung kommen von inneren Bildern mit äußeren Objekten und ihrem Verhalten zu deuten ist. Die Möglichkeit des Verstehens zeigt aufs Neue das Vorhandensein regulierender typischer Anordnungen, denen sowohl das Innen wie das Außen des Menschen unterworfen ist.*
Wolfgang Pauli

The introduction of Planck's *constant h into mathematics*!
Jean Leray

To every rotation in the group $\mathbf{O}_3^+(\mathbb{R})$ there corresponds two *opposite quaternions of norm 1. There is no "reasonable" way of "selecting" one of these*

two quaternions for $O_3^+(\mathbb{R})$ as a whole. Nevertheless the quaternions and the rotation are readily linked "geometrically". The claim that four-dimensional spaces are quite exceptional is no idle talk—at least from the point of view of the Euclidean rotation group $O_3^+(\mathbb{R})$. The greatest difference to the case of an indefinite quadratic form is the existence of isotropic vectors. Classically, this name is only used in the geometry over the complex number field $\mathbb{C}$, a geometry, which all throughout the 19th century and even in many more recent books, has been so hopelessly and ridiculously muddled up with the geometry over $\mathbb{R}$ that it is practically impossible to say, at any given moment, what space one is working in.
Jean A. Dieudonné

Quantum mechanics is a bizarre theory, invented to explain atoms. As far as we know today it is capable of explaining everything about ordinary matter (chemistry, biology, superconductivity), sometimes with stunning numerical accuracy. But it also says something about the occurrence of the most spectacular event in the cosmos—the supernova. The range is 57 orders of magnitude!
Elliott H. Lieb

Nur Beharrung führt zum Ziel, nur die Fülle führt zur Klarheit, und im Abgrund wohnt die Wahrheit.
Niels Bohr's favourite Schiller aphorism

1. Introduction

The physicist-philosopher Michael Polanyi emphasized the fact that no other area of mathematical physics has changed our sense of intellectual satisfaction to such an extent as the discovery of quantum theory ([37]). Although basically anti-intuitive, "quantum reality" captures the imagination like relativity theory *and* cosmology. It is remarkable, however, that most discussions of the conceptual, as well as philosophical, issues of quantum physics are still based on the very first attempts to mathematically formalize this theory which has been worked out more than 60 years ago ([51]). These restrictions of the mathematical armamentarium have considerably slowed down the development of quantum information processing and quantum cosmology. It is even more remarkable that the Lorentz invariant *projectivization* $P_2(\mathbb{C})$ of the *complexification* of the coadjoint orbit picture of the Heisenberg nilpotent Lie group G and the internal antisymmetries of the Hopf bundle model representing the *spin* structure on the tangent bundle TS_2 of the unit sphere $S_2 \hookrightarrow \mathbb{R}^3$, as well as the Heisenberg nilmanifold have not been applied to the central cognitive problems of non-local quantum reality such as quantum teleportation to a remote location in the universe via EPR information transmission channels, and clinical magnetic resonance imaging (MRI) of medical diagnosis, although the *pictorial* turn forming a temporal successor of the paradigms of the linguistic turn, has been *projectively* implemented by the *semantics* of computer vision. A serious application of the Heisenberg group G and the projectivization $\mathbf{P}_2(\mathbb{C})$ of its complexified coadjoint orbit picture to quantum theory would have obviated any sincere attempt to construct an artificial conflict between the non-locality of quantum realism on the one hand, and special relativity on the other hand. Actually a conflict such as the relativistic EPR channel dilemma ([11] and [19]) proposed by the advocates of an epistemologically idealistic view on quantum effects never existed. Harmonic analysis of the Heisenberg nilpotent Lie group G and its symmetries provide a scenario for quantum information processing that is compatible with quantum physics as well as relativity theory. "Scritto in lingua matematica" (Galileo Galilei) it does not leave room for contradictions to the principle of causality, paradoxes, and mysteries, but opens the window to the non-locality of quantum information processing and the Heisenberg Lie group approach to relativity and cosmology.

Because quantum theory does *not* directly deal with relativity, in quantum field theory there is *no* concept of photon trajectory. The distance independent concept of internal antisymmetry in circle bundle models of spin-drift fields provides the fundament of the spinorial *spectral* approach to the non-locality phenomenon of quantum information processing, as well as the third Kepplerian fundamental law of planetary motion which allows to conceive mass as a spectral object ([44]). Thus quantum theory is not restricted to microphysical systems. Due to its inherent Lorentz invariance, the algebraic *projective* approach to quantum theory is concerned with *relativistic* large scale macro systems as well. It is the projective aspect of the coadjoint orbit picture of G *and* the Kaluza-Klein model of the Minkowskian space-time manifold which permits the unification of both aspects in the spirit of the de Broglie spectral unveiling procedure of relativistic *matter* modes, and quantum information processing. Specifically, the communication channelling aspects of quantum cosmology are exhibited by the methods of algebraic projective geometry.

Concerning the *neurofunctional* microsystems, it should be observed that the diagnostic part of medicine has gone through a revolution in the last few decades, primarily due to the improvements in computer technology. The computationally supported diagnostic facilities have culminated in clinical MRI. Few techniques involving sophisticated instrumentation have made so rapid an impact on medical diagnosis as has clinical MRI. The MRI modality has replaced very invasive and less diagnostic methods such as pneumoencephalography, myelography, and nuclear medicine brain scans. MRI is the premier clinical method for the diagnosis of disease in the central nervous system, because of its speed, noninvasiveness, and capability to create tomograms with exquisite contrast

([3], [16], [26], [53] and [56]). Presently clinical MRI scanners are ubiquitous and able to cover the full range of all organs of the human body ([14]). The step from anatomic to neurofunctional MRI (fMRI) is small, as the magnet itself is identical for both purposes, and the remaining hardware usually requires merely an upgrade, if that. So structural details of the brain can be reconstructed non-invasively, even as three-dimensional images, and small, task-related changes of the cerebral blood flow can be displayed, even in the deepest recesses of the brain.

One of the intriguing insights which emerged from the decade of the brain is that synchronous neuronal firing plays a crucial role in brain function. Neurobiological research showed that the synchronous firing can define assemblies of cortical cells, the synchrony providing the resonance glue that *semantically* binds distributed neuronal activity into *coherent* presentations of objects and events ([50]). By including the time-line as an additional dimension in which patterns of neuronal activity are *spatio-temporally* organized within the glia, spatio-temporal coding helps to solve a series of fundamental organizational problems in understanding how the brain works, for instance, how figure-ground segregation and scene segmentation are achieved. In the visual system, which forms the pictorial gate, temporal coding permits the integration of the responses of excited neurons that are widely distributed over the cortex without requiring the anatomical convergence of information transmission channels onto a single target region. The temporal code allows multiple presentations to coexist in the same cortical region. The spatio-temporal excitation patterns corresponding to each presentation remain distinguishable as long as they have a fixed phase relationship to each other.

Beyond the neuroimaging modality fMRI, a particular important application of quantum holography resides in the imaging of brain dynamics in terms of internal spatio-temporal signal patterns. The fact that the firing rate of neurons in peripheral structures of the central nervous system reflects the intensity of sensory stimuli introduced the concept of frame-rate encoding. The evidence of a close correlation between the location of a neuron within the brain and its functional properties led to the concept of position coding. The message conveyed by a neuron was thought to be defined entirely by the amplitude of the neuronal response and its provenance. As a consequence, in single unit recordings, the temporal state received relatively little attention as a dimension for the encoding process. Temporal codes were either ignored or undiscovered with the commonly applied methods of single unit analysis. Recently, however, a dramatic change in attitude and interest occurred. Attention has shifted towards *time* as an encoding variable of the functional organization of the brain. Phase relations between the responses of distributed neuron assemblies are recognized as important a code variable as relations between the response amplitudes. The oscillatory activity exhibits to be the result of dynamic interactions within the cortical network itself ([17], [31] and [44]). The neuroimaging fMRI modality represents one of the roots of the renaissance of interest in the spatio-temporal excitation aspects of nervous activity.

Quantum holography allows for neuroimaging the cooperativity of *coherently* organized assemblies of excitatory and inhibitory neurons in the living brain. The point to note is that the neuronal network cooperativity is not just one of the many concepts within the vast area of brain research. It actually represents the missing link between the peripheral events at the cellular level and the behaviour of the entire organism. These peripheral events in the environment are encoded by the spatio-temporal excitation patterns of higher brain neuro-functional networks. The planar coadjoint orbits of G form the *synchronized* hypostasis for the organization of the spatio-temporal signal patterns.

Equally important as these purely practical and cognitive aspects is the unique set of physical quantities that can be measured by the MRI modality. Substantial efforts have been made to increase the temporal resolution in clinical MRI. In this process, the notion of real-time imaging has emerged, denoting the continuous database acquisition of entire images at a frame-rate sufficiently high to resolve dynamic processes such as neuronal activity, the motion of tissues in cardiac imaging or interventional devices. Continuous heart imaging with high frame-rates could provide unprecedented insight into cardiac dynamics. Owing to its highly accurate diagnostic ability combined with its minimal risk and non-invasive nature, the use of MRI has skyrocketed ([3], [16], [40], [44], [53] and [56]). Like synthetic aperture radar (SAR) image formation and cooperatively organized neuronal networks, the semantics of MRI is based on the spatio-temporal principles of spectral coding and *not* on the deformation of wavefront sets. Laguerre's elegant projective *phase* formula in terms of the orthogonality involution defined by the zero-dimensional *absolute* projective quadric

$$\Phi_\infty^\circ \hookrightarrow \mathbf{P}_2(\mathbb{C})$$

emphasizes the importance of algebraic projective geometry over the *extended* ground field $\mathbb{C}$ for quantum information processing, magnetic resonance tomography, and relativity theory. Combined with the spin structure on the tangent bundle $\mathbf{TS}_2$, the Laguerre projective phase formula is compatible with the transactional interpretation of the concept of quantum physical non-locality ([11]).

As firstly discovered by Jean Victor Poncelet, the projective quadric

$$\Phi_\infty^\circ = \{\bar{i}, i\}$$

consists of the conjugate *cyclic* points. They are joined by the line at infinity of the complex projective plane $\mathbf{P}_2(\mathbb{C})$ under its standard Kähler metric. The role they are playing for the non-locality of quantum information processing ([11]) and the large scale geometry of the Minkowskian space-time manifold indicates towards applications to cosmology ([21]).

The conjugate cyclic points are $\{\bar{i}, i\}$ common to all circles of the symplectic plane $\mathbb{R} \oplus \mathbb{R}$ having the compact Kähler manifold $\mathbf{P}_2(\mathbb{C})$ as the projectivization of its complexification. Because they give rise to the *isotropic* lines of any pencil of lines in the complex projective plane $\mathbf{P}_2(\mathbb{C})$, they are capable to serve as a *phase reference*. The compact unit sphere $\mathbf{S}_5 \hookrightarrow \mathbb{R}^6$ forms a circle bundle over $\mathbf{P}_2(\mathbb{C})$ and, by passing to the quotient, cuts down to the fundamental four-dimensional central Plücker quadric $\Psi \hookrightarrow \mathbf{P}_5(\mathbb{R})$ of the Kaluza-Klein model of the Minkowskian space-time manifold. Due to the line generators of $\mathbf{S}_2 = \mathbf{P}_1(\mathbb{C})$ there exists a Segre embedding of the manifold of Kepplerian orbits into the four-dimensional Plücker quadric

$$\mathbf{S}_2 \times \mathbf{S}_2 \hookrightarrow \Psi \hookrightarrow \mathbf{P}_5(\mathbb{R}).$$

Switching from large scale geometry to the geometric control of spin dynamics, it should be noticed that MRI forms the most valuable application of the projectivization $\mathbf{P}_2(\mathbb{C})$ of the complexification of the coadjoint orbit picture of the Heisenberg Lie group G which itself forms, from the non-holonomic differential geometric and control theoretic point of view, a *paradigm* of sub-Riemannian geometry ([45]). The sub-Riemannian metric is defined as the Legendre transform of a semi-elliptic symbol. Due to an application of the Kustaanheimo-Stiefel transformation of celestial mechanics and insertion of the correct gravitational potential, the sub-Riemannian metric gives rise to the Schwarzschild metric of cosmology in the Kruskal coordinatized two-fold covered Minkowskian space-time manifold ([21]) of real Clifford algebra $\mathcal{C}\ell(3, 1, \mathbb{R})$, and the information transmission channels implemented by the Kruskal diagram in which the radial light rays everywhere have the slope ∓ 1.

In the roughly 15-year history of clinical MRI, the number of applications of MRI, and the progress made in discovering and implementing new means of geometrically controlling image contrast based on a wide range of chemical and physiological bases, has been nothing short of astounding. With open-access magnet designs now available ([40]), the application of MRI to oncology say, will rapidly expand to include numerous novel interventional procedures. Undoubtedly, this rate of development does not show any signs of slowing in the near future.

As another spectral encoding based information processing system representing the pictorial turn, SAR remote sensing is an airborne or spaceborne echo-mode array imaging system ([34]). Similar to other array imaging systems such as clinical MRI, the SAR imaging modality acquires a multi-dimensional database that can be manipulated via signal processing algorithms to acquire an image that carries information on the remote target under study. The synchronized hypostasis of the database which has been coherently acquired of remote targets on a terrain or planet is formed by spectral variation of the radar frequency and spatio-temporal signal pattern in the target area. High resolution

SAR remote sensing offers one of the prime examples of the role of digital image formation which is heavily dependent on the computing power available for rapidly processing discrete Fourier analysis. The capability of the SAR imaging modality to holographically exploit signal phase has given rise to completely new tools for glaciology and the study of tectonic activity.

A significant development in computing has been the discovery that the computational power of quantum computers exceeds that of Turing machines. Central to the experimental realization of quantum information processing is the discovery of fault-tolerant quantum logic gates which has greatly improved the long-term prospects for quantum computing technology. Their operation requires conditional quantum dynamics, in which one subsystem undergoes a coherent evolution that instantaneously depends on the quantum state of another spatially separated subsystem. In particular, the spin-drift field of the spectral object within the evolving subsystem may acquire a conditional phase shift from a subsystem installed at a remote location. Conditional geometric phases of the spectral object depend only upon the non-local geometry of the photon trajectory relativistically performed, and are therefore resilient to a variety of error types. Because the spin-echo technique of magnetic resonance allow to protect spatio-temporally encoded quantum information from errors that arise owing to decoherence through uncontrolled interaction with the environment ([27]), it suggests the possibility of an intrinsically fault-tolerant implementation of quantum gate operations. This is the quantum theoretical reason why the discovery of the spin-echo phenomenon by Erwin Louis Hahn changed so dramatically the face of nuclear magnetic resonance ([2] and [27]). For the design of quantum computing hardware, magnetic resonance techniques have already been used to demonstrate quantum information processing by using controlled phase shift gates. An alternative is to invoke magnetic Aharonov-Bohm interactions to non-locally process quantum holography within an electron microscope by means of parallel displacements in the fiber bundle of phase shifts in spin-drift fields over the circle $\mathbf{S}_1 \hookrightarrow \mathbb{R} \oplus \mathbb{R}$.

2. The crystalline micro system

A central role in the study of the cotangent bundle $T^{\star}\mathbb{R}^2$ is played by the real Heisenberg Lie group G. The three-dimensional connected and simply connected Heisenberg nilpotent Lie group G forms the non-split *central* group extension

$$\mathbb{R} \lhd G \longrightarrow \mathbb{R} \oplus \mathbb{R}.$$

The concept of central extension is of fundamental importance in the theory of loop groups ([38]). Lebesgue measure of $\mathbb{R}^3$ provides a Haar measure of G.

The standard presentation of G is given by the multiplication law $(g_1, g_2) \rightsquigarrow g_1 g_2$ of unipotent matrices with real entries

$$\begin{pmatrix} 1 & x_1 & z_1 \\ 0 & 1 & y_1 \\ 0 & 0 & 1 \end{pmatrix} \begin{pmatrix} 1 & x_2 & z_2 \\ 0 & 1 & y_2 \\ 0 & 0 & 1 \end{pmatrix} = \begin{pmatrix} 1 & x_1 + x_2 & z_1 + z_2 + x_1 y_2 \\ 0 & 1 & y_1 + y_2 \\ 0 & 0 & 1 \end{pmatrix}.$$

All the eigenvalues of the elements $g \in G$ are equal to 1, the neutral element of G is given by the unit matrix

$$\begin{pmatrix} 1 & 0 & 0 \\ 0 & 1 & 0 \\ 0 & 0 & 1 \end{pmatrix},$$

and the inverse $g^{-1} \in G$ obviously reads

$$\begin{pmatrix} 1 & x & z \\ 0 & 1 & y \\ 0 & 0 & 1 \end{pmatrix}^{-1} = \begin{pmatrix} 1 & -x & -z + xy \\ 0 & 1 & -y \\ 0 & 0 & 1 \end{pmatrix}.$$

The point to note is that, in the standard representation, G forms a linear superposition of *transvections* which allow to adjust the world lines or photon trajectories of the experimental setup to the convex light cone. The moment presentation of G is given by the equation

$$\begin{pmatrix} 1 & x_1 & z_1 \\ 0 & 1 & y_1 \\ 0 & 0 & 1 \end{pmatrix} \begin{pmatrix} 1 & x_2 & z_2 \\ 0 & 1 & y_2 \\ 0 & 0 & 1 \end{pmatrix} = \begin{pmatrix} 1 & x_1 + x_2 & z_1 + z_2 + \dfrac{1}{2} \det \begin{pmatrix} x_1 & x_2 \\ y_1 & y_2 \end{pmatrix} \\ 0 & 1 & y_1 + y_2 \\ 0 & 0 & 1 \end{pmatrix},$$

so that in the moment presentation the inverse simply reads

$$\begin{pmatrix} 1 & x & z \\ 0 & 1 & y \\ 0 & 0 & 1 \end{pmatrix}^{-1} = \begin{pmatrix} 1 & -x & -z \\ 0 & 1 & -y \\ 0 & 0 & 1 \end{pmatrix}.$$

The inclusion of the moment

$$\det \begin{pmatrix} x_1 & x_2 \\ y_1 & y_2 \end{pmatrix}$$

suggests the introduction of homogeneous coordinates, the Plückerian coordinates of the generic world line or photon trajectory of relativity, in the three-dimensional real projective spinor space $\mathbf{P}_3(\mathbb{R})$. In both presentations of G, the one-parameter group of dilation operators $(s.I)_{s \in \mathbb{R}^\times}$ acts non-isotropically on G according to the assignment

$$s.I : \begin{pmatrix} 1 & x & z \\ 0 & 1 & y \\ 0 & 0 & 1 \end{pmatrix} \rightsquigarrow \begin{pmatrix} 1 & sx & s^2z \\ 0 & 1 & sy \\ 0 & 0 & 1 \end{pmatrix} \quad (s \in \mathbb{R}^\times).$$

It will be shown that the infinitesimal version of the dilation operators play an important role in quantum cosmology. The moment presentation reveals that there is a real *symplectic* structure inherent to the Heisenberg Lie group G ([25] and [57]). The natural symplectic structure is an essential ingredient of the coadjoint orbit picture of G which permits to classify the equivalence classes of irreducible unitary linear representations of G and to model a synchronized hypostasis for the spatio-temporal signal patterns of the wavelets of *simultaneity* of quantum information processing by mirrored *pairs* of planar coadjoint orbits of G ([29], [30] and [45]). The associated orthogonality involution is defined by the zero-dimensional absolute projective quadric $\Phi_\infty^\circ \hookrightarrow \mathbf{P}_2(\mathbb{C})$ consisting of the conjugate cyclic points $\{\bar{i}, i\}$ which are joined by the line at infinity of the complex projective plane $\mathbf{P}_2(\mathbb{C})$ under its standard Kähler metric.

The Heisenberg group $G = \mathcal{U}G_0$ represents the universal covering group of the *reduced* Heisenberg Lie group G_0 defined by the central extension

$$\mathbb{T} \lhd G_0 \longrightarrow \mathbb{R} \oplus \mathbb{R}.$$

The group law of G_0 is defined by the circular symmetry action

$$(x_1, y_1, \zeta_1) \cdot (x_2, y_2, \zeta_2) = (x_1 + x_2, y_1 + y_2, e^{2\pi i x_1 y_2} \zeta_1 \zeta_2),$$

so that the inverse admits the expression

$$(x, y, \zeta)^{-1} = (-x, -y, e^{2\pi i xy} \bar{\zeta}).$$

Of course, the overbar denotes complex conjugation. The center $C \hookrightarrow G$ is given by the set of unipotent matrices

$$\begin{pmatrix} 1 & 0 & z \\ 0 & 1 & 0 \\ 0 & 0 & 1 \end{pmatrix} \quad (z \in \mathbb{R}).$$

Clearly C is isomorphic to the real line $\mathbb{R}$ and the quotient G/C is group isomorphic to the real symplectic plane $\mathbb{R} \oplus \mathbb{R}$. The symplectic structure of the plane $\mathbb{R} \oplus \mathbb{R}$ is induced by the orthogonality involution which leaves fixed the isotropic lines of any pencil of lines in the complex projective plane $\mathbf{P}_2(\mathbb{C})$ under its standard Kähler metric. Due to the spin-drift resonance modes supported by

C which implies the *axisymmetry* of G, the center $C \hookrightarrow G$ plays a significant role. It contains the infinite cyclic group

$$\begin{pmatrix} 1 & 0 & m \\ 0 & 1 & 0 \\ 0 & 0 & 1 \end{pmatrix} \quad (m \in \mathbb{Z})$$

as a discrete, hence closed subgroup of G which allows to curl up C in a circle. Performing the quotient group therefore allows to compactify G by *periodization* modulo $\mathbb{Z}$ along the longitudinal direction. Similarly, the center

$$\{(0, 0, \zeta) \mid \zeta \in \mathbb{T}\}$$

of the reduced Heisenberg Lie group G_0 is isomorphic to the one-dimensional compact torus group $\mathbb{T} = \mathbb{R}/\mathbb{Z}$. Its Haar measure is the normalized push-down of Lebesgue measure of the real line $\mathbb{R}$.

The normal subgroup $\mathbf{SL}(2, \mathbb{R}) \hookrightarrow \mathbf{GL}(2, \mathbb{R})$ coincides with the real symplectic group $\mathbf{Sp}(2, \mathbb{R})$ and therefore forms the group of continuous automorphisms of G which keep the center $C \hookrightarrow G$ pointwise fixed. This follows from the classical Stone-von Neumann theorem of quantum physics asserting that, up to unitary isomorphy, there is only one irreducible unitary linear representation ρ_1 of G_0 whose central character is the identity subduced on $\mathbb{T}$. The continuous automorphisms of G keeping the center $C \hookrightarrow G$ pointwise fixed give rise to the metaplectic representation ω of the metaplectic group $\mathbf{Mp}(2, \mathbb{R})$ referred to below. It is the action of ω which reflects the internal *global* symmetries of the Heisenberg Lie group G and recovers the theory of spherical harmonics with extra structure. Denoting by $\{\ \}^\circ$ the neutral connected component of a Lie group, the spin isomorphy

$$\mathbf{SL}(2, \mathbb{R}) = \mathbf{Spin}^\circ(2, 1, \mathbb{R})$$

identifies the metaplectic representation ω as a spin representation.

For $m = n = 0$, the center $\mathbb{T}$ of G_0 forms a subgroup of the Abelian subgroup $L_0 \hookrightarrow G_0$ given by the triples

$$\{(m, n, \zeta) \mid (m, n) \in \mathbb{Z} \oplus \mathbb{Z}, \ \zeta \in \mathbb{T}\}.$$

Actually L_0 is a *maximal* Abelian subgroup of G_0, and the coset space $L_0/\mathbb{T}$ for L_0 is isomorphic to the planar lattice

$$L = \mathbb{Z} \oplus \mathbb{Z}\tau \hookrightarrow \mathbb{C} \quad (\Im \tau > 0).$$

The quotient manifold G_0/L_0 is homeomorphic to a two-dimensional flat torus foliated by the spherically linked Villarceau circles of parataxy in three-dimensional elliptic non-Euclidean geometry. These are stereographic projections of the Clifford parallels of the first and second kind referred to below.

The planar lattice L is represented by the array of Dirac measures

$$\{\varepsilon_{(m+n\tau)} \mid (m, n) \in \mathbb{Z} \oplus \mathbb{Z}\}.$$

The compact homogeneous manifold $G/(L \times \mathbb{R})$ is conformally equivalent to the compact homogeneous manifold $G/(L' \times \mathbb{R})$ if and only if the planar lattices L, L' satisfy the condition

$$L' = a \cdot L \quad (a \in \mathbb{C}^\times),$$

where the subgroup $\mathbb{C}^\times \hookrightarrow \mathbf{GL}(2, \mathbb{R})$ is diffeomorphic to the cylindrical direct product $\mathbf{S}_1 \times \mathbb{R}$.

This justifies to adopt the canonical form of the planar lattices $L = \mathbb{Z} \oplus \mathbb{Z}\tau$ and $L' = \mathbb{Z} \oplus \mathbb{Z}\tau'$. They are conformally equivalent if and only if the discrete elliptic modular subgroup $\mathbf{SL}(2, \mathbb{Z}) \hookrightarrow \mathbf{SL}(2, \mathbb{R})$ generated by the matrices

$$\begin{pmatrix} 1 & 1 \\ 0 & 1 \end{pmatrix}, \quad J = \begin{pmatrix} 0 & -1 \\ 1 & 0 \end{pmatrix}$$

transforms τ in the Poincaré upper half-plane $\Im\tau > 0$ onto τ' contained in the complex open upper half-plane by Möbius linear fractional transformations. The two generators are conjugate in $\mathbf{SL}(2, \mathbb{C})$ but *not* in $\mathbf{SL}(2, \mathbb{R})$ and therefore not in the discrete, hence closed subgroup $\mathbf{SL}(2, \mathbb{Z}) \hookrightarrow \mathbf{SL}(2, \mathbb{R})$. At the Lie algebra level of real forms the isomorphies

$$\mathrm{Lie}\left(\mathbf{SL}(2, \mathbb{R})\right) \otimes \mathbb{C} = \mathrm{Lie}\left(\mathbf{SL}(2, \mathbb{C})\right) = \mathrm{Lie}\left(\mathbf{SU}(2, \mathbb{C})\right) \otimes \mathbb{C}$$

hold.

Clearly the subgroup $\mathbf{SO}(2, \mathbb{R}) \hookrightarrow \mathbf{O}(2, \mathbb{R})$ is isomorphic to the compact torus group $\mathbb{T}$. It forms a maximal compact connected subgroup of $\mathbf{SL}(2, \mathbb{R})$ and has the property that any compact subgroup of $\mathbf{SL}(2, \mathbb{R})$ may be conjugated to become a subgroup of the group of planar rotations $\mathbf{SO}(2, \mathbb{R})$. The action of the compact subgroup $\mathbf{SO}(2, \mathbb{R}) \hookrightarrow \mathbf{SL}(2, \mathbb{R})$ subduced from the projectivization of the usual linear representation of $\mathbf{SL}(2, \mathbb{R})$ on the real symplectic plane $\mathbb{R} \oplus \mathbb{R}$ stabilizes the cyclic point $i \in \mathbb{C}$ so that the open upper half-plane can be identified with the homogeneous manifold

$$\mathbf{SL}(2, \mathbb{R}) / \mathbf{SO}(2, \mathbb{R}).$$

Every orbit of $\mathbf{SL}(2, \mathbb{Z})$ admits one element in the module figure

$$\mathscr{D} = \left\{ \tau \in \mathbb{C} \mid |\tau| \geq 1, |\Re\tau| \leq \frac{1}{2}, \Im\tau > 0 \right\}$$

and two elements of $\mathscr{D}$ belong to the same orbit of the Möbius action of $\mathbf{SL}(2, \mathbb{Z})$ if and only if they lie on the boundary of $\mathscr{D}$. Boundary identification of the module figure $\mathscr{D} \hookrightarrow \mathbb{C}$ establishes that the double coset space

$$\mathbf{SL}(2, \mathbb{Z}) \backslash \mathbf{SL}(2, \mathbb{R}) / \mathbf{SO}(2, \mathbb{R})$$

is homeomorphic to the complex plane $\mathbb{C}$. It is well known that the modular function provides an explicit homeomorphism.

In view of the geometry of light cones or cones of causality, it is natural to let the special linear group $\mathbf{SL}(2, \mathbb{R})$ act on the compact subset $\mathbf{S}_1 \hookrightarrow \mathbb{R}^2$ of truncated rays in the real plane $\mathbb{R}^2$. The open positive half-line $\mathbb{R}_+^\times$ forms an abelian Lie group with respect to the multiplication law $(s_1, s_2) \rightsquigarrow s_1 \cdot s_2$, and admits a Haar measure

$$\frac{\mathrm{d}s}{s} \quad (s > 0).$$

The stabilizer of the truncated horizontal ray $\mathbb{R}_+^\times$ is given by the Borel subgroup

$$B_\mathbb{R} \hookrightarrow \mathbf{SL}(2, \mathbb{R})$$

consisting of the triangular matrices

$$\left\{ \begin{pmatrix} s & r \\ 0 & \dfrac{1}{s} \end{pmatrix} \mid s > 0, r \in \mathbb{R} \right\}.$$

The transitive action of the group of planar rotations $\mathbf{SO}(2, \mathbb{R})$ on the compact set $\mathbf{S}_1$ of rays in $\mathbb{R}^2$ provides the homeomorphism

$$\mathbf{SO}(2, \mathbb{R}) \times B_\mathbb{R} \longrightarrow \mathbf{SL}(2, \mathbb{R}).$$

The associated homeomorphism of $\mathbf{SL}(2, \mathbb{R})$ onto the fiber space

$$\mathbb{T} \times \mathbb{R} \times \mathbb{R}$$

implements a logarithmic scale along the longitudinal direction of the geometry of the circular energy cylinder $\mathbb{T} \times \mathbb{R}$. It will be emphasized below that this logarithmic scale is also at the stochastic edge of quantum theory and special relativity. The coaxial circular cylinders $\mathbb{T} \times \mathbb{R} \times \mathbb{R}$ can be regarded as the fiber bundle of directions associated to the compact unit sphere $\mathbf{S}_2 \hookrightarrow \mathbb{R}^3$.

For $s = 1$ the subgroup of unipotent matrices

$$\left\{ \begin{pmatrix} 1 & r \\ 0 & 1 \end{pmatrix} \mid r \in \mathbb{R} \right\}$$

of the Borel subgroup $B_\mathbb{R} \hookrightarrow \mathbf{SL}(2, \mathbb{R})$ allows to parametrize every element of $\mathbf{SL}(2, \mathbb{R})$ in terms of a product of planar rotations, dilations, and transvections

$$\begin{pmatrix} \cos\theta & \sin\theta \\ -\sin\theta & \cos\theta \end{pmatrix} \begin{pmatrix} s & 0 \\ 0 & \dfrac{1}{s} \end{pmatrix} \begin{pmatrix} 1 & r \\ 0 & 1 \end{pmatrix} \quad (\theta \in]-\pi, +\pi], s > 0, r \in \mathbb{R}).$$

For $\theta \in]-\pi, +\pi]$, the rotation defines a spin-drift mode. The transition from **SL**$(2, \mathbb{R})$ to its two-fold covering group **Mp**$(2, \mathbb{R})$ will afford the phase angle restriction $\theta \in]0, \pi]$. The group formed by the diagonal matrices

$$\mp \begin{pmatrix} e^t & 0 \\ 0 & e^{-t} \end{pmatrix} \quad (t \in \mathbb{R})$$

is ismorphic to **Spin**$°(1, 1, \mathbb{R})$. Due to the LU-decomposition, the transposed matrices

$$\left\{ \begin{pmatrix} 1 & 0 \\ u & 1 \end{pmatrix} \mid u \in \mathbb{R} \right\}$$

together with the transvections

$$\left\{ \begin{pmatrix} 1 & r \\ 0 & 1 \end{pmatrix} \mid r \in \mathbb{R} \right\}$$

in the aforementioned fiber space $\mathbb{T} \times \mathbb{R} \times \mathbb{R}$ form a set of generators of the special linear group **SL**$(2, \mathbb{R})$.

Due to the two-step nilpotency of the Heisenberg Lie group G, the *commutator* mapping $(g_1, g_2) \rightsquigarrow [g_1, g_2]$ throws the product group $G \times G$ into the center $C \hookrightarrow G$ according to the identity

$$\left[\begin{pmatrix} 1 & x_1 & z_1 \\ 0 & 1 & y_1 \\ 0 & 0 & 1 \end{pmatrix}, \begin{pmatrix} 1 & x_2 & z_2 \\ 0 & 1 & y_2 \\ 0 & 0 & 1 \end{pmatrix} \right] = \begin{pmatrix} 1 & 0 & \det \begin{pmatrix} x_1 & x_2 \\ y_1 & y_2 \end{pmatrix} \\ 0 & 1 & 0 \\ 0 & 0 & 1 \end{pmatrix}.$$

Hence the concept of stratigraphic time scale is defined by the commutators of G which provide at the Lie algebra level the Heisenberg uncertainty relation under the normalization $h = 1$. The point to note is that in the center $C \hookrightarrow G$ the moment

$$\det \begin{pmatrix} x_1 & x_2 \\ y_1 & y_2 \end{pmatrix}$$

is doubled in comparison to the center in the moment presentation of G. In the standard presentation of G, the center C supports an *alternating* bilinear form. The associated tracial Plancherel measure $(1/2\pi)v dv$ is dual to the longitudinal logarithmic scale. It exhibits that the homogeneous Pfaffian polynomial of G ([35]) is a *linear* density form.

Let the real vector space $\mathbb{R}^3$ be endowed with the sub-Riemannian metric, also called Carnot-Carathéodory metric in geometric control theory ([28]). As in quantum cybernetics ([19]), introduce the vector fields

$$X_1 = \begin{pmatrix} 1 \\ 0 \\ 0 \end{pmatrix}, \quad X_2 = \begin{pmatrix} 0 \\ 1 \\ x \end{pmatrix} \quad (x \in \mathbb{R})$$

corresponding to

$$\frac{\partial}{\partial x}, \qquad \frac{\partial}{\partial y} + x\frac{\partial}{\partial z} \quad (x \in \mathbb{R}).$$

Setting

$$X_3 = [X_1, X_2] = \begin{pmatrix} 0 \\ 0 \\ 1 \end{pmatrix},$$

corresponding to the timelike Killing vector field

$$\frac{\partial}{\partial z},$$

it follows

$$[X_1, X_3] = [X_2, X_3] = 0.$$

Therefore, the real Heisenberg Lie algebra Lie(G) is isomorphic to the real Lie algebra spanned by $\{X_1, X_2\}$ and the distance with respect to the Carnot-Carathéodory metric defined by X_1 and X_2 becomes left-invariant with respect to the standard presentation

$$\exp_G \left(x X_1 + y X_2 + z X_3 \right)$$

of the elements G. The one-dimensional center of Lie(G) is spanned over $\mathbb{R}$ by the vector field $\{X_3\}$.

As a consequence of the Campbell-Hausdorff formula, the moment presentation

$$\exp_G \left(x X_1 + y X_2 + z[X_1, X_2] \right)$$

of the elements of G arises by choosing the vector fields

$$X_1 = \begin{pmatrix} 1 \\ 0 \\ -\dfrac{x}{2} \end{pmatrix}, \qquad X_2 = \begin{pmatrix} 0 \\ 1 \\ \dfrac{x}{2} \end{pmatrix} \quad (x \in \mathbb{R}),$$

and

$$X_3 = [X_1, X_2].$$

Due to the classical Stone-von Neumann theorem of quantum physics, the real symplectic structure of G gives rise to the chronology of the stratigraphic time scale which is defined by the *central* characters curling up the timelike center $C \hookrightarrow G$. The stratigraphic time scale therefore parametrizes the spin-drift modes. Because there is a canonical bijective correspondence between stationary symmetry actions and timelike Killing vector fields, the timelike center $C \hookrightarrow G$ gives rise to the parametrization of the Heisenberg helices

$$
\begin{pmatrix}
1 & \dfrac{-x(1 - \cos 2\pi vt) - y \sin 2\pi vt}{2\pi v} & \dfrac{(x^2 + y^2)(2\pi vt - \sin 2\pi vt)}{2(2\pi v)^2} \\
0 & 1 & \dfrac{x \sin 2\pi vt - y(1 - \cos 2\pi vt)}{2\pi v} \\
0 & 0 & 1
\end{pmatrix} \quad (t \in \mathbb{R})
$$

which are, as axisymmetric geodesics of the sub-Riemannian geometry of G starting at the origin, the solutions of the associated Hamiltonian equations ([44] and [45]). The unit sphere is then the image of a geodesically *incomplete* Lorentz invariant circular cylinder of radius $r = \sqrt{x^2 + y^2}$. The minimizing geodesics supported by the circular energy cylinder derive from the classical isoperimetric problem of the Euclidean plane $\mathbb{R}^2$. The pull-back of the Einstein cylinder world lines or photon trajectories of relativity to the spheres in space-time define the Kruskal coordinatized two-fold covered Minkowskian space-time manifold of Clifford algebra $\mathscr{C}\ell(3, 1, \mathbb{R})$.

Applications: The Coriolis term attached to the longitudinal axis of the Heisenberg helices should be noticed. It is due to the rotating frame induced by the natural symplectic structure of the planar coadjoint orbits of G. As an application, the relativistic perihelion advance of Keppler's favourite planet Mars is 1.35" per century in good agreement with the calculated value. An analogous reasoning holds outside the solar system.

As an application of the planar signal pattern recognition technique of quantum holography and the transactional interpretation of the concept of quantum physical non-locality ([11]), the phase shift of a light ray or radar beam bent by the sun mass is two times as large as the curvature gauged according to the Newtonian reference for a mass grazing the sun with the velocity of light c. The quantum holographically gauged deflection angle coincides with the relativistic curvature angle which has been experimentally confirmed. The *phase* correction by the factor 2, gauged by halving the *spatial* frequency via the Laguerre projective phase formula, played an important historical role in the approval of Einstein's relativistic extension of the Newtonian gravitation theory: Einstein's range of computation was from the deflection angle of 8.4" he predicted in a letter dated October 14, 1913 to the correct deflection angle of 1.68". The deflection angle he corrected in 1915 was experimentally confirmed by Sir Arthur Stanley Eddington in May 1919, three years after Karl Schwarzschild's untimely death in Potsdam 1916.

The gravitational lens effect induced by the light ray deflection has been discovered by Einstein already in 1912 but not published before 1936. The deflection effect generates the axis of real *focal* points. For the first time, the gravitational lens effect was confirmed in 1979. It suggests an application of quantum information theory to the relativistic astrophysics of gravitational waves. Specifically the gravitational waves emitted by the binary radio *pulsar* PSR 1913+16 and indirectly detected by Russell A. Hulse and Joseph H. Taylor, are of special interest for the reconfirmation of the predictions of general relativity.

For the Einstein field equations, an exact solution in terms of the energy-momentum tensor is only available in spaces of rather high symmetry. Birkhoff's theorem states that the Schwarzschild metric is the unique spherically symmetric solution of the Einstein field equations in the exterior vacuum region of spherically symmetric distributions of matter ([21]). As a historically note, this exact solution was discovered by Karl Schwarzschild (1873–1916) in December 1915 while serving as a soldier of the Germany army on the eastern front.

The Schwarzschild solution suggests to remove the non-intrinsic singularity of the longitudinal logarithmic scale associated to the Borel subgroup $B_{\mathbb{R}} \hookrightarrow \mathbf{SL}(2, \mathbb{R})$ by resonance. Isometric Lorentz centralisation of the spinors defined by the Heisenberg helices via the resonant *homogeneous* linear differential operators of order ≤ 1 in $\omega(\mathrm{Lie}(\mathbf{O}(3, 1, \mathbb{R}))) \hookrightarrow \mathrm{Lie}(\mathbf{SL}(2, \mathbb{R}))$ is a different method to deduce the Schwarzschild metric in the Kruskal coordinatized two-fold covered Minkowskian space-time manifold of Clifford algebra $\mathscr{C}\ell(3, 1, \mathbb{R})$. The Kruskal extension is the *maximal* singularity-free extension of the Schwarzschild metric of cosmology. This may be seen by a direct examination of the geodesics: Every geodesic either runs into the intrinsic spherical singularity, or can be extended. The Kruskal diagram provides the complete description of a spherically symmetric *black hole* which is the end state of a gravitational collapse.

The real symplectic group $\mathbf{Sp}(4, \mathbb{R})$ of linear Poisson isomorphisms of $\mathrm{T}^{\star}\mathbb{R}^2$ is connected but not simply connected. At the Lie algebra level, $\mathrm{Lie}(\mathbf{Sp}(4, \mathbb{R}))$ is anti-isomorphic to the Poisson algebra of quadratic polynomials on $\mathrm{T}^{\star}\mathbb{R}^2$. The maximal compact subgroup of $\mathbf{Sp}(4, \mathbb{R})$ is given by $\mathbf{Sp}(4, \mathbb{R}) \cap \mathbf{O}(4, \mathbb{R}) = \mathbf{U}(2, \mathbb{C})$. In complex coordinates the Poisson bracket reads

$$\{f, g\} = i \left(\frac{\partial f}{\partial w} \frac{\partial g}{\partial \bar{w}} - \frac{\partial f}{\partial \bar{w}} \frac{\partial g}{\partial w} \right),$$

where f, g are holomorphic functions on the complex vector space $\mathbb{C}^2$.

The theory of reductive dual Lie group pairs within $\mathbf{Sp}(4, \mathbb{R})$ provides mutually *commuting* restrictions of ω to the full Lorentz isometry group $\mathbf{O}(3, 1, \mathbb{R})$ and its centralizer in $\mathbf{Sp}(4, \mathbb{R})$. The commutant associates to quantum cosmology via the projectivization $\mathbf{P}_2(\mathbb{C})$ of the complexification of the coadjoint orbit picture of the Heisenberg nilpotent Lie group G the concepts of sub-Riemannian geometry as well as the Schwarzschild radius R of general relativity theory.

Their asymptotic sectors determine two quantum cosmological external worlds for which there exist no information transmission channels between.

The concept of stratigraphic time scale implements the Keppler II law. It allows to resolve the geometric eidetics of the Keppler I law via the concept of square integrable representation ([35]). The relativistic chronology performs a rescaling of the stratigraphic time scale of the fiber bundle of directions associated to the compact unit sphere $S_2 \hookrightarrow \mathbb{R}^3$.

The quantum holographic approach to the magnetic Aharonov-Bohm effect is based on an analogous curvature reasoning of displacements in fiber bundles.

Strong Stone-von Neumann Theorem. There exists, up to unitary isomorphy, a unique irreducible unitary linear representation ρ of G which is square integrable mod C and subduces the principal *central* character

$$C \ni \begin{pmatrix} 1 & 0 & z \\ 0 & 1 & 0 \\ 0 & 0 & 1 \end{pmatrix} \rightsquigarrow e^{2\pi i z}$$

as a central (tempered) distribution of positive type and spin-drift mode of normalized temporal frequency $\nu = 1$.

Corollary 1. The complex vector space of $\mathscr{C}^\infty$ vectors of ρ is the Schwartz space $\mathscr{S}(\mathbb{R})$ of complex-valued $\mathscr{C}^\infty$ functions on the real line $\mathbb{R}$ rapidly declining at infinity.

Corollary 2. The standard representation ρ of G gives rise to a linear isomorphism

$$f \rightsquigarrow K_\rho(f)$$

of the convolution algebra $\mathscr{S}(\mathbb{R} \oplus \mathbb{R})$ with convolution structure inherited from G, onto the algebra of *all* kernel operators on $\mathscr{S}(\mathbb{R})$ with kernel in $\mathscr{S}(\mathbb{R} \oplus \mathbb{R})$, which are of trace class. It extends to an isometric isomorphism of $L^2(\mathbb{R} \oplus \mathbb{R})$ onto the algebra of all Hilbert-Schmidt operators on $L^2(\mathbb{R})$ realized as linear kernel operators with kernels belonging to $L^2(\mathbb{R} \oplus \mathbb{R})$.

It follows that the strong Stone-von Neumann theorem is equivalent to the uniqueness of the irreducible representation of the $C^\star$-algebra of compact operators.

The spinorial *symbol calculus* allows to invert the linear isomorphism $f \rightsquigarrow K_\rho(f)$ in terms of the symbol of K_ρ in the sense of pseudodifferential operators and the *symplectic* Fourier transformer $\mathscr{F}_{\mathbb{R} \oplus \mathbb{R}}$ operating on the complex Schwartz space $\mathscr{S}(\mathbb{R} \oplus \mathbb{R}) \hookrightarrow L^2(\mathbb{R} \oplus \mathbb{R})$ on the real symplectic plane $\mathbb{R} \oplus \mathbb{R}$.

Applications: High resolution filter bank implementation by means of quantum holography via the Lévy inversion formula displayed below.

Image formation and symplectic Fourier transform reconstruction in synthetic aperture radar (SAR) remote sensing. With synthetic aperture image formation techniques, resolution hundreds, or even thousands, of times finer than the diffraction limit of the *actual* receiving antenna can be achieved, particularly in the centimeter wavelength region. Due to projective geometry, there is *no* conflict between SAR remote sensing and the special theory of relativity ([41]).

Image reconstruction by application of a symplectic Fourier transformer $\mathscr{F}_{\mathbb{R}\oplus\mathbb{R}}$ from the quantum hologram in clinical magnetic resonance tomography (MRI). MRI techniques have been in widespread clinical use now for more than fifteen years. This non-invasive imaging modality helped to achieve unprecedented precision in visualizing human morphology, including an exact display of normal anatomy of the patient *in vivo*, as well as of the gross pathology of organs and diseases ([44]).

Signal processing, conceived as the spinorial symbol calculus of the distributional wave equation in the sense of pseudodifferential operators.

Associated to the simply connected Lie group G is the three-dimensional real Heisenberg Lie algebra $\mathrm{Lie}(G)$ of nilpotent matrices

$$\begin{pmatrix} 0 & a & c \\ 0 & 0 & b \\ 0 & 0 & 0 \end{pmatrix} \quad (a, b, c \in \mathbb{R}),$$

where the matrix entries a, b are related to the complex coordinates $w, \bar{w} \in \mathbb{C}$ according to the identities

$$a = \frac{w + \bar{w}}{\sqrt{2}}, \qquad b = -i\frac{w - \bar{w}}{\sqrt{2}}.$$

Connecting $\mathrm{Lie}(G)$ and G there is the exponential map

$$\exp_G : \mathrm{Lie}(G) \longrightarrow G$$

which forms a global diffeomorphism such that

$$\exp_G\left(\begin{pmatrix} 0 & a & c \\ 0 & 0 & b \\ 0 & 0 & 0 \end{pmatrix}\right) = \begin{pmatrix} 1 & a & c + \frac{1}{2}ab \\ 0 & 1 & b \\ 0 & 0 & 1 \end{pmatrix}$$

holds. The group multiplication in G can then be expressed via the Campbell-Hausdorff formula which in the case of the moment presentation of G is quite simple ([42]):

$$\exp_G(X_1)\exp_G(X_2) = \exp_G\left(X_1 + X_2 + \frac{1}{2}[X_1, X_2]\right).$$

The adjoint action of G on the Heisenberg Lie algebra $\mathrm{Lie}(G)$ is defined by the adjoint representation

$$\mathrm{Ad}_G : G \longrightarrow \mathbf{GL}(\mathrm{Lie}(G)).$$

The *contragredient* representation to the adjoint representation

$$\mathrm{CoAd}_G : G \longrightarrow \mathbf{GL}\left(\mathrm{Lie}(G)^\star\right)$$

gives rise to the coadjoint action of G on the vector space dual $\mathrm{Lie}(G)^\star$ of the Heisenberg Lie algebra. Thus, the cotangent bundle $\mathrm{T}^\star G$ of G admits the representation

$$\mathrm{T}^\star G = G \times \mathrm{Lie}(G)^\star.$$

For any element $g \in G$ and linear form $\nu \in \mathrm{Lie}(G)^\star$, the element $\mathrm{CoAd}_G(g)(\nu) \in \mathrm{Lie}(G)^\star$ is determined by the rule

$$\mathrm{CoAd}_G(g)(\nu)(X) = \nu\left(\mathrm{Ad}_G\left(g^{-1}\right)(X)\right)$$

for all vector fields $X \in \mathrm{Lie}(G)$. The classification of the unitary dual $\hat{G}$ then follows from the coadjoint orbit picture of G in the vector space dual of the Heisenberg Lie algebra $\mathrm{Lie}(G)^\star$. Explicitly,

$$\mathrm{CoAd}_G \begin{pmatrix} 1 & x & z \\ 0 & 1 & y \\ 0 & 0 & 1 \end{pmatrix} = \begin{pmatrix} 1 & 0 & -x \\ 0 & 1 & y \\ 0 & 0 & 1 \end{pmatrix}.$$

An analogous method will be used to determine the cotangential equations of projective quadrics.

Identify the linear forms $\nu \in \mathrm{Lie}(G)^\star$ ($\nu \neq 0$) with their central control gradients. Due to the Pfaffian linear form which defines the Plancherel measure of G, the natural symplectic form

$$\nu \cdot dx \bigwedge dy$$

of the planar coadjoint orbits

$$\mathcal{O}_\nu = \mathrm{CoAd}_G(G)(\nu) \quad \left(\nu \in \mathbb{R}^\times\right)$$

of G exhibits the real variable x as a spatial frequency. Its symplectic conjugate is then the phase variable y of the polarization state. Both variables characterize a local accessible quantum state ([59]).

The planar coadjoint orbits in

$$\mathrm{CoAd}_G(G)\left(\mathbb{R}^\times\right) \hookrightarrow \mathrm{Lie}(G)^\star$$

coincide with the symplectic *leaves* of the Poisson structure of $\mathscr{C}^\infty(\mathrm{Lie}(G)^\star)$ with Poisson bracket

$$\{f, g\} = c \left(\frac{\partial f}{\partial a} \frac{\partial g}{\partial b} - \frac{\partial f}{\partial b} \frac{\partial g}{\partial a} \right) \quad (a, b, c \in \mathbb{R})$$

for complex-valued functions $f, g \in \mathscr{C}^\infty(\mathrm{Lie}(G)^\star)$. The symplectic leaves are slicing up the real dual $\mathrm{Lie}(G)^\star$ and form the symplectic spinorial substrate or synchronized hypostasis of the spatio-temporal *filter bank* implementation of quantum information processing. The slices come in *mirrored* pairs

$$\left(\mathbb{O}_\nu, \mathbb{O}_{-\nu}\right) \quad \left(\nu \in \mathbb{R}^\times\right)$$

according to the following non-locality theorem which refers to the concept of complex anti-Hilbert space ([44], [46] and [47]). Each single point in $\mathrm{T}^\star\mathbb{R} \times \{0\}$ forms a *leaf*.

Antisymmetric Entanglement Theorem. Let G denote a simply connected *nilpotent* Lie group. The following properties of the element $\rho \in \hat{G}$ are equivalent:

(i)   The coadjoint orbit $\mathbb{O}_\rho$ is a symplectic affine *linear* variety in the vector space dual $\mathrm{Lie}(G)^\star$;

(ii) If the one-dimensional identity representation I is *vaguely* contained in the tensor product representation $\rho \hat{\otimes} \sigma$ for $\sigma \in \hat{G}$, then $\sigma = \check{\rho}$ is the contragredient representation of $\rho \in \hat{G}$ acting on an everywhere dense vector subspace of the complex anti-representation Hilbert space of ρ.

Corollary 3. The element $\rho \in \hat{G}$ is *square integrable* modulo its kernel if and only if each element $\sigma \in \hat{G}$ for which I is vaguely contained in $\rho \hat{\otimes} \sigma$ must be equal to $\check{\rho}$.

The following comments are now in order: The concept of vague containment is standard in the theory of unitary group representations and (tempered) distributions of positive type ([15]). The non-locality of quantum physics is performed by global antisymmetry coupling of positive type. It was from early on this kind of coupling the principal reason for Einstein's reservations about quantum physics, namely the then-current treatment of the spatial non-separability of the quantum physical account of interactions. The antisymmetric entanglement theorem forms a geometric *extension* of Born's concept of quantum probability which has been accepted by Einstein ([24]). It includes Bell's theorem on Lorentz invariant expectation values ([1], [23], [27], [49] and [51]).

For the three-dimensional Heisenberg Lie group G the counting measure of the planar lattice

$$L \hookrightarrow G$$

is an invariant distribution under the Fourier transformer. It follows via the Poisson summation formula ([25] and [54]) that I is vaguely contained in $\rho \hat{\otimes} \sigma$ for elements $\rho \in \hat{G}$ and $\sigma \in \hat{G}$ of dimensions > 1. Thus the antisymmetric entanglement theorem can be applied to the quantum physics of the solid state where the role of the action of I is well known from the theory of Bloch functions ([10]).

For the three-dimensional Heisenberg nilpotent Lie group G, the sandwich configuration of mirrored pairs of planar coadjoint orbits implements the smooth action

$$\mathbb{T} \ni \zeta \rightsquigarrow (\mathbb{C}_0 \times \mathbb{C}_1 \ni (w_0, w_1) \rightsquigarrow (\zeta w_0, \zeta w_1) \in \mathbb{C}_0 \times \mathbb{C}_1)$$

which gives rise to a partition of the compact unit sphere $\mathbf{S}_3 \hookrightarrow \mathbb{R}^4$ into circles. This circle bundle is obtained by two stereographic projections with *antipodal* centers. It is exactly the Hopf fibration

$$\mathbf{S}_1 \longrightarrow \mathbf{S}_3$$
$$\downarrow$$
$$\mathbf{S}_2$$

where $\mathbf{S}_1 = \mathbf{SO}(2, \mathbb{R})$ is isomorphic to the one-dimensional compact torus group $\mathbb{T}$. The quantum rotation group

$$\mathbf{S}_3 = \mathbf{SU}(2, \mathbb{C}) = \mathbf{Spin}(3, \mathbb{R})$$

forms the two-fold covering group of the Euclidean rotation group $\mathbf{SO}(3, \mathbb{R})$, whereas the spin group

$$\mathbf{Spin}(4, \mathbb{R}) = \mathbf{Spin}(3, \mathbb{R}) \times \mathbf{Spin}(3, \mathbb{R})$$

forms the two-fold covering of $\mathbf{SO}(4, \mathbb{R})$. The lifting of the structure group $\mathbf{SO}(3, \mathbb{R})$ of the tangent bundle $T\mathbf{S}_2$ to $\mathbf{Spin}(3, \mathbb{R})$ is then the spin structure represented by the Hopf bundle. Its compact base manifold

$$\mathbf{S}_2 = \mathbb{C}_0 \cup_\varphi \mathbb{C}_1$$

regarded as the union of two copies $\{\mathbb{C}_0, \mathbb{C}_1\}$ of $\mathbb{C}$ supposes that the complex planes are *coherently* arranged so that their positive real axes are parallel. The orientation reversing identification map φ of the union of the complex planes $\{\mathbb{C}_0, \mathbb{C}_1\}$ is defined by the assignment

$$\varphi : \mathbb{C}_0^\times \ni w_0 \rightsquigarrow \frac{1}{w_0} = w_1 \in \mathbb{C}_1^\times.$$

It may be regarded as arising from *two* stereographic projections with antipodal centers of $\mathbf{S}_2$ onto the tangential planes located in $T\mathbf{S}_2$ at diametrically opposite points. In this way, the Hopf fibration

$$\mathbf{S}_3$$
$$\downarrow$$
$$\mathbf{S}_2$$

together with the bundle homomorphism $S_3 \to SO(3, \mathbb{R})$, is the *uniquely* determined spin structure on the tangent bundle TS_2 of the compact unit sphere S_2.

The point to note is that switching the orientation of the replica $\mathbb{C}_0$ which is defined by the cyclic points as a phase reference, *instantaneously* reverses the orientation of the replica $\mathbb{C}_1$, and vice versa. The violation of the chronology of the stratigraphic time scale is independent of the distance of the remote subsystems.

The submersion $\pi : S_3 \to S_2$ is given by the assignment

$$\pi : w \rightsquigarrow \begin{cases} \dfrac{w_1}{w_0} \in \mathbb{C}_1 & \text{if } w_0 \neq 0, \\[2mm] \dfrac{w_0}{w_1} \in \mathbb{C}_0 & \text{if } w_1 \neq 0. \end{cases}$$

Notice that if $w_0 \cdot w_1 \neq 0$, the two definitions agree. It is easy to verify that $\pi(w) = \pi(w')$ if and only if $w = \zeta \cdot w'$ for some element $\zeta \in \mathbb{T}$. The local trivializations of π are

$$\pi^{-1}(\mathbb{C}_0) = \mathbb{C}_0 \times S_1$$
$$(w_0, w_1) \rightsquigarrow \left(\frac{w_1}{w_0}, \frac{w_0}{|w_0|} \right)$$

and

$$\pi^{-1}(\mathbb{C}_1) = \mathbb{C}_1 \times S_1$$
$$(w_0, w_1) \rightsquigarrow \left(\frac{w_0}{w_1}, \frac{w_1}{|w_1|} \right).$$

The bundle chart change

$$\mathbb{C}_0^\times \times S_1 \longrightarrow \mathbb{C}_1^\times \times S_1$$

between tori is seen to be

$$(w_0, \zeta) \rightsquigarrow \left(\frac{1}{w_0}, \zeta \frac{w_0}{|w_0|} \right).$$

Since the bundle chart change is of the form

$$(w_0, \zeta) \rightsquigarrow (\varphi(w_0), \zeta \cdot h(w_0)),$$

the Hopf fibration forms a circle bundle over the quotient manifold S_2.

The compact base manifold $S_2 \hookrightarrow \mathbb{R}^3$ conically projects onto the one-dimensional absolute projective quadric Φ_∞ of the three-dimensional real projective spinor space

$$P_3(\mathbb{R}) = P\left(\mathbb{R}^3 \times \mathbb{R} \right) = SU(2, \mathbb{C})/\{+I, -I\}.$$

Historically, it was the idea of immersing a vector space into a larger space as above that led to the idea of projective geometry. Saying that two linear varieties in $\mathbb{R}^3$ have the same direction translates into the condition that their images under the canonical injection $\mathbb{R}^3 \hookrightarrow \mathbf{P}_3(\mathbb{R})$ should generate projective linear varieties which intersect the hyperplane at infinity in the same set. Let the hyperplane at infinity of $\mathbf{P}_3(\mathbb{R})$, which actually defines boundary conditions for the immersed vector space $\mathbb{R}^3$, be equipped with a symplectic structure. Then there exists a uniquely determined two-dimensional non-degenerate central projective quadric $\Phi \hookrightarrow \mathbf{P}_3(\mathbb{R})$ homeomorphic to the unit sphere $\mathbf{S}_2$ such that its trace on $\mathbb{R} \oplus \mathbb{R}$ defines the one-dimensional absolute projective quadric $\Phi_\infty \hookrightarrow \mathbf{P}_2(\mathbb{C})$. The trace of Φ_∞ itself on the line at infinity of the complex projective plane $\mathbf{P}_2(\mathbb{C})$ coincides with the zero-dimensional absolute projective quadric Φ_∞° consisting of the conjugate cyclic points $\{\bar{i}, i\}$. As a phase reference, they are common to all circles of the symplectic plane $\mathbb{R} \oplus \mathbb{R}$. The isotropic lines stemming from the center of a circle in $\mathbb{R} \oplus \mathbb{R}$ are tangent to the circle at the conjugate cyclic points $\{\bar{i}, i\} \hookrightarrow \mathbf{P}_2(\mathbb{C})$. They form the circle's asymptotes and the set of circles form a three-dimensional real projective space with the conjugate cyclic points $\{\bar{i}, i\}$ as base points of all pencils of circles in $\mathbb{R} \oplus \mathbb{R}$.

Because the quadratic form of Φ has signature

$$\begin{pmatrix} 1 & & \\ & 1 & \\ & & 1 \end{pmatrix}$$

and the Lorentzian form $r_{3,1}^2$ on $\mathbb{R}^4$ with normalization of the velocity of light $c = 1$ is defined by the diagonal matrix

$$\begin{pmatrix} 1 & & & \\ & 1 & & \\ & & 1 & \\ & & & -1 \end{pmatrix}$$

of rank 4, the orthogonal projective group $\mathbf{PO}(3, \Phi)$ is given by $\mathbf{O}(3, 1, \mathbb{R})/\{+I, -I\}$, where

$$\mathbf{O}(3, 1, \mathbb{R}) \hookrightarrow \mathbf{GL}(3, 1, \mathbb{R})$$

denotes the full Lorentz isometry group of the four-dimensional Minkowskian space-time manifold of Clifford algebra $\mathscr{C}\ell(3, 1, \mathbb{R})$. It leaves the Lorentzian form $r_{3,1}^2$ on $\mathbb{R}^4$ invariant and preserves the Poincaré sphere of symplectic spinorial geometry ([8]). The isotropic cone associated to the quadratic form $r_{3,1}^2$ spans the real vector space $\mathbb{R}^4$, realized by $\mathbb{C}^2$. By scaling the Lorentzian form $r_{3,1}^2$ on $\mathbb{R}^4$ with the velocity of light c, standardization to the gauge circle $t = 1$ as the gauge circle group forming the intersection of the convex light cone with the fiber bundle of directions associated to the compact unit sphere $S_2 \hookrightarrow \mathbb{R}^3$ will allow the implementation of the Lorentz dilation ([18])

$$\frac{1}{\sqrt{1-(v^2/c^2)}} \quad (0 \le v < c)$$

associated with the velocity gradient

$$\frac{v}{c}$$

which has been measured by an inertial observer. Because the Minkowskian space-time manifold of Clifford algebra $\mathscr{C}\ell(3, 1, \mathbb{R})$ implements the relativistic curvature of the Heisenberg helices supported by the circular energy cylinders, the projectivization $\mathbf{P}_2(\mathbb{C})$ of the complexification of the coadjoint orbit picture of the Heisenberg Lie group G permits an application of the Laguerre projective phase formula to compute the phase shift of a radar beam just grazing the sun mass in such a way that the angle of deflection is in *complete* agreement with quantum holography. Furthermore, it provides the relativistic perihelion precession of the planetary motions, especially pronounced in the case of Mercury, and hence a relativistic adjustment of the first Kepplerian fundamental law.

The orthogonal projective group $\mathbf{PO}(3, \Phi)$ can be realized by the Möbius group generated by all inversions of the compact unit sphere $S_2 \hookrightarrow \mathbb{R}^3$. The orientation preserving conformal group $\mathbf{PO}^+(3, \Phi)$ is naturally isomorphic to the projective group $\mathbf{PGL}(2, \mathbb{C})$ of automorphisms of the Riemann sphere $S_2 = \mathbf{P}_1(\mathbb{C})$. Then $\mathbf{P}_1(\mathbb{C}) - \mathbb{C}^\times = \{0\} \cup \{\infty\}$. Owing to the *antiprojectivities*, the projective group $\mathbf{PGL}(2, \mathbb{C})$ forms a subgroup of index 2 of the circular group of the complex projective line $\mathbf{P}_1(\mathbb{C}) = (\mathbb{C}^2 - \{0\})/\mathbb{C}^\times$ which is not homeomorphic to the real projective plane $\mathbf{P}_2(\mathbb{R})$ under its standard Kähler metric since the former is orientable but *not* the latter.

An application of the two-fold covering group $\mathbf{O}(3, 1, \mathbb{R})$ of the orthogonal projective group $\mathbf{PO}(3, \Phi)$ is formed by Bell's theorem on Lorentz invariant expectation values ([23] and [49]). Consider the elliptic non-Euclidean displacements in $\mathbf{PO}^+(3, \Phi)$ which are the images of rotations of the form

$$M_{Q,+I} : R \rightsquigarrow Q \cdot R \quad (\|Q\| = 1)$$

or

$$M_{+I,Q'} : R \rightsquigarrow R \cdot Q' \quad (\|Q'\| = 1)$$

in the Clifford algebra $\mathbb{H}$ for $\mathbb{R}^2$ obtained by left or right multiplication of $Q \in \mathbb{H}$ and $Q' \in \mathbb{H}$, respectively. They form the Clifford translations of the first and second type, respectively. The Clifford translations form groups PG_1 and PG_2, respectively, of *projectivities* of the real projective spinor space $\mathbf{P}_3(\mathbb{R})$ which are isomorphic to $\mathbf{SO}(3, \mathbb{R})$. The groups PG_1 and PG_2 act as simply transitive groups on $\mathbf{P}_3(\mathbb{R})$. Therefore the Clifford translations $\neq I$ do not admit fixed points.

Consider a non-Euclidean line D, hence a homogeneous plane $F \hookrightarrow \mathbb{H}$. Then F is isomorphic to the complex plane $\mathbb{C}$. The Clifford translations of the first and second type, respectively, which leave D globally invariant form *Abelian* subgroups $\mathrm{T}_1(D) \hookrightarrow \mathrm{PG}_1$ and $\mathrm{T}_2(D) \hookrightarrow \mathrm{PG}_2$, both isomorphic to the planar rotation group $\mathbf{S}_1 = \mathbf{SO}(2, \mathbb{R})$. The same argument also shows that, for every Clifford translation $v_1 \in \mathbf{PG}_1$, either $v_1(D) = D$ or $v_1(D) \cap D = \emptyset$. The lines

$$\left\{ v_1(D) \mid v_1 \in \mathbf{PG}_1 \right\}$$

form the Clifford parallels to D of the first kind ([9]). Clearly, Clifford parallelism of the first kind is an equivalence relation. In the pencil of lines through every point $A \in \mathbb{P}_3(\mathbb{R})$ as a base point there is one and only one non-Euclidean line

$$\left\{ v_1(D) \mid v_1 \in \mathrm{T}_1(D) \right\}$$

which is Clifford parallel to D.

Let $D' \neq D$ be a line in the three-dimensional elliptic non-Euclidean spinor space $\mathbf{P}_3(\mathbb{R})$ which is Clifford parallel of the first kind to D. Let A be a point of D, A' a point of D', and $v_1 \in \mathbf{PG}_1$ the unique element such that $v_1(A) = A'$. Let D_1 denote the non-Euclidean line connecting A and A'. Then, as $v_1(D)$ meets D_1 at the point A', it follows $v_1(D_1) = D_1$. Because the action of the orthogonal projective group $\mathbf{PO}(3, \Phi)$ preserves non-Euclidean angles, choices of suitable orientations provide the identity of the angles of D and D_1, and of D' and D_1.

Consider a non-Euclidean plane F containing D in the three-dimensional elliptic non-Euclidean spinor space $\mathbf{P}_3(\mathbb{R})$. Then F meets D' in a unique point A' because D and D' are *not* located in the same projective plane. Thus the image $v_1'(F)$ of F under $v_1' \in \mathrm{T}_1(D)$ is a plane containing D and meeting D' in the point $v_1'(A')$. On the other hand D' makes equal angles with the normals to the two planes F and $v_1'(F)$ at the points A' and $v_1'(A')$, respectively. This property has no analogue in Euclidean geometry and therefore exhibits one of the most serious blows for the Kantian theory.

Of course, there are analogous results for the equivalence relation of Clifford parallelism of the second type, defined by the group PG_2. The concept of Clifford parallelism leads from three-dimensional elliptic non-Euclidean geometry to the geometry of the open unit ball of three-dimensional Euclidean space $\mathbb{R}^3$. Under this transition, non-Euclidean lines become circles whose plane passes through the origin, and for which the point 0 has power -1 again. Then Clifford parallels correspond to Villarceau circles of parataxy ([9]).

The space of fibers of the Hopf bundle fit together topologically to form the compact unit sphere $\mathbf{S}_2 \hookrightarrow \mathbb{R}^3$. The preimage of a circle on $\mathbf{S}_2$ in $\mathbf{S}_3$ is a flat torus which can be considered as homeomorphic to a surface of revolution $S_1 \times S_1$ with *two* axes

$$\left\{ (w_0, w_1) \in \mathbb{C}^2 \mid |w_0| = \cos\frac{\theta}{2}, \ |w_1| = \sin\frac{\theta}{2} \right\},$$

where

$$\tan\frac{\theta}{2} = \Im\tau, \quad \tau \in \mathfrak{D}.$$

As an intersection of the compact unit sphere $S_3 \hookrightarrow \mathbb{R}^4$ with the circular diffraction cone in $\mathbb{R}^4$ of angle $\theta \in [0, \pi]$ with vertex at the origin of $\mathbb{C}^2$, it reads

$$S_3 \cap \left\{ (w_0, w_1) \in \mathbb{C}^2 \mid |w_0|^2 \sin^2\frac{\theta}{2} - |w_1|^2 \cos^2\frac{\theta}{2} = 0 \right\}.$$

The bundle of *bitangential* planes of the flat torus $S_1 \times S_1$ gives rise to the *non-orientable* real projective plane

$$\mathbf{P}_2(\mathbb{R})$$

which contains Möbius strips S as $\mathbb{Z}/2\mathbb{Z}$ bundles over the compact base manifold S_1. It is the *twist* of the concept of Möbius strip $S \hookrightarrow \mathbf{P}_2(\mathbb{R})$ with typical fiber $[0, 1]$ given by

$$\mathbb{Z}/2\mathbb{Z} \longrightarrow S$$
$$\downarrow$$
$$S_1$$

which geometrically represents the antisymmetric entanglement process non-locally realized by antisymmetrically polarization-entangled photon pairs. The entanglement does not contradict the principle of causality.

A cross-sectional disc for the corresponding solid torus meets every fiber inside the torus just once. The complementary disc meets each fiber outside the torus once. The fibers on the torus meet each disc once in its bounding circle and provide a diffeomorphism between these two circles. Thus the space of fibers is made up of two-discs glued together by a diffeomorphism of their frontiers, which is just the two-sphere $S_2 \hookrightarrow \mathbb{R}^3$.

Similarly, $S_3 \hookrightarrow \mathbb{R}^4$ can be looked at as two solid tori in $\mathbb{R}^3$ glued along their Villarceau circles such that the meridians of the one torus corresponds to the parallels of latitude of the other one.

Notice the intersection properties of the Villarceau circles as the two kinds of Clifford parallels on the flat torus to an arbitrary great circle of $S_3 \hookrightarrow \mathbb{R}^4$ through

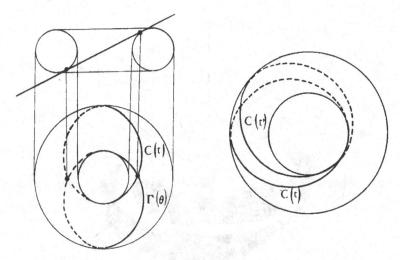

Figure 1. The bundle of Villarceau circles $C(t)$ and $\Gamma(\theta)$ of parataxy in three-dimensional elliptic non-Euclidean geometry is found as intersections of a flat torus with a bitangential plane. The parameters t and θ denote time scales. The circle bundle over the flat torus forms the Heisenberg nilmanifold.

the poles ([9]). By Debeye's cut-off procedure, the compact base manifold $\mathbf{S}_2$ implements the Lévy inversion formula

$$\lambda([a, b]) = \lim_{T \to \infty} \frac{1}{2T} \int_{-T}^{T} \frac{e^{2\pi i b v} - e^{2\pi i a v}}{i v} \cdot \mathscr{F}_{\mathbb{R}} \lambda(v) dv \quad (a < b)$$

for any bounded Radon measure λ with Fourier transform

$$v \rightsquigarrow \mathscr{F}_{\mathbb{R}} \lambda(v).$$

The averaging is performed with respect to the density $\mathscr{F}_{\mathbb{R}} \lambda(v)$ of the Lebesgue measure dv on the real line $\mathbb{R}$ such that the boundary conditions

$$\lambda(\{a\}) = \lambda(\{b\}) = 0$$

are satisfied ([4]). The Lévy inversion formula visualizes the Kepplerian conchoid construction which is spectral theoretic in nature ([43]).

For the three-dimensional Heisenberg nilpotent Lie group G, each infinite dimensional irreducible unitary linear representation is unitarily equivalent to the linear Schrödinger representation ρ_v acting on the complex Schwartz space $\mathscr{S}(\mathbb{R}) \hookrightarrow L^2(\mathbb{R})$ according to the rule

$$\rho_v \left(\begin{pmatrix} 1 & x & z \\ 0 & 1 & y \\ 0 & 0 & 1 \end{pmatrix} \right) \psi(t) = e^{2\pi i v(z+yt)} \psi(t + x) \quad (t \in \mathbb{R}),$$

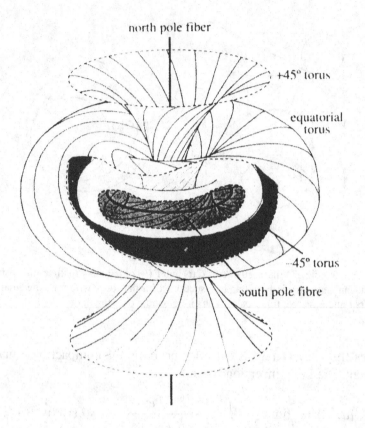

Figure 2. Cutaway perspective view in $\mathbb{R}^3$ of the Hopf fibration: Visualization via stereographic projection of the foliation of $\mathbf{S}_3 \hookrightarrow \mathbb{R}^4$ by nested flat tori. The Hopf fibration together with the bundle homomorphism $\mathbf{S}_3 \rightarrow \mathbf{SO}(3, \mathbb{R})$ forms a fiber bundle model of the spin structure on the tangent bundle $\mathbf{TS}_2$ of $\mathbf{S}_2$. The traces of the bitangential planes on the flat tori are displayed as a bundle of Villarceau circles of parataxy. The Hopf bundle is tautologous in the sense that a point in $\mathbf{P}_1(\mathbb{C})$ represents a complex line in the vector space $\mathbb{C}^2$, and to this point its complex line is associated. By fiberwise multiplication with the time-line $\mathbb{R}$, the four-dimensional Taub-NUT manifold arises which forms a cylinder bundle over $\mathbf{S}_2$.

where the spectral object $\psi \in \mathcal{S}(\mathbb{R})$ is complex-valued and $\nu \neq 0$ denotes its temporal frequency. Then

$$\rho = \rho_1$$

denotes the *generic* infinite dimensional irreducible unitary linear representation of G associated with the symplectic planar coadjoint orbit

$$\mathcal{O}_1 = \text{CoAd}_G(G)(1).$$

By duality, the symplectic monomial matrix

$$J^{-1} = -J = \begin{pmatrix} 0 & 1 \\ -1 & 0 \end{pmatrix}$$

defines the infinitesimal spin-drift mode.

In correspondence to the zero-dimensional absolute projective quadric Φ°_∞ $\hookrightarrow \mathbf{P}_1(\mathbb{C})$ which consists of the conjugate cyclic points $\{\bar{i}, i\}$ joined by the line at infinity of $\mathbf{P}_2(\mathbb{C})$ and common, as a phase reference, to all circles of the symplectic plane $\mathbb{R} \oplus \mathbb{R}$, the traceless matrices J^{-1} and J introduce the Fourier transformer $\mathscr{F}_{\mathbb{R}}$ *and* the Fourier cotransformer $\bar{\mathscr{F}}_{\mathbb{R}}$, respectively. These transformers are acting on the complex Schwartz space $\mathscr{S}(\mathbb{R}) \hookrightarrow L^2(\mathbb{R})$ on the real line $\mathbb{R}$. They are associated to the metaplectic representation ω of the metaplectic group $\mathbf{Mp}(2, \mathbb{R})$ via the normalization relations

$$\omega(J^{-1}) = \mathscr{F}_{\mathbb{R}}, \qquad \omega(J) = \bar{\mathscr{F}}_{\mathbb{R}}.$$

The intertwiners exhibit x as a phase variable and y as a spatial frequency variable ([33], [12], [13], [39] and [42]). The symplectic Fourier transformer

$$\mathscr{F}_{\mathbb{R} \oplus \mathbb{R}} = \mathscr{F}_{\mathbb{R}} \bigwedge \bar{\mathscr{F}}_{\mathbb{R}}$$

is of order 2 instead of order 4 because only two isotropic tangents drawn from the conjugate cyclic points $\Phi^\circ_\infty = \{\bar{i}, i\}$ to the central projective quadric $\Phi_\infty \hookrightarrow \mathbf{P}_2(\mathbb{C})$ are activated.

The symplectic Fourier transformer $\mathscr{F}_{\mathbb{R} \oplus \mathbb{R}}$ allows one to holographically implement the high resolution filter banks of the SAR imaging and the MRI modality ([34] and [44]).

At the Lie algebra level,

$$\omega(\mathrm{Lie}(\mathbf{Sp}(2, \mathbb{R})))$$

consists of the linear differential operators of total order 2 spanned over $\mathbb{R}$ by the expressions

$$\left\{ \pi i \xi^2, \pi i \xi \eta, \pi i \eta^2, \frac{i}{4\pi} \frac{\partial^2}{\partial \xi^2}, \frac{i}{4\pi} \frac{\partial^2}{\partial \xi \partial \eta}, \frac{i}{4\pi} \frac{\partial^2}{\partial \eta^2}, \right.$$
$$\left. \xi \frac{\partial}{\partial \xi} + \frac{1}{2}, \xi \frac{\partial}{\partial \eta}, \eta \frac{\partial}{\partial \xi}, \eta \frac{\partial}{\partial \eta} + \frac{1}{2} \right\}.$$

They display the obstruction of order 2 to linearizing ω as a group representation of $\mathbf{Sp}(2, \mathbb{R}) = \mathbf{SL}(2, \mathbb{R})$, as opposed to the metaplectic group $\mathbf{Mp}(2, \mathbb{R})$.

The tensor product representation

$$\rho \hat{\otimes} \check{\rho}$$

associated with the mirrored pair of coadjoint orbits

$$\left(\mathbb{O}_1, \mathbb{O}_{-1} \right)$$

gives rise to antisymmetrically *entangled* quantum states and the sandwich principal bundle. As a projective unitary linear representation of the metaplectic group $\mathbf{Mp}(2, \mathbb{R})$ in the projective space $\mathbf{P} L^2(\mathbb{R})$, the metaplectic representation ω acts on $\rho \hat{\otimes} \check{\rho}$ by conjugation. In terms of quantum information physics, it generates an extremal entropy.

Applications: The spin-echo technique of magnetic resonance as the workhorse of clinical MRI ([2], [20] and [44]). As illustrated by the figure, the timing diagram includes an excitation pulse, typically using a flip angle of $\alpha = 90°$, followed by one or more refocusing pulses, typically using a flip angle of $\alpha = 180°$ to perform a quantum gate operation geometrically supported by the mirrored pair of planar coadjoint orbits $(\mathbb{O}_\nu, \mathbb{O}_{-\nu})$ where $\nu \in \mathbb{R}^\times$. Conventional spin-echo imaging is a steady-state technique devoted almost exclusively to the organizations of two-dimensional multisection acquisitions. Its typical hologram trajectory collects the spectral database sequentially, line by line, moving from one extreme to the opposite extreme. Each line is defined by the spatio-temporal control of phase encoding gradients. Each position along the rows is defined by the spatio-temporal controlling of the frequency encoding gradients. The speed with which the hologram is traversed determines the scan time. Fast spin-echo imaging sequences provide scan times routinely two to 32 times faster than conventional spin-echo sequences.

The gradient-echo control technique of clinical magnetic resonance tomography with flip angle α ([20] and [44]). Comparing the timing diagram with the spin-echo pulse sequence, the most important difference to note is the absence of a refocusing pulse in the gradient-echo control method. Because the spatio-temporal encoding effect of a gradient pulse is reversed by a refocusing pulse, the gradient pulse controls along the phase-encoding and readout directions that occur before the refocusing pulse in the spin-echo imaging sequence are reversed in polarity for the gradient-echo pulse sequence. Gradient-echo pulse sequences are so named because it is the reversing polarity of the readout gradient control that creates the echo, not a refocusing pulse that overcomes the decoherence tendency of relaxation ([44]). The decoherence implies the delocalization of phase relations ([27] and [60]).

The spin-echo echo-planar imaging technique which is among the oldest MRI pulse sequence techniques ([20] and [44]). The timing diagram for a spin-echo EPI pulse sequence is analogous to that for a spin-echo imaging sequence. The key feature of the EPI technique is the rapidly oscillating readout gradient that collects repeated gradient echoes.

The phenomenon of quantum teleportation establishing the non-locality of quantum realism: *"Quantum spookiness wins, Einstein loses in photon test"*. It

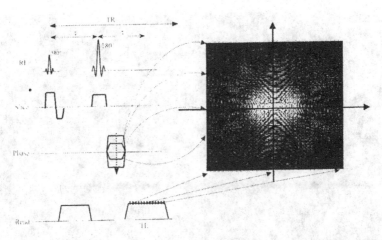

Figure 3. The spatio-temporal phase encoding gradient controls which row of the hologram is collected, whereas the sampling of the induced voltage determines which data of each row is collected. The figure displays the spatio-temporal phase encoding gradient that controls the row, the data sampling during the spatio-temporal frequency encoding procedure that determines the position along the row, and the appearance of the MRI raw database forming the hologram before its symplectic Fourier transform reconstruction.

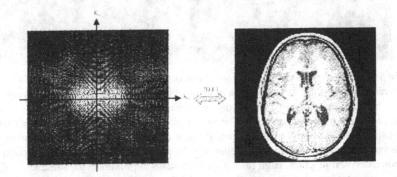

Figure 4. A full sampling of the hologram allows a symplectic Fourier transform reconstruction: The application of the symplectic Fourier transformer $\mathcal{F}_{\mathbb{R} \oplus \mathbb{R}}$ generates the final image.

is based on the Hopf fibration of the compact unit sphere $\mathbf{S}_3 \hookrightarrow \mathbb{C}^2$ given by the diagram

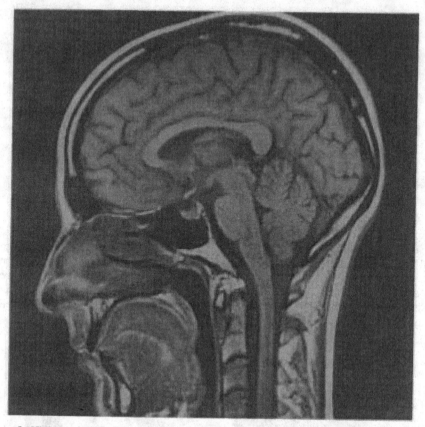

Figure 5. High resolution clinical magnetic resonance tomography: Sagittal cross-section of the neurocranium along the falx cerebri within the longitudinal interhemispheric fissure demonstrating midline sagittal neuroanatomy of the outwardly rounded gyri and inwardly invaginating fissures and sulci of the human brain. The various portions of the corpus callosum shown include the rostrum, genu, body and splenium, pineal gland, quadrigeminal plate, infundibulum, third and fourth ventricle, pituitary gland, cerebellar vermis, pons, aqueduct of Sylvius prepontine space, and craniocervical junction. High resolution MRI scans approximate the same level of detail as cut specimens to depict neuroanatomy even in the deepest recesses of the brain.

where the circle $\mathbf{S}_1$ is homeomorphic to the one-dimensional compact torus group $\mathbb{T} = \mathbf{U}(1, \mathbb{C})$. The compact base manifold

$$\mathbf{S}_2 = (\mathbb{C}^2 - \{0\})/\mathbb{C}^\times = \mathbf{P}_1(\mathbb{C})$$

is homeomorphic to the homogeneous manifold $\mathbf{SU}(2, \mathbb{C})/T$ consisting of the quantum rotation group $\mathbf{SU}(2, \mathbb{C})$ factored through the maximal torus

$$T = \left\{ \begin{pmatrix} \zeta & 0 \\ 0 & \bar{\zeta} \end{pmatrix} \mid \zeta\bar{\zeta} = 1 \right\}.$$

The non-degenerate two-dimensional central projective quadric Φ associated to $\mathbf{S}_2$ and defining the isotropic cone of rays in the three-dimensional real projective spinor space $\mathbf{P}_3(\mathbb{R})$ is homeomorphic to the compact subgroup of unit quaternions $\mathbf{SU}(2, \mathbb{C}) \hookrightarrow \mathbb{H}^{\times}$. Obviously the direct product identity

$$\mathbf{H}^{\times} = \mathbb{R}_+^{\times} \odot \mathbf{S}_3$$

holds. The left cosets

$$\{Q \cdot T \mid Q \in \mathbf{SU}(2, \mathbb{C})\}$$

are great circles of the quantum rotation group $\mathbf{S}_3 = \mathbf{SU}(2, \mathbb{C})$ which avoid the poles and provide the Hopf partition of $\mathbf{S}_3 \hookrightarrow \mathbb{R}^4$ via the submersion $\pi : \mathbf{S}_3 \to \mathbf{S}_2$ of the circle bundle over the compact unit sphere $\mathbf{S}_2$.

The two coverings $\mathbf{SU}(2, \mathbb{C}) - \{+I, -I\}$ and $\mathbf{S}_2 \times (\mathbf{S}_1 - \{1\})$ of $\mathbf{SO}(3, \mathbb{R}) - \{I\}$ are equivalent. To see this, observe that the hyperplane $q_0 = 0$ of the four-dimensional real vector space $\mathbf{H}$ consists of the *pure* or traceless quaternions $E \hookrightarrow \mathbf{H}$. The real vector space E of those quaternions which are equal to their *vector* parts admits as its canonical basis the *associated* Pauli spin matrices of the theory of Clifford algebras

$$\begin{pmatrix} i & 0 \\ 0 & -i \end{pmatrix}, \quad \begin{pmatrix} 0 & 1 \\ -1 & 0 \end{pmatrix}, \quad \begin{pmatrix} 0 & i \\ i & 0 \end{pmatrix}.$$

They represent the real form $\mathrm{Lie}(\mathbf{SU}(2, \mathbb{C}))$ on $\mathbb{C}^2$. Then the real vector space $\mathbf{H}$ of quaternions admits the direct sum decomposition

$$\mathbf{H} = \mathbb{R} \cdot I \oplus E,$$

where I spans the central subfield

$$\{(q_0, 0, 0, 0) \mid q_0 \in \mathbb{R}\}$$

of the skew field $\mathbf{H}$. Thus $\mathbf{H}$ should be considered as a central extension of the field $\mathbb{R}$.

The compact Lie group whose complexified Lie algebra is $\mathrm{Lie}(\mathbf{SL}(2, \mathbb{C}))$ is the special unitary group in two variables $\mathbf{SU}(2, \mathbb{C})$. It acts on the three-dimensional real vector space $E \hookrightarrow \mathbf{H}$ of traceless antihermitian matrices

$$\begin{pmatrix} q_1 i & q_2 + q_3 i \\ -q_2 + q_3 i & -q_1 i \end{pmatrix} \quad (0, q_1, q_2, q_3) \in \mathbb{H}$$

by rotations. The rotations keep invariant the natural scalar product

$$E \times E \ni (Q, Q') \rightsquigarrow -\frac{1}{2} \mathrm{Tr}(Q \cdot Q')$$

and hence the compact unit sphere $S_2 \hookrightarrow E$ of equation

$$q_1^2 + q_2^2 + q_3^2 = 1.$$

The Pauli spin matrix

$$\begin{pmatrix} i & 0 \\ 0 & -i \end{pmatrix}$$

forming the north pole of the sphere S_2 is associated to the zero-dimensional absolute quadric of conjugate cyclic points $\Phi_\infty^\circ = \{\bar{i}, i\}$ joined by the line at infinity of the complex projective plane $\mathbf{P}_2(\mathbb{C})$ under its standard Kähler metric. The conjugation actions of $\mathbf{H}$ on $\Phi_\infty^\circ \hookrightarrow \Phi_\infty$ provide the Kustaanheimo-Stiefel transformations mentioned above.

Conjugation of the canonical basis of $E \hookrightarrow \mathbf{H}$ by the elements of the real Lie group $\mathbf{SU}(2, \mathbb{C})$ provides the orthogonal representation of the quantum rotation group $\mathbf{SU}(2, \mathbb{C})$ in the Euclidean rotation group $\mathbf{SO}(E) = \mathbf{SO}(3, \mathbb{R})$ which is given by the surjective homomorphism

$$\begin{pmatrix} w_1 & w_2 \\ -\bar{w}_2 & \bar{w}_1 \end{pmatrix}$$

$$\rightsquigarrow \begin{pmatrix} w_1\bar{w}_1 - w_2\bar{w}_2 & i(\bar{w}_1w_2 - w_1\bar{w}_2) & \bar{w}_1w_2 + w_1\bar{w}_2 \\ i(w_1\bar{w}_2 - w_1w_2) & \frac{1}{2}(w_1^2 + \bar{w}_1^2 + w_2^2 + \bar{w}_1^2) & \frac{1}{2}(w_1^2 - \bar{w}_1^2 - w_2^2 + \bar{w}_1^2) \\ -(w_1\bar{w}_2 + w_1w_2) & \frac{1}{2}(\bar{w}_1^2 - w_1^2 + \bar{w}_2^2 - w_2^2) & \frac{1}{2}(w_1^2 + \bar{w}_1^2 - w_2^2 - \bar{w}_1^2) \end{pmatrix}$$

of $\mathbf{SU}(2, \mathbb{C})$ onto $\mathbf{SO}(E)$. Its kernel keeps the hyperplane $E \hookrightarrow \mathbf{H}$ pointwise fixed and is therefore given by the subgroup $\{+I, -I\}$ of order 2. Furthermore, it factors through the homeomorphism

$$\mathbf{SU}(2, \mathbb{C}) - \{+I, -I\} \longrightarrow S_2 \times (S_1 - \{1\})$$

composed with the projection

$$S_2 \times (S_1 - \{1\}) \longrightarrow \mathbf{SO}(3, \mathbb{R}) - \{I\}.$$

In terms of quaternions, the Hopf bundle has as its submersion $\pi : S_3 \to S_2$ the conjugation

$$\pi : \mathbf{SU}(2, \mathbb{C}) \ni Q \rightsquigarrow Q \cdot \begin{pmatrix} i & 0 \\ 0 & -i \end{pmatrix} \cdot Q^{-1}$$

because its fibers are the left cosets $\{Q \cdot T \mid Q \in \mathbf{SU}(2, \mathbb{C})\}$ indicated above. Due to the sandwich configuration of antipodal stereographic projections and the conjugation action of the metaplectic representation ω, the Hopf fibration model of the spinor structure on the tangent bundle TS_2 of S_2 forms the *geometric* background of the quantum teleportation phenomenon (Einstein-Podolsky-Rosen (EPR) experiment: Parametric down conversion of fluorescence type II). Thus there is actually neither mystery nor telepathy with *"the actions at a distance"* and Bell's theorem.

A companion of the orthogonal representation of the compact subgroup of unit quaternions $SU(2, \mathbb{C}) \hookrightarrow \mathbf{H}^{\times}$ in $SO(E)$ is the parametric representation of the Euclidean rotation group $SO(\mathbb{H}) = SO(4, \mathbb{R})$ of the four-dimensional real vector space of quaternions $\mathbf{H}$. Then

$$SU(2, \mathbb{C}) = SO(E) \cap GL(2, \mathbb{C}).$$

Every rotation belonging to $SO(\mathbb{H})$ can be written in the form

$$M_{Q,Q'} : R \rightsquigarrow Q \cdot R \cdot Q',$$

where $Q \in \mathbb{H}$ and $Q' \in \mathbb{H}$ are non-zero quaternions such that their norms satisfy

$$\|Q\| \cdot \|Q'\| = 1.$$

The consistent normalization

$$\|Q\| = \|Q'\| = 1$$

ensures that the assignment $(Q, Q') \rightsquigarrow M_{Q,Q'}$ defines a surjective homomorphism

$$SU(2, \mathbb{C}) \times SU(2, \mathbb{C}) \longrightarrow SO(\mathbb{H})$$

whose kernel is the two-element subgroup

$$N = \{(+I, +I), (-I, -I)\}.$$

of $SU(2, \mathbb{C}) \times SU(2, \mathbb{C})$. The center of the real product group $SU(2, \mathbb{C}) \times SU(2, \mathbb{C})$, however, is given by the product

$$S_2 = \{+I, -I\} \times \{+I, -I\},$$

hence a subgroup containing N of order 4. Thus the Euclidean rotation group $SO(\mathbb{H}) = SO(4, \mathbb{R})$ is isomorphic to the quotient

$$(SU(2, \mathbb{C}) \times SU(2, \mathbb{C}))/N.$$

It contains two normal subgroups $H_1 = M_{SU(2,\mathbb{C}),+I}$ and $H_2 = M_{+I,SU(2,\mathbb{C})}$ both isomorphic to the real Lie group $SU(2, \mathbb{C})$. It follows

$$H_1 \cap H_2 = S_2/N = \{+I, -I\}.$$

Toward quantum information processing: EPR information transmission channels ([22], [24] and [58]) of the quantum teleportation phenomenon of antisymmetrically polarization-entangled photon pairs. The antisymmetrically polarization-entangled photon pairs are generated via *equirepartitioned* sequences mod $\mathbb{Z} \oplus \mathbb{Z}$ in crystal lattices acting as EPR sources (BBO in Figure 6). Bohl's theorem on equirepartition then establishes that the slope $\tan \frac{\theta}{2} = \Im\tau$ of the incident laser beam direction, where $\tau \in \mathfrak{D}$, and the periodicity direction of L must be an *irrational* number in order to achieve a full torus.

Boundary identification provides the Keppler II law for the laser beam trajectory. The diffraction cone of the lattice lies within the convex light cone associated to the non-degenerate two-dimensional projective quadric Φ which is embedded into the three-dimensional real projective spinor space $\mathbf{P}_3(\mathbb{R})$.

When the slope $\Im\tau$ of the incident laser beam direction, where $\tau \in \mathfrak{D}$, is rational, the trajectory on the flat torus is periodic.

There is *no* collision of the phenomenon of quantum teleportation realized by an EPR experiment with the special theory of relativity ([11] and [19]). Niels Bohr, typically antiplatonic, did not like the von Neumann quantum theory and finally lost the war of quantum relativity ([22]).

The fundamental theorem of planetary dynamics exhibits the spin-drift resonance modes of the Keppler III spinorial spectral theorem. It exhibits the *twisted cubic* curve associated to the non-degenerate null polarity of the three-dimensional real projective spinor space $\mathbf{P}_3(\mathbb{R})$. The one-dimensional projective geometry of the twisted cubic is inherited from $\mathbf{P}_3(\mathbb{R})$. From any of its points, the twisted cubic projects onto a two-dimensional projective quadric in $\mathbf{P}_3(\mathbb{R})$. If the twisted cubic meets the plane at infinity of $\mathbf{P}_3(\mathbb{R})$ at two conjugate cyclic points on the isotropic lines of any pencil of lines in $\mathbf{P}_2(\mathbb{C})$, and a third point to boot, the planar trace of the projective quadric on the tangent plane of the absolute projective quadric $\Phi_\infty \hookrightarrow \mathbf{P}_2(\mathbb{C})$ is a central one-dimensional projective quadric, the elliptic planetary trajectory of the zeroth and first Keplerian laws. The conjugate cyclic points $\{\bar{i}, i\}$ forming the zero-dimensional projective quadric Φ_∞° and joined by the line at infinity of the complex projective plane $\mathbf{P}_2(\mathbb{C})$ enter the Laguerre projective phase formula as a phase reference. The associated orthogonality involution defines the *four* focal points of the Keppler ellipse via the conjugate isotropic tangents $\{T, \bar{T}\}$ and $\{S, \bar{S}\}$ from Φ_∞°, and the eccentricity. Indeed, the contragredient non-degenerate bilinear form defining the push-forward cotangential equation of Φ_∞ establishes that the two real or imaginary focal points determine the other two, as the non-cyclic intersection points of the isotropic lines going through the given foci. Reflection through the axes of the central projective Φ_∞ exchanges the conjugate cyclic points $\{\bar{i}, i\} \hookrightarrow \mathbf{P}(2, \mathbb{C})$ as well as the tangents drawn from these points. Passing to the intersection points $T \cap \bar{T}$ and $S \cap \bar{S}$, it follows that the real focal points lie on one of the two axes, and the imaginary focal points on the other. Both pairs are symmetric with respect to the center. In Keppler's helio*centric* system, one of the real focal points is occupied by the sun. The Laguerre projective phase formula then defines the relativistic perihelion precession of the planetary motions as well as the correct deflection effect of radar beams grazing the sun mass.

The two-fold covering group $\mathbf{Mp}(2, \mathbb{R})$ of the real symplectic or special real linear group $\mathbf{Sp}(2, \mathbb{R}) = \mathbf{SL}(2, \mathbb{R})$ exists because the fundamental group of $\mathbf{SL}(2, \mathbb{R})$ is the cyclic group $\mathbb{Z}$ ([30]). It contains the central subgroup C_2 isomorphic to the cyclic group $\mathbb{Z}/2\mathbb{Z}$ of order 2 such that

$$\mathbf{Mp}(2, \mathbb{R})/C_2 = \mathbf{Sp}(2, \mathbb{R})$$

and $\mathbf{Mp}(2, \mathbb{R})$ is a non-trivial extension of $\mathbf{Sp}(2, \mathbb{R})$ by C_2 in the sense that it is *not* the direct product $\mathbf{Sp}(2, \mathbb{R}) \times C_2$. One may regard the standard representation ρ of G as a representation of the semidirect product of $\mathbf{Mp}(2, \mathbb{R})$ and G.

The metaplectic representation ω of the metaplectic group

$$C_2 \longrightarrow \mathbf{Mp}(2, \mathbb{R})$$
$$\downarrow$$
$$\mathbf{Sp}(2, \mathbb{R})$$

acts on the standard complex Hilbert space $L^2(\mathbb{R})$ as an irreducible unitary linear representation. Its normalization action is defined via the covariance identity

$$\omega(\tilde{\alpha})\rho(g)\omega(\tilde{\alpha})^{-1} = \rho(\alpha(g))$$

for all elements $g \in G$ where the basepoint projection $\tilde{\alpha} \rightsquigarrow \alpha$ maps the metaplectic group $\mathbf{Mp}(2, \mathbb{R})$ onto the base manifold $\mathbf{Sp}(2, \mathbb{R}) = \mathbf{SL}(2, \mathbb{R})$. The point to note is that it is this mapping which indicates in the quantum teleportation problem that antisymmetrically polarization-entangled photon pairs can reproduce quantum information transmitted *instantaneously* to a remote location via EPR information transmission channels because $\omega(\tilde{\alpha})$ where α are phases in the subgroup $\mathbf{SO}(2, \mathbb{R})$ of the group of automorphisms $\mathbf{SL}(2, \mathbb{R})$ of G allows to perform instantaneously changes of the polarization state by antisymmetric continuous phase shifts in the relativistic trajectories of a photon pair interferometer which preserves the moments (Figure 6). The fact that

$$\mathbb{Z}/2\mathbb{Z} \longrightarrow \mathbf{Mp}(2, \mathbb{R})$$
$$\downarrow$$
$$\mathbf{SL}(2, \mathbb{R})$$

forms a two-fold covering group of the real symplectic group $\mathbf{Sp}(2, \mathbb{R}) = \mathbf{SL}(2, \mathbb{R})$ implies via the Maslov index ([12], [13], [33], [39] and [44]) a violation of Bell's inequality for certain states of polarization ([23] and [49]).

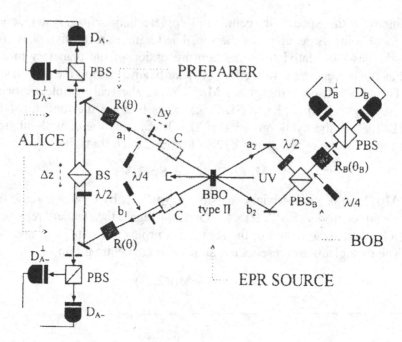

Figure 6. Quantum teleportation in a photon pair interferometer which preserves the moments. The EPR source spatially separates the quantum teleportation system into two remote subsystems. The photon trajectories of the left hand subsystem represent the preparing procedure of the spectral object. It allows to perform phase shifts by the action of the metaplectic representation ω. The left hand side subsystem forms the cohomological aspect of the chirality experiment. Calcite crystals (C) are acting as double refracting polarizers which separate horizontally and vertically polarized photons. The photon trajectories on the right hand side represent the sandwich configuration of mirrored pairs of planar coadjoint orbits which permit to acquire the phase shifts. The right hand side subsystem forms the cohomotopic aspect of the experiment. Both sides are non-locally coupled by the antisymmetric entanglement process of quantum reality (BS = beam splitter, D = detector).

However, antisymmetrically polarization-entangled photon pairs cannot perfectly clone the information which has been transmitted ([5] and [7]) so that there is an aspect of *undecidability* in the transmission of quantum information to remote locations via EPR information transmission channels. Because the 2-cocycle

$$c : (\tilde{\alpha}_1, \tilde{\alpha}_2) \rightsquigarrow c(\tilde{\alpha}_1, \tilde{\alpha}_2)$$

associated to ω is ∓ 1 valued ([12], [13], [30], [33] and [39]), the quantum non-cloning property is of cohomological nature. As a consequence, quantum information and classical information cannot be separated.

The Quantum Non-Cloning Theorem. The perfect reconstruction of quantum information transmitted to a remote location needs the transmission of at

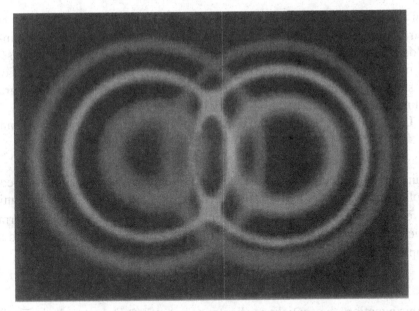

Figure 7. The generation of antisymmetrically polarization-entangled photon pairs for EPR information transmission channels. Three frequencies of the spectral object are spatially filtered out. Note the reversed ordering of the spinorial spectra in the two families of paratactic circles. Two spatially different circles from the same family not only do not intersect but they are linked in the sense of homotopy theory. Two circles from different families intersect in exactly two points.

least two bits of external classical information through a classical information transmission channel.

The conventional operator theoretic formulation of quantum physics does not cover the circle bundle models of spin-drift fields. However, due to the quantum non-cloning theorem ([5]), Einstein's reservations about quantum physics ("The more successes the quantum theory has the sillier it looks.") are made more precise ([24]).

3. The planetary macro system

The non-locality of the internal symmetry techniques applied to the Heisenberg nilpotent Lie group G allows to switch from the crystalline micro system to the planetary and even interstellar macro system. A simple but nevertheless important methodologic observation is that the discovery of the first Kepplerian fundamental law was the solution of a planar signal pattern recognition problem. The solution was based on the second Kepplerian law which was chronologically the first one in the list of the fundamental laws and created the chronology of the stratigraphic time scale. The present paper establishes that, in the spirit of the de Broglie spectral unveiling procedure of matter modes, the third Kepplerian law of planetary motion is a consequence of the symbolic calculus of the distributional wave equation which governs circle bundle models of spin-drift fields by means of Fourier transformers. The symbolic calculus is based on the metaplectic representation. The alternate name oscillator representation for the same representation of the metaplectic group remembers of the holographic technique of Kepplers libration theory ([6]).

Let $\mathbb{T} = \mathbf{U}(1, \mathbb{C})$ denote the one-dimensional compact torus group and

$$G \hookrightarrow \text{Loop}(\mathbb{T}^2)$$

again the Heisenberg nilpotent Lie group. Note that the preceding embedding implies the wavelet condition of vanishing integral ([38]). The distributional wave equation of the generic Fourier transformed propagator $\hat{P}$ in terms of which the Cauchy (initial value) problem is solved reads in the coordinates adapted to the unitary dual $\hat{G}$ as follows:

$$\left(\xi^2 + \eta^2 - t^2\right)\hat{P} = 0.$$

In this equation, the spatial coordinates $(\xi, \eta) \in \mathbb{R} \oplus \mathbb{R}$ denote dual symplectic coordinates of the generic Fourier transformed propagator $\hat{P}$ and

$$r_{2,1}^2(\xi, \eta, t) = \xi^2 + \eta^2 - t^2$$

denotes the square of the restricted Lorentz metric $r_{2,1}(\xi, \eta, t)$ of the vector $(\xi, \eta, t) \in \mathbb{R}^3$ of three-dimensional space-time manifold ([8]). Note that one has carefully to distinguish between the two opposite orientations of the real symplectic plane $\mathbb{R} \oplus \mathbb{R}$ having $\mathbf{P}_2(\mathbb{C})$ as the projectivization of its complexification, and the two different time orderings by virtue of the effect of time reversal.

The $r_{3,1}^2$-adapted Fourier transformed propagators

$$\hat{P}_0, \hat{P}_1$$

of the Cauchy problem for the distributional wave equation are compactly supported distributions invariant under the compact group $\mathbf{O}(2, \mathbb{R})$. Owing to the internal symmetry properties under infinitesimal dilations they admit the explicit form

$$\hat{P}_0 = c_0 t^2 \cdot m_{2,1}, \qquad \hat{P}_1 = c_1 t \cdot m_{2,1}$$

with suitable real constants c_0, c_1 of normalization. Here $m_{2,1}$ denotes the invariant measure of the Lorentz isometry group of the three-dimensional space-time geometry

$$\mathbf{O}(2, 1, \mathbb{R}) \hookrightarrow \mathbf{GL}(2, 1, \mathbb{R})$$

acting on the convex light cone of truncated rays

$$\{(\xi, \eta, t) \in \mathbb{R}^3 \mid r_{2,1}(\xi, \eta, t) = 0\} - \{(0, 0, 0)\}.$$

Each of its two connected components, the forward and backward light cone of truncated rays, is stabilized by the determinant one subgroup of index two

$$\mathbf{O}^+(2, 1, \mathbb{R}) \hookrightarrow \mathbf{O}(2, 1, \mathbb{R}).$$

Notice that the isometry group $\mathbf{O}^+(2, 1, \mathbb{R})$ is the image of the natural homomorphism $\mathbf{SL}(2, \mathbb{R}) \to \mathbf{GL}(3, \mathbb{R})$. If $\mathbb{R}^3$ with its canonical basis is identified with the real vector space of traceless matrices spanned by

$$\begin{pmatrix} 1 & 0 \\ 0 & -1 \end{pmatrix}, \quad \begin{pmatrix} 0 & 1 \\ 1 & 0 \end{pmatrix}, \quad \begin{pmatrix} 0 & -1 \\ 1 & 0 \end{pmatrix}$$

and equipped with the natural scalar product

$$(A, A') \rightsquigarrow \frac{1}{2} \mathrm{Tr}(A \cdot A'),$$

then the conjugation by elements of $\mathbf{SL}(2, \mathbb{R})$ preserves the restricted Lorentzian form $r_{2,1}^2$ of diagonal matrix

$$\begin{pmatrix} 1 & & \\ & 1 & \\ & & -1 \end{pmatrix}$$

of rank 3, and the isomorphy

$$\mathbf{O}^+(2, 1, \mathbb{R}) = \mathbf{SL}(2, \mathbb{R})/\{+I, -I\}$$

holds. If the associated $\mathbf{O}^+(2, 1, \mathbb{R})$ invariant measures $m_{2,1}^{\mp}$ on the forward and backward light cone of truncated rays are consistently normalized, then

$$\mathrm{supp}(m_{2,1}^{\pm}) = \{(\xi, \eta, t) \mid r_{2,1}(\xi, \eta, t) = 0\}$$

and the decomposition

$$m_{2,1} = m_{2,1}^{-} + m_{2,1}^{+}$$

holds. Notice also the subgroup inclusion

$$\mathbf{O}(2, \mathbb{R}) \times \mathbf{O}(1, \mathbb{R}) \hookrightarrow \mathbf{O}(2, 1, \mathbb{R})$$

preserving $t = 0$. The non-trivial element in the factor $\mathbf{O}(1, \mathbb{R})$ represents the effect of time reversal.

Let ρ denote the generic linear Schrödinger representation of G with generic orbit under the coadjoint action

$$\mathbb{O}_1 = \mathrm{CoAd}_G(G)(1)$$

in the vector space dual $\mathrm{Lie}(G)^\star$ of the real Heisenberg Lie algebra $\mathrm{Lie}(G)$. Then ρ acts trivially on the dual of the longitudinal center $C \hookrightarrow G$ and, according to the theorem of Stone-von Neumann ([42]), this action determines uniquely the equivalence class of ρ within the unitary dual $\hat{G}$.

Due to the fact that the nilpotent Lie group G is an exponential Lie group ([42]), the global diffeomorphism

$$\log_G : G \longrightarrow \mathrm{Lie}(G)$$

allows to identify conceptually distinct objects by means of one parameter subgroups of G. The reciprocal diffeomorphism is given by the surjective mapping

$$\exp_G : \mathrm{Lie}(G) \longrightarrow G,$$

alluded above. It is only by ignoring the spin-drift modes, hence by factoring out the center $C \hookrightarrow G$, that $\exp_G$ forms the *inverse* diffeomorphism of $\log_G$. Due to the transference property, the logarithmic map $\log_G$ provides the real Heisenberg Lie algebra $\mathrm{Lie}(G)$ as well as its vector space dual $\mathrm{Lie}(G)^\star$ with a logarithmic scale.

Toward quantum cosmology: The discovery of the logarithmic scale can be traced back to Keppler's first publication, the famous *Mysterium Cosmographicum* of 1595. In quantum cosmology, the intrinsic time appears as the logarithm of the expansion parameter of the Friedman model ([60]).

The point to note is that the Hopf bundle is *tautologous* in the sense that a point in $\mathbf{P}_1(\mathbb{C})$ represents a complex line in $\mathbb{C}^2$, and to this point its complex line is associated. By performing the fiberwise product with the real line $\mathbb{R}$, the four-dimensional Taub-NUT manifold ([21]) arises as a cylinder bundle over $\mathbf{S}_2$. A fifth dimension can be augmented either to make space-time geometry into a *curved* manifold which is embedded into the five-dimensional real projective space $\mathbf{P}_5(\mathbb{R})$, or to make space-time geometry into a three-dimensional real projective spinor space $\mathbf{P}_3(\mathbb{R})$. The choice of a particular Minkowskian space-time manifold of Clifford algebra $\mathscr{C}\ell(3, 1, \mathbb{R})$ in a five-dimensional space amounts to breaking the symmetry from the spin isomorphy

$$\mathbf{Spin}^\circ(3, 2, \mathbb{R}) = \mathbf{Sp}(4, \mathbb{R})$$

to the spin isomorphy

$$\mathbf{Spin}^\circ(3, 1, \mathbb{R}) = \mathbf{Sp}(2, \mathbb{C}).$$

The set of projective lines of the three-dimensional real projective spinor space $\mathbf{P}_3(\mathbb{R})$ forms the Plückerian *line* geometry $\bigwedge^2 \mathbb{R}^4$. Let the six-dimensional real Grassmannian manifold of projective lines of the three-dimensional real projective spinor space $\mathbf{P}_3(\mathbb{R})$ be defined by

$$\mathbf{G}_{4,2}(\mathbb{R}) = \bigwedge^2 \mathbb{R}^4 = \mathrm{Lie}\big(\mathbf{Spin}(4, \mathbb{R})\big),$$

and embedded into the five-dimensional real projective space $\mathbf{P}_5(\mathbb{R})$. In order to identify the projective lines of $\mathbf{P}_3(\mathbb{R})$ under their line coordinates or Plückerian homogeneous ray coordinates in the six-dimensional real vector space

$$\bigwedge^2 \mathbb{R}^4 = \bigwedge^2 \mathbb{R}^3 \oplus \bigwedge^2 \mathbb{R}^3$$

with a point set in $\mathbf{P}_5(\mathbb{R})$, the Kaluza-Klein model of the Minkowskian space-time manifold of Clifford algebra $\mathscr{C}\ell(3, 1, \mathbb{R})$ introduces the fundamental four-dimensional central Plücker quadric $\Psi \hookrightarrow \mathbf{P}_5(\mathbb{R})$. It is determined by the square root of the Pfaffian of the skew symmetric matrix of line coordinates and has signature

$$\begin{pmatrix} 1 & & & & & \\ & 1 & & & & \\ & & 1 & & & \\ & & & -1 & & \\ & & & & -1 & \\ & & & & & -1 \end{pmatrix}.$$

One fascinating aspect of the theory of projective quadrics is the behaviour of the vector spaces lying *on* them. The fundamental four-dimensional central Plücker quadric $\Psi \hookrightarrow \mathbf{P}_5(\mathbb{R})$ has two sets of plane generators and is bijectively mapped under the Plücker correspondence onto the *linear* line complex or *thread* consisting of self-polar lines for a null polarity of the three-dimensional real projective spinor space $\mathbf{P}_3(\mathbb{R})$. It is the existence of the Plücker correspondence and the natural *realization* of the Grassmannian manifold of projective lines

$$\mathbf{G}_{4,2}(\mathbb{R}) \hookrightarrow \mathbf{P}_3(\mathbb{R})$$

by the non-degenerate four-dimensional projective quadric

$$\Psi \hookrightarrow \mathbf{P}_5(\mathbb{R})$$

which justifies Einstein's and Pauli's vague statement that one *"cannot help feeling that there is some truth in Kaluza's five-dimensional theory"*.

The thread is associated to the non-degenerate *null* polarity or inclusion reversing *dualizing* correlation

$$\delta : \mathbf{G}_{4,k+1}(\mathbb{R}) \longrightarrow \mathbf{G}_{4,3-k}(\mathbb{R}) \quad (k \in \{0, 1, 2\})$$

inside the three-dimensional real projective spinor space $\mathbf{P}_3(\mathbb{R})$. It is uniquely determined up to factor $\neq 0$ by a twisted cubic curve in $\mathbf{P}_3(\mathbb{R})$ which is the intersection of three two-dimensional projective quadrics in $\mathbf{P}_3(\mathbb{R})$, and induced by the canonical symplectic form of the four-dimensional real vector space $\mathbb{R}^4$, realized by $\mathbb{C}^2$. The associated symplectic basis of $\mathbb{R}^4$ admits the representation by traceless matrices

$$E_1 = \begin{pmatrix} & -1 & & \\ 1 & & & \\ & & & 1 \\ & & -1 & \end{pmatrix}, \qquad E_2 = \begin{pmatrix} & & -1 & \\ & & & -1 \\ 1 & & & \\ & 1 & & \end{pmatrix},$$

$$E_3 = \begin{pmatrix} & 1 & & \\ 1 & & & \\ & & & -1 \\ & & -1 & \end{pmatrix}, \qquad E_4 = \begin{pmatrix} & & 1 & \\ & & & 1 \\ 1 & & & \\ & 1 & & \end{pmatrix}$$

with Clifford product

$$E_5 = \begin{pmatrix} -1 & & & \\ & 1 & & \\ & & -1 & \\ & & & 1 \end{pmatrix}.$$

It completes the set of generators $\{E_1, E_2, E_3, E_4\}$ of $\mathscr{C}\ell(3, 1, \mathbb{R})$ by the element $E_5 \in \mathscr{C}\ell(3, 1, \mathbb{R})$ which anticommutes which each of the generators, and satisfies the Clifford identity

$$E_5^2 = - \begin{pmatrix} 1 & & & \\ & 1 & & \\ & & 1 & \\ & & & 1 \end{pmatrix}.$$

In view of

$$\dim_{\mathbb{R}} \mathscr{C}\ell(3, 1, \mathbb{R}) = 16,$$

the scalar 1 together with the pentad $\{E_1, E_2, E_3, E_4, E_5\}$ and the set of bivectors $\{E_j E_k | 1 \leq j < k \leq 5\}$ form a real Clifford basis of $\mathscr{C}\ell(3, 1, \mathbb{R})$.

The skew symmetric matrix $E_1 E_2$ provides the null polarity δ with the symplectic normal form

$$\begin{pmatrix} & & & 1 \\ & & -1 & \\ & 1 & & \\ -1 & & & \end{pmatrix}.$$

Obviously, each point of $\mathbf{P}_3(\mathbb{R})$ is self-conjugate under the non-degenerate null polarity δ, and each line of $\mathbf{P}_3(\mathbb{R})$ passing through it and lying in its polar plane forms a null line, hence a self-conjugate line with respect to δ. The twisted cubic in $\mathbf{P}_3(\mathbb{R})$ is self-conjugate under the dualizing correlation δ with respect to the skew symmetric matrix

$$\begin{pmatrix} & & & 1 \\ & & -3 & \\ & 3 & & \\ -1 & & & \end{pmatrix}.$$

It forms the locus of the focal points of the osculating planes to the twisted cubic in $\mathbf{P}_3(\mathbb{R})$.

The thread associated to the dualizing correlation δ forms a three-dimensional projective quadric located on Ψ. As the trace of Ψ on a non-tangent hyperplane in $\mathbf{P}_5(\mathbb{R})$, it contains the non-degenerate two-dimensional projective quadric $\Phi \hookrightarrow \mathbf{P}_3(\mathbb{R})$ and admits *no* double point. It follows the peeling-off sequence of *spherical* inclusions at one by one stepwise increasing projective quadric dimensions

$$\Phi_{\infty}^{\circ} \hookrightarrow \Phi_{\infty} \hookrightarrow \Phi \hookrightarrow \Psi_{\infty} \hookrightarrow \Psi.$$

Two lines of the thread embedded into $\mathbf{P}_3(\mathbb{R})$ intersect if and only if the corresponding points of the real projective space $\mathbf{P}_5(\mathbb{R})$ are conjugate with respect to the non-degenerate projective quadric $\Psi \hookrightarrow \mathbf{P}_5(\mathbb{R})$. It follows that the set of lines of the Plückerian line geometry of the real projective spinor space $\mathbf{P}_5(\mathbb{R})$ is in bijective correspondence to the four-dimensional Laguerre *sphere* geometry of signature

$$\begin{pmatrix} 1 & & & \\ & 1 & & \\ & & 1 & \\ & & & -1 \end{pmatrix}$$

associated to the Lorentzian form $r_{3,1}^2$ on $\mathbb{R}^4$.

A line of the three-dimensional real projective spinor space $\mathbf{P}_3(\mathbb{R})$ is isotropic if and only if it is in bijective Plücker correspondence to a point of the three-dimensional *absolute* projective quadric $\Psi_{\infty} \hookrightarrow \mathbf{P}_5(\mathbb{R})$. As indicated by the connotation, Ψ_{∞} is the intersection of Ψ with the hyperplane at infinity of $\mathbf{P}_5(\mathbb{R})$. The absolute projective quadric Ψ_{∞} is a central quadric with the center as the pole of the hyperplane at infinity of $\mathbf{P}_5(\mathbb{R})$, located on the axis of the thread. All spheres of $\mathbf{P}_4(\mathbb{R})$ contain Ψ_{∞}, and the absolute projective quadric is invariant under the action of the orthogonal projective group $\mathbf{PO}(3, \Phi)$. If the line is non-isotropic, it determines a *biaxial involution* of $\mathbf{P}_3(\mathbb{R})$ which provides a map from the trace of the fundamental four-dimensional central Plücker quadric Ψ on $\mathbf{P}_3(\mathbb{R})$ onto the thread.

Due to the fact that the minimizing property of the Heisenberg helices derive from the natural symplectic structure of the planar coadjoint orbits of G, the transition to the spinorial spectra is performed by the Fourier transformer

$$\omega\left(J^{-1}\right) = \mp \exp_{\mathbf{SL}(2,\mathbb{R})} \left(\frac{\pi i}{2} \left(\frac{\pi r_{3,1}^2}{2} - \frac{\Box_{3,1}}{4\pi} \right) \right)$$

which is $r_{3,1}^2$-adapted to the full Lorentz isometry group $\mathbf{O}(3, 1, \mathbb{R})$ covering the orthogonal projective group $\mathbf{PO}(3, \Phi)$ of the four-dimensional Minkowskian space-time manifold of Clifford algebra $\mathscr{C}\ell(3, 1, \mathbb{R})$. As in the distributional wave equation, it provides the spinorial symbol calculus of the linear differential operators belonging to

$$\omega(\mathrm{Lie}(\mathbf{O}(3, 1, \mathbb{R}))) \hookrightarrow \mathrm{Lie}(\mathbf{SL}(2, \mathbb{R})).$$

The spinorial symbol calculus exhibits the relativistic Klein-Gordon equation

$$\Box_{3,1}\psi = m^2\psi \quad (m > 0)$$

for the complex-valued spectral object $\psi \in \mathscr{S}(\mathbb{R}^4)$, invariant under the action of the Lorentz group $\mathbf{O}(3, 1, \mathbb{R})$ ([46] and [47]). Explicitly, in terms of the Laplacian operator, the Klein-Gordon equation ([32] and [33]) reads

$$\left(\frac{\partial^2}{\partial t^2} + m^2 \right) \psi = \Delta_3\psi.$$

The d'Alembertian wave operator $\Box_{3,1} = \Delta_3 - (\partial^2/\partial t^2)$ is self-adjoint and admits two fundamental solutions that are of particular importance for the gravitational waves. They are, respectively, the advanced and retarded potentials of ψ, evaluated at the origin.

Taking the space-time potential in the complexified Clifford algebra

$$\mathscr{C}\ell(3, 1, \mathbb{R}) \otimes \mathbb{C}$$

leads by polarization to the Dirac equation for spin-$1/2$ particles. The point to note is that there are *two* different real structures on the same complex Clifford algebra.

The Schwarzschild radius R of general relativity theory, which is included into the Schwarzschild metric of cosmology as a spherical singularity of the Kruskal coordinatized two-fold covered Minkowskian space-time manifold of Clifford algebra $\mathscr{C}\ell(3, 1, \mathbb{R})$, is determined by the *rest-mass* m_0 (Figure 8). Any particle having a spherically symmetric gravitational field is contained in a black hole.

The logarithmic scale can be *spectrally* visualized by antisymmetrically polarization-entangled photon pairs generated to establish the quantum teleportation phenomenon via quantum holography. Due to the non-commutativity, the

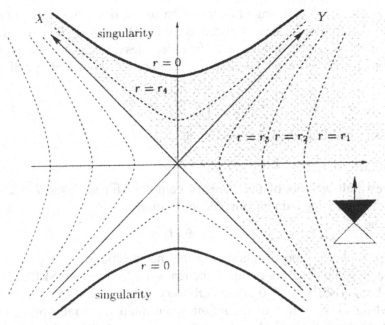

Figure 8. The Kruskal coordinatized two-fold covered Minkowskian space-time manifold displaying radial null geodesics as straight lines inclined at 45°. The region in the Kruskal diagram covered by the Schwarzschild coordinate chart is shaded. The metric is regular not only in the shaded area but in the entire area between the two branches of the hyperbola $r = 0$. This comprises two images of the exterior of the spherical singularity and two of its interior.

group law of G is *not* linearized under the logarithmic map although the image $\log_G(G)$ is a vector subspace of $\text{Lie}(G)$.

Infinitesimal dilations in all three directions preserve the space of solutions of the distributional wave equation. They act along truncated rays in the real plane $\mathbb{R}^2$ on the Fourier transformed propagators $\hat{P}_0$, $\hat{P}_1$ by topological vector space duality as follows:

$$\omega(s \cdot I)\hat{P}_0 = s^{-3/2} \cdot \hat{P}_0 \quad (s > 0),$$

and

$$\omega(s \cdot I)\hat{P}_1 = s^{-1/2} \cdot \hat{P}_1 \quad (s > 0).$$

The point to note is that the action of the infinitesimal dilation operator $\omega(s \cdot I)$ along truncated rays in the real plane $\mathbb{R}^2$ particularly includes the infinitesimal version of dilation along the longitudinal direction of the coadjoint orbit picture of the unitary dual $\hat{G}$. This *partial* action of the infinitesimal dilation operator $\omega(s \cdot I)$ allows spin-drift resonance adjustment by means of parallel displacement in the fiber bundle of phase shifts over the compact base manifold $\mathbf{S}_1$ along the longitudinal axis of $\hat{G}$ in order to remove the non-intrinsic singularity of the longitudinal logarithmic scale associated to the Borel subgroup $B_{\mathbb{R}} \hookrightarrow \mathbf{SL}(2, \mathbb{R})$.

The point to note is that the preceding identities are distributional *eigenvalue* equations for the infinitesimal dilation operator $\omega(s \cdot I)$ of lowest half-weights

$$\left\{ \frac{3}{2}, \frac{1}{2} \right\}$$

with associated Lie($\mathbf{SL}(2, \mathbb{R})$) submodules

$$M_{3/2}, M_{1/2}$$

of lowest half-weights of the standard complex Hilbert space $L^2(\mathbb{R})$. These submodules provide a direct sum decomposition

$$M_{3/2} \oplus M_{1/2}$$

of the metaplectic representation ω of the metaplectic group $\mathbf{Mp}(2, \mathbb{R})$ which forms a two-fold covering group of the real symplectic group $\mathbf{SL}(2, \mathbb{R})$.

If the real scalar $s \neq 0$ admits arbitrary sign, one has to replace s by $|s|$ everywhere in the action of the infinitesimal dilation operator $\omega(s \cdot I)$ along truncated rays in the real plane $\mathbb{R}^2$ and to insert $\text{sign}(s)$ into the sign sensitive action of the operator $\omega(s \cdot I)$ on the reference eigendistribution $\hat{P}_1$ of *lowest* half-weight $1/2$. In any event, the ratio of the lowest half-weights is an integer.

As in the Kepplerian libration theory ([52]), let $a > 0$ denote the length of the major semi-axis at the aphelion of the elliptical orbit, and $T > 0$ the period of the planetary orbit. The unit circle $t = 1$ given by

$$\xi^2 + \eta^2 = 1$$

on the intersection of the affine coadjoint orbit

$$\mathbb{O}_1 \hookrightarrow \text{Lie}(G)^\star$$

associated to ρ with the convex light cone serves as the gauge circle group. By identification of Lie($\mathbb{T}$) with the real line $\mathbb{R}$, the gauge circle group $t = 1$ can be parametrized by the spin-drift resonance modes of temporal frequencies located on the dual of the longitudinal center $C \hookrightarrow G$. In this way, the circular energy cylinder of screw motions

$$\mathbb{T} \times \mathbb{R}$$

arises as a Lorentzian manifold. The time compactification admits closed timelike curves.

By torus standardization to the unit circle traced out by spin-drift resonance modes on the dual of the longitudinal center $C \hookrightarrow G$, the logarithmic scale defined by

$$s = -\frac{1}{2} \log a$$

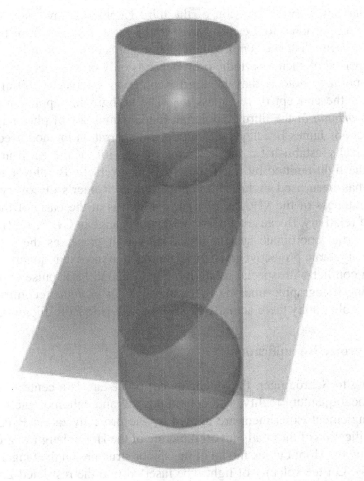

Figure 9. The symplectic spinorial geometry of the Keppler III spinorial spectral theorem: Visualization of a bitangential plane in the circular energy cylinder $\mathbb{T} \times \mathbb{R}$ which represents a coherent wave tube. The symplectic spinorial manifold allows also to establish the magnetic Aharonov-Bohm effect of electron holography by means of parallel displacements in the fiber bundle of phase shifts in spin-drift fields over the circle $\mathbf{S}_1 \hookrightarrow \mathbb{R} \oplus \mathbb{R}$.

provides the *ratio orbium* of the third Kepplerian spinorial spectral theorem.

The Keppler III Spinorial Spectral Theorem. In terms of gravitation geometrically implemented by the choice of the Borel subgroup $B_\mathbb{R} \hookrightarrow \mathbf{SL}(2, \mathbb{R})$, the 1–2–3 equation holds:

$$(\text{mass of the center of attraction})^1$$
$$= \left(\frac{2\pi}{T}\right)^2 \times (\text{major semi-axis of the elliptical orbit})^3.$$

A familiar but *incorrect* version of the third Kepplerian law asserts that the period is proportional to the $3/2$ power of the mean distance from the origin. In fact the mean distance is *not* equal to the major semi-axis a. It is noteworthy that the period of such a periodic orbit depends only on the energy.

The point to note is that the third Kepplerian spinorial spectral theorem introduces the concept of rest-mass m_0. The holographic approach to derive the *ratio orbium* of the third Kepplerian fundamental law of planetary motion remembers of James Bradley's elegant optical aberration method used in 1728 to empirically establish for the first time the motion of the earth around the sun by the light emitted by fixed stars ([6]). Conversely, Bradley's aberration method has been used in 1728 to confirm Olaus Römer's discovery in 1676 of the finiteness of the velocity of light c which is at the basis of the special theory of relativity. Because a linear coordinate change in $\mathbf{GL}(3, 1, \mathbb{R})$ provides a new inertial coordinate system if and only if it preserves the set of light rays, the algebraic projective geometric approach assures that quantum physics is *not* in conflict with special relativity. Femtosecond laser pulse experiments performing holographic imaging by capturing *light in flight* reconfirm that in quantum holography there actually exists no relativistic EPR dilemma.

4. The projective unification

According to Schrödinger, the entanglement procedure is a central ingredient of non-local quantum reality. The cohomological and cohomotopic aspects of the entanglement experiment are unified by the projectivization $\mathbf{P}_2(\mathbb{C})$ of the complexification of the coadjoint orbit picture of the Heisenberg Lie group G as realized by the Hopf bundle model of the spinor structure on the tangent bundle $\mathbf{T}\mathbf{S}_2$ of $\mathbf{S}_2$. Let the velocity of light c be inserted into the restricted Lorentzian form $r_{2,1}^2$ on $\mathbb{R}^3$ so that

$$r_{2,1}^2(\xi, \eta, t) = \xi^2 + \eta^2 - c^2 t^2,$$

and observe the explicit form of the Fourier transformed propagator of the Cauchy problem $\hat{P}_1$ associated to the lowest half-weight $1/2$ in terms of the invariant measure $m_{2,1}$ of the Lorentz isometry group $\mathbf{O}(2, 1, \mathbb{R})$. Notice that

$$\operatorname{supp}(\hat{P}_1) = \{(\xi, \eta, t) \in \mathbb{R}^3 \mid \xi^2 + \eta^2 - c^2 t^2 = 0\}$$

holds and that the compact support of the spin-drift propagator P_1 lies inside the convex light cone

$$\{(\xi, \eta, t) \in \mathbb{R}^3 \mid \xi^2 + \eta^2 - c^2 t^2 \leq 0\}.$$

Standardization of the total energy $\mathscr{E}$ onto the gauge circle group of the convex light cone associated to the non-degenerate projective quadric $\Phi \hookrightarrow \mathbf{P}_3(\mathbb{R})$ is performed by the infinitesimal dilation operator $\omega(s \cdot I)$, $s \neq 0$ along truncated rays in the real plane $\mathbb{R}^2$. Its partial action along the longitudinal axis allows the spin-drift resonance adjustment of the expansion in the logarithmic time scale by Lorentz contraction to the matter modes of the de Broglie spectral unveiling approach in information physics. For the total energy, it follows ($h =$ Planck's constant)

$$\mathscr{E} = h\nu = \frac{h\nu_0}{\sqrt{1 - (v^2/c^2)}} = \frac{m_0 c^2}{\sqrt{1 - (v^2/c^2)}},$$

where ν_0 denotes the temporal frequency implemented by the nested flat tori of the Hopf fibration of maximal bitangential gradient *bounded away* from the critical gradients

$$\mp \frac{\pi}{2}.$$

The gradients are in correspondence to the extremal rays of the convex light cone

$$\mathbb{R}_+^\times \ni s \rightsquigarrow \mp c \cdot s$$

which are parametrized by the Borel subgroup $B_{\mathbb{R}} \hookrightarrow \mathbf{SL}(2, \mathbb{R})$. Furthermore, $m_0 > 0$ denotes the rest-mass implemented by the third Kepplerian spinorial spectral theorem. The point to note is that the last identity reflects the sandwich configuration of mirrored pairs of planar coadjoint orbits of Chern class ∓ 1 supporting the total or relative energy $\mathscr{E}$, the momentum

$$\frac{h\nu}{c},$$

as well as the proper or rest-energy $\mathscr{E}_0$.

As a consequence of the antisymmetry of the relativistic Doppler effect, the Einstein equivalence of mass and energy ([18])

$$\mathscr{E} = mc^2 \quad \left(m = \frac{m_0}{\sqrt{1 - (v^2/c^2)}} \right),$$

and the rest-energy rest-mass identity of the special theory of relativity

$$\mathscr{E}_0 = m_0 c^2$$

follows. The relativistic kinetic energy associated to the velocity

$$v = c \cdot \tanh \nu$$

is then given by the expression

$$\mathscr{E}_{\text{kin}} = m_0 c^2 \left(\frac{1}{\sqrt{1 - (v^2/c^2)}} - 1 \right).$$

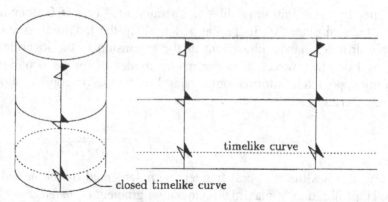

Figure 10. Einstein's cylinder world lines: Visualization of the three-dimensional cylindrical space-time manifold $\mathbb{T} \times \mathbb{R}$. The Einstein equivalence of mass and energy $\mathscr{E} = mc^2$ follows by spin-drift resonance adjustment to the sectional geometry of light cone and circular energy cylinder. Cylindrical gravitational waves are coherent solutions of the Einstein field equations and therefore accessible to quantum holography.

As a further consequence of the projective unification, the identity for the de Broglie wavelength λ_{dB} of the matter modes of the Bohr-Sommerfeld quantum condition follows ([8] and [27]) in terms of the impulse

$$p_0 = \frac{m_0 v}{\sqrt{1 - (v^2/c^2)}}$$

reads

$$\lambda_{dB} = \frac{h}{mv} = \frac{h\sqrt{1 - (v^2/c^2)}}{m_0 v}.$$

Then the total energy reads in terms of the impulse p_0 and the rest-mass m_0 defined by the Keppler III spinorial spectral theorem

$$\mathscr{E} = c\sqrt{m_0^2 c^2 + p_0^2}.$$

The last equation is called the *relativistic* energy theorem. It can be translated into the relativistic Klein-Gordon equation for the complex-valued spectral object $\psi \in \mathscr{S}(\mathbb{R}^4)$ by the spinorial symbol calculus

$$\Box_{3,1}\psi = m^2 \psi \quad (m > 0).$$

The relativistic Klein-Gordon equation can be regarded as a generalization of the Laguerre projective phase formula asserting that the orthogonality involution of $\mathbf{P}_2(\mathbb{C})$ under its standard Kähler metric has the isotropic lines through the pole with respect to $\Phi_\infty \hookrightarrow \mathbf{P}_2(\mathbb{C})$ of the line at infinity as self-corresponding or double lines. Thus the algebraic projective geometric approach to quantum physics which assures that the non-locality of quantum physics is *not* in conflict with special relativity ([11] and [19]), leads via the fundamental four-dimensional central Plücker quadric $\Psi \hookrightarrow \mathbf{P}_5(\mathbb{R})$ of the Plückerian line geometry of the real projective spinor space $\mathbf{P}_5(\mathbb{R})$ to the Klein-Gordon equation instead of the Schrödinger equation. Because the Schrödinger equation treats time and space coordinates differently, it is *not* of relativistic character and therefore not compatible with the transactional interpretation of the concept of quantum physical non-locality ([11]).

Two final comments are in order. Firstly, a relativistic consequence concerning the first Kepplerian fundamental law should be mentioned: Due to the relativistic rescaling of the chronology of the stratigraphic time scale, the perihelion of the planetary motions is phase shifted. The relativistic perihelion precession can be read off from the explicit form of the Heisenberg helices given above and the Laguerre projective phase formula. Thus the Keppler III spinorial spectral theorem permits a relativistic adjustment of the first Kepplerian planetary law. Its focal points admit the same projective base as the Laguerre phase formula. Similar arguments hold for the change of the periastron of the binary radio pulsar PSR 1913+16. Due to the dense matter of the companion star, the rotation is abpout 4 degrees per year.

Secondly, the spin-drift resonance adjustment is performed at the critical time T_c when the photon trajectory is confined to the convex light cone with probability larger than $1 - \varepsilon$ for any given small $\varepsilon > 0$. The critical time T_c is defined in terms of the dual symplectic coordinates $(\xi, \eta) \in \mathbb{R} \oplus \mathbb{R}$ of the critical vertex of a radial Lévy stochastic process ([4] and [36]) parametrized by the logarithmic scale of the Borel subgroup $B_{\mathbb{R}} \hookrightarrow \mathbf{SL}(2, \mathbb{R})$.

5. Concluding remarks

Carl Gustav Jacob Jacobi invented theta functions in the 1820s. They come into full flower in Jacobi's *"Fundamenta nova theoriae functionarum ellipticarum"* of 1829. Since then theta functions have been used in numerous investigations by generations of number theorists. They are involved in many fascinating identities of number-theoretical and combinatorial import, and they provide one of the most effective ways to construct automorphic forms. In the early 1960s André Weil, inspired by the work of Carl Ludwig Siegel on quadratic forms, provided a representation-theoretic foundation for the theory of theta functions ([57]). He discovered that theta functions were intimately connected with a most singular projective unitary representation ω of the real symplectic group $\mathbf{Sp}(2, \mathbb{R}) = \mathbf{SL}(2, \mathbb{R})$. This representation ω arises by virtue of the existence of a smooth action by automorphisms of the real symplectic group $\mathbf{SL}(2, \mathbb{R})$ on the two-step nilpotent Lie group G.

Whereas Hermann Weyl overlooked the natural occurrence of the Heisenberg group, Weil was working in the more general adelic framework of number theory, so he actually considered symplectic groups and Heisenberg groups with coefficients in a general local field when he showed that the metaplectic representation ω of the metaplectic group $\mathbf{Mp}(2, \mathbb{R})$ is intimately related to the law of quadratic reciprocity. When the ground field is $\mathbb{R}$, both groups are Lie groups and so have associated Lie algebras over $\mathbb{R}$. The commutation relations of the real Heisenberg Lie algebra are, as its name suggests, essentially the Heisenberg canonical commutation relations under the normalization $h = 1$. These form the foundations of quantum physics. Motivated by quantum field theory, David Shale, a student of Irving Segal, had several years earlier constructed the metaplectic representation ω of the real symplectic group ([48]). The remarkable coincidence that the internal global symmetries expressed by ω are involved in a fundamental way both in modern number theory and physics was discovered very soon, even before the publication of Weil's seminal paper ([57]), and is one of the most attractive of the many highly fascinating properties of the metaplectic representation ω.

Quantum physics and special relativity have more in common than might at first appear. The global symmetries inherent to the Heisenberg nilpotent Lie group G, represented by the metaplectic representation ω of the metaplectic group

allows to explain the entanglement phenomenon of quantum physics and to geometrically deduce the Einstein equivalence of mass and energy $\mathcal{E} = mc^2$ in relativity via the Keppler III spinorial spectral theorem. The underlying sub-Riemannian geometry of G combined with the radial Lévy stochastic process parametrized by the Borel subgroup $B_{\mathbb{R}} \hookrightarrow \mathbf{SL}(2, \mathbb{R})$ permits also to establish the magnetic Aharonov-Bohm effect of electron holography ([55]) by means of parallel displacements in the fiber bundle of phase shifts in spin-drift fields over the circle $\mathbf{S}_1 \hookrightarrow \mathbb{R} \oplus \mathbb{R}$. It confirms the non-locality phenomenon of quantum physics in an electron microscopical manner as convincing as the laser optical experiment and establishes the inherent Lorentz invariance of the projectivization $\mathbf{P}_2(\mathbb{C})$ of the complexification of the coadjoint orbit picture of the Heisenberg Lie group G which serves as a synchronized hypostasis for the spatio-temporal signal patterns. As a consequence, the global symmetries inherent to the Heisenberg nilpotent Lie group G permits to deduce the Schwarzschild metric in the coordinatized two-fold covered Minkowskian space-time manifold of Clifford algebra $\mathcal{C}\ell(3, 1, \mathbb{R})$. In analogy to the sandwich configuration of mirrored pairs of planar coadjoint orbits of G, the Kruskal coordinatized manifold can be thought of as being formed by two copies of asymptotically flat Schwarzschild manifolds joined at a throat which is determined by the Schwarzschild radius R. It is impossible to transmit information through the throat in such a way as to contradict the principle of causality.

The Kruskal solution is compatible with the Laguerre projective phase formula which permits to compute the relativistic precession of the perihelion of the planetary motions as well as the deviation of laser beams grazing the sun mass. As a maximal singularity-free extension of the Schwarzschild metric of cosmology it provides the complete description of a spherically symmetric black hole. Thus G is a fundamental structure for the large scale geometry of quantum cosmology, as well. The projectivization $\mathbf{P}_2(\mathbb{C})$ of the complexification of its coadjoint orbit picture is the mathematical base of quantum information processing. It allows to deduce its different microphysical and macrophysical system components from a single geometric object ([44]).

Acknowledgments

The authors are grateful to Professors George L. Farre and Maurice de Gosson, Dr. Peter Marcer, and Professor Jürgen Potthoff for valuable discussions and continuous support.

References

[1] Aspect, A. and Grangier, P., Experiments on Einstein-Podolsky-Rosen-type correlations with pairs of visible photons, In *Quantum Concepts on Space and Time*, R. Penrose and C.J. Isham (eds.), Oxford University Press, Oxford, New York, Toronto 1986, 1–15.

[2] D.M.S. Bagguley (ed.), *Pulsed Magnetic Resonance-NMR, ESR, and Optics: A Recognition of E.L. Hahn*, Oxford University Press, Oxford, New York, Toronto 1992.

[3] Bauer, R., van der Flierdt, E., Mörike, K., and Wagner-Manslau, C., *MR tomography of the central nervous system*, 2nd edition, Gustav Fischer Verlag, Stuttgart, Jena, New York 1993.

[4] Berton, J., *Lévy processes*, Cambridge University Press, Cambridge, New York, Melbourne 1996.

[5] Binz, E. and Schempp, W., Quantum teleportation and spin echo: A unitary symplectic spinor approach, In *Aspects of Complex Analysis, Differential Geometry, Mathematical Physics and Applications*, S. Dimiev and K. Sekigawa (eds.), World Scientific, Singapore, New Jersey, London 1999, 314–365.

[6] _____, *Vector fields in three-space, natural internal degrees of freedom, signal transmission and quantization*, Results in Math. 37 (2000), 226–245.

[7] _____, Quantum systems: From macro systems to micro systems. The holographic technique, In *Cybernetics and Systems 2000*, vol. 1, R. Trappl (ed.), Austrian Society for Cybernetic Studies and University of Vienna, Vienna 2000, 123–128.

[8] _____, *Spacetime geometry and information transmission*, Preprint (in progress).

[9] Bloch, A., Sur les circles paratactiques et la cyclide de Dupin, *J. de Math.* **3** (1924), 51–78.

[10] Callaway, J., *Quantum theory of the solid state*, 2nd edition, Academic Press, Boston, San Diego, New York 1991.

[11] Cramer, J.G., The transactional interpretation of quantum mechanics, *Rev. Mod. Phys.* **58** (1986), 647–687.

[12] de Gosson, M., Maslov indices on the metaplectic group $Mp(n)$, *Ann. Inst. Fourier, Grenoble* **40** (1990), 537–555.

[13] _____, *Maslov classes, metaplectic representation and Lagrangian quantization*, Akademie Verlag, Berlin 1997.

[14] Edelman, R.R., Hesselink, J.R., and Zlatkin, M.B., *Clinical Magnetic Resonance Imaging*, vol. 1 and 2, 2nd edition, W.B. Saunders (ed.), Philadelphia, London, Toronto 1996.

[15] Dixmier, J., C^*-Algebras. North Holland, Amsterdam, New York, Oxford 1977.

[16] Duvernoy, H.M., *The human brain: Surface, three-dimensional sectional anatomy with MRI, and blood supply*, 2nd edition, Springer-Verlag, Wien, New York 1999.

[17] Elbert, T. and Keil, A., Imaging in the fourth dimension, *Nature* **404** (2000), 29–31.

[18] Einstein, A., Elementary derivation of the equivalence of mass and energy, *Bull. (New Series) Amer. Math. Soc.* **37** (1999), 39–44; Reprinted from *Bull. Amer. Math. Soc.* **41** (1935), 223–230.

[19] Grössing, G., Nonlocality and the time-ordering of events, In *Cybernetics and Systems 2000*, vol. 1, R. Trappl (ed.), Austrian Society for Cybernetic Studies and University of Vienna, Vienna 2000, 185–188.

[20] Haacke, E.M., Brown, R.W., Thompson, M.R., and Venkatesan, R., *Magnetic Resonance Imaging: Physical Principles and Sequence Design*, Wiley-Liss, New York, Chichester, Weinheim 1999.

[21] Hawking, S.W. and Ellis, G.F.R., *The large scale structure of space-time*, Cambridge University Press, Cambridge, New York, Melbourne 1973.

[22] Held C., *Die Bohr-Einstein-Debatte: Quantenmechanik und physikalische wirklichkeit*, Mentis, Paderborn 1999.

[23] Horne, M.A., Shimony, A., and Zeilinger, A., Introduction to two-particle interferometry, In *Sixty-Two Years of Uncertainty: Historical, Philosophical and Physical Inquiries into the Foundations of Quantum Mechanics*, A.I. Miller (ed.), Plenum Press, New York, London 1990, 113–119.

[24] Howard, D., "Nicht sein kann was nicht sein darf," or the prehistory of EPR, 1909–1935: Einstein's early worries about the quantum mechanics of composite systems. In *Sixty-Two Years of Uncertainty: Historical, Philosophical and*

Physical Inquiries into the Foundations of Quantum Mechanics, A.I. Miller (ed.), Plenum Press, New York, London 1990, 61–111.

[25] Igusa, J., *Theta functions*, Springer-Verlag, Berlin, Heidelberg, New York 1972.

[26] Jinkins, J.R., *Atlas of neuroradiologic embryology, anatomy, and variants*, Lippincott Williams & Wilkins, Philadelphia, Baltimore, New York 2000.

[27] Joos, E., Decoherence through interaction with the environment, In *Decoherence and the Appearance of a Classical World in Quantum Theory*, D. Guilini, E. Joos, C. Kiefer, J. Kupsch, I.-O. Stamatescu, and H.D. Zeh (eds.), Springer-Verlag, Berlin, Heidelberg, New York 1996, 35–136.

[28] Jurdjević, V., *Geometric control theory*, Cambridge University Press, Cambridge, New York, Melbourne 1997.

[29] Kirillov, A.A., Merits and demerits of the orbit method, *Bull. (New Series) Amer. Math. Soc.* **36** (1999), 433–488.

[30] Kostant, B., Symplectic spinors, In *Geometria Simplettica e Fisica Matematica*, *Symposia Mathematica*, vol. XIV, Academic Press, London, New York 1974, 139–152.

[31] J. Krüger (ed.), *Neuronal cooperativity*, Springer-Verlag, Berlin, Heidelberg, New York 1991.

[32] Leray, J., Hyperbolic differential equations, The Institute of Advanced Studies, Princeton, New Jersey 1952.

[33] _____, *Lagrangian analysis and quantum mechanics: A mathematical structure related to asymtotic expansions and the Maslov index*, The MIT Press, Cambridge, Massachusetts 1981.

[34] Leith, E.N., Synthetic aperture radar, In *Optical Data Processing*, D. Casasent (ed.), Springer-Verlag, Berlin, Heidelberg, New York 1978, 89–117.

[35] Moore, C.C. and Wolf, J.A., Square integrable representations of nilpotent groups, *Trans. Amer. Math. Soc.* **185** (1973), 445–462.

[36] Nagasawa, M., *Stochastic processes in quantum physics*, Birkhäuser Verlag, Basel, Boston, Berlin 2000.

[37] Polanyi, M., *The tacit dimension*, Doubleday & Company, Garden City, New York 1966.

[38] Pressley, A. and Segal, G., *Loop groups*, Oxford University Press, Oxford, New York, Toronto 1990.

[39] Ranga Rao, R., The Maslov index on the simply connected covering group and the metaplectic representation, *J. Funct. Anal.* **107** (1992), 211–233.

[40] Rothschild, P.A. and Reinking Rothschild, D., *Open MRI*, Lippincott Williams & Wilkins, Philadelphia, Baltimore, New York 2000.

[41] Sachs, R.K. and Wu, H., General relativity for mathematicians, Springer-Verlag, New York, Heidelberg, Berlin 1977.

[42] Schempp, W., *Harmonic analysis on the Heisenberg nilpotent Lie group, with applications to signal theory*, Longman Scientific and Technical, London 1986.

[43] _____, Zu Kepplers conchoid-konstruktion, *Results in Math.* **32** (1997), 352–390.

[44] Schempp, W.J., *Magnetic resonance imaging: Mathematical foundations and applications*, Wiley-Liss, New York, Chichester, Weinheim 1998.

[45] Schempp, W., Sub-Riemannian geometry and clinical magnetic resonance tomography, *Math. Meth. Appl. Sci.* **22** (1999), 867–922.

[46] Schwartz, L., Sous-espaces hilbertiens et noyaux associés; applications aux représentations des groupes de Lie, In *Deuxième Colloq. l'Anal. Fonct.*, Centre Belge Recherches Math., Librairie Universitaire, Louvain 1964, 153–163.

[47] ———, *Application of distributions to the theory of elementary particles in quantum mechanics*, Gordon and Breach, New York, London, Paris 1968.

[48] Shale, D., Linear symmetries of free boson fields, *Trans. Amer. Math. Soc.* **103** (1962), 149–167.

[49] Shimony, A., An exposition of Bell's theorem, In *Sixty-Two Years of Uncertainty: Historical, Philosophical and Physical Inquiries into the Foundations of Quantum Mechanics*, A.I. Miller (ed.), Plenum Press, New York, London 1990, 33–43.

[50] Singer, W., Synchronization of cortical activity and its putative role in information processing and learning, *Annu. Rev. Physiol.* **55** (1993), 349–374.

[51] Stapp H.P., *Mind, matter, and quantum mechanics*, Springer-Verlag, Berlin, Heidelberg, New York 1993.

[52] Stephenson, B., *Kepler's physical astronomy*, Princeton University Press, Princeton, New Jersey 1994.

[53] Tamraz, J.C. and Comair, Y.G., *Atlas of regional anatomy of the brain using MRI: With functional correlations*, Springer-Verlag, Berlin, Heidelberg, New York 2000.

[54] Terras, A., *Fourier analysis on finite groups and applications*, Cambridge University Press, Cambridge, New York, Melbourne 1999.

[55] Tonomura, A., *Electron holography*, 2nd edition, Springer-Verlag, Berlin, Heidelberg, New York 1999.

[56] Truwit, C.L. and Lempert, T.E., *High resolution atlas of cranial neuronanatomy*, Williams & Wilkins, Baltimore, Philadelphia, Hong Kong 1994.

[57] Weil, A., Sur certains groupes d'opérateurs unitaires, *Acta Math.* **111** (1964), 143–211. In *Collected Papers*, vol. III, Springer-Verlag, New York, Heidelberg, Berlin 1980, 1–69.

[58] J.A. Wheeler and W.H. Zurek (eds.), *Quantum theory and measurement*, Princeton University Press, Princeton, New Jersey 1983.

[59] Wootters, W.K., Local accessibility of quantum states, In *Complexity, Entropy and the Physics of Information*, W.H. Zurek (ed.), Santa Fe Institute Studies in the Sciences of Complexity, Addison-Wesley, Redwood City, California, Menlo Park, California, Reading, Massachusetts 1990, 39–46.

[60] Zeh, H.D., Quantum mesasurements and entropy, In *Complexity, Entropy and the Physics of Information*, W.H. Zurek (ed.), Santa Fe Institute Studies in the Sciences of Complexity, Addison-Wesley, Redwood City, California, Menlo Park, California, Reading, Massachusetts 1990, 405–422.

SUR LE CONTRÔLE DES ÉQUATIONS DE NAVIER–STOKES

J.L. LIONS

À la mémoire de Jean Leray

Contents

1   Résumé 545
2   Introduction et conjectures 545
3   Équations de Navier–Stokes linéarisées 551
    3.1   Énoncé du résultat 551
    3.2   Premières estimations *a priori* 552
    3.3   Système d'Optimalité (S.O.) 552
    3.4   Estimations pour l'état adjoint 553
    3.5   Dérivation en μ 553
    3.6   Estimations 554
4   Equations de Navier–Stokes (et autres modèles et problèmes) 556
    4.1   Equations de Navier–Stokes 556
    4.2   Autres modèles 557
    4.3   Problèmes 557

M. de Gosson (ed.), Jean Leray '99 Conference Proceedings, 543-558.

1. Résumé

On étudie dans cet article le contrôle de l'écoulement d'un fluide incompressible dans un ouvert Ω de $\mathbb{R}^3$. L'état du système est donné par les *solutions faibles* (ou *turbulentes* dans la terminologie introduite en 1933 par Jean Leray) des équations de Navier–Stokes. On peut agir sur la vitesse (et la pression) de l'écoulement par un contrôle distribué (dans une partie de Ω) ou frontière (sur une partie du bord de Ω). On considère ici le cas de contrôles distribués, pour un peu simplifier l'exposé. La situation, pour les conjectures et les résultats disponibles est la même pour le cas des contrôles frontères.

Dans cette situation —et *dans toutes les situations de contrôle de systèmes turbulents ou chaotiques*— on a trois conjectures principales:

Conjecture 1.1 ([12]). Il y a contrôlabilité approchée.

Des résultats importants ont été obtenus dans ces directions, notamment par J.–M. Coron, A. Fursikov et O. Yu. Imanuvilov.

Conjecture 1.2 (E. Zuazua et l'Auteur). Le coût du contrôle, convenablement défini, est majoré par une constante indépendante de la viscosité μ, lorsque $\mu \to 0$.

On formule et on étudie pour la première fois ici la Conjecture 1.3:

Conjecture 1.3. Il existe un contrôle optimal dépendant de façon différentiable de μ. La norme de la dérivée en μ de ce contrôle tend vers l'infini comme $1/\mu$ lorsque $\mu \to 0$.

L'analogue de cette conjecture est démontré dans la section 3 par les équations de Navier–Stokes *linéarisées*.

2. Introduction et conjectures

Soit Ω un ouvert borné de $\mathbb{R}^3$. On considère dans Ω un écoulement de fluide Newtonien visqueux. La vitesse de l'écoulement est *l'état* du système étudié. Elle est représentée par
$$y = \{y_1, y_2, y_3\},$$
fonction de $x \in \Omega$, $t \in (0, T)$ et *d'un contrôle* $v = \{v_1, v_2, v_3\}$.

Ce contrôle v est 'appliqué' sur un *ouvert* $\mathbb{O} \subset \Omega$. On dit que c'est un *contrôle distribué* (il est distribué sur $\mathbb{O}$).

Remarque. Dans les applications, aéronautiques notamment, le contrôle est (le plus souvent) appliqué sur une partie de la frontière Γ de Ω. Tout ce qui va être dit est valable dans ce cas, au prix de complications techniques que nous essayons de réduire dans le présent article.

L'état y est donc «donné» par 'la' solution des équations de Navier–Stokes

$$\left|\begin{array}{l} \dfrac{\partial y}{\partial t} + y\nabla y - \mu\Delta y = v1_{\mathbb{O}} - \nabla p, \quad \mu \text{ donné} > 0, \\[2mm] \text{div } y = 0 \end{array}\right. \tag{2.1}$$

avec la condition (le fluide adhère aux parois)

$$y = 0 \text{ sur } \Gamma \times (0, T) \tag{2.2}$$

et la condition

$$y|_{t=0} = y(x, 0) = y^0(x). \tag{2.3}$$

Dans (2.1) p désigne la pression et $1_{\mathbb{O}}$ la fonction caractéristique de $\mathbb{O}$. *On veut «agir sur le système» —C-á-d, choisir le contrôle v —de façon à ce qu'il se comporte conformément à nos souhaits.*

Avant de préciser cela, il convient de préciser ce que l'on entend par «solution» de (2.1), (2.2) et (2.3). On utilise les solutions faibles «à la Leray» ([8], [9] et [10]). Pour les définir, on introduit les espaces fonctionnels suivants

$$V = \left\{\varphi \mid \varphi = \{\varphi_1, \varphi_2, \varphi_3\}, \ \varphi_k \in H^1(\Omega), \ \varphi_k = 0 \text{ sur } \Gamma, \ \text{div } \varphi = 0\right\}$$

où $\varphi_k \in H^1(\Omega)$ signifie $\varphi_k, \partial\varphi_k/\partial x_i \in L^2(\Omega)$, $i = 1, 2, 3$), $H = $ fermeture de V dans $(L^2(\Omega))^3$.

On pose, pour $\varphi, \widehat{\varphi} \in V$

$$(\varphi, \widehat{\varphi}) = \sum_{i=1}^{3} \int_{\Omega} \varphi_i \widehat{\varphi}_i \, dx,$$

$$a(\varphi, \widehat{\varphi}) = \sum_{i,j} \int_{\Omega} \frac{\partial\varphi_i}{\partial x_j} \frac{\partial\widehat{\varphi}_i}{\partial x_j} \widehat{\varphi}_i \, dx$$

et pour $\varphi, \psi, \widehat{\varphi} \in V$,

$$b(\varphi, \psi, \widehat{\varphi}) = \sum_{i,j} \int_{\Omega} \psi_j \frac{\partial\varphi_i}{\partial x_j} \widehat{\varphi}_i \, dx.$$

La formation faible variationelle de (2.1), (2.2) et (2.3) est alors:
trouver y avec

$$y \in L^2(0, T; V) \cap L^\infty(0, T; H) \tag{2.4}$$

satisfaisant à

$$\left(\frac{\partial y}{\partial t}, \widehat{y}\right) + \mu a(y, \widehat{y}) + b(y, y, \widehat{y}) = (v1_{\mathbb{O}}, \widehat{y}) \quad \forall \widehat{y} \in V \tag{2.5}$$

avec en outre la condition (2.3) où y^0 est donné dans H.

Remarque. Il faut en fait préciser cela en remplaçant dans (2.5) $\widehat{y}$ par une fonction $\widehat{y}(t)$ dépendant de t de façon régulière et satisfaisant à $\widehat{y}(T) = 0$. On intègre alors par parties en t [11].

On démontre alors *l'existence d'une solution* —ce que Leray appelait «*solution turbulente* » de (2.4), (2.5) et (2.3).

On montre que y a la propriété supplémentaire [11]

$$\frac{\partial y}{\partial t} \in L^2 \left(0, T; V^{-3/2} \right) \tag{2.6}$$

(où $V^{-3/2}$ utilise les espaces $H^{-3/2}(\Omega)$, obtenus par dualité de $H_0^{3/2}(\Omega)$, [14]). Beaucoup d'autres propriétés ont été obtenues [7] et [19] mais le *problème de l'unicité est toujours ouvert depuis Leray*.

Remarque. On considère ici des fluides incompressibles (d'où la condition div $y = 0$). Pour les fluides compressibles, la situation connue actuellement est présentée par [20].

Dans la suite, on désigne donc par $y(x, t; v) = y(v)$ *l'une quelconque des solutions (éventuelles) de* (2.4), (2.5) *et* (2.6).

C'est *l'état* $y(v)$, ainsi défini (*peut-être* donc s'agit-il d'un *ensemble*), que l'on veut «contrôler».

On va considérer ici le problème de la *contrôlabilité*, le plus important (et le plus difficile) dans ces directions.

On désigne par $y(T; v)$ «l'ensemble» des fonctions $x \to y(x, T; v)$. Ce sont des éléments de H. Le système est dit *approximativement contrôlable si l'ensemble parcouru par tous les* $y(T; v)$ *lorsque* v *parcourt l'espace de tous les contrôles*

$$v \in \mathcal{U} = L^2(\mathbb{O} \times (0, T))^3 \tag{2.7}$$

est dense dans H.

On a formulé en 1989 la conjecture suivante:

Conjecture 2.1. Le système régi par les équations de Navier–Stokes est approximativement contrôlable.

Remarque. Dans cette conjecture $\mathbb{O}$ est un ouvert quelconque contenu dans Ω, donc arbitrairement petit.

Si $\mathbb{O} = \Omega$ le problème est essentiellement trivial. On a alors contrôlabilité exacte. Etant donné y^T dans H, il existe un contrôle v tel que

$$y(T; v) = y^T.$$

Remarque. La même conjecture est formulée dans l'Auteur [12] pour des contrôles frontières.

Remarque. En fait nous avons formulé dans [12] une conjecture plus générale, pour tous les systèmes qui peuvent être turbulents ou chaotiques, tels les fluides compressibles et tels aussi les systèmes de la météorologie ou de la climatologie.

Remarque. Au cours d'un exposé sur [12] à l'Académie des Sciences d'Israël, le Professeur Dvoretsky a signalé une remarque attribuée à J. von Neumann «Le climat est plus facile à contrôler qu'à prévoir ». (Voir l'exposé de J. von Neumann sur le contrôle du climat [21].)

Remarque. La conjecture —si elle est vraie!—, voir Remarques 2 à 2 ci-dessous, signifie que, étant donné y^T dans H, et étant donné $\beta > 0$ arbitrairement petit, il existe v tel que

$$y(T; v) \in y^T + \beta B \tag{2.8}$$

(pour l'une des solutions de la famille éventuelle des solutions) où B désigne la boule unité de H.

Il est peu probable que la conjecture soit valide pour une topologie beaucoup plus fine que celle de H, mais cette question est largement ouverte.

Remarque. Le cas $\mu = 0$ correspond aux équations d'Euler. On peut formuler le même problème (on cherche alors $y \in L^\infty(0, T; H)$, y ne peut être à valeurs dans V) et la même conjecture. En fait il y a ici peut-être beaucoup plus. J.–M. Coron [1] a en effet démontré la contrôlabilité exacte, dans les espaces de vecteurs indéfiniment différentiables, mais, pour l'instant, seulement en dimension 2 ($\Omega \subset \mathbb{R}^2$).

Remarque. La conjecture a été démontrée [2] et [4] en dimension 2, à quelques difficultés techniques près (conditions sur la topologie de Ω, contrôle frontière «non standard»).

Dans le cas de la dimension 3, [1] et [6] ont démontré (2.8) (avec β arbitrairement petit) dans le cas où y^T 'n'est pas trop éloigné' de y^0.

Remarque. E. Zuazua et l'Auteur ont démontré la conjecture [16] pour le cas des approximations en dimension finie du système (2.5) (ceci par la méthode de Galerkin, pour un espace «convenable» d'approximation), et aussi pour le contrôle frontière.

Tout cela rend raisonnable de s'attaquer aux questions fondamentales suivantes:

(i) admettant possible la réalisation de (2.8), *à quel coût*? Cela est précisé ci-dessous, pour aboutir à la Conjecture 1.2.

(ii) ayant trouvé un contrôle $v = v(\mu)$ réalisant (2.8) au «meilleur coût », *comment ce contrôle* (optimal) *dépend-t-il de* μ? Cela est précisé ci-dessous, pour aboutir à la Conjecture 1.3, présentée ici pour la première fois et étudiée dans les sections 3 et 4 ci-dessous.

Précisons d'abord ce que l'on entend par *coût de contrôle*.
La définition correspond naturellement à (2.8) est

$$\text{Coût} = \inf \frac{1}{2}\|v\|_{\mathcal{U}}^2, \qquad y(T; v) \in y^T + \beta B, \tag{2.9}$$

où $\mathcal{U}$ est l'espace introduit dans (2.7), munie de sa norme notée $\| \ \|_{\mathcal{U}}$.

Remarque. Dans (2.9) le coût dépend, bien entendu de β, de y (et de $\mathbb{O}$), mais surtout *de* μ (la viscosité). On va, dans la suite, étudier la dépendance du coût en μ.

La définition (2.9) est évidemment ambigüe, puisque son énoncé même repose sur une conjecture. C'est pourquoi on modifie la notion de coût de la manière suivante. On introduit

$$J(v) = \frac{1}{2}\|v\|_{\mathcal{U}}^2 + \frac{\alpha}{2}\left\| y\left(T; v - Y^T\right)\right\|_H^2 \tag{2.10}$$

et *on définit*

$$\text{Coût} = \mathcal{C}(\mu) = \inf J(v), \quad v \in \mathcal{U}. \tag{2.11}$$

Remarque. On peut montrer qu'il existe au moins un élément $v(\mu)$ tel que

$$J(v(\mu)) = \inf J(v) = \mathcal{C}(\mu). \tag{2.12}$$

Ce sont les fonctions $\mu \to \mathcal{C}(\mu)$ et $\mu \to v(\mu)$ que l'on étudie ci-après, *lorsque* $\mu \to 0$.

Remarque. Même dans les situations où la contrôlabilité (approchée ou exacte) est établie, il est habituel de remplacer (2.9) par (2.10), sachant que pour un choix convenable de α (dépendant de β), (2.9) et (2.10) sont 'équivalents'. Ceci est précisé (pour des problèmes linéaires) dans R. Glowinski et l'Auteur [5], la formation (2.10) étant celle utilisée de préférence dans les approximations numérique

On peut maintenant formuler la:

Conjecture 2.2. Lorsque $\mu \to 0$, $\mathscr{C}(\mu)$ demeure borné. Autrement dit, il existe $v(\mu)$ avec (2.12) et

$$\|v(\mu)\|_{\mathcal{U}} \leq \text{constante lorsque } \mu \longrightarrow 0. \tag{2.13}$$

Remarque. La Conjecture 1.2 a été démontrée par E. Zuazua et l'Auteur [17] et [18] pour les problèmes analogues où l'état est approché en dimension finie, comme plus haut.

Ayant montré l'existence d'un contrôle $v(\mu)$ «optimal» (dans le sens (2.12)), *quelle est la «robustesse» de $v(\mu)$ par rapport à des variations de μ?*

Conjecture 2.3. On peut trouver $v(\mu)$ dépendant de façon différentiable de μ, tel que

$$\left\| \frac{\partial v(\mu)}{\partial \mu} \right\|_{\mathcal{U}} \leq \frac{C}{\mu} \tag{2.14}$$

lorsque $\mu \longrightarrow 0$.
En fait on peut, probablement, quelque peu améliorer (2.14) comme suit:
On introduit

$$\mathcal{U}_0 = \left(L^2 \left(\mathcal{O} \right) \right)^3 \tag{2.15}$$

de sorte que

$$\mathcal{U} = L^2 \left(0, T; \mathcal{U}_0 \right). \tag{2.16}$$

On conjecture que

$$\left\| \frac{\partial v(\mu)}{\partial \mu} t \right\|_{\mathcal{U}_0} \leq \frac{C}{\mu} \quad \text{p.p en } t. \tag{2.17}$$

Remarque. On va démontrer, dans la section 2 suivante, les estimations (2.14) et (2.17) dans le cas des équations de Navier–Stokes linéarisées.

Des remarques sur le cas général (ouvert) des équations de Navier–Stokes sont données au section 4.

Remarque. Nous ignorons si le facteur $1/\mu$ peut être amélioré dans (2.14). La présence d'un tel facteur, si elle ne pouvait être significativement améliorée serait une démonstration quantitative de la difficulté, bien connue, du contrôle effectif de la turbulence.

Remarque. Naturellement $v(\mu)$ dépend aussi de β, y^T et de $\mathcal{O}$. La sensitivité de $v(\mu, \mathcal{O})$ par rapport à $\mathcal{O}$ (en fait, de la région de la frontière de Ω où s'exerce le contrôle) est un problème fondamental mais non abordé ici.

3. Équations de Navier–Stokes linéarisées

3.1. ÉNONCÉ DU RÉSULTAT

On se donne

$$\eta \in L^2(0, T; V). \tag{3.1}$$

On lui associe l'équation linéaire

$$
\left|
\begin{aligned}
&\left(\frac{\partial y}{\partial t}, \widehat{y}\right) + \mu a\,(y, \widehat{y}) + b\,(\eta, y, \widehat{y}) \\
&\qquad = (v1_{\mathcal{O}}, \widehat{y}) \quad \forall \widehat{y} \in V,\; y \in L^2(0, T; V) \cap L^\infty(0, T; H), \\
&y(0) = 0 \text{ pour fixer les idées.}
\end{aligned}
\right. \tag{3.2}
$$

On désigne par $y(v)$ la solution de (3.2), sans expliciter la dépendance en η.

Remarque. Tout point fixe de l'application $\eta \to y$ (v étant fixé) donne une solution des équations de Navier–Stokes.

On considère maintenant, comme en (2.10) mais avec la nouvelle signification (3.2) pour $y(v)$,

$$\mathcal{J}(v) = \frac{1}{2}\|v\|_{\mathcal{U}}^2 + \frac{\alpha}{2}\left\|y(T; v) - y^T\right\|_H^2. \tag{3.3}$$

Le problème

$$\inf \mathcal{J}(v), \quad v \in \mathcal{U} \tag{3.4}$$

admet une solution unique, notée $v(\mu)$.

On va démontrer que

$$\left\|\frac{\partial}{\partial \mu}v(\mu)\right\|_{\mathcal{U}_0} \leq \frac{C}{\mu} \quad \text{p.p en } t \in (0, T) \tag{3.5}$$

où C désigne, ici et dans la suite, des constantes diverses.

Remarque. Il va résulter des démonstrations qui suivent que:

$$\begin{aligned} &\text{on peut choisir dans (3.5) une} \\ &\text{constante } C \text{ indépendante de } \eta. \end{aligned} \tag{3.6}$$

Remarque. Naturellement, l'estimation (3.5) montrera aussi l'existence de la dérivée $(\partial/\partial \mu)v(\mu)$ dans l'espace $\mathcal{U} = L^2(0, T; \mathcal{U}_0)$.

3.2. PREMIÈRES ESTIMATIONS *A PRIORI*

Pour $v = 0$, on a $y = 0$ et donc

$$\mathcal{J}(0) = \frac{\alpha}{2} \left\| y^T \right\|_H^2.$$

Donc

$$\mathcal{J}(v(\mu)) \leq \frac{\alpha}{2} \left\| y^T \right\|_H^2$$

de sorte que

$$\|v(\mu)\|_{\mathcal{U}} \leq C, \qquad \|y(T, v(\mu))\|_H \leq C. \tag{3.7}$$

Prenant $\widehat{y} = y = y(v(\mu))$ dans (3.2), on obtient (car $b(\eta, y, y) = 0$)

$$\frac{1}{2} \frac{d}{dt} \|y\|_H^2 + \mu a(y) = (v(\mu) 1_{\mathcal{O}}, y) \tag{3.8}$$

(où l'on écrit, ici et dans la suite, $a(\varphi, \varphi) = a(\varphi)$).

On déduit de (3.7) et (3.8) que, pour $y = y(v(\mu))$ on a

$$\left| \begin{array}{l} \|y(t)\|_H \leq C, \\ \displaystyle\int_0^T a(y)\, dt \leq C/\mu. \end{array} \right. \tag{3.9}$$

3.3. SYSTÈME D'OPTIMALITÉ (S.O.)

(Pour les S.O. dans le contrôle des systèmes distribués, [13]).

L'équation d'Euler (du Calcul des Variations) pour (3.4) est donnée par

$$(v, \delta v)_{\mathcal{U}} + \alpha \left(y(T; v) - y^T, \delta y(T, v) \right) = 0 \quad \forall \delta v \in \mathcal{U}. \tag{3.10}$$

On introduit l'état adjoint φ donné par

$$\left| \begin{array}{l} -\left(\dfrac{\partial \varphi}{\partial t}, \widehat{\varphi} \right) + \mu a\left(\varphi, \widehat{y} \right) + b\left(\eta, \widehat{\varphi}, \varphi \right) = 0 \quad \forall \widehat{\varphi} \in V, \\[2mm] \varphi(T) = \alpha \left(y(T; v) - y^T \right), \\[2mm] \varphi \in L^2(0, T; V) \cap L^\infty(0, T; H). \end{array} \right. \tag{3.11}$$

(Noter la place de $\widehat{\varphi}$ dans $b(\eta, \widehat{\varphi}, \varphi)$).

Prenant dans (3.11) $\widehat{\varphi} = \delta y$ (correspondant à δv) et intégrant par parties sur $(0, T)$, il vient

$$\alpha \left(y(T, v) - y^T, \delta y(T) \right)$$

$$= \int_0^T \left[\left(\frac{\partial}{\partial t} \delta y, \varphi \right) + \mu a(\delta y, \varphi) + b(\eta, \delta y, \varphi) \right] dt$$

$$= \int_0^T (\delta v 1_{\mathbb{O}}, \varphi) \, dt \quad \text{(d'après 2.2),}$$

de sorte que (3.10) s'écrit

$$\int_0^T (v + \varphi 1_{\mathbb{O}}, \delta v)_{\mathcal{U}_0} \, dt = 0 \quad \forall \delta v,$$

C-á-d,

$$v + \varphi 1_{\mathbb{O}} = 0. \tag{3.12}$$

Le (S.O.) est alors donné par (3.2) où $v = -\varphi 1_{\mathbb{O}}$ et (3.11).

3.4. ESTIMATIONS POUR L'ÉTAT ADJOINT

Prenant $\widehat{\varphi} = \varphi$ dans (3.11), on en déduit que

$$\frac{1}{2} \|\varphi(t)\|_H^2 + \mu \int_t^T \alpha(y) \, dt = \frac{\alpha^2}{2} \left\| y(T) - y^T \right\|_H^2 \tag{3.13}$$

(où $y(T) = y(T; v(\mu))$ donc le deuxième membre de (3.13) est $\leq C$.

On en déduit les estimations analogues à (3.9):

$$\left| \begin{array}{l} \|\varphi(t)\|_H \leq C, \\[2mm] \displaystyle\int_0^T \alpha(\varphi) dt \leq C/\mu. \end{array} \right. \tag{3.14}$$

Remarque. On déduit de (3.12) et (3.14$_1$) que

$$\|v(\varphi)\|_{\mathcal{U}_0} \leq C \quad \text{p.p en } t. \tag{3.15}$$

3.5. DÉRIVATION EN μ

On commence par un calcul formel, en supposant l'existence des dérivées

$$\frac{\partial y}{\partial \mu} = \dot{y}, \qquad \frac{\partial \varphi}{\partial \mu} = \dot{\varphi} \tag{3.16}$$

(où y, φ est *la* solution du (S.O.)). On a alors, d'après (3.12)

$$\dot{v} = -\dot{\varphi} 1_{\mathbb{O}} \tag{3.17}$$

Dérivant donc μ le (S.O.), on obtient

$$\left| \begin{array}{l} \left(\dfrac{\partial \dot{y}}{\partial t}, \widehat{y} \right) + \mu a \left(\dot{y}, \widehat{y} \right) + b \left(\eta, \dot{y}, \widehat{y} \right) + \left(\dot{\varphi} 1_{\mathbb{O}}, \widehat{y} \right) = -\alpha \left(y, \widehat{y} \right), \\[2mm] \dot{y}(0) = 0. \end{array} \right. \tag{3.18}$$

$$\left| \begin{array}{l} -\left(\dfrac{\partial \dot{\varphi}}{\partial t}, \widehat{\varphi} \right) + \mu a \left(\dot{\varphi}, \widehat{\varphi} \right) + b \left(\eta, \widehat{\varphi}, \dot{\varphi} \right) = -\alpha \left(\varphi, \widehat{\varphi} \right), \\[2mm] \dot{\varphi}(T) = \alpha \dot{y}(T). \end{array} \right. \tag{3.19}$$

On va vérifier que (3.18) et (3.19) est le (S.O.) d'un nouveau problème de contrôle. Cela montrera l'existence et l'unicité de la solution (3.18) et (3.19) et justifiera les dérivations en μ.

On introduit un nouveau contrôle $w \in \mathcal{U}$. L'état $z = z(w)$ est donné par

$$\left| \begin{array}{l} \left(\dfrac{\partial z}{\partial t}, \widehat{y} \right) + \mu a \left(z, \widehat{y} \right) + b \left(\eta, z, \widehat{y} \right) = \left(w 1_{\mathbb{O}}, \widehat{y} \right) - \alpha \left(y, \widehat{y} \right), \\[2mm] z(0) = 0. \end{array} \right. \tag{3.20}$$

La nouvelle fonction coût est donnée par

$$K(w) = \frac{1}{2} \| w \|_{\mathcal{U}}^2 + \frac{\alpha}{2} \| z(T) \|^2 - \int_0^T a(\varphi, z) \, dt, \tag{3.21}$$

où φ est donné par la solution du S.O.

Le problème

$$\inf K(w), \quad w \in \mathcal{U}, \tag{3.22}$$

admet une solution unique $w = w(\mu)$.

On vérifie que le (S.O.) pour (3.22) est l'ensemble des équations (3.18) et (3.19) avec $w(\mu) = -\dot{\varphi} 1_{\mathbb{O}}$.

Par conséquent, le système (3.18) et (3.19) admet une solution unique et cela permet de justifier les dérivations formelles faites en (3.5). En outre $\dot{v} = w(\mu)$.

3.6. ESTIMATIONS

On a

$$K(w(\mu)) \le K(0). \tag{3.23}$$

Pour $w = 0$, $z(w = 0) = \zeta$ est donné par

$$\left| \begin{array}{l} \left(\dfrac{\partial \zeta}{\partial t}, \widehat{y} \right) + \mu a\,(\zeta, \widehat{y}) + b\,(\eta, \zeta, \widehat{y}) = -\alpha\,(y, \widehat{y})\,, \\[2mm] \zeta(0) = 0. \end{array} \right. \tag{3.24}$$

et

$$K(0) = \frac{\alpha}{2} \|\zeta(T)\|^2 - \int_0^T a(\varphi, \zeta)\, dt. \tag{3.25}$$

On déduit de (3.24) que

$$\left| \begin{array}{l} \dfrac{1}{2} \|\zeta(T)\|^2 + \mu \displaystyle\int_0^T a(\zeta)\, dt \\[4mm] \qquad = \displaystyle\int_0^T a(y, \zeta)\, dt \\[4mm] \qquad \leq \dfrac{\mu}{2} \displaystyle\int_0^T a(\zeta)\, dt + \dfrac{1}{2\mu} \displaystyle\int_0^T a(y)\, dt. \end{array} \right.$$

Utilisant (3.9), on en déduit que

$$\left| \begin{array}{l} \|\zeta(T)\|^2 \leq C/\mu^2, \\[3mm] \mu \displaystyle\int_0^T a(\zeta)\, dt \leq C/\mu^2. \end{array} \right. \tag{3.26}$$

Alors

$$\int_0^T a(\varphi, \zeta)\, dt \leq \left(\int_0^T a(\varphi)\, dt \right)^{1/2} \left(\int_0^T a(\zeta)\, dt \right)^{1/2} \leq C\mu^{-2}$$

(par (3.14) et (3.26)), donc

$$K(0) \leq C/\mu^2. \tag{3.27}$$

Il résulte alors de (3.23) et (3.24) que

$$\|w(\mu)\|_{\mathcal{U}} \|v\|_{\mathcal{U}} \leq C/\mu. \tag{3.28}$$

Il est possible d'améliorer (3.28) pour obtenir (3.5). Comme $w(\mu) = \dot{v} = -\dot{\varphi} 1_{\mathcal{O}}$, il suffit d'estimer $\|\dot{\varphi}(t)\|_H$.

On peut maintenant utiliser (3.19) avec $\widehat{\varphi} = \dot{\varphi}$, d'où

$$\frac{1}{2} \|\dot{\varphi}(t)\|_H^2 + \mu \int_t^T a(\dot{\varphi})\, dt$$
$$= -\int_t^T a(\varphi, \dot{\varphi})\, dt + \frac{\alpha^2}{2} \|\dot{y}(T)\|_H^2. \tag{3.29}$$

Le premier terme du second membre est majoré par

$$\frac{\mu}{2} \int_0^T a(\dot{\varphi})\,dt + \frac{1}{2\mu} \int_t^T a(\varphi)\,dt$$

et comme on a (3.14), on déduit que

$$\|\dot{\varphi}(t)\|_H^2 + \mu \int_t^T a(\dot{\varphi})\,dt \le \frac{C}{\mu^2} + \frac{\alpha^2}{2}\|\dot{y}(T)\|_H^2. \tag{3.30}$$

On estime maintenant $\|\dot{y}(T)\|_H^2$. Prenant $\widehat{y} = \dot{y}$ dans (3.18), on en déduit que

$$\left|
\begin{aligned}
&\frac{1}{2}\|\dot{y}(t)\|^2 + \mu \int_0^t a(\dot{y})\,dt \\
&= -\int_0^t a(y,\dot{y})\,dt - \int_0^t (\dot{\varphi}1_{\mathbb{O}},\dot{y})\,dt \\
&\le \frac{\mu}{2}\int_0^t a(\dot{y})\,dt + \frac{1}{2\mu}\int_0^t a(y)\,dt + C\int_0^t \|\dot{\varphi}\|_H\|\dot{y}\|_H\,dt.
\end{aligned}
\right.$$

Utilisant (3.9), on en déduit que

$$\|\dot{y}(t)\|_H^2 \le \frac{C}{\mu^2} + \int_t^t \|\dot{\varphi}\|_H^2.$$

Utilisant cette inégalité dans (3.30) on en déduit le résultat désiré.

4. Equations de Navier–Stokes (et autres modèles et problèmes)

4.1. EQUATIONS DE NAVIER–STOKES

Remarque. Les estimations de la section 2 sont indépendantes de η. Comme, pour v fixé, une solution de l'équation de Navier–Stokes est donnée par tout point fixe de l'application $\eta \to y$, cela donne un élément favorable pour la Conjecture 1.3, mais pas une démonstration.

Remarque. On peut considérer directement le (S.O.) pour le problème (2.10) et (2.11), $y(T; v)$ désignant une solution de Navier–Stokes. Mais le (S.O.) donne des conditions nécessaires pour l'optimalité, dont la suffisance n'est pas établie (faute de convexité).

On peut ensuite dériver formellement en μ. Les estimations ainsi obtenues (c'est possible!) semblent difficile à justifier rigoureusement. Elles sont, finalement, moins convaincantes que la remarque du début du section 3.

4.2. AUTRES MODÈLES

Tout ce qui a été présenté dans les sections précédentes s'étend à «*tous*» les modèles de la Physique Mathématique où interviennent des fluides incompressibles (ou non). Par exemple, les modèles de la Climatologie, où l'on peut mettre «toutes» les équations sous une forme qui généralise les équations de Navier–Stokes [15] (et la bibliographie de ce travail).

4.3. PROBLÈMES

De très nombreuses questions restent ouvertes, signalons en deux.

1) Peut-on obtenir un *développement asymptotique* plus précis de la fonction $\mu \to v(\mu)$, lorsque $\mu \to 0$ faisant intervenir *les couches limites*?

2) Peut-on améliorer l'estimation (ou la conjecture) (2.14) ou (2.17)?

Naturellement, toutes les considérations faites ici ne donnent pas d'*algorithme* pour calculer une approximation de $v(\mu)$, question qui n'a pas été abordée ici.

Références

[1] Coron, J-M., On the controllability of 2D imcompressible perfect fluids, *J.M.P.A.* **75** (1996), 155–188.

[2] _____, On the controllability of the 2D imcompressible Navier–Stokes with the Navier slip boundary conditions, *ESAIM–COCV1* (1996), 35–75.

[3] Coron, J-M. and Imanuvilov, O.Yu., A paraître.

[4] Fursikov, A. and Imanuvilov, O.Yu., *Controllability of evolution equations*, Lecture Notes Series 34, Seoul National University 1996.

[5] Glowinski, R. and Lions, J-L., Exact and approximate controllability for distributed parameter systems, *Acta Numerica* (1994), 269–378; (1996), 159–333.

[6] Imanuvilov, O.Yu., On exact controllability for Navier–Stokes equations, *ESAIM–COCV3* (1998), 83–117.

[7] Ladyzenskaya, O.A., *La théorie mathématique des fluides visqueux imcompressibles*, Moscou 1961.

[8] Leray, J., Etude de diverses équations intégrales non linéaires et de quelques problèmes que pose l'hydrodynamique, *J.M.P.A.* t. XII (1933), 1–82.

[9] _____, Essai sur le mouvement plan d'un liquide visqueux que limitent des parois, *J.M.P.A.* t XIII (1934), 331–418.

[10] _____, Sur le mouvement d'un liquide visqueux emplissant l'espace, *Acta Math.* **63** (1934), 193–248; *Voir aussi Oeuvres Complètes*, Vol. 3, P. Malliavin (ed.), Springer 1998.

[11]   Lions, J.L., *Quelques méthodes de résolution des problèmes aux limites non linéaires*, Dunod Gauthier-Villars, Paris 1969.

[12]   _____, *Are there connections between turbulence and controllability? in Analyse et Optimisation des Systèmes*, Lecture Notes in Control and Information Sciences 144, Springer 1990.

[13]   _____, *Contôle optimal de systèmes gouvernés par des équations aux dérivées partielles*, Dunod Gauthier-Villars, Paris 1968.

[14]   Lions, J.L. and Magenes, E., *Problèmes aux limites non homogènes et applications*, Dunod, Paris 1968.

[15]   Lions, J.L., Temam, R., and Wang, S.R., Models for the coupled atmosphere and ocean, *Computational Mechanics*, Vol. 1, n° 1, 1993, 3–120.

[16]   Lions, J.L. and Zuazua, E., Exact boundary controllability of Galerkin's approximations of Navier–Stokes equations, *Ann. Scuola Normale Sup.* Pisa XXVI (1998), 605–621.

[17]   _____, On the cost of controlling unstable systems: the case of boundary controls, *Journal d'Analyse Mathématique* **73** (1997).

[18]   _____, *Contrôlabilité exacte des approximations de Gakerkin des équations de Navier–Stokes*, C.R.A.S. **234** (1997), 1015–1021.

[19]   Lions, P.L., *Mathématical topics in fluid Mechanics*, Vol 1, Oxford. Univ. Press 1996.

[20]   _____, *The compressible case*, Vol 2, 1998.

[21]   von Neumann, J., Can we survive technology?, *Fortune. Juin* (1955); *Voir auusi Oeuvres Complètes*, Vol. 6, Pergamon 1963.

ADDRESSES

VINCENZO ANCONA
University of Firenze, Department of Maths
67/A Viale Morgagni
Firetize, Italy

RICHARD BEALS
Yale University, Department of Maths
10 Hillhouse Ave
New-Haven CT 06520 8283, USA

B. BENDIFFALAH
University of Montpellier II,
Department of Maths
Montpellier, France
e-mail: ben@darboux.math.univ-montp2.fr

BERNHELM BOOSS-BAVNBEK
Universty of Roskide, Department of Maths
IMFUFA, Box 260
4000 Roskilde, Danmark
e-mail: BOOSS@mmf.ruc.dk

LOUIS BOUTET DE MONVEL
University of Paris VI, Department of Maths
4 Pl Jussieu
75 252 Paris cedex 05, Prance

A. BOVE
University of Bologna
Piazza di porta S. Donato 5
40127 Bologna, Italy
e-mail: bove@leondegrange.dm.unibo.it

ENRICO BERNARDI
University of Bologna
Piazza di porta S. Donato 5
40127 Bologna, Italy
e-mail: bernardi@dm.unibo.it

YVONNE CHOQUET-BRUHAT
University of Paris VI
Department of Maths
4 Pl Jussieu
75 252 Paris cedex 05, France

PHILIPPE G. CIARLET
City University of Hongkong and
Université Pierre et Marie Curie
Intstitut de Mathématiques
4, place Jussieu
75252 Paris cedex 05, France

FREDERICO COLOMBINI
University of Pisa, Department of Maths
Via F Buonarroti
56127 Pisa, Italy
e-mail: Colombini@dm.unipi.it

PIERRE DAZORD
University of Lyon I, Department of Maths
7 rue Franklin
69002 Lyon, France

HÉLÈNE DÉLQUIÉ
University of Paris VI, Department of Maths
4 Pl Jussieu
75 252 Paris cedex 05, France

MAURICE DE GOSSON
Blekinge Institute of Technology
Campus Annebo
371 79 Karlskrona, Sweden
e-mail: mdg@ihn.bth.se

SERGE DE GOSSON
University of Växjö, Department of Maths
35 195 Växjö, Sweden
e-mail: sergedegosson@hotmail.com

KENRO FURUTANI
Science University of Tokyo
Noda, Chiba 278–8510, Japan
e-mail: furutani_kenro@ma.noda.tus.ac.jp

BERNARD GAVEAU
University of Paris VI, Department of Maths
4 Pl Jussieu
75 252 Paris cedex 05, France

DANIEL GOURDIN
University of Paris VI, Department of Maths
4 P1 Jussieu
75 252 Paris cedex 05, France
e-mail: Daniel.Gourdin@math.jussieu.fr

PETER GREINER
University of Toronto, Department of Maths
Toronto, Ontario M5S3G3, Canada
e-mail: greiner@math.toronto.edu

Y. HAMADA
61-63 tatekura-cho, Shimogamo, Sakyo-ku
Kyoto 606-0806, Japan
e-mail: yhamada@oak.ocn.ne.jp

CHRISTIAN HOUZEL
11, rue Monticelli
75014 Paris, France

NAIL IBRAGIMOV
Blekinge Institute of Technology
Campus Annebo
371 79 Karlskrona, Sweden
e-mail: nib@ihn.bth.se

SÖREN ILLMAN
University of Helsinki, Department of Maths
Helsinki, Finland
e-mail: illman@cc.helsinki.fi

K. KAJITANI
University of Tsukuba, Department of Maths
Ibaraki 305-8571, Japan
e-mail: kajitani@math.tsukaba.ac.jp

ANDREI KHRENNIKOV
University of Växjö, Department of maths
35 195 Växjö, Sweden
e-mail: Andrei.khrennikov@masda.vxn.se

T. KOBAYASHI
Science University of Tokyo
Noda, Chiba 278-8510, Japan
e-mail: Takao@ma.noda.sut.ac.jp

PASCAL LAUBIN
University of Liège, Department of Maths
Grande Traverse 12
Liège, Belgium
e-mail: P.Laubin@ugl.ac.be

PAULETTE LIBERMANN
116 Avenue du Général Leclerc
75014 Paris, France

JACQUES-LOUIS LIONS
Collège de France
3 rue d'Ulm
75005 Paris, France

CHARLES-MICHEL MARLE
Université Pierre et Marie Curie
Intstitut de Mathématiques
4, place Jussieu
75252 Paris cedex 05, France
email: marle@math.jussieu.fr

VLADIMIR NAZAIKINSKII
Institute for Problems
in Mechanics, RAS
117526 Moscow, Russia

JEAN-PIERRE NICOLAS
École Polytechnique
91 128 Palaiseau, France
e-mail: Nicolas@math.polytechnique.fr

T. NISHITANI
Science University of Osaka,
Department of Maths
Machikaneyama 1-16 Toyonaka
Osaka 560, Japan

YUJIRO OHYA
Koshien University
665-0006 Momijigaoka 10-1
Kyoto, Japan
e-mail: y-ohya@koshien.ac.jp

ANTON SAVIN
Moscow State University,
Department of Maths
Vorobe'vy Gory
119 899 Moscow, Russia
e-mail: antonsavin@mtu-net.ru

WALTER SCHEMPP
Lehrstuhl fuer Mathematik I
University of Siegen
Walter Flex Strasse 3
57068 Siegen, Germany
e-mail: Schempp@mathematik.uni-siegen.de

BENGT-WOLFGANG SCHULZE
University of Potsdam
Postfach 601553
14415 Potsdam, Germany
e-mail: schulze@math.uni-potsdam.de

J. SNIATYCKI
University of Calgary
2500 University Drive NMW
Calgary AB T2N IN4, Canada
e-mail: Sniat@math.ucalgary.ca

S. SPAGNOLO
University of Pisa, Department of Maths
Via F Buonarroti
56127 Pisa, Italy
e-mail: Spagnolo@dm.unipi.it

BORIS STERNIN
Independent University of Moscow
Moscow, Russia
e-mail: sternin@mtu-net.ru

G. TAGLIALATELA
University of Paris VI, Department of Maths
4 P1 Jussieu
75 252 Paris cedex 05, France
e-mail: Taglialatela@math.jussieu.fr

JOACHIM TOFT
Blekinge Institute of Technology
Campus Annebo
371 79 Karlskrona, Sweden
e-mail: Toft@ihn.hk-r.se

N. TOSE
Keioo University, Department of Maths
Hiyoshi Campus
4-1-1 Hiyshi, Kohoku
Yokohama, Japan
e-mail: Tose@math.hc.keio.ac.jp

NIKOLAI TARKAHANOV
University of Potsdam
Department of Maths
Potsdam, Germany
e-mail: tarkhan@math.uni-potsdam.de

GIJS TUYNMAN
University of Lille, Department of Maths
UST
59655 Villeneuve d'Ascq, France
e-mail: Gmt@gat.univ-lille.fr

JEAN VAILLANT
University of Paris VI, Department of Maths
4 P1 Jussieu
75 252 Paris cedex 05, France
e-mail: Vaillant@math.jussieu.fr

CLAUDE WAGSCHAL
University of Toulouse
Department of Maths
118 Route de Narbonne
31062 Toulouse cedex, France
e-mail: Wagschal@math.toulouse.fr

KRZYSTOF P. WOJCIECHOWSKI
Department of Mathematics
IUPUI (Indiana University/Purdue University)
Indianapolis, IN 46202-3216, USA
email: kwojciechowski@math.iupui.edu

Mathematical Physics Studies

1. F.A.E. Pirani, D.C. Robinson and W.F. Shadwick: *Local Jet Bundle Formulation of Bäcklund Transformations.* 1979 ISBN 90-277-1036-8
2. W.O. Amrein: *Non-Relativistic Quantum Dynamics.* 1981 ISBN 90-277-1324-3
3. M. Cahen, M. de Wilde, L. Lemaire and L. Vanhecke (eds.): *Differential Geometry and Mathematical Physics.* 1983 Pb ISBN 90-277-1508-4
4. A.O. Barut (ed.): *Quantum Theory, Groups, Fields and Particles.* 1983 ISBN 90-277-1552-1
5. G. Lindblad: *Non-Equilibrium Entropy and Irreversibility.* 1983 ISBN 90-277-1640-4
6. S. Sternberg (ed.): *Differential Geometric Methods in Mathematical Physics.* 1984 ISBN 90-277-1781-8
7. J.P. Jurzak: *Unbounded Non-Commutative Integration.* 1985 ISBN 90-277-1815-6
8. C. Fronsdal (ed.): *Essays on Supersymmetry.* 1986 ISBN 90-277-2207-2
9. V.N. Popov and V.S. Yarunin: *Collective Effects in Quantum Statistics of Radiation and Matter.* 1988 ISBN 90-277-2735-X
10. M. Cahen and M. Flato (eds.): *Quantum Theories and Geometry.* 1988 ISBN 90-277-2803-8
11. Bernard Prum and Jean Claude Fort: *Processes on a Lattice and Gibbs Measures.* 1991 ISBN 0-7923-1069-1
12. A. Boutet de Monvel, Petre Dita, Gheorghe Nenciu and Radu Purice (eds.): *Recent Developments in Quantum Mechanics.* 1991 ISBN 0-7923-1148-5
13. R. Gielerak, J. Lukierski and Z. Popowicz (eds.): *Quantum Groups and Related Topics.* Proceedings of the First Max Born Symposium. 1992 ISBN 0-7923-1924-9
14. A. Lichnerowicz, *Magnetohydrodynamics: Waves and Shock Waves in Curved Space-Time.* 1994 ISBN 0-7923-2805-1
15. M. Flato, R. Kerner and A. Lichnerowicz (eds.): *Physics on Manifolds.* 1993 ISBN 0-7923-2500-1
16. H. Araki, K.R. Ito, A. Kishimoto and I. Ojima (eds.): *Quantum and Non-Commutative Analysis.* Past, Present and Future Perspectives. 1993 ISBN 0-7923-2532-X
17. D.Ya. Petrina: *Mathematical Foundations of Quantum Statistical Mechanics.* Continuous Systems. 1995 ISBN 0-7923-3258-X
18. J. Bertrand, M. Flato, J.-P. Gazeau, M. Irac-Astaud and D. Sternheimer (eds.): *Modern Group Theoretical Methods in Physics.* Proceedings of the Conference in honour of Guy Rideau. 1995 ISBN 0-7923-3645-3
19. A. Boutet de Monvel and V. Marchenko (eds.): *Algebraic and Geometric Methods in Mathematical Physics.* Proceedings of the Kaciveli Summer School, Crimea, Ukraine, 1993. 1996 ISBN 0-7923-3909-6

Mathematical Physics Studies

20. D. Sternheimer, J. Rawnsley and S. Gutt (eds.): *Deformation Theory and Symplectic Geometry.* Proceedings of the Ascona Meeting, June 1996. 1997
ISBN 0-7923-4525-8
21. G. Dito and D. Sternheimer (eds.): *Conférence Moshé Flato 1999.* Quantization, Deformations, and Symmetries, Volume I. 2000
ISBN 0-7923-6540-2 / Set: 0-7923-6542-9
22. G. Dito and D. Sternheimer (eds.): *Conférence Moshé Flato 1999.* Quantization, Deformations, and Symmetries, Volume II. 2000
ISBN 0-7923-6541-0 / Set: 0-7923-6542-9
23. Y. Maeda, H. Moriyoshi, H. Omori, D. Sternheimer, T. Tate and S. Watamura (eds.): *Noncommutative Differential Geometry and Its Applications to Physics.* Proceedings of the Workshop at Shonan, Japan, June 1999. 2001
ISBN 0-7923-6930-0